The Study of Time II

Proceedings of the Second Conference of the
International Society for the Study of Time
Lake Yamanaka — Japan

Edited by

J. T. Fraser and N. Lawrence

With 80 Figures

Springer-Verlag
New York Heidelberg Berlin 1975

Library of Congress Cataloging in Publication Data

International Society for the Study of Time.
  The study of time.

  English and German.
  Vol. 2 edited by J.T. Fraser and N. Lawrence.
  First conference held in 1969 at Oberwolfach,
Germany; 2d held in 1973 near Lake Yamanaka, Japan.
  Includes bibliographies.
  1. Time—Congresses.   I. Fraser, J.T., ed.
II. Haber, Francis C., ed.   III. Müller, Gert
Heinz, 1923-   ed.   IV. Lawrence, Nathaniel
Morris, 1917-   ed.   V. Title.
QB209.I55 1972   529   72-80472

ISBN 978-3-642-50123-4   ISBN 978-3-642-50121-0 (eBook)
DOI 10.1007/978-3-642-50121-0

Foreword

The Second Conference of the International Society for the Study of Time was held at Hotel Mt. Fuji, near Lake Yamanaka, Japan, on July 1 to 7, 1973. The present volume is the proceedings at that Conference and constitutes the second volume in *The Study of Time* series.*

At the closing session of our First Conference in Oberwolfach, Germany, in 1969, I was honored by being elected to the Presidency of the Society, following Dr. J. G. Whitrow, our first President. My mandate was to organize a Second Conference, consistent with the aim of the Society, which is the holding of interdisciplinary conferences for the presentation and discussion of papers on various aspects of time. Several participants expressed to me their wish to have a second conference held in Japan so as to emphasize the international and intercultural dedication of this Society. Dr. Fraser carefully evaluated this and many other suggestions, weighed the possible conference sites and our chances of raising the necessary funds to convene a meeting at such sites, and concurred with my conclusions that we should go ahead with the plans for a Japanese meeting.

For the difficult and complicated task of raising funds and organizing a conference in Japan, I had to select and rely heavily on somebody both capable and reliable and also living in Japan. Thus, I asked the Reverend Michael Mutsuo Yanase, S.J., Professor of Sophia University, Tokyo, to undertake this arduous task. He graciously accepted and flawlessly carried out this demanding assignment. Without his powerful dedication and silent tenacity the conference would have never come about. He organized the "P.C.J.", the Preparatory Committee in Japan for the Second World Conference of ISST, with Professor Takahiko Yamanouchi as Chairman, and Professor Yoichiro Murakami as Local Secretary. Other members of P.C.J. were Professors Kodi Husimi, Tarow Indow, Masao Mutsumoto, Shozo Ohmori and Professor Yanase himself.

The self-effacing tireless efforts of the P.C.J. Secretary, Dr. Murakami, with the wise advice of Chairman Yamanouchi and close collaboration of other members of P.C.J. deservedly paid off, and they could raise sufficient funds to defray part of the travel expenses of speakers, and room and board of participants. The behind-the-scene support of Professor Seishi Kaya and Professor Kankuro Kaneshige opened many doors that would have otherwise been closed to us. P.C.J. also took care of all the details of hotel arrangements, transportation and entertainment. Dr. Husimi and Dr. Yanase went out of their way to promote the "cause" of the study of time in Japan, giving lectures, organizing small local symposia on time and urging a scientific magazine** to publish a series of articles on time.

The contributors of funds were as follows: Asia Foundation, Commemorative Association for the Japan World Exposition, Federation of Electric Power Companies Japan, Fuji Bank and its affiliated

---

*The proceedings of our First Conference is *The Study of Time*, v.1. Berlin, Heidelberg, New York: Springer-Verlag 1972.
**"Shizen" (in Japanese).

enterprizes, Japan Iron and Steel Federation, Japan Medical Association, K. Hattori and Company, Mitsubishi Foundation, Mitsui Bank and its affiliated enterprizes, Nippon Telegraph and Telephone Public Corporation, Sumitomo Bank and its affiliated enterprizes.

Apart from these benefactors, Bulova Watch Company, Inc. of New York and Bulova-Citizen Watch Company made a separate contribution to sponsor a special session on *Timekeepers and Time* which took place on the fourth day of the Conference. The papers of that symposium are collected at the end of this volume, and will be separately reprinted in paperback, for distribution by Bulova Watch Company, Inc. of New York.

Dr. Fraser, our Founder and Secretary, with his typical efficiency, erudition and devotion, acted as a one-man program committee for the Conference. Based on his worldwide correspondence with scholars and scientists active in the study of time, and his seemingly endless familiarity with pertinent literature, he enlisted the array of fine specialists whose papers are collected in this volume. I am convinced that the reader will share my joy in reading them. Inevitably, there have been several prospective speakers who were prevented from attending the Conference for one reason or another. I hope they will participate in our future conferences.

Professional societies do not come about simply because some people wished that they would exist. A successful conference does not just happen unless there is a well-coordinated collaboration of sponsors, organizers and participants. I would, therefore, like to close by expressing my most sincere thanks and appreciation to Dr. Fraser and to Dr. Helen B. Green, Treasurer of I.S.S.T., to the Chairman, Secretary and members of P.C.J., to all those organizations which made generous contributions and, above all, to all the participants who provided the conference with the substance. If anyone of these people or organizations had not done what they did, the Conference would indeed have been impossible.

Honolulu, Hawaii, July 1974.

Satosi Watanabe
Second President
International Society for the
Study of Time.

Contents

LIST OF PARTICIPANTS

I. AGING

Temporal Stages in the Development of the Self

H.B. GREEN

This paper will propose that there is a self, that the self is imbedded in time, and that the self develops by solving a succession of eleven temporal problems encountered in its lifespan. The approach to these propositions will include consideration of five psychological givens. In the context of this paper, the term 'psychological given' refers to the nature of the psychophysical inheritance which determines to some extent the potentials for behavior throughout the individual's lifetime. These psychological givens underlie the continuous functioning of the self in time. It will be shown that they permit the self to become increasingly selective, cumulative, unique, consistent, and time-aware.

PSYCHOLOGICAL GIVENS ON WHICH THE SELF IS BASED

The first psychological given is the selectivity of the human organism due to its rhythmic nature. It changes and moves because it has life and it does this with a momentum which alternates thrust and recovery. As it changes it comes into contact with new surroundings. Out of the complexity of these external conditions, it selects a few characteristics as stimuli. The number selected is limited because each of the various neural mechanisms reacting within the organism has a rhythmic process which is able to conduct only when it synchronizes with the others. A peak of synchrony determines which few stimuli will arouse the entire organism and mobilize it for action.

Maturation and learning interact in a process which is the second psychological given. The understanding of this process rests upon accepting the inseparability of psychophysical structure and function. Inherent in this unity is an epigenetic groundplan (to use Erikson's term) for development which unfolds only as it interacts with its milieu. Maturation causes engagement, and engagement causes learning which is an internal modification of the initial behavioral thrust. Maturation provides the capacity for arousal and for holistic action. Learning is based on the maturing organism, which has an incredible plasticity for the accumulation of experience in the form of internal patterns corresponding to external conditions. Initially, these·external conditions are experienced as percepts related to what is out there. But the human capacity to store, retrieve and combine fragments of this experience gives rise to concepts. Eventually the individual can react to these concepts by internal stimulation in the absence of external events. The accumulation of these concepts, and the habits of repeated reactions to them affect the steering and on-going behavioral potential of the person.

Unique individual differences are the third psychological given. By these are meant the innate predispositions that determine which aspects of the external surroundings will be approached and avoided. They therefore start the self on its life course of selection because they determine the compatibility of that particular self with certain features of its environment. Personality may be viewed as the sum-total of the various innate predispositions which are measurable as behavior and subject to conditioning. The self is the integrating force which has a dynamic for holding together these predispositions and their learned modifications. This force exerts internal pressure toward an overall form which is different from all others, a Gestalt, and never duplicated. This gives the cohesion and articulation seen in the distinction which characterizes an individual's life.

Soon after birth, several distinct types of personality can be observed and measured. Under uniform conditions, newborn babies can be classified according to the following behaviors: activity level, attention span, rhythmicity, sensory threshold, tendency to approach or withdraw from new stimuli, mood, distractibility, and adaptability. It must be assumed that these early differences are the manifestations of innate predispositions to exhibit certain kinds of behavior. The list of categories is growing, as more psychological research focuses on the initial character of infants. Longitudinal studies have correlated many of these first attributes of babies with their personality measures taken at adolescence or at adulthood and have found lasting characteristics.[1,2,3,4,5,6] The fourth psychological given, then, is this lifelong general consistency.

One of the innate dispositions of humans is an awareness of passing time and the ability to recollect or anticipate it as a concept which can be expressed.[7] It appears as soon as the child is old enough to be a subject who can be tested. References to the present occur first in children's speech, and then references to a short future are soon followed by references to a short past.[8] These periods grow longer as the child gets older but the primitive ability to estimate time shows very little improvement with practice, and remains stable into old age.[7] This awareness of time is the fifth psychological given.

For the self, time is entirely subjective, and subjective time may have little relation to external or clock time. Time does not exist as an independent philosophical or physical entity. It is a result of the reciprocal inter-relation of experience, maturation, learning and innate personality predispositions. The time which the self knows has quality as well as extent. As Whitrow has said, we do not experience time, *per se,* but only what goes on in time, and the experience of time has both quality and quantity. For example, the child will experience as very long the time it takes to fill a pail at a well, or to wait and then run at a traffic stop light. This is time's extent. Habituation later changes this childhood perception of time by levelling certain aspects and sharpening other aspects, thus altering its quality. The self grows with these time changes. Furthermore, the self finds itself only in relation to time. For example, to ascertain whether one knew Harry before Dick, one compares oneself at various time periods: 'I knew Dick when I was still a student, but did I know Harry? Later on I took that job; yes, that's when I met Harry.'

People who are disoriented in their former relation to time are maladjusted, for they have no past time. Amnesia for time cripples the self, causing it to feel empty and unable to function in the present. The confusion of schizophrenes concerning present time is a crucial aspect of their disorder.[9,10] For example, schizophrenes sometimes think they are children when they are grownups, or otherwise forget themselves in past time. Cultures which have had their past forcibly cut off from their present existence — such as the American Indians or the ex-slave Blacks — experience a loss of identity. Returning soldiers, who have participated in a period of shocking war, report confusion as to who they really are,[11,12] because the nature of the self needs to be re-established and related to its time.

The inner environment of the individual constantly seeks a cognitively compatible relationship with the objective world. When new developments in that world present some difficulty in adjustment, temporary imbalance occurs in the internal process. This necessitates an internal reorganization which may be termed a small crisis. At critical periods, problems occur in time oreintation which cause upset to the cognitive-emotional structure of the individual and necessitate rebalancing. That the resulting behavior change is fundamental to the state of the self is indicated by the extent and saliency of any new action pattern — a whole new self in a new time emerges.

Psychologists have often described the development of behavior as a series of qualitatively different stages, each linked to an age for its first appearance. Freud described psychosexual development in periods;[13] Gesell divided motor development into stages;[4] Piaget looks at cognitive development as qualitatively different at various ages;[14] Erikson writes of psycho-social growth in stages;[15] and Kohlberg has found progressively higher steps to moral development.[16] While all these theorists maintain that development is basically continuous, they also agree that distinct spurts occur, and that these tend to have a universal sequence in the order of their emergence. These spurts are referred to as stages which are tied to ages because growing capacities lead to more engagement and the perception of more complex situations. Learning to easily use one set of concepts and to express them in behavior frees the person to concentrate upon other stimulative aspects of his environment. Thus ability to walk or talk changes a child's contacts as his ability to learn from them enlarges his store of concepts. Maturation increases with engagement, and engagement increases maturation. It is an interaction process.

This paper will propose eleven temporal problems. The solution of each one forms a new state in the development of the self in time. Although these states are introspectively familiar to all of us and are described in terms of temporal problems which everyone experiences, most of their aspects can now be substantiated by American psychological research.

It is here proposed that the self forms in a progression of stages, each of which involves a previous organization of time, grows out of a problem in that time organization, and is resolved by reaching a new and more complex level of time orientation. The self is inextricably bound to time because neither exists without the other. The self is the agent upon which time is recorded, but the agent cannot grow or act without the experience of time. The lifespan of the self-in-time can be divided into stages, each one normally occurring within an age-range when the new self *first* begins to change through achieving a new time orientation.

I. PERMANENCE OF OBJECTS AND PERSONS (DURING YEAR 1)

The infant's first time problem is permanence or the continued existence of objects and persons. Upon the establishment of this concept rests the security of the self.[17,18,19,20,21] When early experience indicates an untrustworthy or capricious environment, a resulting anxiety foreshortens the future time sense.[22,10]

American research is producing evidence which makes us honor the unlearned readiness of infants to test permanence. At birth or very soon afterward, babies are able to respond and are alert to a wide range of stimuli: touch, movement, taste, sound, color, shape, size, pattern and distance. Newborn babies orient to any of these stimuli by a slowly coordinated whole response of the whole body. This has been measured by finding correlations in such behavior as turning, increased activity, reaching or withdrawing, vocalization, tracking with the eyes, surprise reactions, rising heart and respiration rates and skin temperature.[23] Psychologists, therefore, infer from these reactions that newborn infants can form percepts or encoded correspondencies to outer reality. Furthermore, they can learn, that is, discriminate differences in percepts, as well as retrieve and combine percepts. For example, one-day-old babies have learned at the sound of a bell to turn left for a bottle, and at the sound of a buzzer to turn right for a bottle.[24,25] The motor response is slow but the learned intent is evident.

Infant behavior suggests that their first perceptions are singles, and when repeated, are at first regarded by them as discrete events. Continuing recurrences present a problem to the infant who, as an active creator of hypotheses about the functioning of his world,[26,27,14] must soon progress to a new concept that objects continue to exist — for example, that his mother exists when she is out of sight. Research indicates that startle reactions occur when consistent attributes are changed. A three-week-old infant is upset when his mother, seen talking, has her voice displaced to seem to appear to emerge three feet to the side of her mouth.[28] If a favorite object such as a teddy bear or bottle is repeatedly raised until it disappears behind the left side of an overhead screen and then reappears down from the right side, a three-month old baby soon learns to look to the right immediately after the object disappears on the left. However, a five-month old baby will slowly track with its eyes an imaginary arc across the overhead screen, showing that it has assumed the continuity of the object's motion during the time when it was invisible. If the object does not reappear finally on the right, a six-month old baby, unlike a younger one, will look back to the left to check its initial perception which failed to meet expectations. Infants show surprise when a familiar object, seen to be covered, is not there when the cover is removed. Four-month old babies will accept a favorite toy or 'security' object as the *same* one when it reappears immediately in a different location. But eight-month old babies will show startle, hesitation in accepting the object and continue to examine the spot where it disappeared. [23,29] It can be seen from these examples that the permanent existence of objects and persons is the first time concept underlying behavior, and that it becomes a time crisis sometime between four and eight months.

When the early experience of babies indicates an untrustworthy environment, self-security suffers. Psychosomatic symptoms often appear and basic adjustment is usually delayed or distorted.[17,18] Reactions of anxiety, non-communication and shallow social relations are typical of the later personality structure of infants who have been separated for a length of time from their familiar caretaking person.[19,20,21] Erikson and other psychoanalysts have suggested that inconsistent caretaking in infancy may shorten the sense of future time in later years. Data from time studies with juvenile delinquents indirectly indicates this relation.[7,10]

These findings have important ramifications: Impermanence in the family setting is a concomitant of the rising American divorce rate. The security derived from the permanent and supportive presence of the father until the child is at least four years old has been found to modify the later capacity of women to reach orgasm[30] and of men to form the conventional attributes of masculinity.[31] These examples demonstrate the significance of a stable environment for forming the first foundations of a non-distorted time sense for the infant who experiences parental continuity.

II. CLOCK TIME (1 - 3 YEARS)

The next time adjustment for the child is based upon the establishment of routines for eating, sleeping, etc., which must appear to the toddler to relate to an arbitrary or clock time.[32] The fact that there is another time necessitates some reorganization of his subjective time and increases his sense of the saliency of cultural time.[33] For instance, his mother may say "Why are you so slow eating your supper tonight?", whereas he may have thought he was making fair progress. Children's difficulty in adjusting to schedules at variance with their internal rhythms is relfected in the decreasing number of angry outbursts after the child reaches two years.[23] By this time, feeding, toilet training, and sleeping patterns are usually established. His behavior, resulting from his constructs about how to anticipate events has brought his activity into line with adult time. He has learned that 'there is a time as well as a place' for everything. Furthermore, his time concepts of past, present, and future begin to have extent. The child is then no longer out-of-step with daily events and his temporal understanding contributes to his self-regard as adequate. He has thereby forced his internal rhythms to accommodate to external cycles so that, in a sense, time has become a discipline.[34] The psychoanalytical literature associates the pleasure principle with timelessness which is coerced by frustration toward a sense of time which is more structured and differentiated.[35]

American mothers, when compared cross-culturally, are excessively strict with cleanliness and toilet training.[2] Firm scheduling and emphasis on cleanliness in the care of children is a product of bottle-feeding programs advocated by Dr. Holt and other influential pediatricians in the first quarter of the century. The psychology of vehaviorism, or the view that baby personalities are the result of total conditioning and shaping, was also begun at that time and is still widely believed by the American public. Fifty percent of the children referred to child guidance clinics for behavior problems have been forced at earlier ages to conform to meticulous schedules:[2] one factor in their maladjustment may have been that these children needed more time to learn the ground rules of socialization.

III. RESTRICTION TIME (3 - 5 YEARS)

The child of this age finds there are times when he can safely expand his interests, and times when he must stop, obey, and accept a dependent status. Because of this dichotomy of time, he must learn to tolerate larger and larger periods of frustration. For example, to wait while others finish eating, shopping, chatting, etc.. In return, he feels the pride of being more like grownups and hence acceptable to himself and to others. Also, he learns to handle the aggression which rises in reaction to frustration and discipline. The child evolves strategies within time such as sulking until his anger wears off, sublimating it by pitching rocks, kicking the chair rung, or waiting until he can aggress against a more helpless child or animal.

American discipline can be divided into three kinds: (i) *authoritarian* discipline, based on force or threat, which research indicates produces conforming, and responsible, if somewhat conventional and anxious, older children and adults. Their time perception is well related to external time because the unhappy consequences of not complying with conventional time have been forcibly taught them. They become docile though unwilling conformists to the time requirements of society.[36,37](ii) *permissive* discipline, where the parent deliberately refrains from punishing in order to allow the child to learn for himself or from his peers. By comparison with children trained forcibly, this method produces more independent, aggressive, and creative types. These children have a more elestic sense of time with more imaginative fantasies about the past and future, and a less rigid adherence to external time.[38] Time is viewed as more flexible, and its users may be somewhat 'timeless' in the absorption of the moment. At maturity, this group has more idlers and more innovators. (iii) The third method of discipline is the *reasoning* method, which is based on an appeal to self-understanding and democratic procedures. This generally produces a higher level of moral behavior and more self-generated guilt from transgression. These children are apt to manipulate time for their own ends because their exposure to reasoning and analysis helps them to use time as a means.[39,40] As they grow older, they are apt to see the relationship between delay of gratification or long-term planning and their access to goals and achievement.[36,41]

IV. CAUSAL SEQUENCES (6 - 11 YEARS)

Busy trying, doing, reality-seeking, and muscle-flexing, this age is exploring larger and larger causal chains or sequences in the time units necessary for the accomplishment of tasks. The individual begins to tackle problems in a certain workable order. A systematic approach and orderly method for reaching goals is seen to effect an economy in time.

Having understood that one thing leads to another, the youngster is ready to learn another secret of time which is that certain sequences in his behavior cause others to have feelings which he has experienced. The self begins to empathize. To do this, he has to gain the ability to project himself into another's place and foretell in time the experience that another person will have. Thus, to a playmate in baseball he can say 'Good try' and know that this remark will console his friend for missing the ball. Or he can say to himself 'Mother is irritable tonight because she is tired from shopping all day: I will wait awhile before asking to go to the movies.' He has learned to employ time as causal to his interpersonal relationships.

Pre-adolescents use their greater verbal fluency to argue or arbitrate their problems, seeking over a period of time for a desired solution. Instead of the short angry episodes of younger years, their interchanges last longer and aim to maintain continuities with others rather than break these off.[23] Future gains begin to take more precedence over immediate reactions.

Until recently, it was believed that better schooling at this age would increase the long-term aspects of achievement in children from Black, Chicano, and migratory sub-cultures. Recent findings are that improvement in the quality of the education offered these children brings no significant amount of improvement in their achievement level.[42] Psychologists believe that their negative self-concepts account for these findings. Insecurity about their early care and support, lack of training in time routines and little explanation of the reason for discipline — all of which are necessary to self-growth in time — may have combined with general deprivation to cause negative self-concepts among these underprivileged school children.

V. PERSONAL TIME (12 - 15 YEARS)

The problem here is the acceptance of the configuration of the self.[43,44,5,11,15] The adolescent opens inwardly, giving much time to assessing his person — its capacities and inadequacies, new body and aspects of masculinity-feminity. He does the same with other persons. In so doing, the person forms a conscious self-image of who he is. Previous positive or negative self-images were unconscious. Now the adolescent can see himself in past time and project his potentialities in the future. He can think about his thoughts. Until now, the self has been an unconscious cohesive force which begins to be conscious of the self in time at adolescence. However, the time perspective, like the self-perspective does not project very far into the future.

Jungian psychoanalysts find that innate personality predispositions group into one of four different time oreintations as the individual approaches maturity.[45,46] The efficient *Sensation* type, who is held by concrete immediate experience, is absorbed by interest in present time. The behavior of the logical *Thinking* type is dominated by the past which is conceived as flowing along a linear line into the present and toward the future. For the *Feeling* type, time is circular because the emotional values of the personal past illumine the present, and the future revives the emotional impact of the past through the memory of it. For the *Intuitive* type, future time is unrelated to past or present time because he is relatively free of time constraints as he acts upon his hunches and unverbalizable cues.

Eysenck and Orme find experimental evidence for the greater involvement of extroverts in the present and introverts in the past and future.[10] Research findings indicate that for extroverts present time seems longer than it does to introverts.[7,10] Any further generalizations about the estimate of time cannot be made with confidence because of the anarchy of techniques, the diversity in subject samples, the interplay of cultural variables, the conflicting results, and the lack of replication which has characterized the experimental work on this aspect of time.[7]

In the American culture, adult society is often ineffective in relating to the adolescent age group. Tacit acceptance of the right of adolescents to reject parental control, less isolation and increased contact between the sexes in school and out, plus the unavailability of early work are all reflected in the adolescent rift. Parents nostalgically try to keep the nuclear family tight whereas the adolescent demands much free time to isolate himself from the family or to gather without supervision with other adolescents. These uses of time are disapproved by the older generation who in turn are resented by the young for attempts to fill their time productively. The difficulty lies partly in the differing concepts of time: adolescents find so-called unfilled time necessary for orientating themselves, whereas adults consider that do-nothing time is wasted time. Thus the adults are oriented toward the useful immediate futures of their offspring at a period when the latter are blase about futures or practical matters. It may be surmised that they are concentrating on their identity in time as a prerequisite to their later employment of time.

VI. MUTUAL TIME (15 - 25 YEARS)

The problem here is the search for intimate compatibility with another person for no other purpose than the complete sharing in time of experience. It may be the mutuality in time of two lovers or of two intimate friends, or the rapport of a younger person with a significant adult. The sense of personal isolation is forever reversed. Even if the loved person disappears, the experience remains and the self is more. It now becomes clear to the individuals involved that sharing time with someone else accelerates one's self development. Time is perceived by them as benign under circumstances where it is open-ended and continuous. Future such relationships are anticipated with eagerness – a new step forward for the self.

Psychiatrists know how many people are unable to reach a level of mutual time. Encounter groups also aim to help individuals through group pressure to re-direct their behavior toward better inter-personal relationships of depth. In these sessions, group aid is directed toward building each person's non-discursive abilities, i.e., the development of more sensitivity, more honesty, more expressiveness together with greater understanding and sympathy. These programs vary with the method and skill of the trainer. Research indicates that these sessions make little long-range change in behavior, but do produce peak experiences or moments of extreme self-expansion which are highly valued by those members who experience them.[47]

There is observable in the United States a new solution to the old need for belongingness. The migra-tion from rural regions and the dissolution of stable communities, even in urban areas, has created a new way to form the ordinary affiliations broken by frequent changes of job and region. The average American has twelve different jobs in a lifetime[48] and some of the population moves every five years. In some circles, anything over two years is considered a long-term career commitment.[48] As a result of this high-technology condition, associations become temporary, and many Americans have develo-ped techniques for a relaxed approach to new acquaintances. The result is a capability for forming and re-forming groups rapidly and in a manner which affords inclusion of newcomers. These groups are characterized by much sharing of activities and personal problems, and the building of quick loyalties. These interpersonal relations reflect an easy, non-rigid use of time. The self is typically undefensive about inroads upon his own time-routines, dropping activities in order to be approachable for others or to initiate some acquaintanceship.

VII. ALTERNATIVES IN TIME (18 - 25 YEARS)

The time span of a person's life may be considered as an arc. In the beginning one has only a momen-tum for change and some innate potentials for selecting from the rapidly expanding world. During the fast crescendo of early life there is little consciousness of past or future. However, early adulthood forces a new time perspective because it is a choice point. First of all, the individual is now old enough to evaluate his own past in its various emphases on socialization. He is also old enough to analyze the culture heritage in general. It is a crucial period during which a youth examines and ponders upon the nature of each of society's institutions: education, occupation, religion, government, class structure, and the family. He knows he must choose a future to which he will belong. The nature of his future depends on one of three alternatives: (i) He can continue in the societal mold to which he has been poured or (ii) he can reject it for one of his own making or (iii) he can join others to try to change the forms of society. He is at the top of the arc of time and he pauses, taking a moratorium to decide where he wants to stand.[49,50]

This is a new period in the American lifespan.[50,51,52] Formerly, young people grew imperceptibly through a prolonged adolescence to an early adulthood of jobs and responsibility. Now there is time for a youth period because the preceding adolescent stage arrives sooner by both social acquiescence and in physical puberty. For example, the menarche is three years earlier in the United States than it was one hundred years ago in Europe.[53]

It is frequently difficult for American youth to find work in this highly-industrialized society. Also, this age group has discovered that the affluence of their middle- and upper-class parents will float them for awhile. Much of this aid filters down indirectly to help maintain lower-class youth as a consequence of their mutual agreement to share, have minimal needs, and postponement of it, is manifested in youth's different dress, language, music, health foods, night rather than day living, mystic cults, exercise and meditation routines, biofeedback training, unplanned travel, etc.. In short, they explore and cultivate every conceivable use for time that contributes to their consciousness expansion, that is, their self-growth. Future time and past time are weighted evenly in a poised present time.

This period of youth has had one lasting effect in their new insistence on the new right of man to more leisure. In the United States, the work hours of the year are becoming roughly equal to the free time (exclusive of sleeping). By developing their ability to use leisure in a personally creative way, the young enrich their later life in variety and depth.

Youth's explorations of consciousness include the manipulation of time through extreme slowing, repetition or acceleration of it and by the denial of time, as in the unstructured day. These experiments toward increasing the significance of time are reflected in some new art forms in American cinema, choreography, music and theatre. Such innovations show that behavior which is usually overlooked or considered repetitious can become, by time-extension, new or different or significant.

On an individual level, many youths are pursuing meditative controls of time through the voluntary regulations of their beta, alpha and theta brain waves. These studies have shown consciousness of time to alter with the dominant neural activity. Beta waves hasten time and make it more vivid. Alpha waves relax time, permitting more merging of past, present and future. Theta waves produce lack of consciousness of time, that is, a timeless feeling.[54]

VIII. THE USES OF TIME (25 - 40 YEARS)

This is the period of greatest convergence between external or objective time and subjective time. One comes to grips with reality, so to speak, by no longer upholding the dichotomy between personal time and clock time. Behavior is brought into line with the consensual validation of group time in order to get things done. Commitment to job, marriage, children and property finally focuses the use of time toward establishing and securing these goals. Productive lives become a necessity, and the personality unifies its various capacities toward the demands of reality.

With the increased responsibilities of maturity, time becomes a greater problem in differentiation and articulation in order to impose more on the environment than the environment imposes on the person. In a high-technology society, this means viewing "time as an energized, expendable and directional phenomenon ... not as an encompassing medium lacking dynamic properties."[36] Minor effects of this attitude are that the more highly motivated persons tend to underestimate time[55] and to experience more time anxiety.[56] Time becomes more restrictive with increased responsibilities and drudgery. Less rebellious, sensuous and self-indulgent than in the youth period, the mature individual no longer toys with alternative self-images, but settles primarily on one life-style.

The irregularity of an individual's synchrony between his internal rhythms and the external rhythms to which he is exposed shows in the rise and fall of correlations between seasonal periodicities and psychological events. There is evidence that the time of the year causes variation in measures of intelligence and abilities,[57] length of menstrual periods,[59] labor,[58] the menarche,[58] mental breakdown, [57,60] suicide[61] and crime.[62] While these findings do not apply to every individual, they tend to show up in studies of large populations. Women are more affected than men because of their reproductive cycles, and the relationship of these to their mood shifts.[10]

Cultures vary in the amount of time-conformity which is concentionally required of their members. Many cultures are relatively tolerant of individual variations and also recognize more legitimate priorities over commitments of time such as one's state of health or the necessity for unhurried interpersonal relations. Cultures also vary in the amount of time which individuals are expected to invest in recognizing their past (as in homage to their ancestors), or toward earning their immortality (as in the Hindu *dharma*).

By comparison, the typical American adult is tremendously time-bound into meticulous observance and concentration upon the present. The average American of this age group is harried by time. He measures his character by it ("I've saved $6,000 over the last four years"); he measures his job by it (destroy rather than dis-assemble old equipment because "time is money"). He is paid in terms of time. His work programs are all time-limited. He unconsciously budgets his free time, expecting to use so many minutes to phone a friend, to read the newspaper, or to play with his son. In other cultures, these behaviors tend to come to a natural close as a period of unstructured time. But for the adult American only when time is under his control, that is, properly parceled out in relation to his needs, does a person feel he has himself in control. Women are as pressured as the men to utilize a regulation of time. The diversity of roles expected of an American woman demands that she be ready for anything anytime thus meeting new contingencies rapidly.[63] This flexibility segments any bulk-time for activity she may attempt.

Time-budget studies indicate some psychological derivatives of the fact that the working day, the working week, and the working life are becoming shorter. Lower-level employees still view their pay as buying their hours at the plant plus a supervised minimum or time-measured productive units. However, management in large organizations attempts to build a company morale which is independent of time by giving the upper-level personnel more trust and less supervision in their use of time. It has been found that a sense of corporate belonging can supplant the coercion of time as a drive to raise the production indices.[64]

The average American spends roughly the same amount of time for work commitments as for other activities. Free time — that amount left out of the twenty-four hours after work time and sleeping time are deducted — is, of course, divided between maintenance duties and leisure. Each maintenance activity is becoming shorter in duration because of more efficient services: supermarkets center the shopping requirements, cars need fewer reapirs, laundry and cleaning are rapid, foods are prepared and pre-cooked, and short-order meals are available everywhere. However, the rising standard of living for the great mass of Americans introduces a subtle demand for more *kinds* of maintenance.

The nature of one's use of leisure is now considered to be the criterion of one's life-style, one's significant way of life. One trend in the use of this free time externalizes the psychological focus through

tangible involvements which range from the interest in brightening one's physical appearance and surroundings to brightening one's view by casual shopping or driving about or by longer trips.

The other trend in the use of leisure is toward internalizing the psychological focus through new forms
of individualism by way of innovations in the intellectual and affective side of life.[65] People are investing less time in the large institutions of religion, politics, and schooling,[50] with the result that these
are losing psychological force for the individual. To offset this vacuum, there is a widespread movement toward more personal forms of self-enrichment. In 1965, twenty-five million Americans, or one
out of every five in the population, participated in some form of voluntary education. This indicates
an extraordinary rise in the free time which is used for directed purposes, being four times greater
than it was twenty years earlier.[50] The commonest form of leisure activity in the U.S. is watching
television. The attraction, of course, lies in psychic variation and change because the screen satisfies
the need to laugh, to be emotionally involved or to experience sensory color. Andre Malraux said that
television is the 'industrialization of dreams.'[48] As an art form it may permit the viewer to be briefly
translated into a different time and reality.

IX. RECONSIDERED TIME (40 - 50 YEARS)

This is a time of re-evaluation of life's course while it is in full swing. In this period the individual is
haunted by thoughts of performing the major aspects of his life again differently. Doubts surface as to
the value of decisions made, stations won, or the status aimed for. One asks oneself, 'Do I really want
to do what I am doing or should I change before it is too late?' It is a time of re-evaluation when reflection illumines dissatisfactions. The fantasy of reliving time by abandoning things as they are and
starting over can become a major decision or a matter for repression. Minor adjustments for men
frequently take the form of a change in job or career whereas women often seek an arena of interest
outside the home. Whether alternatives are acted upon or only imagined, they create for the individual a crisis in the resolution of time. Whatever the outcome, this midway pause incolves some adjustment before it is possible to settle down once again to practical ways and means.

These uncertainties are known to accompany the early psychological changes of women's menopause
but recent research[66] has established a somewhat parallel period for men. Whether these behaviors are
due to hormonal imbalances in both sexes or whether they are a normal derivative of the self in time
remains open to investigation. All that is known so far is that insecurities about time tend to occur
at this period of life.

The American divorce rate, now nearly one out of three marriages, is heavily weighted by persons in
this age range of 40-50 years. The other large age category for divorce is the one for the early adulthood (18-25 years) when there is also a thoughtful period about the future.

X. THE FORESHORTENED FUTURE (50 - 60 YEARS)

The rugged individualism of late maturity creates a perception of time as overly swift and as possibly
preventing the completion of important work and the final achievemen[of life's goals.[67] In the latter
part of this period, typically around fifty, the self begins to realize that the body has only a limited

store of energy. While people at this age are 'the command generation' at the height of whatever their competence and authority, they are nevertheless faced with waning physical and psychological stamina. Economy in the use of time becomes a source of anxiety because saved time can be used for recovery of energy. The self begins to worry that its life-time is too short ("later than you think").[68] The push forward still exists to achieve goals or, once achieved, to fully appreciate and enjoy them. But the deeper reality for the self is that time is running out.

In the American culture there are more apt to be golden years for men than for women. Whereas the man's work load is still a full one, the offspring may have settled far away leaving his wife to occupy and care for a somewhat empty nest. She has much time to wonder who she really is and how she will fulfill the rest of her life. Without close relatives or outside employment, 'killing time by keeping busy' can be a problem. Furthermore, three women out of every five women are widowed in this period.

To disguise the signs of aging often becomes a preoccupation of both sexes, for it comforts the self to avoid recognizing the foreclosure of time. Thus, older Americans try to feel younger by dieting, exercising to keep in shape, refusing to curtail the frenetic activity of their more vital years. Heart attacks and other signs of hypertension take a higher toll in the United States than elsewhere.

XI. THE RICH PAST (BEGINNING ABOUT 65)

Older people are subject to veridical $d\bar{e}j\grave{a}$-vu in which new experiences are the same yet not the same to them. This requires explication — "I've seen this before but it's not identical" — as well as adjustment to a slight sense of being displaced in time. They are beguiled by time, as though the quintessence of time lay in the return to something known before.[34] This divides their inner life between concentraring on the present and perusing the past, which is the main temporal problem of this period.

This is a time of gradual enfolding and withdrawal during which the self views itself in the long past. Crucial decisions are worked through again in a search for self-justification. The quality of recalled experience is savored which makes any recounting of it slow and filled with detail.[69] New experiences are not desired because there is so little left. The aging person wants to be comfortable with the familiar — old friends, favorite children, places that revive the aura of the past.

Time passes quickly rather than slowly[70] both because it is filled with associations and because even small changes require effort and small rests. Deterioration of the sensori-motor functions are not usually accompanied by intellectual impairment — that is, until the onset of senility. The old person is much less open to new stimuli, slower to react, and subject to increasing defects in hearing, vision, and mobility. Nevertheless, cognitive and emotional functions are usually retained, including sexuality. Studies indicate that persons between 60 and 80 years do not show a decline in general intelligence test scores because what they lose in speed they tend to gain in accuracy. Fifty percent retain their vocabulary and a third increase their word power.[71] (Contradictory findings in early testing of elderly persons were due to the use of 'speed tests' which had a time limit for completion.) The need for social interchange and the need for play still exist, although the old person has less physical and innovative resources for finding the means to satisfy these needs. The effectiveness of the person may be impaired without changing the inner core or self.[72]

The American culture is cruel to this age group and there are multiple underlying reasons for this. First, Americans are intolerant about decline and deterioration. As a new nation built rapidly, they regard everything new as better and anything old as probably unimportant or obsolete. Second, they lack respect for tradition because there is little evidence of a crumbling antiquity or a beautiful former way of living. Third, denial of unpleasant turth can take extreme forms of denial in the U.S. For example, the religion of Christian Science denies the reality of disease, infirmity, suffering and death. 'Pollyanna' attitudes of exaggerating the good in everything have glinded Americans to the difficulties of aging. Also the Horatio Alger myth that everyone can rise from his circumstances 'if he but tries' is often applied thoughtlessly to the poor and the old. While these may be middle class attitudes, the great bulk of Americans are middle class (1½% are upper class and 10% live on a subsistence level). In groups which are marginal in socio-economic status, such as migrant workers, many consider that the pattern of expected events and the consequences of their behavior are unpredictable and not under their control.[73] The aged among this class are especially apt to have this attitude.[74]

The loneliness of old age is a product of peculiarly American conditions: the short generational continuities of an immigrant stock spread over a huge land, the moving about — whether opportunistic or necessary — which reduces the contact-time between generations, and the American tendency to discount genetic pride. Other cultures with less generation gap, less rapid industrialization, fewer facilities for removal and resettling, and firmer class barriers for intermarriage, do not create as much social emptiness in the lives of their older people. This is a melting-pot society in a highly technological style of living, and the secluded old suffer most from it.

Psychology notes that death usually has three phases: first, the social part of the self withdraws and dies. In the United States, ostracism and neglect hasten this process, for the dying are removed to hospitals where they are segregated by sex. Relatives may visit briefly at stated hours. Hospitals for terminal cases do not even keep hope alive. There is little for patients to do except wait. In the second phase, the psychological self withdraws and dies. Surroundings are gradually closed out. Finally, the body dies, as one after the other, the organ functions cease.

Added to the wealth of anecdotal evidence that individuals have postponed their deaths to coincide with the return of a son, the publication of a book, etc., there is statistical evidence that the death rates for large populations drop for birth dates and for events of religious, political, cultural and national importance.[75,76] The implication is that the self has some control over forcing death to defer to time.[77]

Death is the most important *rite de passage* in the majority of world cultures. Elsewhere than in America, it is spoken of with foreboding and looms against life more or less constantly. The actual event is usually preceded by a long series of farewells to significant persons and takes place finally in the presence of a solemn circle. The behavior of the dying person indicates that he meets death as he lived life, with the same hidden resources, the same consistency, and the same insecurities. In short, it appears that the self is still there at this time. In the U.S., death is seldom referred to and then in passing — for example, the death toll on the highways or the need to carry a heavy load of insurance against catastrophes — one way of 'writing off' the future.

When death is imminent, the American medical practice is to administer medication to make this last phase an easy one for the dying. The dosage necessary to produce relief from symptoms also tends to confuse and disorient the patient. At the end death usually happens quietly and without pain. Life goes away softly. Instead of being a final time it is no time, for the self is not there. Nevertheless, many Americans take a self-reliant, if rueful, satisfaction in making advance provision for these conditions of hospitalization and death because it is a quiet departure, without trauma or disruption of others' lives.

CONCLUSION

This paper has proposed that the self in time is a primitive concept because time and the self are inexorably associated. Time exists only when there is a psychological agent or self which perceives it, whether consciously or unconsciously. The self, in turn, is strung out along the thread of its time. It travels back and forth along this thread to search for memories or to project the future. But if the self is not found there, neither is time.

The self is the invisible binder of the visible personality. While it ages in the arc of its lifetime, it demonstrates a selective and unique dynamic. Although the metamorphosis is continuous, it is also uneven, depending on the self's ability to resolve the temporal problems and to go forward at each stage in a new matrix of its own time perception.

Everyone according to his age is familiar with some of the temporal problems. However, this paper sets them forth as though they were the successive rises of a stairway. In this sense, to surmount them forms stages for the development of the self. Some of the research in psychology which supports this view has been indicated, as well as modifying factors in American cultural conditioning.

NOTES

1. Kagan J. and Moss, H.A.: *Birth to Maturity*. New York: Wiley 1962.

2. Sears, R.R.; Maccoby, E. and Levin, H.: *Patterns of Child Rearing*. New York: Harper & Row 1957.

3. Schaefer, E.S. and Bayley, N.: "Maternal Behavior and Personality Development – Data from the Berkeley Growth Study." *Child Development, 13* (1960), 155-73.

4. Gesell, A.: *First Five Years of Life*. New York: Harper & Row 1940.

5. Gesell, A.; Ilg, F.L. and Ames, L.B.: *Youth: The Years from Ten to Sixteen*. New York: Harper 1956.

6. Kagan, J. and Moss, H.A.: "The Stability of Passive and Dependent Behavior from Childhood through Adulthood." *Child Development, 31* (1960), 577-91.

7. Doob, L.: *The Patterning of Time*. New Haven: Yale University Press 1971.

8. Holme, R.: *Developmental Psychology Today*. Palo Alto, Cal.: Communications Research Machines 1971.

9. Dahl, M.: "A Singular Disturbance of Temporal Orientation." *American Journal of Psychology, 115* (1958), 146-54.

10. Orme, J.E.: *Time, Experience and Behavior*. London: Iliffe 1969.

11. Erikson, E.H.: "The Problem of Ego-Identity." *Journal of American Psychoanalytic Association, 4* (1964), 56-121.

12. Bettleheim, B.: *The Informed Heart*. Glencoe, Ill.: The Free Press 1960.

13. Freud, S.: *Collected Papers*. London: Hogarth 1925 (Vols. 1-4).

14. Piaget, J.: *Construction of Reality in the Child*. New York: Basic Books 1954.

15. Erikson, E.H.: *Childhood and Society*. New York: Norton 1963.

16. Kohlberg, L. and Turiel, E.: *Research in Moral Development*. New York: Holt, Rhinehart & Winston 1971.

17. Fries, M.C.: "Psychosomatic Relationships between Mother and Infant." *Psychosomatic Medicine, 6* (1968), 159-62.

18. Spitz, R.A.: *The First Year of Life*. New York: International Universities Press 1965.

19.   Rheingold, H.: "The Modification of Social Responsiveness in Institutional Babies." *Monographs of the Society for Research in Child Development,* Vol. 21 (1956), 2.

20.   Bowlby, J.: *Attachment.* London: Hogarth 1969.

21.   Bowlby, J.: *Child Care and the Growth of Love.* Baltimore: Penguin 1965.

22.   Erikson, E.H.: *Identity, Youth and Crisis.* New York: W.W. Norton 1968.

23.   *Developmental Psychology Today.* Del Mar, Cal.: Communications Research Machine 1971.

24.   Lipsitt, L.P.: "The Concepts of Development and Learning in Child Behavior." In Lindsley, D.B. and Lumsdaine, A.A. (Eds.): *Brain Function.* Berkeley: University of California Press 1967. Vol 4.

25.   Lipsitt, L.P.: "Learning in the Human Infant." In Stevenson, H.W. (Ed.): *Developmental Approaches.* New York: Wiley 1967. Pp. 225-47.

26.   Bruner, J.R.; Olver, R. and Greenfield, P.: *Studies in Cognitive Growth.* New York: Wiley 1966.

27.   Inhelder, B. and Piaget, J.: *The Early Growth of Logic in the Child.* New York: Harper & Row 1964.

28.   Aronson, E. and Rosenbloom, S.: "Space Perception in Early Infancy: Perception within a Common Auditory-visual Space." *Science,* Vol. 172, No. 3988 (June 1971), 1161-63.

29.   Mundy-Castle, A.C. and Anglin, J.: "The Development of Looking in Infancy." Unpublished paper presented at S.R.C.D., Santa Monica, California, April 1969.

30.   Fisher, Seymour: *The Female Orgasm: Psychology, Physiology, Fantasy.* New York: Basic Books 1974. (Reference in *McCalls,* Nov. 1973, pp. 56-8).

31.   Wohlford, P. and Liberman, D.: "Effect of Father-absence on Personal Time, Field Dependence and Anxiety." *Proceedings of the American Psychological Association, 5(1)* (1970), 263.

32.   Springer, D.: "Development in Young Children of an Understanding of Time and the Clock." *Journal of Genetic Psychology, 67* (1946), 97-106.

33.   Zern, D.: "The Influence of Certain Child-rearing Factors upon the Development of a Structured and Salient Sense of Time." *Genetic Psychology Monographs, 81(2)* (1970), 197-254.

34.   Knapp, R.H.: Personal communication to the author.

35.   Bergler, E. and Roheim, G.: "Psychology of Time Perception." *Psychoanalytic Quarterly, 15* (1946), 190-206.

36.   McClelland, D.C.: *The Achieving Society.* Princeton, N.J.: Van Nostrand 1961.

37. McClelland, D.C.: "The Importance of Learning in the Formation of Motives." In Atkinson, J.W. (Ed.): *Motives, Action and Society*. Princeton, N.J.: Van Nostrand 1958.

38. Loehlin, J.C.: "The Influence of Different Activities on the Perception of Time." *Psychological Monographs, 474* (1959), 4, 73.

39. Klineberg, S.L.: "Future Time Perspective and the Preference for Delayed Reward." *Journal of Personality and Social Psychology, 8* (1968), 253-57.

40. McClelland, D.C.: *Motivating Economic Achievement*. New York: Free Press 1969.

41. Cottle, T.J.: "Temporal Correlates of the Achievement Value and Manifest Anxiety." *Journal of Consulting and Clinical Psychology, 33(5)* (1969), 541-50.

42. Coleman, J.S.; Campbell, E.Q.; Hobson, C.J.; McPartland, J.; Mood, A.M. and York, R.L.: *Equality of Economic Opportunity*. Washington, D.C.: U.S. Government Printing Office 1966.

43. Ausubel, D.F.: *Theory and Problems of Adolescent Development*. New York: Grune & Stratton 1954.

44. McCandless, B.R.: *Adolescents: Behavior and Development*. Hinsdale, Ill.: Dryden 1970.

45. Mann, H.; Siegler, M. and Osmond, H.: "The Many Worlds of Time." *Journal of Analytic Psychology, 18(1)* (1968), 33-56.

46. Osmond, H.; Yakir, H. and Cheek, F.: *The Future of Time*. New York: Doubleday 1971.

47. Lieberman, M.A.; Yalem, I.D. and Miles, M.B.: "Encounter: the Leader makes the Difference." *Psychology Today,* March 1973, 69-76;

48. Kaplan, M. and Besserman, P.: *Technology, Human Values and Leisure*. Nashville, Tenn.: Abingdon Press 1971.

49. Rizzo, A.E.: "The Time Moratorium." *Adolescence, 2(8)* (1968), 469-80.

50. Dumazedier, J.: "Leisure and Post-industrial Societies." In Kaplan, M. and Besserman, P., *op. cit.*

51. Reich, C.: *The Greening of America*. New York: Random House 1970.

52. Keniston, K.: *Young Radicals: Notes on Committed Youth*. New York: Harcourt, Brace and World 1968.

53. Dubos, R.: "Biological Individuality." In *Annual Editions, Readings in Human Development '73-'74*. Guilford, Conn.: Dushkin Publishing Group 1973.

54. Karlins, M. and Andrews, L.M.: *Psychology: What's in It for Us*. New York: Random House 1973.

55. Meade, R.D.: "Effect of Motivation and Progress on the Estimation of Longer Time Intervals." *Journal of Experimental Psychology, 65* (1963), 564.

56. Knapp, R.H. and Green, H.B.: "Time Imagery and the Achievement Motive." *Journal of Personality, 25* (1958), 426-34.

57. Davies, A.D.M.: "Season of Birth, Intelligence and Personality Measures." *British Journal of Psychology, 55* (1964), 475.

58. Malek, J.; Gleich, J. and Maly, V.: "Characteristics of the Daily Rhythm of Menstruation and Labor." *Annals of the New York Academy of Science, 98* (1962), 1042.

59. Pasamanick, B. and Knobloch, H.: "Epidemilogie Studies on the Complications of Pregnancy and the Birth Process." In Caplan, G.(Ed.): *Prevention of Mental Disorders in Childhood.* London: Tavistock 1961.

60. Kellner, R.: "The Seasonal Prevalence of Neurosis." *British Journal of Psychiatry, 112* (1966), 69.

61. Pokorny, A.D.; Davis, F. and Harbenson, W.: "Suicide, Suicide Attempts and Weather." *American Journal of Psychiatry, 120*(1963), 377.

62. Stott, D.H.: "Evidence for a Congenital Factor in Maladjustment and Delinquency." *American Journal of Psychiatry, 118* (1962), 781.

63. Calabresi, R. and Cohen, J.: "Personality and Time Attitudes." *Journal of Abnormal Psychology, 73(5)* (1968), 431-39.

64. Wolman, B.B.: *Handbook of General Psychology.* Englewood, N.J.: Prentice Hall 1973.

65. Obermeyer, C.: "Challenges and Contradictions." In Kaplan, M. and Besserman, P., *op. cit.*

66. Scarf, M.: "Husbands in Crisis." In *Annual Editions, Readings in Human Development, '73-'74.* Guilford, Conn.: Dushkin Publishing Group 1973.

67. Cottle, T.J.: "Temporal Correlates of Dogmatism." *Journal of Consulting and Clinical Psychology, 36(1)* (Feb. 1971), 70-81.

68. Dickstein, L.S. and Blatt, S.J.: "Death Concern, Futurity and Anticipation." *Journal of Consulting Psychology, 30* (1966), 11.

69. Costa, P. and Kastenbaum, R.: "Some Aspects of Memories and Ambitions in Centenarians." *Journal of Genetic Psychology, 110(1)* (1967), 3-16.

70. Aisenberg, R.: "What Happens to Old Psychologists." In Kastenbaum, R. (Ed.): *New Thoughts on Old Age.* New York: Springer 1964.

71. Blum, J.E.; Fosshage, J.L. and Jarvic, L.F.: "Old Age and Intelligence." *Human Behavior,* June 1973.

72. Kastenbaum, R. (Ed.): *New Thoughts on Old Age.* New York: Springer 1964.

73. Nelkin, D.: "Unpredictability and Life-style in a Migrant Labor Camp." *Social Problems, 17(4)* (1970), 473.

74. Curtin, S.R.: *Nobody Ever Died of Old Age.* Boston: Little Brown 1972.

75. Phillips, D.: *Statistics: A Guide to the Unknown.* New York: Holden-Day 1971.

76. Koestler, A.: *Roots of Coincidence.* New York: Random House 1972.

77. Inman, W.S.:"Emotion, Cancer and Time: Coincidence or Determinism." *British Journal of Medical Psychology, 40* (1967), 225-31.

Time, Death and Ritual in Old Age

R. KASTENBAUM

INTRODUCTION

From Rigidity to Ritual?

Consider first the young: fresh, spontaneous, varied, unpredictable, profligate of energy.

Consider next the *very* young: reflex-animated, predictable in needs and responses, limited in behavioral repertoire and adaptational flexibility.

Now consider the old, especially the very old: routinized, predictable, spinning out a thin network of repetitious and constricted behavior, cautious, and conserving of energy.

These are stereotypes, of course. One should not accept such statements uncritically. They fail to recognise individual differences at each age level, for example, or the influence of socio-physical circumstances. Nevertheless, the apparent curvilinear relationship between chronological age and spontaneity is owrth our attention here. The very young are not yet free from biological imperatives nor possessed of the mature central nervous system functioning required for resourceful adaptation and subtle experiencing. The very old also tend to appear constricted in behavior. The nature of their experiential states requires sensitive inquiry, but one frequently has the impression that feeling and thought are far more repetitious than spontaneous.

Perhaps life-span human development can be regarded as a movement from bio-social rigidity to socio-biological ritual with an intervening phase of spontaneity and flexibility. The processes ordinarily summarized under rubrics such as "growth" and "socialization" might alternatively be interpreted as the *ritualization* of experience and behavior.

Outcomes and goals are not necessarily identical. Yet it might be heuristic to remind ourselves that there is more than one way to view the constellation of thoughts and behaviors often seen in old age. One can make the usual assumption that it is "better" to be young, fresh, and remarkably open to new experience. From this standpoint, old age represents decline and impoverishment. But one might counter-assume that the outcome *is* the goal. In other words, Nature "wants" to make old men and women of us. It is "better" to be fully-formed and completed beings, an end-product that requires much time experienced and many thoughts and behaviors practiced to perfection. Still again, it could be argued that youth and old age (or spontaneity and predictability) are *different* states of the organism, but not "better" or "worse" states. One might as well philosophize over the relative desirability of being a caterpillar or a butterfly when, in fact, both represent intimately related stages of the "same" identity.

In this paper we call attention to the possibility that the old person's relationship to time and death can be understood in terms of a process of ritualization. This approach will lead us to present some relatively familiar observations in an unfamiliar light, as well as to introduce some observations and perspectives that are not typically considered in this context.

We admit freely that the emphasis upon ritualization is in part a methodological device. The relationship of any person — young or old — to time and death is subtle, shifting, and complex. Only a very limited, simplistic, or arbitrary approach could provide the semblance of integration. We have chosen to develop one particular theme, *ritualization,* as a means of "recruiting" a variety of processes and phenomena that should be included in any thorough exploration of the topic. Hopefully, this will provide a useful starting-point for the thoughts and investigations of others, even if the particular interpretations offered here seem less than convincing.

"Capturing the Moment"

Those who are familiar with the literature on temporality have learned perforce to pick their way through a network of distinctions, definitions, and coordinates. We will have to contend with some of these problems here as well. One distinction appears more fundamental than the others, however: the "raw" moment as contrasted with the "captured" or, one might say, "domesticated" moment.

"Raw temporality" is difficult to describe verbally. One gropes clumsily with terms such as "evanescent," "in passing," "here-it-comes-and-there-it-goes." Perhaps one of the difficulties we have on this point is that language is itself a falsifying or domesticating technique. If words "preserve" the raw moment for us, they do so by adding a kind of preservative. The evanescent quality is translated into a realm of symbolic permanence. "Quick! Look! Run!" These exclamations come somewhat closer to the experience of raw temporality than do more elaborate utterances.

Another aspect of the difficulty may be the close relationship between the *moving* quality of the moment, and its phenomenological *immediacy.* We are more likely to be aware of the moment to the extent that either its transitory or its immediacy characteristics are brought to the fore. The unexpected glimpse of a firefly in the evening calls our attention to the fact that we are having a moment while still having the moment. An episode of pain or pleasure reminds us that our sense of reality is intimately associated with the "right now." Although often related, the transitory and immediacy characteristics of the raw moment are not identical, and so may confound our efforts to formulate a simple, veridical expression of experiential temporality.

Much human intention and invention has been invested in efforts to capture the moment. The history of art and monument-making is, in part, the history of preserving or carrying-forward "time-scenes" that one was loathe to see perish. Contemporary examples include the polaroid camera and photography in general, as well as the proliferation of sound recordings in various modalities. We can have the moment both ways in some instances: the television camera presents the spectacular bit of sports action as it occurs — but we can also view it as a replay, perhaps in slow motion as well. A generation of athletes and viewers has already been influenced by the image of the "replay", and this might prove to become a significant and pervasive concept quite apart from its origins in sports telecasting. The replay lovingly (or obsessively) raises a moment of action beyond the sphere of raw temporality: a "nowness" that can be examined and enjoyed retrospectively, that retains some of its essential qualities but does not perish. Along with the "endless loop" tape cassette, this technical innovation can penetrate into our deepest fantasies about life, death, and identity.

Continuation of the sports example is relevant. The action itself eventually becomes translated into that popular conversational coin: the sports statistic. The relentless fan has a headful of numbers. He knows the average number of yards gained per reception by his favorite wide receiver or the earned run average of his team's leading pitcher. Indeed preoccupation with the statistics often threatens to take precedence over the actions themselves. The statistics have an advantage: they are definitive and available. The actions glowed for their moment, then vanished. And so fans (and participants) may lose themselves in the abstractions, analyses and comparisons that numbers yield, forgetting the daring and graceful actions that provided their basis in experiential reality.

The "captured" moment can be shared. It can be woven into a pattern. It can be replayed at convenience. Perhaps most significantly, it offers a protected experience. The encounter with the raw moment and its dangers is shielded against, insulated. One can have a safe and predictable experience. A unit or series of captured moments also may convey a sense of temporal directionality. The replay or pattern resembles on-going life sufficiently for some purposes. Yet one can feel secure because he knows where /when the pattern is going. Temporality thus becomes something like a man-made stream. Properly managed, it will never overflow its bounds, or run dry.

Those of us who read and write about temporality are mostly neither very young nor very old: therefore, imperfectly ritualized. We are in a position to enjoy both "raw" and "captured" or "processed" temporality. There is no need to insist that we experience what might be called a "pure" moment apart from the patterning imposed by experience. It is sufficient to recognize that we can distinguish a range of temporal experiences that vary in their experiential proximity to the pulsating "moment-of-the-moment."

At a theoretical extreme of ritualization, one would never experience a "live" moment. Phenomenological life would be taken up with the enactment of patterned series. Another way of saying this: every experience has its "place," its "meaning." A sudden intrusion into this powerful framework would be quickly assimilated, rejected, or otherwise worked-upon. The fully-formed, ritualized person might not have to recognize that his pattern had even been challenged, so deft would be the screening/manipulating operation.

The opposite extreme perhaps should be characterized as well. Totally engrossed in the moment, the individual who lacks a coherent patterning seems almost to be a different person at different times. He is at the mercy of unprocessed temporality. Pain explodes him; pleasure sweeps him away; deficit of stimulation drops him into torpor. Little continuity could be observed in his phenomenological life, had we the techniques for such observation. He might be, in a sense, a creature begotten by Hume.[1] This Hume-ian being would exist in a succession of instants, possessing no more continuity than what might be yielded by the structure of his (inner and outer) stimulating fields. So captured by the moment, he might scarcely experience time at all, in the sense of continuous, directional flow.

We have focused almost exclusively upon experience rather than behavior up to this point. Experience will continue to be our primary concern here, but the relationship to behavior and interaction will not be entirely neglected as we proceed. What we hope to have suggested by this point is that a pervasive change in the quality of experience may occur with increased life experience (an index of, but not identical with chronological age). The directionality is toward what we are regarding as *phenomenologic ritualization*. We see this as a rather orderly, systematic process that has much to do with the "old age" constellation that is commonly observed, but which leads us to a different perspective on this constellation.

Succinctly, on this view, to be old is to live within an experiential world that is highly patterned and redundant of meanings, relatively less permeable to the "raw moment." We will examine this state further as well as some of its implications after consideration of other temporal dimensions that are salient in old age.

TIME: INNER DIMENSIONS

We concentrate here upon several dimensions of time that are especially germane to understanding the old person's experience and functioning. These dimensions are also relevant to younger people, of course, but the aged are of prime concern in this paper.

Futurity: Probabilistic and Axiologic Modes

A variety of studies have indicated that old people generally think less of the future, whether compared with their own past orientations, or with the prospection of young adults[2,3,4]. This conclusion sometimes becomes obvious during the interviewing process itself, well before data are obtained and analyzed. The old person may *resist* talking about the future. "What's there to say?" he will ask. "Can't we talk about something more pleasant?" "Future — what future?" "Don't think much about that any more." These are fairly typical responses when the topic is broached.

Close inspection of what the old person says (and does not say) suggests that the quality of the orientation deserves as much attention as its quantitative aspects (e.g., how *many* events forecast, how *far* into the future). We have found it particularly useful to analyze future orientations of the aged in terms of probabilistic and axiologic modes. By *probabilistic*, we mean a scanning of the future that concentrates upon objective estimation of contingencies. "I think it *will* rain tomorrow" is a statement about what is likely to occur, based upon the speaker's best interpretation of the information available to him. It is a future-minus-my-preferences modality. "I *hope* the rain holds off" projects a future relevant to the speaker's likes and dislikes. The *axiologic* mode is concerned with prospections that emphasize either longed-for or dreaded contingencies.[5]

Theoretically, we would expect a mature adult to make appropriate use of both modalities. He does not have to choose between expectation and desire. "I *want* it to be this way," he tells us, "but I *expect* it to turn out otherwise." Or, "I do not have enough information to make a good estimate of what actually will happen, but I am afraid that it might go poorly for me." This approach is consonant with a developmental-field theory of human aging.[6] Developmental-field theory proposes, among other things, that individuals create very particularized "careers" of maturation and aging, and generally are more effective if they have achieved versatile frameworks that can support diverse and contradictory experiential trends. In both clinical practice and research we have encountered a number of old people who steadfastly refrain from one modality or the other. Some appear reconciled to a future that is irrelevant to their own hopes and fears. The old man will adopt the perspective of a disinterested observer. He will speak willingly enough about the future, provided one does not press him to indicate how his own needs and preferences will be served. Others appear swamped by their needs. The future is projected only in terms of terrors-to-be-avoided or delights-to-be-savored. Questions about the likelihood of these contingencies actually coming to pass are "not heard" or misunderstood.

A simple technique often helps to clarify the respondent's future orientation in this regard. He can be asked an obvious question, "How does the future look to you (today, at this time)?" In his reply, the elements of an axiologic and a probabilistic orientation may both be evident. If not, one can pursue with a non-directive probe, ("Can you tell me a little more about that ... ?") If one of the modalities is still absent, the inquiry becomes more specific, (e.g., "I understand now what you hope will happen: what do you think the chances are that your hope will be fulfilled," or words to that effect). A detailed set of coding categories has been developed for this simple line of inquiry, and adequate interrater reliability can be achieved with proper experience and training.[7]

The probabilistic/axiologic modalities can be taken as illustrative of qualitative aspects of the older person's orientation toward futurity, although other dimensions also merit attention.

Retrospective Modalities and Their Functional Significance

Directionality and Social Values. In the Western technological world we are accustomed to value positively those who are "forward-looking." Tomorrow is more relevant and important than yesterday. It is idle and regressive to "live in the past," but constructive to harness one's energies and thoughts to the service of the future. This orientation introduces a bias into observations and interpretations of retrospection. Psychologists and other students of human behavior often share the cultural prejudice that the plans and dreams of the young are somehow more useful or valuable than the reflections of the old. The prejudice is not without its implicit contradictions. We tend to complain about the elder's reputed engrossment in his past, and use this as a rationalization for taking his views less seriously, or even for classifying him as a "fossil", a "has-been", or the psychiatric equivalent of these terms. However, what the old person takes as the subject of his retrospections may be a record of accomplishments that the young person still must project as future hopes and aspirations. It is a peculiar society that urges people to plow the present in order to bring in a future harvest, but which then looks reproachfully or condescendingly when the fruits of the labor are enjoyed.

Furthermore, could it be that we have a vested interest in the old person's past orientation? If his thoughts flow backwards, then he is not in a position to interfere with our schemes and ambitions which have futurity as referent. We do not always enjoy having a person with the vast experience of the aged looking over our shoulder and offering opinions; perhaps we feel more comfortable if he looks the other way. Additionally, it is possible that we are miffed because the past he contemplates is a private world from which we are at least partially excluded. His more remote retrospections feature people and situations alien to our own experiences. The young person may feel snubbed, left out without comprehending that in his own future projections he tends to exclude the aged. Perhaps enough has been said along these lines to remind ourselves that cultural values reach deep into our interpretations of past and future and are likely to influence even our "objective" assessments of the role of retrospection.

Assumptions and Evidence. It is commonly assumed that old people think "more"about the past. We do not doubt that there is something to this proposition. Yet neither the assumption nor the evidence are as clear as one might be tempted to believe. Without unfurling a thorough review of the data in this field, we can nevertheless indicate that the proposition is neither simple nor entirely resolved at this time. Often, for example, we are careless in defining what is meant by the "more." Consider two illustrative propositions: (a) An old person thinks more about the past than does a young person; (b) As a person grows older, he thinks more about the past than he did when he was young. Most statements fail to differentiate between these propositions, yet the statements are not identical and require different types of evidence to be tested properly. Most available research compares the time perspectives of independent populations of young and old.[8,9] The findings apply much more directly to the first proposition than to the second. Differences between those who are young and old, in the 1970's, possibly are attributable to many factors other than age differences per se – e.g., the different life circumstances both groups face today, or critical differences in early experience which might have led to one generation having developed in a more past-oriented or future-oriented style from the beginning.

There are other ways to interpret the "more" as well, for example: (a) Old people tend to think less of the *future,* therefore, retrospection dominates their mental life by default as it were; it is not intensified, but simply has little competition; (b) The old person continues to think about the present and the future in the sense of taking current demands and opportunities into account for daily survival and adaptation. However, he thinks "more" about the past in the specific sense that this is where his deepest emotions and personal themes are to be found; (c) The old person truly "lives in the past," i.e. is so engrossed in by-gone situations that he is virtually out of contact with the present and the future. It is seldom that we make the kind of observations and analyses that would provide the basis for accurate discriminations among these (and other) alternative propositions. Yet old people in general or a specific old person might be past-oriented in one of the above senses, but not in the others. We must know quite a bit of the old person's total mental life to appreciate the specific scope and role of his retrospective processes.

Let us take an admittedly extreme example here. Suppose that we have entered into conversation with an old woman who seems most at home in discussing the past. Actually, she is dwelling largely upon a past that antedates her own birth. On and on she talks about long-vanished people, places, and circumstances. It appears that she is providing us with additional evidence that old people think "more" about the past. This particular woman, however, is a professional historian. Her retrospective processes have been highly disciplined and practiced over the years. Had this conversation taken place forty years ago, she would have ranged backward in time at that point also. The woman shows no defect in relating herself to the present or the future. But she does have special abilities and interests that call upon a retrospective turn of mind, and she engages in this form of thought as a matter of preference. If we were overly impressed by her age and did not know of her professional credentials, we might then interpret her retrospection differently. It is not easy to determine the precise relationship between chronological age, career, and personality in this instance. But is it *ever* appropriate to make a facile judgment of this kind? Can we comprehend *any* old person's retrospective processes by focusing entirely upon his age?

Perhaps the most stultifying and misleading assumption about retrospection in the aged is that this characteristic is so uniform and powerful that it provides a safe basis for differentiating young from old. In point of fact, however, there are major differences among the aged with respect to uses of the past.[10,11,12] Individual personality and life circumstances often seem to be more significant than age per se in shaping time perspective. Furthermore, a broad spectrum of psycho-gerontological research has established that individual differences flourish in old age.[13] By maintaining that old people "live in the past" one fails to make adequate differentiations within the population of aged themselves, or between the young and the old.

Memory. A thorough exploration of retrospection in old age should attend to the dynamics of memory. Although we are not attempting a comprehensive, in-depth exploration here, a few points can be made. Use of the past is related to (i) type and range of potentially available memory content, and (ii) the functional estate of all those processes involved in calling upon stored information. The elderly survivor of a concentration camp that took the lives of intimate friends and relatives may have a different retrospective inclination than the elder who has many years of comfortable family life to look back upon. Differences in retrospection among the aged can also be related to differential vulnerability to deficits in the registration, scanning, and retrieval of recent information. Perhaps the person who has difficulty utilizing his recent past must for that reason rely more heavily upon knowledge and experiences that are in "long-term storage." We know little about the dynamics that might be associated with long-term and short-term memory processes with advancing age, but a growing body of research makes it clear that "memory" is a term denoting a set of complex processes which undergo differential change with time and experience.[14] To understand retrospection in old age we must continue to improve our understanding of memory in both its psychobiological and psychosocial aspects.

Retrospective Modalities. The author's clinical experience with the elderly as well as his research efforts have impressed him with a multiplicity of retrospective modalities, rather than a single "living in the past" orientation. In a preliminary and rather clumsy way we will attempt to sketch several of the modalities which appear most relevant to the time-death-ritual thesis.

1. *Life Review.* It has been reported that some old people engage in a "life review"[15] to integrate their experiences and identity in the face of death. While it is reasonably certain that such experiences do occur and can be highly significant for the persons involved, there is no evidence to indicate that the life review is a focal and universal occurrence among the aged (or, for that matter, limited only to the aged). It is possible that some form of life review is relatively common, but reaches the level of clear consciousness and articulate communication for only a few individuals. For sake of clarity it might be best to limit the application of Butler's phrase, life review, to those instances in which an old person is "processing" his entire past experience for self-integrative purposes as death becomes salient on the subjective (and perhaps objective) horizon.

2. *Validation.* There is another form of retrospection that seems to be used by a greater number of old people. Perhaps it is less systematic and profound than the life review, and it is not necessarily linked directly to anticipations of death. The old person may find himself with relatively little to "go on" in his daily encounters with the world. Low in status and current resources, he is subject to self-doubt. Therefore, he makes frequent trips to the past — but for the purpose of returning to the immediate situation with "reinforcements." Use of the past in this sense is a mode of *validating* oneself. "I was a strong, competent, beloved

person once — therefore, I am still a worthwhile person, capable of further survival and adaptation." We do not mean to imply that the validating process is infinitely effective; eventually the old person may return from his retrospective journey with the conviction that who he was is no longer sufficient to bolster who he is now in an untenable situation. In passing, it might be noted that the young person who is in a crisis of "nerve" and self-doubt is more likely to validate himself through a projection into the future: "Who I will be someday establishes ahead of time my value and competency right now."

3.   *Boundary-Setting.* As we move along our personal life-lines, there are numerous occasions for re-thinking where we have been and where we are going. Some of the most critical "turning the corner" experiences are likely to occur before what is concensually recognized as old age.[16] In the approach to old age and the subtle way-stations that are experienced within that area, the person may have difficulty in establishing which aspects of his total identity remain operative. The past is then worked and re-worked in an effort to set appropriate boundaries. "I am no longer a working man" is a conclusion that may separate one zone of total identity into "then" and "now." Entry into age-segregated housing, a nursing home, or geriatric hospital might require still another boundary adjustment. In a dysphoric judgment, an ailing and uprooted elder might decide, "Then, I was a person; now, I am a patient — an old crock!" Whether or not the consequences prove gratifying, the retrospective process of boundary-setting appears to be important and fairly common. For some older people, it is boundary-setting, a functional process, that calls upon most of their retrospective energies.

4.   *Perpetuation of the Past.* "Ritualistic perpetuation of the past" is a concept familiar to us through the observations and insights of cultural anthropologists.[17,18] A kind of ritualistic perpetuation can also be observed in our own society. However, there are important differences between the old person's solitary carrying-forward of an idiosyncratic past and an entire tribe or community's ritualistic, concensual interpenetration of past-present-future. Here we wish to make only the following points: (i) Perpetuation of the past, no matter how obvious, dramatic, or colorful, is only one of the modalities by which an old person can relate himself to the by-gone; (ii) Although perpetuation seems to build up its own momentum it is vulnerable to onslaughts from current reality (including the elder's own mental and physical condition); the individual may receive very little support in his effort to maintain a style of life that has passed from fashion; (iii) As long as the perpetuation "succeeds," the elder may show an impressive range of thought, affect and behavior, all of which bear the imprint of an earlier era; he is competent and adjusted, but to a world other than our own; (iv) Strictly speaking, this is not always a retrospective modality; some elders never have relinquished the past, therefore, it is a process of continuation, not re-establishment; other elders, however, *return* to a more secure and stable life-style, and then attempt to bring it forward in time; (v) Perpetuation (or continuation) is most striking when it consistently dominates the individual's functioning; but it can also appear as a subtle or oscillating component of a more up-to-date life style — for young as well as for old.[19]

5.   *Replay.* One of the most fascinating retrospective modalities has not received the attention it deserves, possibly because it is encountered so frequently that it tends to be taken for granted and to fade into the background of our perceptions. Reference is made here to the highly selective repetition-sequences one hears from many old people. This is the confused,

failing aged person who speaks with great specificity and involvement about a few people and events — the same few — on every occasion you talk with him. He does not seem to possess anything except these remnants of the past which alone are "real" to him. But this is also the fairly alert and active aged person who has fallen into a pattern of repetition. This pattern is not identical with perpetuation or continuation of the past. The speaker may clearly appreciate that he is reaching back into the realm of memory, that current realities are quite different. He is introducing past episodes into the present.

One facet of this modality merits special attention: notice how the original time-context often has been abolished. The remembered and replayed experience no longer is bound by its "place in time." It has become essentially ahistorical. What stamped this episode and gave it special personal significance remains; the surrounding fabric of time, the before and after, hangs in shreds. The listener may be impressed by the old person's involvement with his past. However, the speaker may prove unable or unwilling to present a systematic biography. Whole chapters, even volumes, of his personal history are missing. There is the impression that the old person does not experience a sense of intimate continuity with his total past, nor does he possess a more abstract rationalistic framework that arranges his life in proper and uninterrupted sequence. Instead, he has developed a "theater of the mind" in which selected scenes can be played and replayed as the occasion requires.

The needs and functions served by this modality are not well understood at present; indeed, rather little curiosity has been expressed about this form of behavior. Is it failing memory that requires an unfailing source of mental stimulation and conversation? Or is it a disintegration of whatever holds the time-scenes of our lives together, and which then liberates a few episodes for purposes of replay? Again, could it be that the "real" purpose of replay is to ritualize death out of the picture? Perhaps the inner dialogue has something of the following character:

"I am afraid of death, don't even want to think about it. Help me find a way out, will you?"

"Of course. But there is a price you will have to pay."

"You name it."

"Very well. Death can find you in only one place: the future. You were alive yesterday; you are alive today. If we dismantle the apparatus that takes you from today to tomorrow, then we also throw death off the track. Understand?"

"You mean, I can stay alive and not worry about death as long as I don't go any place."

"That's it. But for the operation to succeed, it must be thorough, radical. The entire time-line must be disrupted. Now you go ahead and pick out a headful of memories, scenes that will entertain and nourish you, scenes that will provide the illusion of biography. Fine! Now, I will sever the connections among these scenes, and between these scenes and your general mental fabric. There ... doesn't that feel better? You have all that you need of the past. The remainder is dispensable. Best of all, you have broken off the time-track. You cannot be lured or coerced steadily back in time — you stop at only those stations you have selected; and you are not impelled or compelled to move forward in time. What do you have to say now?"

"That reminds me of ... "

We offer the hypothesis, then, that replaying selected episodes from the past is (i) related to a disintegration of the total time fabric, and (ii) to a subtle but perhaps very effective insulation from the threat of death: time cannot be stopped nor can it run out if it is already frozen and contained within the self.

Other Selected Time Dimensions

Subjective Age. It is customary in the United States to take chronological age (CA) very seriously as a basis for classifying individuals, preparing statistical information, and making decisions of many types. Gerontologists recently have intensified their efforts to supplement CA with dimensions that bear more relationship to the individual's actual capacities. Functional age (FA) is a general concept that, in turn, has started to be fractioned into more specific concepts that are operationally defined.[20] One can speak of biological age as one realm of FA, but, even more specifically, of retinal, epidermal, or bone-marrow age. In other words, alongside the familiar, uniform, concensual increments of CA, one may now introduce a variety of specific time-lines whose properties remain to be determined. A person may be "biologically young" and "psychologically old", for example.

As part of this development, renewed interest is being shown in subjective age: how old the person is to himself. Even within the realm of subjective age, it appears that differentiations must be acknowledged. How old a person believes he looks to others, and how old he himself feels from the inside prove to be two distinct variables, for example.[21] Exploratory research also indicates that the tendency to maintain a subjective age appreciably younger than the CA begins early in adult life (and, incidentally characterizes trainees in gerontology as much as anybody else).

Recognition that most people (young and old) seem to have their private estimations and interpretations of their own age can be helpful in avoiding overgeneralizations based only upon CA and externalistic measures. It remains to be learned how much of the older person's subjective age is based upon such contributing factors as (a) low status of being old, (b) actual inner feeling of youthfulness, and (c) desire to see oneself at a comfortable distance from death ("Only old people die," parents often tell their children in the United States).

Movement. The sense of movement through time is different from other dimensions both logically and empirically.[22] A person might be quite future-oriented, for example, yet have the sense of being trapped in the present. Another person may feel that he is hurtling along through time, and yet possess only an ill-defined and limited outlook on the future. Published research suggests that old people tend to feel that time is moving rapidly for them.[23,24] Although this trend may seem to contradict the impression that time hangs heavy for many of the aged, it is also consistent with this writer's clinical and research experience.[25] But general conclusions are probably less useful here than the attempt to understand how the tempo of time is experienced by old men and women as individuals. We also have much to learn about the experience of time's passage in people whose general framework for the integration of experience has been altered radically. Is it possible, for example, that the "aimless agitation" sometimes observed among the aged is the way temporal experience expresses itself when the continuity of life past/present/future has been so disturbed that there are no more fixed points along which orderly movement can proceed?

Thematics. By *thematics of time* we are referring to the expression of purposes and stylistic characteristics. The old person waiting to retire may "put in" time. In a more active response to a period of duration that holds one captive, the individual may "kill" time. Some elders are able to relax and "savor" time, etc.. There is not much advantage to listing all possible thematics of time. The value resides in learning how an individual continues to express his relationship to life through his use of time as he ages, and what new themes, if any, emerge. In our own experience, we have observed that a *reunion* theme of some kind is fairly common. Time is seen as a curving road that leads increasingly away from everyday life (which becomes increasingly barren of meaning) and closer to people and values known at an earlier time. This curvilinear conception of time may express itself in people who previously had been advocates of a strict linear view. Now, however, "forward" also seems to be "backward." At an extreme, the individual may feel that he is "marking" time until he returns to the psychological starting place, i.e., the time that "counts" is not today, but the future time-to-come which itself will prove to be saturated if not identical with the past.

TIME: OUTER DIMENSIONS

The individual's time dimensions and systems can also be seen in relationship to dimensions and systems that function external to himself. As our aim here is merely to acknowledge the existence of these outer dimensions and suggest a few possible interrelationships, the treatment will be quite schematic.

Environmental Tempo

The speed at which events transpire in the environment is one of the more obvious external variables. Environmental tempo can be analyzed in absolute terms, or in relation to the tempo of the individuals involved. It is possible, for example, that a particular environmental setting will appear "too slow" to a young person and "too fast" to an old person. Further, the environment may be more or less differentiated with respect to tempo. In other words, it may function as though all components were regulated by a single master clock, or there may be a variety of tempo relationships at work within the same environmental setting.

Environmental Stability

Discussions of tempo usually imply a rate of movement within a context that is relatively stable. Even in an environment that is fairly complex and has a multiplicity of tempos, it may be possible to treat the environment itself as constant. But environments are not necessarily stable, as we know. The old person may be required to adjust not only to a variety of tempo in his environment, but also to changes in the basic socio-physical parameters. The new director of a nursing home, for example, may keep things functioning at the accustomed tempo. However, differences in her personality, skills, and expectancies (as well as the simple fact that she is "new" to this environment) can result in both objective and subjective changes. "How rapidly is the environment itself changing? Is it still the same environment to which we have adapted ourselves? Is time bringing change to the world as well as ourselves?" These are among the questions that occur to one who notices external time systems.

Timing and Scheduling

How important is it to do things at a certain time in this environment? How closely and thoroughly must one's activities be scheduled? Is the individual free to use time as his various appetites and purposes propose to him, or must he interdigitate with the socio-physical system at all times? Does he awake, eat, bathe, and move about at his own convenience, or is he under pressure to stay in step with a master schedule? The environments in which older people live vary widely from the tightly timed and scheduled to the "wide open." Additionally, the timing and scheduling dimensions can be established either in such a way as to accommodate well with the individual's own preferences and capabilities, or to provide constant stress because of their unsuitability. A more sophisticated analysis would also consider the complexity and possible contradictory nature of timings and schedulings in a particular environment. From a psychological standpoint it can be important to learn whether or not the individual feels that he has any control of influence over the prevailing time system. Will the environment go its own way, no matter what, or is it responsive to his initiative? The old woman who has taken up residence in an age-segregated facility might be relieved that she no longer has responsibility for preparing three meals a day. Yet she may also feel somewhat at a loss and resentful that meal times are determined and processed by others; she has lost more control over the temporal routines of daily life than she had anticipated, or misses these controls more.

Social Thematics

What other people think about and do with time constitutes a significant aspect of the old person's environment. Is this a place where everybody is largely "marking time?" Or is everybody except the old person fiercely involved with a future-thrusting project? Is the environmental "time climate" perhaps split between "drifters" and "surgers"? Does the daylight staff of a geriatric facility treat the patients with condescension as "back numbers" who do not fit into their own view of life, while the evening staff seeks out the same elders as the source of information, companionship, and linkage to a valued past? The possibilities are many. The important point is that the old person's time system is regularly in contact (and perhaps in contest) with the time systems maintained by other individuals in his environment as well as by any "time climate" that might characterize the environment as a whole.

INNER AND OUTER DIMENSIONS: A FEW RELATIONSHIPS

We will now explore briefly a few of the possible relationships between inner and outer dimensions of time in old age.

Whose Time Is It?

Perhaps this writer's observations are highly skewed, but it is seldom that he has heard an old person ask, "What time is it?" The more relevant question might actually be, *"Whose* time is it?" Although not characteristic of all old people in all situations, the following factors occur frequently enough to be worthy of notice:

1.   The old person did not expect to live to his present age. Furthermore, he has out-lived many of the people and circumstances that provided a sense of embeddedness for him.

2. The socio-physical environment in which the old person currently finds himself may "belong" to others, either officially or by tacit understanding. The old woman, for example, may have become a lodger in her own home, now occupied by married children, and feels that what takes place is mostly out of her control.

3. Events both internal and external can have the effect of seeming outside of the individual's self-system – and in a conflicting manner. Slow-downs and other temporal irregularities in the biosphere may alarm the old person: his body has shifted to another level of functioning. At the "same" time, environmental systems may be changing, accelerating, moving increasingly away from the stability to which he had been accustomed. Both types of change, the somatic and the environmental, can threaten the old person's view of himself. "Where/ when is my time? Can I prevail upon the somatic and the environmental to adjust to my time system, or must I adjust to the other systems, and how could I adjust to such contradictory and complex changes at once, anyhow?"

As a more general formulation, we propose that the sense of identification with and possession of time constitute significant dimensions that are best approached by considering individual and environmental systems simultaneously. (We would apologize for the sketchiness of this formulation, except that such apologies would need to be reiterated tediously in the formulations that follow as well.)

Age-Role and Environmental Systems

The concept of functional age (FA) has already been introduced. Evidence is accumulating to indicate that number-of-years-lived is not at all equivalent with current functional status. Let us add a pair of related concepts:

1. The environmental system can also be assessed in terms of FA. This is true in two senses: (a) the environment as a system that functionally ages individuals at a greater or lesser rate; (b) the environment's own rate of growth or aging. Eventually it should be possible to compare environmental systems in both senses that have been differentiated here. Environment A, for example, may be found to hasten the aging of all the individuals within its compass; while those who operate within Environment B appear remarkably immune to those effects usually interpreted as age-indices. Environment C may resemble in certain formal characteristics as well as in "spirit" an adolescent who is still "getting himself together"; systematization, integration of components has not yet taken place as new components or "ingredients" continue to emerge. Environment D has become highly predictable, resistant to change and "ritualized." The environment itself appears to have grown old.

2. An individual can enact a role within the environment that seems appropriate for a particular age-rank without necessarily being of a particular chronological age. Again, we offer two lines of thought here: (i) The individual is "old" or "young" in socially relative terms. A thirty-five year old athlete, for example, may be the "old man" of the team and treated as such; the sixty year old resident of a geriatric facility may be regarded as "the kid." In a number of social situations there appears to be a group need for establishment of senior and a junior echelon, no matter what the actual range of chronological ages involved; (ii) The physical appearance, performance level, or style corresponds to expectations of a certain age, somewhat independent of the person's age relative to others in the situation. A

youthful face, an agile step, a receptivity to new experiences: characteristics such as these tend to create a different impression than weak tonus, faltering gait, and closed-mindedness although specimens of each type may have precisely the same chronological age.

We propose in this context (as we have on other occasions,[26,27]) that much of what currently passes as the outcome of inevitable and universal aging can be understood more appropriately in terms of personality, learning, and properties of the socio-physical system. Miniature experiments suggest that young adults can be induced to "behave old" and old adults to "behave young" when socio-physical dimensions are modified. Time dimensions are of special importance, e.g., tempo of stimula and demands relative to the individual's tempo, and ability to "use" the time that one is experiencing.[28,29]

On this view, at least some "aging" phenomena are reversible. Time can be traversed in both directions. It remains for future research to determine the potentialities and limitations of this approach. But at least some indices of "old age" might be interpreted rather as products of social ritualization. People have been molded into roles, activities, and life-styles that represent age stereotypes, and which do not necessarily represent their individual relationship to time.

Coming and Going

The Trojans, one of the most magnificent of operas, was delayed and sabotaged many times before its first production finally took place. As the audience streamed in for the premier, a close friend turned to the aging and ailing composer and exulted: "They're coming, Hector, they're coming!" Berlioz' morose reply was, "Yes, ... but I am going."

This incident exemplifies but does not exhaust the relationship between the temporal vantage of the old person and those who share his socio-physical environment. It is common knowledge that individuals occupying the same region in space at the same point in concensual time do not necessarily experience their situations in identical fashion. Critical individual differences may exist even if all are of similar chronological age, are participating in the same activity, and linked by a single theme. Each individual in this hypothetical situation (e.g., residents of a nursing home taking a communal meal) will have brought with him a highly particularized past as well as a current psychobiological mode of functioning that is not precisely the same as anybody else's. The bare fact of public or concensual relatedness in some particular portion of time-space raises easily as many questions as it answers about their inner states of being.

The differences in temporal vantage between an old person and a younger person would seem to include the following:

1. The event or development that is interpreted as "new" by the young person may be interpreted as "the same old stuff" by his senior. Part of the differential in enthusiasm or general responsiveness between young and old may pivot upon the relative size of the "apperceptive mass" that each brings to the situation.

2. The same unit of time is radically different when viewed as proportion-of-life-remaining.[30] One day is a tiny fraction of (assumed) future time for a young adult, but a more substantial "piece" of life for the aged person. From this standpoint, the old person cannot afford to "waste" time as the prodigal youth might do — perhaps a clue to the sense of rapid time

passage that several researchers have found among the aged. Paradoxically, perhaps, the young person often seems to be in a more favorable position to "do something" with "his" time, while the middle-aged person is so thoroughly engaged with social obligations and demands that time and energy tend to be overcommitted.[31]

3.   Another way of looking at the same fact induces a different set of implications: each passing day brings the older person closer to death. The same is true of the younger person, of course, but the "cushion" of unexpended time differs appreciably. Our confidence in the underlying rationality of the universe can be shattered when a death comes in the wrong sequence. The subjective "pecking order of death"[32] helps to support our faith not only that the old will precede the young to the grave, but that somebody is looking after the universe and making sure that the "rules" are diligently obeyed. What we wish to emphasize here is not just the relative difference in proportion-of-life-remaining between the young and the old, but the sense of movement, of acceleration ("Time's winged chariot hurrying near"). As one old man told this writer, "You want to know how I see death? I don't *see* it: I *feel* it and I *hear* it. Death is like a vacuum cleaner that is trying to swoosh a little louder. I'm still here (laughs), but tomorrow or the next day I may be in the bag."

The old person, like the composer of *The Trojans,* may feel that he is "going," while others occupying the same time-space manifold are still "coming." But his feeling-tone may be also permeated with the sense of "coming *to*." In other words, his temporal directionality may be attuned only partially to what he is leaving behind. He is going away from a form of life that has already become depleted in some of its essentials, but "coming to" a conclusion, resting-place, just reward, or reunion.[33] This suggests a state of being that departs from the customary view of the old person as slow and static. Instead he seems to be embarked on a more adventurous "flight plan" than the younger adult who moves from point to point within a limited, socially-controlled compass.

If the older person does appear a bit removed from us, perhaps it is because his future course in time takes him at a different rate and altitude to a destination that we do not expect to reach until we have made a number of reassuring, local intermediary stops.

ON THE RITUALIZATION OF EXPERIENCE

Early in this paper we proposed an alternative approach to interpreting the course of lifespan human development, if only for heuristic purposes. Customarily, we think of people as learning and maturing until they reach a period of prime functioning. This is characterized by vigor, productivity, and a certain "freshness." For a few years we are then treated to the "rising young executive," the "promising artist," the "life of the party," the "nonconforming seeker-after-experiences," etc. — but usually, only for a few years. In the later adult decades the same person may seem to have gone stale and to have reached a plateau in most spheres of functioning. We are then supposed to regret the fade-out of youthful enthusiasm and promise, and its replacement by the somber facade of age. Interpreted in this vein, the story of human development frequently appears to reach a quick and disappointing conclusion.

Alternatively, we might regard the surge of youthful energy and experience-seeking as means rather than end, Society requires a certain amount of innovation, even the most snug, smug   tradition-oriented society. Society also requires a certain maturity and competence among its members that can be achieved only through the application of energy, the exploration of possibilities, and, to some extent, the taking of risks. For innovation and replenishment, it is socially useful to encourage a phase in lifespan development where the individual is relatively free to "collect" experiences and test himself out in a variety of roles and realities. The anxiety, pain, and failure that accompanies some of this youthful activity can be redressed to an appreciable extent by the experience of novelty, pleasures in the use of the body, social approbation simply for being youthful, etc..

Note, if you will, that a society might in principle so arrange itself that the old rather than the young take much of the responsibility for innovations and perhaps even for replenishment. Or, as other possibilities, either a cyclical or a selective system might be devised which would replace the age-dependent relationship we often find with innovation/freshness-of-experience.

But society needs continuity as well as change. The strength of tradition must manifest itself not only at cross-sections in time, but also through points of transition, including intergenerational transmission. Being young, innovative, energy-expending, and experience-gathering might "just" be a vehicle for transforming the immature into dependable conservators. We over-glorify youth, then, but this is a useful maneuver to insure that the important work of becoming an old conservator is accomplished.

By implication, the period of feisty youth could be too long as well as too brief. An efficiency expert might recommend adjustments here or there, depending upon conditions. Youth might be extended if the task of becoming a conservator were deemed more complex and demanding; youth might be shortened if more effective ways were found to induce old age (in the positive sense), or the demands for experience lessened.

But let us return to the individual. When he was young, almost everything was new to him. It required much energy to go through a day filled with discoveries and challenges; fortunately, the necessary drive and resiliency were his. He both enjoyed and suffered acutely. Time without appropriate stimulation and outlet for activity could be sheer torture: the boredom that many youth experience in the midst of their vitality and exuberance should not be forgotten. On more occasions than one, life seemed to come apart, to fragment into competing needs, impulses, and aspirations. Youth was a rich learning experience, but it was not by any means a garden of unalloyed delights.

Now, however, he has reached the sanctuary of patterned experience. The spur of the moment does not impel him to reckless action, nor cause him to doubt his identity or overturn his beliefs. At his best, he can both "eat" his moment, and "keep" it, which perhaps is the hope that underlies the cliche. The moment is acknowledged, but filters through a meaning-system until it has taken on the individual's own "flavor." If the moment contributes nothing of consequence, very little energy is expended upon it. But should the moment have something to contribute, it is then smoothly processed into the enduring pattern.

In his own way, perhaps the old person is a Platonist. From the stream of ephemeral experience he has selected and shaped ideas of truth and beauty. New experiences prove irrelevant to these enduring ideas that sustain him, and are thus discarded; or, the experiences confirm and renew his appreciation of the ideas and so are cherished.

Death? One person is acutely aware of time's unrelenting passage and the transcience of life. For him, death is the cessation of experience, the coin pocket of time emptied out.

Another person, however, no longer resides fully in the stream of time. He is alive, which means that experience registers upon him. But what is most significant in his experience is not the moment by moment jostle. He lives with the patterned ideas, the acted-upon, memorialized symbols of experience. Significant and sustaining experiences are replayed (or preplayed). Death itself takes a place of honor and respect, perhaps also of terror and awe: but as one idea among others. The idea of death is part of the mental fabric, not an alien force dedicated to ripping the fabric apart.

In this sense, the old person (the old person we ourselves have idea-ized for this purpose) does not die to himself. He can and does continue to live with death. Ritual-master that he is, the old person enacts and re-enacts his deeply personal appreciations of life, now embellished, now bare, now repeated with precision, now varied with delicacy. Then, at a certain moment, both ritual and ritual-master fade. Performer and audience exit by the same door.

Perhaps to be old is to be a more developed and perfected human, exercising those virtues of reflection and symbolic integration that distinguish this species from our fellow-travelers in time.

NOTES

1.   D. Hume: *A Treatise of Human Nature*, 1740.

2.   R. Kastenbaum: "Cognitive and Personal Futurity in Later Life," *J. Indiv. Psychol.* 19 (1963): 216-19.

3.   R. Kastenbaum and N. Durkee: "Elderly People View Old Age," *New Thoughts on Old Age,* ed. R. Kastenbaum (New York: Springer 1964), pp. 250-64.

4.   R. Cavan: *Personal Adjustment in Old Age* (Chicago: Science Research Associates 1949).

5.   R. Kastenbaum and B.K. Kastenbaum: "Hope, Survival, and the Caring Environment," *Prediction of Life Span,* eds. E. Palmore and F.C. Jeffers (Lexington, Mass.: Heath 1971), pp. 249-72.

6.   R. Kastenbaum: "Engrossment and Perspective in Later Life," *Contributions to the Psychobiology of Aging,* ed. R. Kastenbaum (New York: Springer 1965), pp. 3-18.

7.   S. Sherwood and R. Kastenbaum: "Open Future": Coding Manual (Boston: Hebrew Center for Rehabilitation of the Aged, mimeo.)

8.   S. Goldstone, W.K. Boardman and W.T. Lhamon: "Kinesthetic Cues in the Development of Time Concepts," *J. Genetic Psychol.* 93 (1958): 185-90.

9.   K. Pollock and R. Kastenbaum: "Delay of Gratification in Later Life: An Experimental Analog," *New Thoughts on Old Age,* ed. R. Kastenbaum (New York: Springer 1964), pp. 281-90.

10.   H.F. Fink: "The Relationship of Time Perspective to Age, Institutionalization, and Activity," *J. Geront.* 12 (1957): 414-17.

11.   P.T. Costa and R. Kastenbaum: "Some Aspects of Memories and Ambitions in Centenarians," *J. Genet. Psychol.* 110 (1967): 3-16.

12.   S. Tobin: "Earliest Memory as Data for Research in Aging," *Research, Planning,* and *Action for the Elderly,* eds. D.P. Kent, R. Kastenbaum and S. Sherwood (New York: Behavioral Publications 1972), pp. 252-78.

13.   C. Eisdorfer and M.P. Lawton, eds.: *The Psychology of Adult Development and Aging* (Washington, D.C.: American Psychological Assoc. 1972).

14.   D. Arenberg: "Cognition and Aging: Verbal Learning, Memory, Problem Solving, and Aging," *The Psychology of Adult Development and Aging,* eds. C. Eisdorfer and M.P. Lawton (Washington, D.C.: American Psychological Assoc. 1972), pp. 74-92.

15. R.N. Butler: "The Life Review: An Interpretation of Reminiscence in the Aged," *Psychiatry* 26 (1963): 65-76.

16. R. Kastenbaum and B.K. Kastenbaum: *Humans ... Developing* (Boston: Holbrook Press, in press.

17. S.C.F. Brandon: *Time and Mankind* (London: Hutchinson 1951).

18. M. Eliades: *The Myth of the Eternal Return* (New York: Pantheon 1954).

19. S. Anthony: *The Discovery of Death in Childhood and After* (New York: Basic Books 1972).

20. B. Bell: "Significance of Functional Age for Interdisciplinary and Longitudinal Research in Aging," *Aging and Human Development* 3 (1972): 145-7.

21. R. Kastenbaum, V. Derbin, P. Sabatini and S. Artt: "'The Ages of Me': Toward Personal and Interpersonal Definitions of Functional Aging," *Aging and Human Development* 3 (1972): 197-212.

22. R. Kastenbaum: "The Structure and Function of Time Perspective," *J. Psychol. Researches* (India) 8 (1954): 1-11.

23. J. Tejmar: "Age Differences in Cyclic Motor Reaction," *Nature* 195 (1962): 813-14.

24. W.W. Surwillo: "Age and the Perception of Short Intervals of Time," *J. Geront.* 19 (1964): 322-24.

25. R. Kastenbaum: "As the Clock Runs Out," *Mental Hygiene* 50 (1966): 332-36.

26. R. Kastenbaum: "Perspectives on the Development and Modification of Behavior in the Aged: A Developmental Perspective," *Gerontologist* 8 (1968): 280-84.

27. R. Kastenbaum: "What Happens to the Man who is Inside the Aging Body? An Inquiry into the Developmental Psychology of Later Life," *Duke University Council on Aging and Human Development, Proceedings of Seminars, 1965-1969*, ed. F.C. Jeffers (Durham, N.C.: Duke University Press 1969), pp. 99-112.

28. R. Kastenbaum: "Getting There Ahead of Time," *Psychology Today* (December 1971).

29. R. Kastenbaum: "On the Experimental Induction of 'Old Behavior,' N.I.M.H. Research Project Proposal (Detroit: Wayne State University, unpub.).

30. P. Sabatini and R. Kastenbaum: "The Do-it-yourself Death Certificate as a Research Technique," *Life Threatening Behavior* 3 (1973): 20-32.

31. E. Cumming and W. Henry: *Growing Old* (New York: Basic Books 1961).

32. R. Kastenbaum and R.B. Aisenberg: *The Psychology of Death* (New York: Springer 1972).

33. R. Kastenbaum: "Two-Way Traffic on the River Styx," Presented at annual meetings of American Psychological Association, Washington, D.C., 1971.

II. BIOLOGICAL RHYTHM

Astronomical References in Biological Rhythms

C.P. RICHTER

This paper concerns three periodic phenomena which because of their roles as timing devices in the lives of animals may be .called clocks:

> 24 -Hour Clock
> 29 - 30 Day Clock
> Yearly Clock

Clocks of squirrel monkeys and Norway rats — the two most commonly used animals for these studies — keep time with great accuracy for months or even years. These biological clocks can be quite as accurate as our mechanical clocks.

How do we know that animals have clocks? How do we know that the clocks keep such accurate time? What significance do the clocks have in the lives of the animals? What is their relation to various astronomical influences.

Biological clocks of all types manifest themselves best in records of spontaneous gross bodily acitvity of animals; but also in eating and drinking times. Methods of recording activity will now be described. This should help in understanding the records.

METHODS

Fig. 1 shows a side view of a stand of cages used in recording spontaneous running activity of rats and other small rodents. Each cage consists (1) of a living compartment containing a non-spillable food cup and a graduated 100 milliliter inverted drinking bottle; (2) a revolving drum to which the rat has free access through a hole in the central partition. A cyclometer attached to a lever on the front end of the axle of the drum registers daily total number of revolutions of the drum. Daily records are made of running activity, food and water intake. Fig. 2 shows the rear view of a stand of 16 activity cages. It shows an eccentric cam on the end of the axle of each drum. With each revolution of the drum the cam opens and closes a microswitch and so sends a current through the connecting wires to an event recorder in an adjoining room. Fig. 3 shows two operation recorders. Twenty ink writing pens on each recorder register revolutions of the drums for 20 rats. The paper moves at a rate of 18 inches per 24 hours. The 18 -inch sheets for each day are cut into strips, one for each rat. The strips are transferred to large charts. These records show the distribution of activity over each 24 hours. (Richter 1965).

Fig. 1. Side view of stand of cages used in measuring daily spontaneous running activity. Each cage has a living compartment with a water bottle and food cup and a revolving drum.

Fig. 2. Rear view of stand of 16 cages showing revolving drums and eccentric cams at ends of axles of drums; also wooden cross bars that hold microswitches in proper position with relation to cams.

Fig. 3. Photograph of two 20–pen operation recorders – one pen for each animal. Paper moves at a rate of 18 inches per 24 hours. Barographs show running activity.

Fig. 4. Cages used in recording daily total spontaneous or gross bodily activity of squirrel monkeys. The device in this cage records activity in a narrow vertical plane around the central axle. Every 80th of a revolution around the axle in either direction is recorded on one of the two cyclometers.

Fig. 4 shows a photograph of an activity cage used for monkeys. All of the monkey's activity in a narrow vertical plane in either direction around axle is recorded on two cyclometers and also on an operation recorder. (Richter 1968).

The laboratory was illuminated from 6 AM to 6 PM and was totally dark from 6 PM to 6 AM. Fig. 5 shows the activity-distribution record of a normal rat. It extends from noon to noon. Records for successive days are placed one under the other. This rat was almost totally inactive in the 12 -hour light period from 6 AM to 6 PM and almost constantly active during the 12 -hour dark period from 6 PM to 6 AM. Here onsets of daily running activity serve as a 'hand' of the clock — telling that the clock is running and at what rate. This record shows that the clock was entrained to start with the dark period at 6 PM. Existence and functioning of the 24 -hour clock can best be seen in blinded animals — that is, after the clock has been freed from the entraining effects of light.

This is illustrated in Fig. 6. Before blinding (EE) onsets of daily active periods occurred quite promptly at 6 PM at start of the dark; after blinding, onsets of daily active periods occurred ten minutes earlier each day with great accuracy. The clock was now freed from entrainment by light. In blinded animals onsets of the main daily active phases may occur earlier in one animal, later in another — that is, the clock may have a period shorter than 24 hours in one animal and longer than 24 hours in another.

Great accuracy of the clock is shown in the record in Fig. 7 of another blinded rat. Onsets of daily active phases occurred 20 minutes later each day during an eight-month period with incredible accuracy in spite of all the many disturbances in the animal's environment: daily record-taking, weighing and filling of food cups, reading and filling of water bottles, weekly weighing, conversation of assistants, etc.. It will later be seen that the clock is independent of all external and internal disturbances. Fig. 8 shows that drinking times also serve as a 'hand' of the clock and with the same accuracy that characterizes spontaneous running activity. The water bottle and cages were wired so that each lap of the water by the rat completed a circuit that registered on a recorder.

The activity-distribution record in Fig. 9 shows that this squirrel monkey was active only in the light. Our observations show that the squirrel monkey is a strictly light active animal. It is active only in the light.

Fig. 10 shows a five-month activity-distribution record of a blinded squirrel monkey. Here onsets of daily active periods occurred 42 minutes later each day over this period. With the same great accuracy shown by onsets of active phases of the rat in Fig. 7. The ink line drawn on the record almost exactly follows the line of onsets of daily active phases throughout the entire period.

The 24 -hour clock of rats is independent of all external and internal disturbances: treatment with all kinds of drugs, or alcohol, interference with the endocrine glands, or removal of the glands, anemia, severe stress, etc. None have any effect on the clock but the clock regulates functioning of every organ of the body. There is no feedback. It is thus a strictly non-homeostatic mechanism — which, of course, makes it such a reliable timing device. (Richter 1965; 67A; 67B).

The 24 -hour clock also measures 12 -hour periods as was seen in Fig. 6, 7 and 10. It was shown also

SPONTANEOUS RUNNING ACTIVITY
RAT #180 CO♀

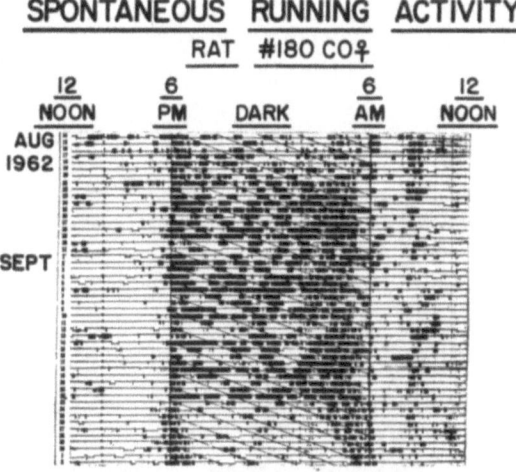

Fig. 5. Forty–eight day excerpt of spontaneous activity distribution of a normal rat. Each line carries the activity record for one day. It extends from noon to noon. On-sets of daily active phase were sharply entrained to start of the 12-hour dark period.

SPONTANEOUS RUNNING ACTIVITY
RAT #55 CO♂

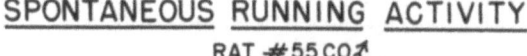

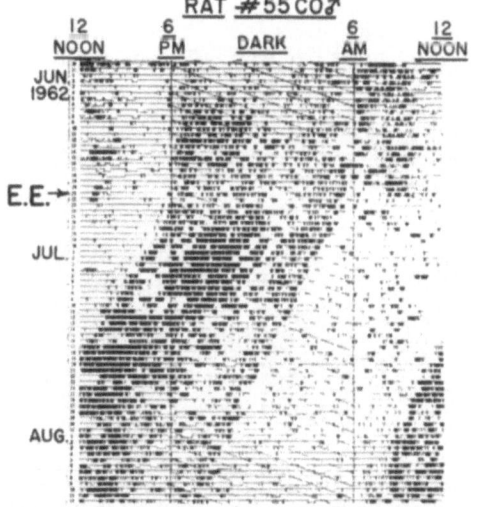

Fig. 6. Activity–distribution record of a rat before and after blinding (EE) showing freeing of the clock after blinding.

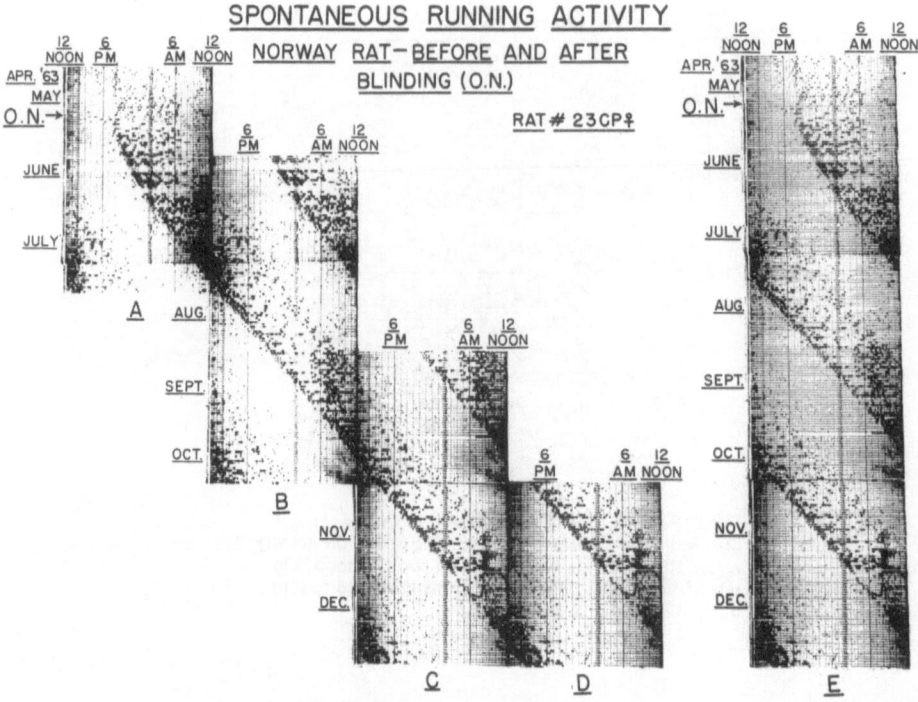

Fig. 7. Activity-distribution record of rat before and after blinding (ON) shows great accuracy of the clock over an eight-month period.

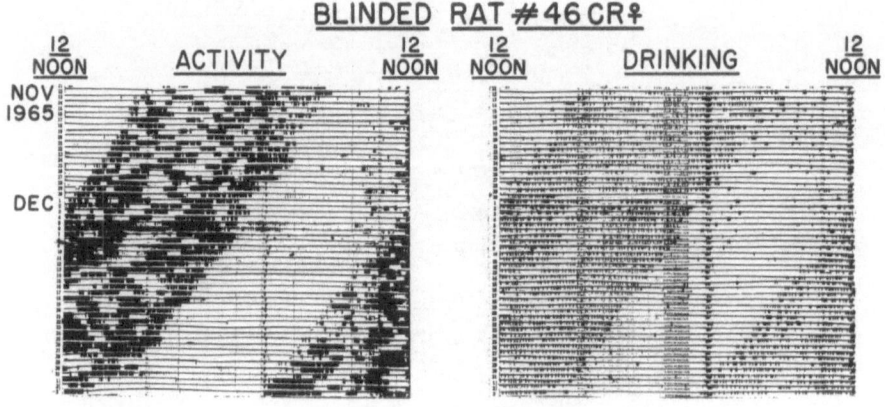

Fig. 8. Fifty-two day excerpt of activity and drinking times of a blinded rat. Each contact of the rat's tongue with the water was registered on an operation recorder.

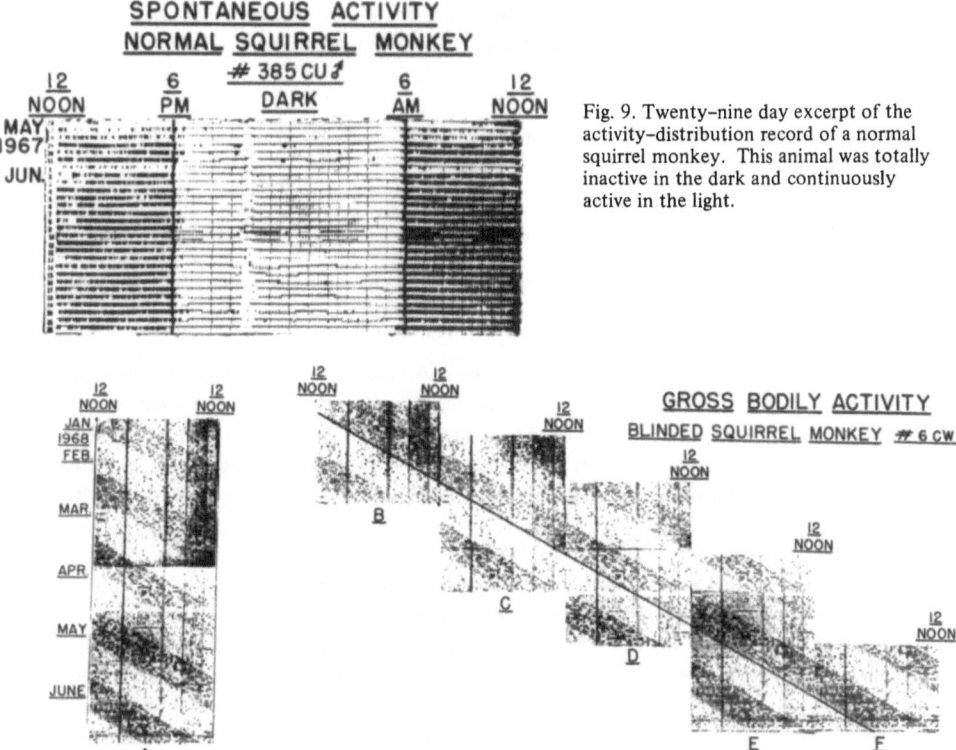

Fig. 9. Twenty–nine day excerpt of the activity–distribution record of a normal squirrel monkey. This animal was totally inactive in the dark and continuously active in the light.

Fig. 10. Five and one–half month excerpt of activity–distribution record of a blinded squirrel monkey. Onsets of daily active phases occurred 42 minutes later each day with great accuracy.

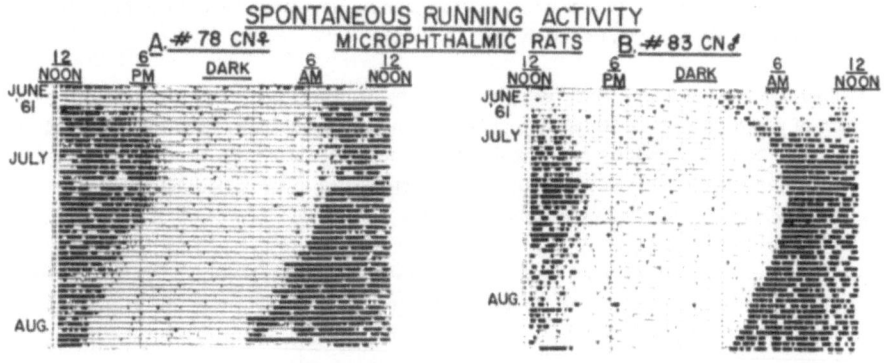

Fig. 11. Excerpts of activity–distribution records of two congenitally blind microphthalmic rats.

by presence of 12 -hour active and 12 -hour inactive phases in early blinded rats. Fig. 11 shows record:
of two congenitally blind rats whose clocks could not have been entrained by light since their optic
nerves were entirely missing.

Our observations indicate that the 24 -hour clock must have originated in the tropics where day and
night have the same length, namely, 12 hours. (Richter 1971).

29 - 30 DAY CLOCK

Experimental observations on blinded squirrel monkeys and rats with brain lesions seem to have est:
blished existence of this clock.

Fig. 12 shows the activity-distribution charts of two squirrel monkeys for a 30 -day period six month
before blinding and for a nine-month period after blinding. Onsets of daily active phases occurre
47-50 minutes later each day. They reached noon at intervals of 28-31 days in the first animal and
29-34 days in the second animal. Over a 3½ year period these values were 30.47 and 30.91 days
respectively. These values closely agree with the lengths of periods of the moon — namely 29.5 days
on the average.

Fig. 13 summarizes our observations made on these two monkeys from August 1964 to December
1967. Ordinates show time in months and days (days not visible in this chart); abscissas successive
24 -hour periods. Times of onsets of active phases for the two monkeys are shown in interrupted lines.
They are the same lines of onsets shown in Fig. 12 only extending over 3½ years. They show that
onsets occurred later each day just as was shown in Fig. 7 and 10. Fig. 13 also shows the times of appearance of the new moon over the same 3½ year period (The Observer's Handbook, The Royal Astronomical Society of Canada). The moon appeared 50.5 minutes later each day shown by the round
black dots. Over this 3½ year period lines of onsets of active phases of the two monkeys closely followed the lines of appearance of the new moon.

At first glance it would appear that this remarkable chart demonstrates that over this 3½ year period
the moon exerted a direct influence over these two monkeys in spite of the fact that they were blir
ded and kept in a light-proofed laboratory.

A closer inspection, however, of the records of these two monkeys shows (1) that over the 3½ yea
length of periods of their clocks varied from time to time over a wide range — above and below lengt
of period of the moon. Furthermore, for most of the time the clocks of the two monkeys were out-o
phase with one another — even though the animals were housed in adjoining cages. Often one was i
an active phase while the other was in an inactive phase.

All of the evidence indicates that these two monkeys harbored inherent clocks that have the sam
length of period as that of the moon — but that they functioned quite independently of any influenc
of the moon.

This 29-30 -day clock must have been built into the nervous system of these animals far back in evol
tionary eras when survival depended on ability of animals to adjust to the actual appearance of the fu
moon.

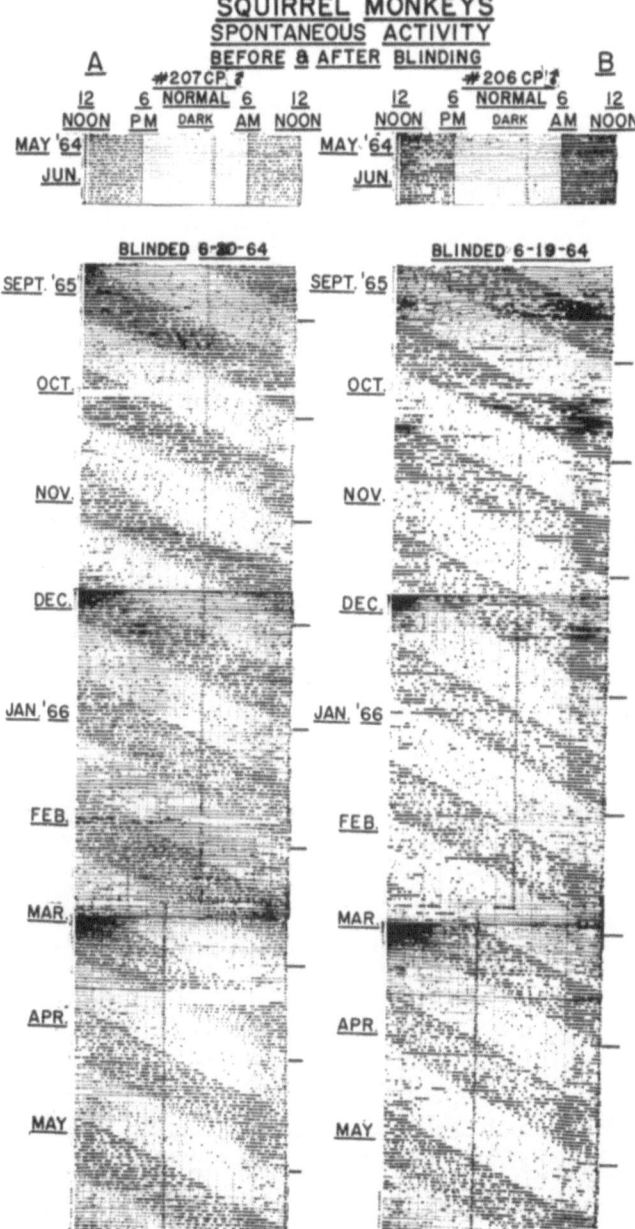

Fig. 12. Activity–distribution records of two squirrel monkeys; for a thirty–day period before blinding and for a nine–month period after blinding. After blinding onsets of daily active phases of the two animals occurred 47–50 minutes later each day respectively.

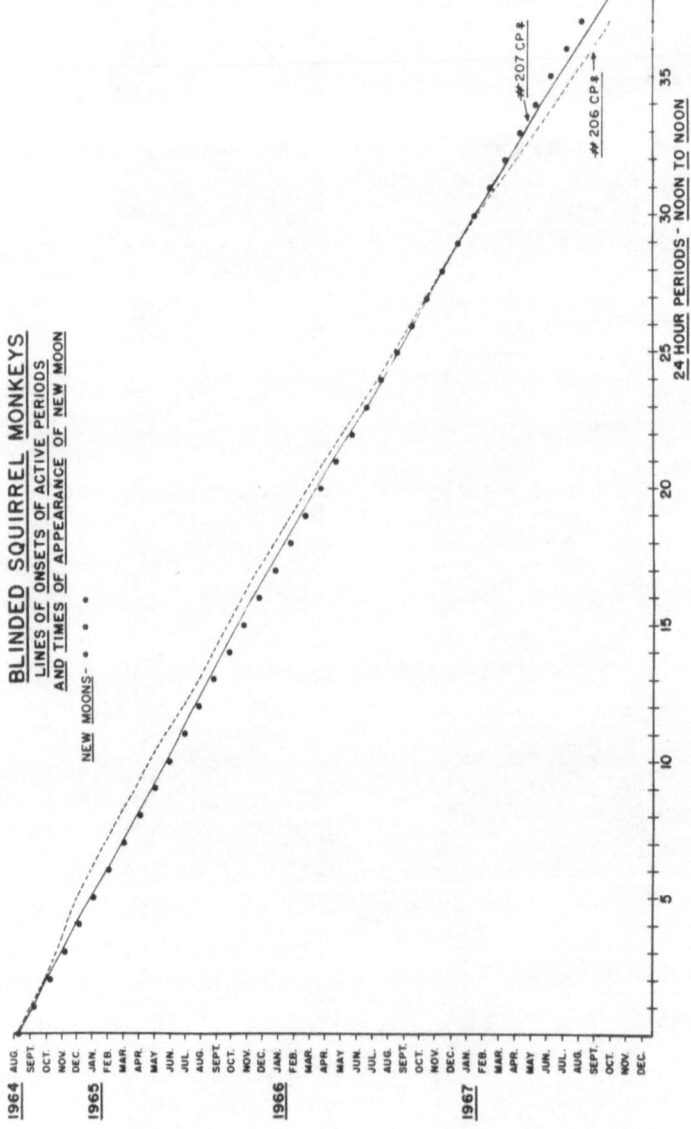

Fig. 13. Graph showing lines of onsets (interrupted lines) of active phases of the two monkeys on record for over 3½ years; it also shows times of appearance of the new moon over the same 3½ year period.

Very accurate 29-30 -day periods were also produced in rats by experimental lesions made in the anterior medial thalami of the brains of both males and remales — bringing further evidence for existence of an inherent 29-30 -day clock in rats as well as in the squirrel monkeys. Fig. 14 shows the remarkable series of 29-30 -day cycles brought out in a rat by a thalamic lesion. (For further examples see Richter 1968).

YEARLY CLOCK

In two species of animals — both hybernators — ground squirrels and chipmunks — definite evidence was found for the existence of a yearly clock. In these animals the clock manifested itself not only in spontaneous running activity, but in daily food and water intake and body weight. These animals were kept in the same activity cages used for rats.

The animals were blinded and in addition were kept in a room that was completely shielded from external light. Furthermore, temperature of the room remained quite constant throughout the year.

These observations were started 15 years ago. In many instances individual animals were under uninterrupted daily observations for periods of 3-4 years.

Fig. 15 shows the record of one of the ground squirrels. The top curve shows three peaks of running activity spaced at intervals of 300, 240 and 230 days. The second curve shows body weight which also manifests quite regular fluctuations but in an inverse relation to activity — that is, the animals were most active when their weight was lowest. The third and fourth curves show food and water intake respectively. They all showed regular cycles in direct relation to activity and indirect relation to body weight.

Intervals between these cycles became shorter from year to year. It is assumed that lengths of cycles became shortened after blinding of the animals just as the periods of the 24 -hour clock of rats and monkeys in many instances became shortened after blinding. The cycles were freed from seasons of the year — maxima of body weight occurring in August, then in April, and finally in November. That is, they were totally freed from the ordinary times of hybernation on the outside. In all of the nine animals used in this experiment lengths of cycles became shortened after blinding.

Fig. 16 shows possibly even a more remarkable record — this time from a blinded chipmunk observed over a four-year period. After blinding, this animal's yearly clock also became freed from external influences — so that three cycles of activity occurred at intervals of 330, 340 and 320 days respectively. Here again the temporal relationship between running activity, body weight and water intake can be clearly followed. Changes in water intake seem to precede changes in all the other functions and length of periods became shortened as in blinded ground squirrels.

The fact that these remarkably regular cycles occurred in these blinded animals independently of any cyclic changes in the external environment must mean that they originated in cyclic mechanisms within in the organism.

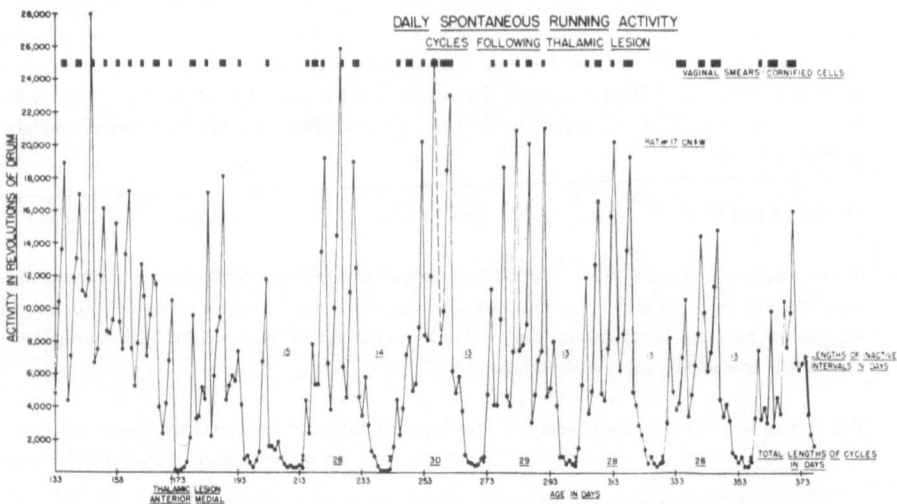

Fig. 14. Record of total daily running activity of rat before and after bilateral lesions in the anterior medial thalamus. Ordinates show daily running activity in number of revolutions of the drum; abscissas, time in days. It also shows lengths of inactive phases and total lengths of entire cycles in days.

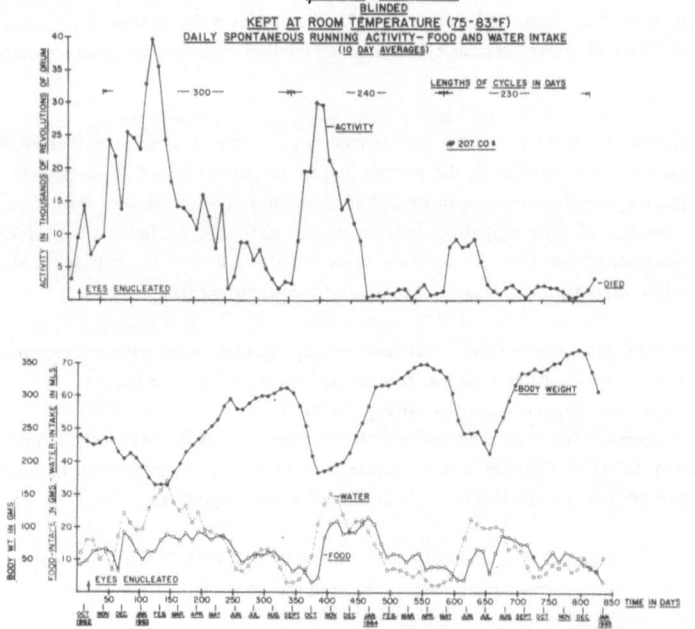

Fig. 15. Graphs for a blinded ground squirrel showing ten–day averages of total daily running activity, body weight, water and food intake over 29 months; it also shows lengths of intervals between cycles. Abscissas show time in days and in months of the year.

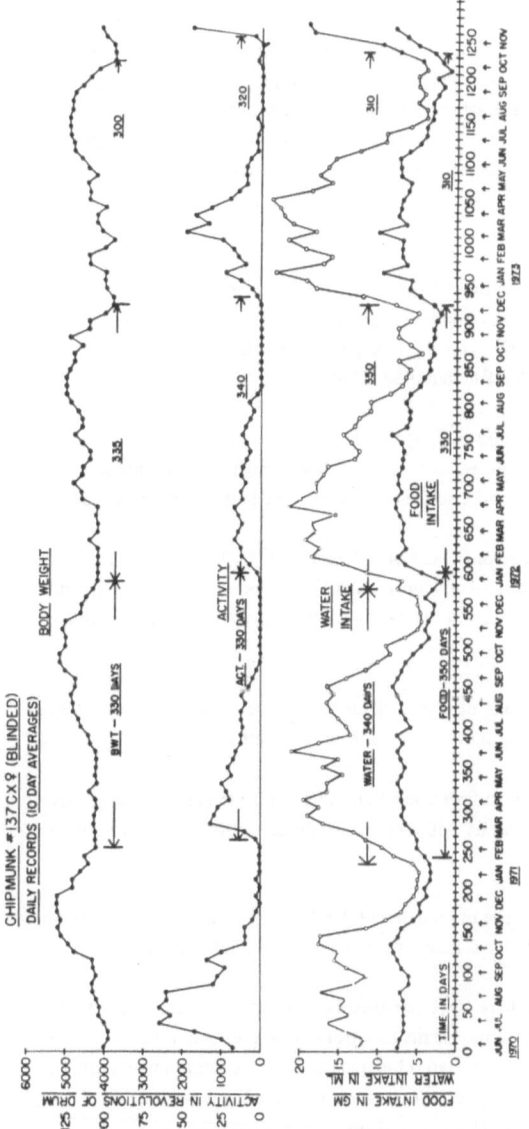

Fig. 16. Graphs for a blinded chipmunk showing ten-day averages of total daily running activity water, and food intake; abscissas show time in days and in months of the year. It also shows lengths of intervals in days between cycles in the various functions.

LOCATION OF CLOCKS

Where are these three clocks located? Most of the evidence at hand shows that the 24 -hour and 29-30 -day clocks are located in fairly specific areas of the brain − in mechanisms that must have been built into the nervous system in the process of evolution many million years ago. The 24 -hour clock must have had its origin in a survival process in the tropics in relation ot the lengths of day and night; and the 29-30 -day cycle must have had its origin also in the tropics with relation to the 29½ day cycles of the moon. The yearly clock must have been built into the brain by-yearly fluctuations in temperature, probably in temperate zones rather than in the tropics.

MAN

Normal man gives little indication of possessing the 24 -hour clock. That, however, he still does possess it but in a submerged state becomes apparent from the fact that under pathological conditions — for instance, after a severe blow on the head, a bout of lethargic encephalitis − the clock may appear quite as clearly as in animals. (Richter 1938, 1967).

Early man up to about 800,000 years ago probably manifested a clock of the same type found in rats, monkeys and apes. It told him when to seek shelter at night to avoid his enemies; and when to wake up to avoid his enemies. The more accurate the clock, the better the chances of survival.

After man's discovery of fire or the use of fire about 800,000 years ago, all this changed. He had light from his hearth fire at night − and protection from wild animals. A premium was now set on his ability to stay awake at night − so his 24 -hour clock gradually became submerged in the process of evolution. (Richter 1966, 67).

Now under normal conditions man gives little indication of possessing a 24 -hour clock as seen in animals − but pathological conditions can still bring it out in full force.

The 29-30 -day clock still manifests itself in normal human females but, otherwise, it too has become submerged to appear only under pathological conditions, chiefly associated with various psychiatric disturbances.

Little is known about existence of an inherent yearly clock in man − except for records of appearance of yearly cycles under pathological conditions in a few psychiatric patients.

In summary, we have seen here three remarkable inherent biological clocks that must originally have been built into organisms by adjustments to astronomical events and that most probably are located in different parts of the brain. Although submerged to a great extent in man during the process of evolution, they undoubtedly still play important though not readily detectable roles in regulation of various physical and mental functions. (Richter 1965).

REFERENCES

Richter, Curt. P.: (1938) *Two Day Cycles of Alternating Good and Bad Behavior in Psychotic Patients.* Arch. Neurol. Psychiat. 39: 587-698.

———. (1965) *Biological Clocks in Medicine and Psychiatry.* Springfield, Illinois: Charles C. Thomas.

———. (1966) *A Hitherto Unrecognized Difference Between Man and Other Primates.* Sci. 154: 427.

———. (1967) *Sleep and Activity: Their Relation to the 24 -Hour Clock. Sleep and Altered States of Consciousness.* William and Wilkens, Vo. 45: 8-29.

———. (1968) *Inherent Twenty-Four Hour and Lunar Clocks of a Primate — the Squirrel Monkey.* Com. Behav. Biol. Part A, 1: 305-32.

———. (1968) *Periodic Phenomena in Man and Animals: Their Relation to Neuroendocrine Mechanisms. (A Monthly or Nearly Monthly Cycle).* Endocrinol. and Human Behav.: 284-309. London, New York: Oxford University Press.

———. (1971) *Inborn Nature of the Rat's 24 -Hour Clock.* J. Comp. & Physiol. Psychol. 75: 1-4.

A great many papers and books on biological clocks or periodic phenomena in general have been published in the last 15-20 years. The appended list contains references to a few books that may be of help to readers who want to learn more about biological clocks.

My own interest in biological clocks started in 1920 during work on my doctoral thesis which disclosed presence of the 24 -hour clock and a 1½ - 2 -hour cycle in gross bodily activity of rats. (A behavioristic study of the activity of the rat. *Comp. Psychol. Monogr. 1*(1922), 1-55.

Aschoff, J.: *Circadian Clocks.* Amsterdam: North-Hollant Publishing Co., 1965.

Bunning, E.: *The Physiological Clock.* New York: Academic Press, 1964.

Conroy, R.T.W.L. and J.N. Mills: *Human Circadian Rhythms.* London: J. & A. Churchill, 1970.

Fisher, K.C.; A.R. Dawe; C.P. Lyman; E. Schonbaum; F.E. South, Jr.: *Mammalian Hibernation III.* Edinburgh, London: Oliver & Boyd, 1967.

Reimann, H.A.: *Periodic Diseases.* Philadelphia: F.A. Davis Co., 1963.

Strughold, H.: *Your Body Clock.* New York: Charles Scribner's Sons, 1971.

Ward, R.R.: *The Living Clocks.* New York: Alfred A. Knopf, 1971.

Cold Spring Harbor Symposia on Quantitative Biology. *Biological Clocks.* Vol. XXV. Cold Spring Harbor, N.Y.: The Biological Laboratory, 1960.

National Institute of Mental Health. *Biological Rhythms in Psychiatry and Medicine.* Chevy Chase, Md.: 1970.

Cyclic States as Biological Space-Time Fields

G. SCHALTENBRAND

1. THE SPACE - TIME PROBLEM

Time is often conceived of as comprising the past, the present, and the future, arranged along a straight line. This conception of time resembles that of the three spatial dimensions with the difference that whereas the direction of time is irreversible, directions in 3 -dimensional space are reversible.

Einstein's relativity theory led to the overthrow of the old concept of this three-dimensional space. According to this theory, every moving object is its own clock which, as seen by an observer at rest, slows down more and more as the relative velocity of the object approaches the speed of light. For the travelling photon, time is arrested in a permanent now.

In the world-view of relativity theory there is no continuous contact between the future and the past. Future and past are thought of, instead, as representable by a cone of world-lines which cross in only one point. Everything which is not future or past lies outside the cone and is considered as belonging to an "absolute elsewhere." The crossing point of world-lines, the "now", must thus be regarded as an illusion since everything in the system is already fixed from the entire past and future history of the world is seen as already there, and we are just meeting it, as it were, "on the way." Whether a permanent movement, such as the rotation of atoms or the vibration of photons are considered as something passing in time or else a permanent present, is seen to depend on the standpoint of the observer. But such a conception of time leads to serious difficulties for any non-abstract observer of the world.

2. THE OBSERVER'S SURFACE

Let us consider such an observer to be a physicist-astronomer for he gave us this idea of the world. Although he receives information from the most distant stars of the world, he must live by interchange with his local world which furnishes him with food, water, oxygen and heat, and which is ready to accept his excretions, and to be moulded by his activities. There exists a surface between the outer world and himself, the surface of his body, creating an artificial separation between his ego and the outside world. The astronomer is a sack of skin containing about 70 kg of dense matter like bones and flesh, semicolloidal substances such as blood, water, salts, and gases, all organized in a peculiar form. In terms of lightyears, an average astronomer will be 6 nano-seconds high and about 5 nano-seconds wide with his arms extended. But such numbers, while true, are misleading because as soon as the skin of the sensory organs of the body are reached by impulses from the outside world, or by touch, or by sound (at a speed of 300 m/sec) or by light (at a speed of 300,000 km/sec), the impulses are coded into other impulses, which travel within the body at a speed from a few cm/sec up to several meters/sec.

There is something that corresponds in the time structure of our being to the conventional limiting surface of the body in space. Since cyclic information which reaches our sense organs is too fast to be perceived, it is coded into a combination of slower rhythms in our ganglion cells as well as into stationary qualities such as colours, images, temperatures, vibrations etc.. These do not allow for the recognition of the external periodical movement as such. The coding system which creates the stationary impressions is something analogous to the skin. It is, as it were, a surface in time, between the internal and external world. The slowing down of the speed of information travel from 300,000 km/sec to about 40 - 60 m/sec is no *relief* for our astronomer! The different world-lines crossing his body and brain are separated not by nano-seconds but by larger fractions of seconds. Is there perhaps a terminal crossing point in the brain where the astronomer sits, like a little monkey, within the web of 13 billion nerve cells and their fibre connections? Even if it were, the same problem would arise again: Is there a microscopic monkey sitting within the astronomer's monkey, where all the impulses would converge to a common "now", one which could not be subdivided any further?

The problem gets even worse when we scrutinize the brain of our astronomer in terms of physics, because his brain consists of atoms molecules in mostly empty space. How does it happen that our astronomer is privileged, among other objects in space, to develop consciousness? Consciousness is certainly connected with the activities of his brain. Medical experience shows that it would disappear with the piece by piece destruction of the brain, yet it cannot be localized in one point within it. Still, consciousness is not a "spaceless" phenomenon: it needs the volume of the brain within which it can represent the entire universe. Consciousness is inseparably connected with the peculiar structure of the brain, with its Gestalt, and with the Gestalt of the astronomer. His Gestalt must be considered as a *Ganzheit,* a totality, as opposed to the sum of its parts. This totality might be described by the Greek word *Holos* (whole, entire). Coining a new word, "holon," we may think of living holons somewhat like Leibniz envisaged his monads, with this difference: a holon has windows, allowing communications with other holons and with the world. Information passes into and leaves through these windows, in contrast with the pre-established physio-psychical parallelism, in which Leibniz believed, when he conceived the idea of his monads.

We now raise the question of whether the properties of such a conscious Gestalt or holon can be sufficiently described by a mathematical analysis.

3. LIMITS OF MATHEMATIZATION

The mathematization of our image of the world is based on a one-sided preference for the optic-haptic sphere of our sensory experience. It is in that sphere that we measure, determine direction, and form number. The qualities of sensory experience such as smell, pain, well-being, sound and colours are embarrassing for mathematization, because they fit only incompletely into the enterprise of the optic-haptic system. Yet, sensory experience exists, and corresponds to the a priori structure of our conscious field; it contains emotional information directly understood even by the new born, such as the agreeability of the sweetness and the attractive smell of mothers' milk. These experiences make one distrust attempts at a pure mathematical conception of science as a *mathesis universalis* even if mathematics remains an extremely useful tool for our intelligence.

Mathematics often developed methods before physicists were able to make use of them. One example is the concept of spaces of four or more dimensions. We cannot visualize such spaces with our a priori space conception, yet they became a useful construct for the understanding of relativity theory, which alone is able to explain certain physical phenomena. Beyond this the general question remains open whether mathematics will ever be in a position to explain the universe of experiential phenomena, the ultimate source of our cognitive experience, and the source of mathematics itself.

4. PHENOMENOLOGICAL APPROACH TO THE PROBLEM OF TIME

Helpful as the mathematization of time may be, it is useless for the elucidation of phenomenological experience: consciousness, becoming, creativity. Therefore, in my opinion, if we want to reach an understanding of time, we will have to start with those elements of experience which appear as overwhelming facts: that we are present in this world, that we have been present, and try to continue to be present, and that we try to change the world so that this continued presentness be possible.

Let us start by distinguishing consciousness, self-awareness, and self-consciousness. In German, self-consciousness and self-awareness are almost the same terms, and both have certain emotional connotations. But *Ich-Bewusstsein* does not quite correspond to self-awareness because *Ich* is considered to be a central point, like the center of gravity in heavy matter, a point in the center of consciousness. But the self is more than such a point: it is the entire conscious person. Hence self-awareness is a preferable expression to the German *Ich-Bewusstsein*. I wish to think of consciousness as a frame of all of our perceptions filled with such Gestalten as we are able to distinguish, and decorated with attributes like colour, sound, smell and taste. Thus, consciousness may exist in animals or in the small child without self-awareness. Self-awareness develops at the time when the child discovers that one of these Gestalten can feel pain when it is hurt. Also, at the same stage, the child experiences certain peculiar double sensations related to touch: when he touches himself he feels the touch with his hands, and feels the same touch with that part of his body which is touched.

Consciousness may still exist when one or many of its components are lost. This is the case for blind persons, or for deaf persons, or even for people who are both blind and deaf, like Helen Keller. Furthermore, consciousness may be limited to feelings of pain, heaviness of a part of the body, or a melody, distinctly but simultaneously, without any other perception, but it has always an extension in space or in time, or in both. Accordingly, I want to think of consciousness, reduced to its most fundamental and irrevocable components, as an extended manifold in space and time. But this extended manifold is perceived within a physical Gestalt and cannot be localized in one of its physical components. Hence, its relation to these physical components is analogous to that which the field of matter bears to the material points, to the atoms and molecules within matter. Thus, I see consciousness as an indivisible presentation of an extended manifold.

It is generally recognized that it is impossible to account for consciousness, mind, emotion and perception using only the meager physical abstraction of matter. This was sufficient for the development of the mechanics of the last century, and of that materialistic conception of a world which regards mental experience as an erroneous phenomenon and, a delusion (but for whom?) or, following Descartes, as a split between mind and matter. Some philosophers, such as Nicolai Hartmann, believed that new levels of function would develop during the evolution of matter into living substance and that mind would emerge from the lower levels like deus ex machina. But many other thinkers hold that we should strip from the organism its own basic properties of consciousness, perception, emotion and will, in order to reach down to the simpler subunits of our world. Such a program would represent an intolerable simplification. It could not even be used to explain, for instance, how I can lift a stone, once I decided to do so. We know from neurophysiology that my will causes a flow of electric charges in my nerve cells, which in their turn, cause, via a well-controlled avalanche of reactions, the motion of the muscles to lift the stone. The crucial point is that first the electric charges have to be set in motion by will power. But, moving of electrons constitute matter moved by mental processes. How can they follow my mental intent, if as it were, they do not understand its command? That the extension of conscious fields of understanding, emotional reactions, or creative actions do not register on macroscopic devices (and senses), and that we cannot decipher coding systems in microphysics, may be due to the fact that their durations and configuration in time and space, and the pulse of events in the microscopical world, are too different from the rather slow space-time structure of our bodies.

5. CONSCIOUSNESS AND THE "DING AN SICH"

The field of our consciousness is filled with many impressions. It is not an assembly of points but of physical Gestalten: plants, animals, men, objects, and things which we distinguish from the background. Since Immanuel Kant the nature of these Gestalten has been questioned because he realized that they can be recognized only within the frame of our innate, a priori, categorical organization. But Hegel has reasoned that we ourselves are the only "Ding an sich", which we actually know. We ourselves have an inside and an outside, and the inside is directly available and known by introspection. It is indirectly knowable by reading the information given to us by other "Ding an sich" which, like ourselves, show growth, motion, maturation, goal-directed activity and are able to exchange information with one another. This is not only true for man, but also for animals and even for plants. We came to recognise that these Gestalten consist of "hardware" and that they use a certain "software". The "software" is communication, which finds its highest development in human language.

There are many objects which were considered as living by primitive men: wells, waterfalls, clouds, thunderstorms, mountains, the earth, the sun and the stars. Today we are inclined to hold that these physical complexes are not alive. Yet, it is us, the observers, who decide what belongs to a cloud or to a mountain and where its boundaries are, whereas the true living Gestalten determine their boundaries themselves. There are regions among Gestalten which organize themselves; these we find similar to ourselves and to the things which we find in nature. There are other Gestalten, however, in which we cannot recognise self-organisation. They include all those objects which are fabricated by man: weapons, gadgets and machinery in general. On this view even a chair seems to have some organic features, though it is unable to repair itself and to multiply. The highly developed computers, however, came to resemble more and more organically developed Gestalten and the miniaturisation of their structure brought them close to the world of the living cell. But there are gaps between the smallest components of our machines and the cellular structure, and further down, the macromolecules of living matter.

What features are most characteristic of ourselves as living and conscious beings? We can detect in ourselves quite a few functions typical of the living individual: goal directedness, creation of order, protection of the self, capacity for self-repair, memory, etc.. Others belong to the supra individual social organisation of life: imitation, representation, information, understanding of information, emotional affectability, esthetic and ethical judgment, and self-reproduction through the fusion of living substance of two individuals into a new unity. From these functions we can descend to abstract differentials of our self. Although we may ignore one or many of our functions, we cannot build an organism out of such "castrated" matter. Therefore, it is appropriate to ask: which of our primordial experiences (as consciousness, emotional affectability, analyzing information and informing, self-repairing and self-reproducing, etc.) may be identified as one descends step by step into the finer structure of the self and of the world.

Many scientists have tried to understand organisms from the standpoint of behaviourism, ignoring certain aspects of their own mental processes. In my opinion such attempts are futile, because interpretations characteristic of living substances creep into the definitions of the "non-living" matter. Even an ordinary brick is made with physical properties so selected that it may be employed for building. The feedback mechanisms which we find in goal-directed machines prove only that such mechanisms can be built by man as tools for human use. They are not alike and conscious but are intended for conscious beings and used by them. Studying living organisms we must begin with the total holon, followed by the interpretation of the operational importance of certain systems within our Gestalt. We may ask, for instance, which systems and which parts are essential for motion, which ones for communication, for sensual perception, for metabolism, for growth, for healing, for self-reproduction? The fact that we are able to build mechanical prostheses which help memory (e.g. tape-recorders), does not mean that the tape-recorders have memory themselves. While some separate functions may thus be imitated, consciousness needs the hierarchical structure of self-organising field and matter, the holons.

6. INFORMATION IN THE HIERARCHY OF LIVING BEINGS

Let us now turn to the "software" that plays such an important role within and among living Gestalten: the means of communication. Animals exchange signals with meaning for sender and receiver alike: the call of the hen for its chicks, the warning calls of birds and mammals, the wooing calls of all kinds of animals. Although most of these calls represent inborn patterns, we should not consider the reaction of the receiver as reflex only. We should not deny a conscious component, that is the communication of an emotional state. In singing birds we see that these signals may be modified by learning and that they show a peculiar feature, the rhythmic arrangement of these sounds and their repetition.

The highest development of communication among the living is human speech, almost entirely acquired by learning. It was probably developed first by the imitation of animal sounds useful to the early hunters for luring animals into traps or chasing dangerous animals away. Our speech consists of sentences, words, syllables and phonemes, each of them occupying a level in a hierarchy of transfer. All of them are generated in our nervous system by reverberating circuits, which may at any time be interrupted, modified and exchanged and which combine to build up an increasingly more complex pattern of information. [1] The development of language from "prefabricated" reverberating components is most easily seen in the language of small children and of primitive people. They tend to repeat the same word or the same sentence again and again. The aboriginal tribes in Africa, on the American continent, and the Eskimos, used to talk and sing repeating phrases, combined with dance and acting. These strongly suggest that in the early beginnings of interpersonal communication in children language cannot be strictly separated from other forms of behaviour. There is no doubt that the same argument also holds for the development of language in the early stages of mankind. Repetition, an old routine of communication, has survived in poetry and music, in rhyme and verse, in rhythm and phrasing, and it is responsible for the lasting appeal of these arts. The tendency toward reverberating utterances is also observable in disturbances of speech. In motor aphasia, for instance, the patient is unable to repeat anything but a few stereotyped words. In the palilalia associated with Parkinson's disease the patient's speech resembles the playing of a scratched gramophone record, where the needle falls back into the same groove again and again.

Let us turn now to the problem of communication *within* the living individual. Analysis of the function of our nervous system by electro-physiological means allows us to register a large set of different periodicities. The slowest periodicities which can be picked up by electrodes are the alpha rhythm (6-8 pulsations per second) and the beta rhythm (about 20-40 pulsations per second). The alpha rhythm has its source in the occipital part of the brain where vision is represented; the beta rhythm is found in the central region of the brain. The alpha rhythm appears when a person is at rest with the eyes closed and disappears as soon as the person is asked to count or listen or think about some problem. What we see now is a desynchronisation as many shapes of waves of different frequency interrupt the alpha rhythm. (Fig. 1)

Studies of the averaging of standardized rhythmical activity of the brain after rhythmical stimulation show that there are specific sources for the different sensory regions of the cortex. Examples are the occipital lobe for vision, the temporal lobe for hearing, and the central region for touch and movement where the impulses arrive in a spatial order that corresponds to the visual field or to the surface of the body. High speed analysis of the topology of these impulses reveals the details of their manner of propagation from their sources towards the other parts of the brain. They show how the signals slow down or speed up depending on the parts of the cortex through which they pass. This suggests that, in addition to the topographical projection of regions of the body there exists chronological organisation in the brain. [2] But the chronological organisation is not recognized by consciousness. Single brain waves of this type allow sufficient time for the nervous impulses to be propagated from one end of the body to the periphery and back several times. Since a re-entrant action is represented in consciousness as a present, or now, it follows that the flow of private time is not unidirectional, and that there is no passing of private time during the "now" of consciousness.

We do not know exactly how the rhythmic electrical discharges develop. Possibly, they are envelopes of more rapid discharges of single ganglion cells which fire at rates of several hundred spikes per second. Or, they may be the results of an integration of the synaptic noise on the surface of their cell bodies, which has an even higher frequency. Thus the nervous system performs coding of higher frequencies into lower ones. However, information which the different parts of the nervous system (page 60) exchange with one another may be more complicated than the spatial organisation and distribution of the electrical pulses. There are also examples of frequency modulation: the intensity of a necessary innervation of a motor unit in the muscle is coded as a change in the frequency of impulses of constant amplitude. (Fig. 2)

We are very far from being able to interpret the complicated interior communication of our nervous system, though in many respects it can be compared with the functions of a complex computer — if we keep in mind that we have no indication that a computer has consciousness. Our nervous system has many properties without parallels in computers. For instance, its activities are influenced by applied electrical fields, hence the functioning of the neurons influences not only the target regions but also their neighborhood.

We are justified to claim that the entire brain functions as a whole, as an extended field of complex electrical pulses functioning as a *Ganzheit*. We know that consciousness disappears when this electrical activity stops, hence we are justified to assume a degree of consciousness when identical electrical activity is recognized in the nervous system of other living beings.

In spite of the fact that we are unable to understand information which is mediated by the electrical activities of the brain and transported from one part of it to others, we must admit meaningful similarity between this type of communication and the rhythm of such interpersonal communication as speech, dance and music. They all represent the cyclic rhythmical order of the "software."

7. THE LANGUAGE OF THE CHROMOSOMES

Microscopes and other scientific tools allow us a descent into the interior of living substance. Wherever we arrive we find subunits which again have the characteristics of Gestalten. These subunits communicate with one another and extend the domain of hardware and software down into the microscopic and submicroscopic levels of our organism. At the lowest stage which we may still consider as alive, we find the double helices of DNA (deoxyribonucleic acid). They carry information for the construction of living beings and use the electrical charges of their own structures to build up their mate by duplication. The function of chromosomes is similar to that of the tape of a tape recorder which at any time can replay a melody.

We see again the peculiar tendency of life to build up the organism from reverberating patterns. Even in the most primitive organisms there is a duplication, triplication, quadruplication, etc. of forms. The most impressive example of this repetitive function of the forming power is the invention of the segmentata (animals which are built up of segments in which a pattern is repeated). This tendency, found already in worms, is continued in the development of the vertebrates. In later stages there is a second periodicity which runs over the first segmental periodicity and induces the development of extremities with longer periods. Still, the anterior and posterior extremities resemble each other. In the construction of extremities we again see rhythm and repetition in the formation of bones and joints succeeding one another from the proximal to the distal end until they reach the finger-tips. Additional alterations add a nice head and tail to the individuum.

I want to illustrate my point with a picture of a segment during this embryological development in man (Fig. 3). We recognize the almost bilateral symmetry of the external shape of the body, with certain deviations of the internal organs from this rule, and the beginning articulation of the buds for the upper extremities right and left, which evidently follow a rhythm of their own. The end result may be illustrated with a figure of the dermatomes in man (Fig. 4). These are the regions of the skin that correspond to the distribution of sensory nerve fibres which belong to one segment of the spinal cord. They still allow us to recognize the primordial law of cyclic development despite all the twists and shifts which have been induced by specialisation, particularly in the arms and legs.

Thus, the weaving of an organism reminds one of the weaving of a sentence in speech, and also of the weaving of a rug. Both are cut off when the pattern is complete; both comprise the spinning off of cyclic states.

8. MAY INFORMATION THEORY BE EXTENDED TO THE ATOMIC LEVEL?

Upon closer examination even amorphic objects such as a cloud, a mountain, a meteor or a planet will, on a certain microscopical level, show a structure similar to that of living organisms. From there down such objects are built up by a continuous series of self-organizing and self-restituting units which exchange information just as the cells of the body do, on a higher level.

The distances between atoms or molecules of a crystal relate mathematically to the frequency of the spins, nuclear rotations, electronic rotations, or to whatever constitute the peculiarities of the components. Thus, there is a direct relation between the spatial structure of a crystal and its temporal structure. The distances between the points of the crystal are not only distances in space but also distances in the time of its private space-time field, determined by its own physical forces. The fact that the steady state of a complicated structure, like a crystal, is "in tune" with itself, shows the absurdity of the conceptions of a point-like "now" applied to the private space-time field of a complex system. The cyclic state in space and time in a stable crystal neither produces nor consumes any energy, but it is full of potential forces which resist any deforming external influences. For instance, the application of pressure on certain crystals will produce a redistribution of its electrical charges; the application of electrical potential will produce changes of the spatial dimension of the crystal. We are talking, of course, of piezo-electric effects. When a piezo-electric crystal is perturbed by a "trauma", that is by an impact which upsets its organization, it shows a tendency to "heal". It reshapes itself; it fills out the gaps within its lattice structure with the correct material, in the correct places. This happens for instance when the crystal is immersed in a solution of its own material. The "healing process", of course, consumes energy, taken from the world around the crystal or from its own reserves (at the expense of a diminution of its size). In stable conditions, when undisturbed, there is no such "metabolism". The sizes of certain crystals, sometimes of several meters, attest to the strength of this organizing power of cyclic states in space and time.

Atoms themselves are complex structures, comprising several nucleons, protons, neutrons and other components, with an exchange of electromagnetic gravitational and, so-called, strong and weak exchange forces. These forces cannot be reduced to a single denominator, thus even the field of the atom is a simple, but manifold structure of the world around it. In the nucleus of atoms we again find a distinction between "hardware," which appears stable, and a "software" which resembles communication. Here we recognize, better than in the macroscopic world, that the software, which is a spinning off of information, corresponds exactly to the hardware, which is frozen information. The Gestalt in space can at any time be changed into a Gestalt in time, carried by radiation for unlimited distances.

It seems then, that the metamorphosis of hardware (corresponding to units of the world in space) into software (corresponding to the temporal Gestalten) is a basic feature of the universe. It reveals itself to the eyes of science as a complete hierarchy of individual Gestalten which share certain properties with us, the observers such as the cyclic architecture, ability to reorganize themselves when disturbed, and ability to communicate with one another by various types of resonances. Through these functions they partake in the creation of a hierarchy of integrative organization from the elementary units of matter to atoms, molecules, crystals, macromolecules, and living cells and up to the organisms which are ourselves.

9. CYCLIC STATES AND THE BLACK BOX PROBLEM

We have seen that recurring, cyclic states are not limited to macroscopic beings. They are also the bases of many enduring phenomena which may be classified as *duree*, or being, such as stable circular movements, as particle spins, or the vibrations of the photons. But cyclic states, (the pulsation of elastic system, the rotation of an electronic cloud, or even the revolutions of planets around the sun) are not yet clocks. In order to become clocks they must drive an index, or a counter which sums up the number of rotations. This is not possible on the atomic level without disturbing the cyclic process, though cyclic processes are able and bound to stay in resonance with one another as long as they remain closely coupled. From the point of view of a larger unit, a smaller unit is a black box, the longer unit is unable to count the rapid oscillations in its interior; but it "feels" its existence as that of a static structure with the magnetic, electrical and gravitational tensions within the surrounding field. Time enters the system via the quantum jumps within the black box. Under extreme conditions, these jumps signify beginnings or ends. The peculiar fact that circular motion, like that of the electron around the nucleus of the atom, is perceived by the surrounding world not as a movement, but as a state, namely a magnetic field, is an excellent example of the method of coding between smaller and larger units in a hierarchy of complex systems.

Cyclic states may be considered reversible processes; the direction of time plays no role in them; they may run either way. Contrasted with cyclic processes, ever since Clausius and Boltzmann developed the idea of statistical laws in the service of thermodynamics, we also recognize irreversible processes, associated with the increase of entropy in closed thermodynamical systems, creating an "arrow of five."

10. QUANTUM JUMPS AND THE NETWORK OF TIME

Since the individual rotations of undisturbed cyclic states on the atomic level cannot be counted (externally they have the properties of stable states), and since only quantum jumps produce a progression in time, quantum jumps appear to be the basis for counting events and comparing temporal distances between such events. Counting quantum jumps allows the creation of a statistical network, which in turn permits time measurements; they correspond to becoming and to passing away in life.

11. EXISTENCE AND BECOMING

Let us return to considering man's experience of time and recall our earlier reasoning that a point-like "now" would be essentially timeless, and that the dating of an instant may be done only by an observer. This can be most easily argued for a beginning, which, even for the smallest but sufficiently complex being, signifies a change in its stationary, stable state, a creation, or becoming, a shifting of a state that existed into the irrevocable past. The end of even the smallest complex of being is another change, a passing away, or an integration into a larger complex. Our consciousness and our whole existence consist of an immeasurable number of such components of being, which become and pass away in an endless stream. Some of them are stable for years, others for seconds, after becoming and before passing away. Thus, one may say that being, or *duree*, emerges like an island out of the flow of communal time. Nevertheless, it is not a state of death even if nothing happens. Hence, there is no reason to distinguish between being (*Sein*) and existence (*Dasein*), as it is done by existentialists who associate being with dead objects, becoming with conscious ones. Being, when carefully scrutinized, is always recognized as a self-organizing state, recurring in itself and able to compensate for certain perturbation by means of homoeostasis.*

12. FOUR COMPONENTS OF TIME

Instead of the conventional three categories of temporality, past, future and present, I wish to accept four categories: (i) existence as a cyclic state of undetermined length; (ii) becoming and passing away, which delineate the surface of existence in communal time; (iii) the irreversible past with its destroyed, complex states of existence and (iv) the undetermined future, which appears to man as a challenge or menace.

The "now" of the present appears as a permanent surface of existence in communal time, even if this surface is not smooth. It seems to us that certain parts of it are "more present" than others; they are ahead of communal time while others are behind it. If we push them back too much, they get lost and join the past; if we push them forward, we experience becoming; thus, becoming and passing away are essential properties of the conscious present. Hence the apparent surface plane of the present has its sources and sinks, which, as a rule, have some emotional coloring such as desire, fear, hope, repulsion, love. Has the physicist any use for these primordial experiences? Yes, for there is a fundamental similarity among them and gravitation, and repulsion. All these are of the nature of tensions, and not simply states of motion.

The irreversible past and the open future fill the consciousness of man. He tries to explain the riddles of the irreversible past and to reconstruct what has been definitely lost and to understand how and why the world has developed, how man himself has come into being. But more important is the preoccupation with the future, which he feels to be his field of activity. Since within certain limits he is able to shape the future, he recognizes his responsibility and his freedom of participating in creation. He also recognizes that it is his ability and his duty to decide what will be and that the "now" is more than just a point of encounter on a world-line that has been determined by eternity.

13. PARALLELS BETWEEN THE MIND - BODY RELATION AND THE FIELD - MATTER RELATION

As we remarked, the cyclic states of elementary particles create static magnetic, electrical and gravitational structures of the world around them. These fields are responsible for forces exerted upon other matter. Thus, they force electrons to circle around magnetic lines of force, or attract massive bodies to move around one another so that finally it is hard to say whether it is stationary matter that shapes the field or whether the field shapes the behaviour of the stationary matter. Changes in these stable conditions occur through quantum jumps, such as during the absorption or emission of light quanta, or during the fusion or fission of matter. I suspect that the coupling between field and matter is analogous to the relation between mind and body in living macroorganisms, and that they both represent a degree of consciousness inherent in the world. Both pairs display interdependence, though their cohesion is not a strict one; each displays a degree of intractable uncertainty as to their limits in space and duration in time.

*The term "homoeostasis" means the maintenance of constancy of relations or equilibrium in the bodily processes necessary for life: e.g., the maintenance of body temperature, oxygen supply, blood pressure etc.. The term was first introduced by W.B. Cannon. (W.B. Cannon: *The Wisdom of the Body*, revised edition, 1939).

Observing ourselves, for instance, we realize that we not only see all the furniture, the sky, landscape and the trees around us, but we are at the same time able to listen to a singing bird, to hold a pen in our hand, to breathe, to tap a melody or a rhythm, and to think. We are putting separate things together within the same plane of the present. In physics we see analogous phenomena in the mutual gravitational attraction between masses and the electrical attraction between positive and negative charges. Both may come together without, as it were, a rope to pull them together. Mechanics tells us only about the forces in the outside world, while those within ourselves seem to be a riddle. Physical theory does not address itself to problems of the structure of our consciousness. Yet, even the idea of curved space-time bears a curious parallel to the Hasidic teaching that space is God and God is consciousness. Namely, just as matter is regarded in general relativity theory as a local irregularity in the structure of space-time, so our consciousness may also be regarded as a lump in the general consciousness of the world, separated from that general field by its peculiar structure in space and time. It is restrained by that structure to "resonate" only to certain aspects of the more general field. Thus, we are enclosed to some extent within the black box of our physical Gestalt.

One may wonder whether the appeal to the similarity between our conscious experience and some features of space and time in the physical world does not amount to an invocation of deus ex machina?

I do not believe we have any choice in this matter. In earlier writings on time and consciousness[3] I have tried to explain that the participation of teleological factors, like the goal-directed tendencies of life, carry a reversal of the direction of time in themselves, and that the phenomena of consciousness cannot be understood unless one admits reversible time and an arrest of the flow of time during conscious phenomena. Are there any observations in physics which would permit the acceptance of negative time?

Such a condition was unacceptable for the old mechanical conception which allowed only for deterministic causation. But in microphysics this problem has been discussed again and again. The black box problem arises not only in living organisms: at the atomic level Feynman has argued for a reversal in the direction of time.[4]

The periods of time reversals postulated by Feynman, in macroscopic communal time, are certainly very brief. It is quite evident that the world has a preference for one direction of time. But still these concepts of modern physics allow us to admit a sphere of events where determination in the classical sense can no longer be maintained. The question arises then whether this is not a general rule for quantum states of even larger dimensions, such as for the interior of crystals, for superconducting metals, or superfluid helium? Einstein himself, one of the inventors of quantum mechanics, was opposed to undetermination and remarked once that "Gott würfelt nicht", God does not play dice.[5] He did not consider the possibility that leaving the gate open for indeterminism might mean more than leaving things to chance; the gate might lead to responsibility, free decision and creativity: essentials of human existence and phenomenology. The problem of the irreconciliability of determinism and of phenomenology was one of the antinomies of Immanuel Kant, one he could not resolve. It seems that the developments of modern science open the gateway leading to a unified vision of mind and body, matter and field.

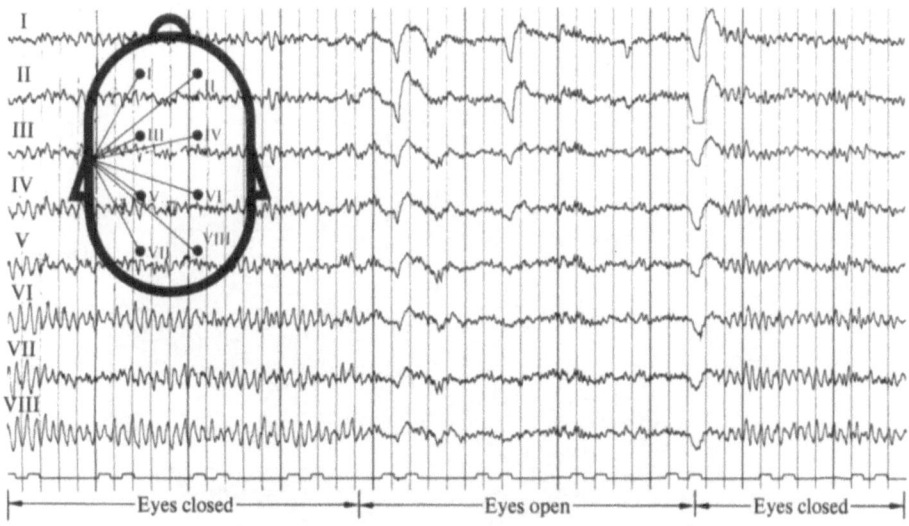

I

II

III

IV

V

VI

VII

VIII

|—————Eyes closed—————|—————Eyes open—————|—————Eyes closed———|

Fig. 1. Electroencephalogram of a healthy person. The template at the left shows the position of the electrodes. The lowest records are from the occipital and temporal lobe and show a regular Alpha-rhythm with 10 waves per second. The other tracings are from the frontal and central region; between some Alpha spikes there are many small Beta spikes. During the opening of the eyes the regular rhythm disappears but returns after closing the eyes.

Fig. 2. Electrical activity of a single motor unit of the human biceps muscle. Uppermost trace shows the time in seconds; the second trace the amount of elongation passively induced; the third trace the increasing muscle tension during and after this elongation; the bottom trace the frequency modulation of the activity of the motor unit which produces the increased tension and counterbalances the passive elongation. Before the elongation takes place the frequency is 8.5 spikes per second, during the elongation for a short period it is about 15 per second, afterwards about 10 per second.

Time (seconds)

Amount of stretch (20°)

Tension

Firing speed of motor unit

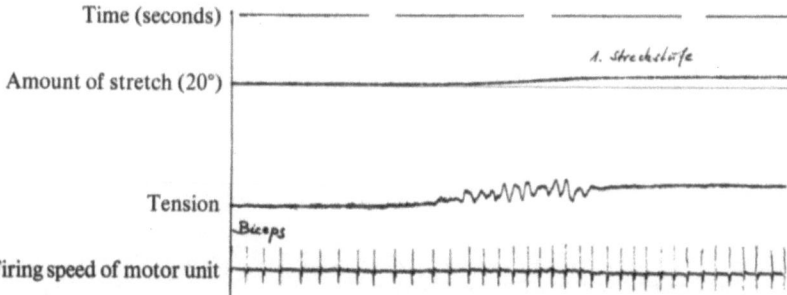

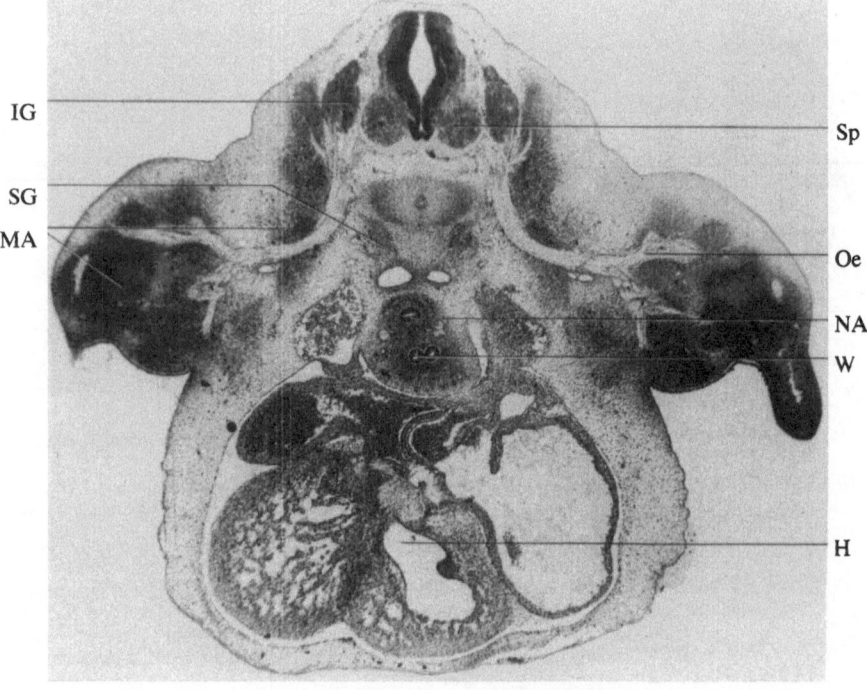

IG

SG

MA

Sp

Oe

NA

W

H

Fig. 3. Cross section of the lower neck region of a human embryo. The buds of the arms may be recognized on both sides. Heart and lung develop in this region, also other organs which at the end of development are located in the chest.

Sp	= spinal cord	Oe	= oesophagus
IG	= intervertebral	W	= windpipe
	ganglion	MA	= muscular
SG	= sympathetic ganglion		Anlage
NA	= nerves for the arms	H	= heart

67

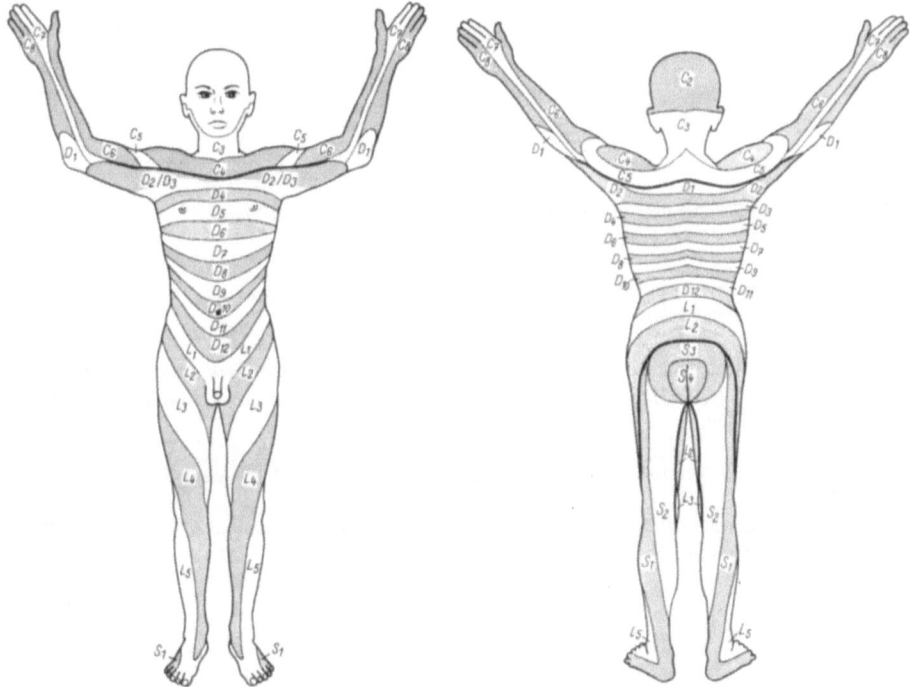

Fig. 4. Dermatomes of adult man. Regions of the skin innervated in sequential order
by the corresponding segments of the spinal cord.

C. -cervical dermatomes belonging to the neck
region
D. -dermatomes belonging to the chest region
L. -dermatomes belonging to the lower trunk
S. -dermatomes belonging to the both parts
of the spinal cord, the sacral region ·

NOTES

1. Georges Schaltenbrand: "Die biologischen Voraussetzungen der Sprache," *Münch. Med. Wochenschrift* 102 (17) (1960): 817-22.

2. This was first postulated by Christian von Monakow in *Die Lokalisation im Grosshirn* (Wiesbaden: Bergmann 1914).

3. Georges Schaltenbrand: "Consciousness and Time," *Annals, New York Academy of Sciences,* 138, Art. 2 (1967): 632-45; "Das Leib-Seele Problem," *Allgemeine Neurologie* (Stuttgart: Thieme 1969).

4. See e.g. G.J. Whitrow: *The Natural Philosophy of Time* (London: Nelson 1961), pp. 280-3.

5. Albert Einstein: "Autobiographical Notes," *Albert Einstein: Philosopher Scientist,* ed. P.A. Schilpp (New York: Tudor 1951).

III. HISTORY OF IDEAS

The Concept of Time in Western Antiquity

P.E. ARIOTTI

My purpose is to trace the development of Western conceptions to Aristotle and to argue that his conception of time is best characterized as *celestial reductionism of time*.

1. TIME RECKONING, CONCRETE, MULTIPLE TIMES

The majority of men, wrote the neoplatonic philosopher Proclus, are aware of time.[1] The ubiquitous and constant alternation of day and night, of the seasons, of birth, growth and death, of wakefulness and sleep, of the lunar phases and circuits of the stars conspire to produce in man a sense of time. The least as well as the most advanced human societies developed ways of reckoning and measuring the passage of time. Natural events were used for timing important human activities. The arrival of the cranes in ancient Greece, Hesiod noted, marks the time for planting, the return of the swallows the end of pruning. Even more reliable are celestial events. The ancient Egyptians recognized the heliacal rising of Sothis (Sirius) as early as the Middle Kingdom (2100-1800 B.C.) and used it to mark the beginning of the year. The South African Bushmen, who lack all systematic knowledge of astronomy, note the rising of Sirius and Canopus and reckon the progress of winter by their movement across the night sky. In *Works and Days*, Hesiod noted the setting of the Pleiades as marking the time for hauling the fragile ships of the Greeks up on the dry land, and the summer solstice for safe seafaring.[2] The cycles of the Moon, i.e. the months, with their clearly demarcating nights of invisibility provided a series of conveniently long time intervals. Human pregnancy lasts approximately nine months. Six moons intervene between sowing and harvesting, between the longest and warmest, and shortest and coldest days of the year. Indeed, ancient man throughout the globe started to reckon the passage of time by counting moons. And he still does, particularly in traditional activities such as religious rites.

The moon cycle, however, is incommensurable with respect to the solar or yearly cycle. The ancient Egyptians recognised such incommensurability early in their history and formulated a yearly calendar of 365 days divided into three seasons and twelve months of thirty days plus five epagomenal days. The ancient Babylonians, on the other hand, persisted in attempts to bring the lunar cycle into synchrony with the solar by means of an occasional intercalation of a thirteenth month. Both the Egyptians and Babylonians developed clocks of considerable sophistication: the *merkhet* or transit slit, shadowclocks, sundials and water or sand clepsydrae as well as the "diagonal" calendar or "star"-clock illustrated by Neugebauer.[3]

And yet, in spite of their sophisticated time measurements, in spite of the Babylonian's constant *nychtemeron* and of the Egyptians' 24 hours day, neither of them, nor indeed any ancient people except the Greeks show any evidence of having conceived time as an isomorphic, continuous quantity or as a frame of reference for events. Ancient peoples, at least in the region of the Mediterranean, did not separate time from its contents. For them time was qualitative, phenomenological, concrete interval or multiple. Time was not a neutral and abstract frame of reference. Time was its own contents. Events were not *in* time, they *were* times. Ancient Hebrew did not have a word for time, but for season, point in time, or eventful duration. Thus ancient Israel could properly say that "this time [or season] is evil," meaning that the event which concretized that particular interval of time was an evil event. For the same token, the Psalmist could meaningfully use the plural and acknowledge to his God that:

"My times [seasons, events, activities] are in thy hands..."[4]

As well as in the famous saying:

"To everything there is a season, and a time [season] to every purpose under the heaven:

2. A time [season] to be born, and a time [season] to die, a time [season] to plant, and a time [season] to pluck up that which is planted..."[5]

And finally, in Job:

"...times [or events] are not hidden from the Almighty..."[6]

2. THE SINGULARIZATION OF TIME

It is a long step to pass from the mere synchronization of natural events and human activities, from the reckoning or simple measurements of time intervals, from the identification of time with its contents, and from multiple times, to the conception of an abstract quantitative time such as we shall find in Plato, and in a more developed state, in Aristotle. A most important stepping stone in such a conceptual development was taken by certain ancient Greeks. While the ancient Egyptians, Babylonians, Hebrews, and probably ordinary Greeks kept to their multiple times, Greek theogonists and poets, perhaps drawing from earlier Iranian sources, hypostatized time into a single entity, cosmological principle or god: Chronos. Such a conception can be exemplified in the view attributed to Pherecydes (middle of the 6th century B.C.) by Damascius:

"Pherecydes of Syros said that Zas always existed, and Chronos and Chtonie, as the three first principles...and Chronos made out of his own seed fire and wind (or breath) and water..."[7]

By the poets of the next century, this cosmological principle, Chronos, was given full divine and anthropomorphic attributes. For Sophocles time was all-mastering, it revealed and hid away all things. For Aeschylus, it taught man all knowledge. For Sophocles, again, and Solon: time was both the supreme teacher and the supreme judge[8] as well as the consoler of man:

"Time is a god who makes rough ways smooth..."[9] And so for Euripides:

"Time will console you..."[10]

From the singularization and hypostatization of time into a universal, all pervading and all-subsuming entity or cosmological principle to abstract time the intermediate step of disvesting it of its anthropomorphic or divine attributes was, in spite of occasional regressions, almost inevitable. Thales, Anaximander, and Anaximenes are rightly regarded as the first true scientists of the West not so much because they hypothesized single, unifying principles or origins for the diversity of the world, for in this they were anticipated by others, e.g. Moses and the Egyptian pharoah Ikhnaton. Rather, the merit of the Milesians rests in having chosen the *will-less, mind-less* water, *Apeiron* or air as the sources or original principles of the universe. It does not matter that the Milesians may have regarded such principles as endowed with intrinsic capacity for action or motion, with *animae* or souls, or even *divine.* The point is that they did not conceive them as willful, mindful entities, as gods. Here is the beginning of the de-deification of nature, the beginning of science. And it is no coincidence that Anaximander, the second of the Milesians, marks the first step in the de-deification and de-anthropomorphization of time. In his extant fragment we read that all coming-to-be from and all ceasing-to-be into the *apeiron* take place not according to the will of time but "according to necessity." Time retains a role in the course of nature. But such a role is not the paramount one of Pherecydes' Chronos or that of necessity. It is, rather, that of setting the time-limit within which the injustice of the opposites must be made good:

> "And the source of coming-to-be for existing things is that into which
> destruction too, happens, "according to necessity; for they pay penalty
> and retribution to each other for their injustice according to the assess-
> ment of time..."[11]

The process of de-deification of nature, and hence of time, continued with Anaximenes, the last of the Milesians. For him air was the ultimate stuff of the world. He did not deny the existence of gods. He did, however, deny them the credit for the creation of the world. For, according to some commentators, even the gods, and hence time, originated from the *aether.*[12] In any case whatever small residue of anthropomorphism may have been left over, it was soon swept away by Xenophanes of Colophon (570-475 B.C.?), the founder of the Eleatic school.

Xenophanes noted that the gods were depicted with all too human traits. Black men depicted their gods black; red-haired men gave them red hair, and all of them attributed few of the virtues and all of the vices of men to the gods. It was not unreasonable, he pointed out, that could beasts conceive of gods of their own, they would make them like themselves. The nature of the divine, he drew from such a *reductio ad absurdum,* did not and could not lend itself to suit the self-centered imagery of a particular species like man, or worse, of particular races of such a species. The truly divine, he concluded, had to be:

> "One god (ἐίς Θεός), greatest among gods and men, in no way similar
> to mortals either in body or in thought."[13]

The process of abstraction from exclusion of anthropomorphic entities continued in Greek philosophy. Heraclitus of Ephesus (fl. 500-480 B.C.), the "obscure", the "dark one" of ancient philosophy is clear enough in not reserving a place or function to the gods of old:

> "...all things happen by strife and necessity..."[14]

and also, even more explicitly:

> "This world-order did none of gods or men make, but always was and is
> and shall be: an everlasting fire..."[15]

But in expressing what he thought time was, Heraclitus did earn his soubriquets. In a second hand report of Sextus Empiricus we find that:

> "Aenesidemus 'according to Heraclitus' stated that time is a body; for it
> does not differ from the existent and the first body."[16]

We are on safer ground when we take Heraclitus as sharing with his contemporaries the traditional view of the occurrences and processes of the world as recurring in cycles like a circle in which "the beginning and the end coincide." In fact, in another fragment of his we read that the soul in death becomes water, water earth, from earth water is generated and thence soul again.[17]

The Pythagorean penchant for finding in number the source and reason of everything is well known. Such a penchant was yet another important step in the conceptual development of abstract time. Like all other Greeks, the Pythagoreans were fascinated by the circle. For them, not only did the celestial bodies move circularly but so did the human soul. Man, however, was unable to "join the beginning and the end", and hence he was fated to die. The celestial bodies, on the other hand, did complete the circle and their motion was thereby eternal.

Given such beliefs, it is not surprising to discover that a later member of the sect, Archytas of Tarentum, an almost exact contemporary of Plato, characterized time mathematically as the movement of the universe:

> "Time is the number of a certain movement, or also, in general, the proper
> interval of the nature of the Universe."[18]

When we shall come to examine Aristotle's notion of time we may be tempted to see in Archytas' *number of a certain movement* the root of the Stagirite's celestial reductionism. It is, rather, the movement or interval of the World-Soul or of the Whole, not of any particular body in it. Rather than being the source of the Aristotelian conception, Archytas' notion of time may be regarded more correctly as the earliest root of the concept of absolute and mathematical time. That such an interpretation is appropriate may be seen in the following additional statement that Simplicus attributed to Archytas:

> "Beyond the circle of fire lay the Unlimited, or the unlimited air from
> which the universe draws its breath. That there must be an Infinite of this
> kind outside the World Archytas had proved. From it time as well as the void
> had entered the world."[19]

Number of the World-Soul movement, interval of the Universe, first substance or body coexistent with it, time was no longer a god. But if time was no longer deified, it was yet to be conceived in a workable manner. What sort of movement, of interval, of substance, or body was it? Ironically, a most important contribution to such a conceptual development was made by Zeno of Elea (born between 490 and 485 B.C.), the pupil and paladin of Parmenides, who like his master denied the reality of motion and of time.

Zeno was concerned with motion rather than time. His four arguments or paradoxes were directed against the reality of motion. But how motion is conceived inescapably depends on how time and space are conceived. By drawing the attention of his contemporaries and successors, e.g. Aristotle, to motion Zeno forced them to examine the question of the nature of time, and of course, space.

Motion, Zeno thought, could be conceived either as continuous, hence infinitely divisible, or as discontinuous, hence ultimately indivisible. In the first case, time and space had to be continuous; in the second case, they had to be discontinuous, with the result that motion had to consist in a succession of discontinuous jerks. The paradoxes of the stadium and Achilles' race with the turtle were directed against the view of continuous motion, and, hence, continuous time and space. Briefly stated, Zeno argued that motion in continuous time and space had to be infinite and hence impossible – even though the distance covered was finite. For before Achilles could reach the end of the race, he had to gain the half-way point, before this point, the quarter-point, the eighth-point, and so on *ad infinitum*.

Equally Zeno claimed that motion was impossible when time and space were regarded as discontinuous and finitely divisible. His arguments were the paradoxes of the flying arrow and of the moving rows. In the former, Zeno argued that, since at any given instant the arrow occupies a space equal to its own dimensions, it would not be moving, and hence, time would not pass. Finally the paradox of the moving rows involves in the words of Aristotle, "the conclusion that half a given time is equal to double that time."[20] Such an absurdity, together with the imports of the previous three paradoxes showed, according to Zeno, the impossibility of motion, and hence of time.

Zeno raised the questions of the reality and of the nature of abstract time. Plato and Aristotle in different ways and with different degrees of success attempted to provide answers.

3. TIME, THE NUMERICAL MOVING IMAGE

In Plato's conception of time both the Pythagorean and Eleatic strands of thought come to a common nexus. Time, for Plato, belongs to the world of appearances or physical world. More precisely, time came into being, and will cease to be, with the ordered universe:

> "...Time came into being together with the heaven, in order that, as they
> were brought into being together, so they may be dissolved together, if ever
> their dissolution should come to pass."[21]

Furthermore, time is mathematical and this mathematical nature is insured, Plato thought, by the motions of the planets:

> "...in order that Time might be brought into being, Sun and Moon and five
> other stars – 'wanderers', as they are called – were made to define and pre-
> serve the numbers of Time."[22]

And yet, in spite of such a genesis and nature, time for Plato is not real; for there are no real "comings-into-being" in the physical world, but merely appearances, copies. Real are only the forms, unchanging, eternal, and essentially static. Time, as a feature of the physical world is one such an appearance, one such copy. More precisely, time for the Athenian is the image of a form. Having reflected the Pythagorean influence by means of his association of time with number and celestial motions, Plato also showed the influence of the Eleatic strand by denying reality to time:

> "...(the demiurge made) of eternity that abides in unity, an everlasting
> likeness moving according to number — that which we have given the
> name Time."[23]

In spite of his denial of a reality for time, Plato's contribution to the development of the concept was important. For he conceived time as the number or, better, series of numbers of specific motions. He thus provided for time a metric, or metrics, understood as the revolutions of the planets. Furthermore, by relating time with number, Plato showed that he was aware of the need for insuring for time not only indefinite, but also uniform progression. Plato, however, conceived numbers as discrete, indivisible wholes.[24] But if so, how could temporal continuity be insured? It could not. Continuity of time however, could be insured by one of those forms which Plato called the "more remote" principles of reality,[25] namely the circle which, in its line or circumference is continuous. Now, time, according to his view, is defined by the planetary motions. But such motions are circular, that is they partake in the attribute of the circle. Therefore, they are in themselves continuous. But if so, time, which they defined, has to be continuous.

And yet, Plato's conception of time left much that was unclear or undecided. In spite of its participation in the continuity of the circle, Plato's time is neither truly continuous or truly uniform. Because it came into being with the ordering of the universe, time has to be discontinuous at that point. But even after such an initial discontinuity, time as an element of the world of appearances, neither is, nor can be perfectly continuous or perfectly uniform. However much those "intricate traceries in the sky", i.e. the celestial orbits, warned Plato, are the "loveliest and most perfect of material things", they do not and cannot partake in full in the perfection of the forms. Therefore, they do not define and preserve time in perfect continuity and uniformity:

> "The genuine astronomer...will think it absurd to believe that these visible
> materials go on forever without change or the slightest deviation..."[26]

4. ARISTOTLE: THE DEFINITION OF TIME

Aristotle's conception of time is clearly indebted to that of Plato. There are, however, major differences between the two views, not the least of which is that for the former time is real. For the Stagirite the physical world, far from being an appearance, is real, and indeed, it is all that there is. Furthermore, this world has a temporal structure, and hence, time is not only real, but important. Nevertheless, the importance of time in Aristotle's science should not be exaggerated; time is not one of his fundamental concepts. For Aristotle explained the comings-into-being and passings away not, as we do, in terms of motion in time and space, but in terms of potentiality and actuality, causes, ends, and change which included motion as we understand it: local motion. These motions, rather than time and space, constituted Aristotle's conceptual framework.

There is, on the other hand, much progress in Aristotle's conception with respect to those of his predecessors. Archytas defined time as the number of the Whole. But what was the nature of such a number and such a motion he left unexplicated. Plato conceived time with a beginning, imperfectly uniform. He related it to number and to the planetary motions. Yet he too, as we saw, left much of his conception unexplicated and unjustified. Aristotle, on the other hand, professional that he was, not only carefully examined their views, but went on to formulate a nearly complete theory of time which he also complemented with clear and explicit rules of application for and to the physical world.

Time, Aristotle thought, is not a form or an image of a form, for there are no forms, only logical classes. Time is indeed real, but Aristotle stated categorically, time does not have an independent existence:

"...time is not defined by time..."

Indeed, he went on to note:

"...time does not exist without change."[27]

For time is perceived only when there is change or motion:

"...when the state of our minds does not change at all, or we have not noticed its changing we do not realize that time has elapsed..."[28]

However, the change or the movement is in the things while time is unique and universal.[29] Moreover, while change is generally not uniform, time is: "...change is always faster or slower, whereas time is not... Clearly then it is not movement."[30] And yet, time and movement, thought Aristotle, are mutually defining:

"Not only do we measure the movement by the time, but also the time by the movement, because they define each other."[31]

What, then is time?

Time, Aristotle agreed with Archytas and Plato, is a number. But, he asked, what is number? An abstract entity, or a material, quasi-atomic extensional component of bodies? It is indeed true, Aristotle remarked, that by conceiving number as inseparable from the objects of the world the Pythagoreans thereby avoided the problems that beset Plato in his theory of the participation of the forms in the world of the sensible. But, he added, to suppose that bodies be composed of number is absurd.[32] For how can sensible objects issue from insensible numbers or forms?

Again, time is a number. But, Aristotle thought, a special kind of number. For while numbers are formed from the discrete, irreducible unit, "...there is no minimum", for time.[33] It must be that time is a number in the sense that it is the numerable aspect of motion:

"...time is not movement, but only movement insofar as it admits of enumeration. A proof of this: we discriminate more or less by number, but more or less movement by time. Time then is a kind of number."[34]

Essentially, time for Aristotle is a magnitude and, like all magnitudes, continuous because its parts are in succession and touch one another. Time, he continued, is also infinite potentially by addition as well as by division, for it has no minimum. And because it is continuous, it cannot have, as Plato thought, a beginning or an end:

"...there was never a time when there was not motion, and there will never be a time there will not be motion."[35]

For, Aristotle argued, had there been a beginning to motion, a prior change or motion would have been required to start it. But this prior motion would itself have required a still prior motion to bring it into being. Clearly, he concluded, motion must regress *ad infinitum,* that is, there must always have been motion. But if motion, so time. Thus time cannot have a beginning, nor for similar reason, can it have an end.[36]

Finally, Aristotle found that Plato's view that the planets, or heavens, define and preserve time, although, as we shall see, not essentially wrong, was incompletely defined and insufficiently justified. For time, he recalled, is unique and universal: "...there is the same time everywhere at once..."[37] But he noted:

> "...if there were more heavens than one, the movement of any of them
> equally would be time, so that there would be many times at the same time."[38]

With the critique of his predecessors completed, Aristotle went on to formulate his own theory of time.

5. ARISTOTLE: CELESTIAL REDUCTIONISM OF TIME

According to Reichenbach and Carnap the concepts used in physics are ultimately *co-ordinated* or *reduced* under stipulated conditions to concrete situation, processes or objects of the world.[39] Aristotle's theory of time, I claim, is an example of such a *coordination or reduction.* For in spite of apparent denial, he *reduced,* as I prefer to say, his concept of time to an actual state of affairs: the moving celestial bodies. Thus his was a *celestial reductionism of time.*

We saw that for Aristotle time is the number of motion. If there is no motion, there is no time. And for him, motion does not exist unless there is at least one actual body in actual motion:

> "...motion and change cannot exist in themselves apart from what
> moves and changes..."[40]

Indeed, outside the bounding sphere of the fixed stars where there is no body, he stressed, there is no motion, hence no time.[41] But if the reality of time rests on the reality of motion, will any moving body do or must there be a body that undergoes a specific kind of motion?

Because time was defined as continuous, uniform, potentially infinite, universal and numerable aspect of motion, the actual motion that reifies time must abide by these conditions.

Of the various kinds of movement or change, locomotion or change of place, Aristotle thought, is the most fundamental.[42] It has to be that time and locomotion are related. But there are, he thought, two types of locomotion: rectilinear and circular. Which of the two is the reifying motion of time?

For Aristotle the universe cannot be infinitely extended.[43] It then follows that rectilinear motion is not appropriate. For, he noted, time is potentially infinite while:

> "The straight line traversed in rectilinear motion cannot be infinite: for
> there is no such thing as an infinite straight line; and even if there were,

it would not be traversed by anything in motion: for the impossible does
not happen and it is impossible to traverse an infinite distance..."[44]

The universe, moreover, is bounded. Hence, even on physical grounds rectilinear motion is finite. Yet, were rectilinear motion to turn upon itself, could it not be potentially infinite? If such were to happen, Aristotle claimed, the overall movement would be discontinuous at the point of change of direction. For going and coming are opposites, and given that:

"...it is not possible for a thing to undergo opposite changes at the same
time, the change will not be continuous, but a period of time will intervene
between the opposite processes..."[45]

Thus, rectilinear motion is neither potentially infinite, nor continuous.

Nor, Aristotle added, is it uniform. For observation shows that:

"In rectilinear locomotion the motion of things in leaving the starting-point
is not uniform with their motion in approaching the finishing-point, since
the velocity of a thing always increases proportionately as it removes itself
further from its position of rest..."[46]

On the other hand, rotary or circular motion, Aristotle believed, is continuous. For since the points on the circumference of the circle are by definition undifferentiated among themselves, there are no starting, middle, or finishing points. It follows that:

"...there is no position on that periphery at which the mobile must rest
as having passed through to the end of its movement, for it is always being
carried round the centre and never to any extreme point;..."[47]

Therefore, circular motion is potentially infinite. It is also, the Stagirite held, uniform. For while rectilinear motion has source, middle, and goal, and hence, acceleration, climax and retardation, circular motion has none of these things:

"...circular motion has in itself neither source nor goals nor middle.
There is no absolute beginning or end or middle-point of it, for in time
it is eternal and in length it returns upon itself and is unbroken. If then
there is no climax to its motion, there will be no irregularity, for irregularity is the result of retardation and acceleration."[48]

Circular motion, then, is continuous and uniform. Indeed

"...rotary motion can be eternal, but no other motion, whether locomotion
or motion of any other kind, can be so, since in all of them rest must
occur..."[49]

Furthermore, circular motion is "...the only one that admits of being regular..."[50] But if so, an actual body that moves in a circle instantiates time:

"This then is what produces continuous motion, namely the body which is
moving in a circle, and its movement makes time continuous."[51]

Time, therefore, is the number or measure of circular motion. For, Aristotle noted:

> "...everything is measured by some thing homogeneous with it, unit by unit,
> horse by horse, and similarly times by some definite time, and as we said
> (220 b 28), time is measured by motion as well as motion by time... If, then
> what is first is the measure of everything homogeneous with it, regular circu-
> lar motion is above all else the measure..."[52]

But which body moves in a circle? The natural motion of terrestrial bodies is accelerated rectilinearly toward or away from the universal center because they are heavy or light. They can be made to move circularly, but only under compulsion and then imperfectly so. On the other hand, Aristotle claimed, that body whose motion is naturally circular:

> "...cannot have either weight (i.e. heaviness) or lightness, for neither
> naturally nor unnaturally can it ever move towards or away from the center."

But if so, such a body will indeed move perfectly circularly, that is, continuously, uniformly and for-ever. The celestial bodies do move in this manner. Hence, Aristotle concluded:

> "...the revolution of the heaven is the measure of all motions, because it
> alone is continuous and unvarying and eternal..."[53]

It does not matter that he avoided making a binding commitment by saying that:

> "(Time) is simply the number of continuous movement, not of any
> particular kind of it."[54]

In the universe that he conceived there is no choice: only the moving celestial bodies describe perfect circles, hence, only such bodies instantiate and reify time because they along observe all the necessary conditions. This is *celestial reductionism of time!*

Finally, all the moving celestial bodies instantiate the same time, not many times simultaneously. What each celestial body instantiates individually is a particular interval of time. For time is one and universal, while there can be many time intervals simultaneously.

From vague and imperfect conceptions of time, number and motion of his predecessors, Aristotle wrought a nearly complete and applicable theory of time. So cogent and useful was his theory that it lasted in science as a living tradition for almost 20 centuries, down, that is to the end of the 17th cen-tury when it was displaced by the conception of absolute time.

REFERENCES

Acknowledgement: I am grateful to Dr. Francis J. Marcolongo of the Department of Philosophy, University of California, Riverside, for translating for me passages from the Greek in Simplicus.

1. Proclus Diadochus: *Commentary of the Timaeus*, IV, III, 8.29.

2. Hesiod: *The Works and Days*, 609 and 663, transl. R. Lattimore, Ann Arbor 1959.

3. O. Neugebauer: *The Exact Sciences in Antiquity*, 2nd ed., New York 1962, p. 83.

4. *Old Testament*, Psalms, XXXI, 15, King James' translation.

5. *ibid.*, Ecclesiastes, III, 1-2.

6. *ibid.*, Job, XXIV, 1.

7. Damascius: *De principiis*, 124 bis. in Kirk and Raven: *The PreSocratic Philosophers*, Cambridge 1966, p. 55.

8. Sophocles: *Oedipus the King*, 613; Solon, frag. 27, Kirk and Raven: p. 120.

9. Sophocles: *Electra*, 179; transl. R.C. Jebb.

10. Euripedes: *Alcestis*, 381; transl. R. Aldington.

11. Simplicus: *In Aristotelis physicorum libros Commentaria*, ed. H. Diels, 24, 17; Kirk and Raven p. 117.

12. Hippolytus: *Refutations*, I, 7 I; Kirk and Raven: p. 144; also Augustine: *De Civitate Dei*, VIII, 2.

13. Xenophanes: frag. 23; Clement: *Stromata*, V, 109, I; Kirk and Raven: p. 169.

14. Heraclitus: frag. 80; Origen: *Contra Celsum*, VI, 42; Kirk and Raven: *ibid.*, p. 195.

15. frag. 30; *ibid.* V, 104, I; *ibid.*; p. 199.

16. Sextus Empiricus: *Adversus mathematicos*,X, 216 & 233; transl. R.G. Bury: *Sextus Empiricus.* Cambridge, Mass. 1936 and 1960, 4 vols.

17. Heraclitus: frag. 36; Clement: *Stromata*, VI, 17, 2; Kirk and Raven: p. 205.

18. Simplicus: *In Aristotelis Categorias Commentarii*, ed. Kalbfleish, Berolini, 1907, p. 350; after F. Marcolongo.

19. E. Zeller: *History of Greek Philosophy*, London 1881, 2 vols., vol. 1, p. 467.

20. Aristotle: *Physics*, VI, 9, 240 a 1, after R.P. Hardie and R.K. Gaye, in R. McKeon: *The Basic Works of Aristotle*, New York 1941.

21. Plato: *Timaeus*, 38B, transl. F.M. Cornford: *Plato's Cosmology*, Cambridge 1937.

22. *ibid.*, 38C.

23. *ibid.*, 37D.

24. Cf. F. Solmsen: *Aristotle's System of the Physical World*, Ithaca, New York 1960, pp. 45-7.

25. *Timaeus*, 53D.

26. *Republic,* 528E-530C; transl. F.M. Cornford: *The Republic of Plato,* Oxford 1941, p. 248.

27. Aristotle: *Physics,* IV, 10, 218, b 16, after Hardie and Gaye.

28. *ibid.,* IV, 11, 218, b 21.

29. *ibid.,* IV, 12, 220, b 5.

30. *ibid.,* IV, 10, 218, b 14-18.

31. *ibid.,* IV, 12, 220, b. 15.

32. *Metaphysics* XIII, 8, 1083, b 11, and XIV, 5, 1092 b 7; also *De Coelo,* III 1, 299 a 24, 300 a 8, 300 a 17.

33. *ibid.,* I, 6, 274 a 9; after J.L. Stocks, in McKeon: *Op.cit.*

34. *Physics,* IV, 11, 219 b 3-5, after Hardie and Gaye.

35. *ibid.,* VIII, 1, 252 b 5; see also 207 b 14.

36. *ibid.,* 251 a 15 - 252 a 4.

37. *ibid.,* IV, 12, 220 b 5.

38. *ibid.,* IV, 10, 218 b 3.

39. H. Reichenbach: *Space and Time,* New York 1958, p. 14; R. Carnap: "Logical Foundations of the Unity of Science", *International Encyclopedia of Unified Science,* Chicago 1955, vol. 1, part 1, pp. 49-50.

40. *Physics,* III 1, 200 b 33; transl. P.H. Wicksteed and F.M. Cornford: *The Physics,* London 1929, 2 vols.

41. *De Coelo,* II, 10, 279 a 12-18.

42. *Physics,* VIII, 7.

43. *De Coelo,* I and II.

44. *Physics,* VIII, 9, 265 a 18 after Hardie and Gaye.

45. *ibid.,* VIII, 7, 261 b 5.

46. *ibid.,* VIII, 9, 265 b 12.

47. *ibid.,* 265 and b 6, after Wicksteed and Cornford.

48. *De Coelo,* II, 6, 288 a 18, transl. W.K.C. Guthrie: *On the Heavens,* Cambridge, Mass., 1939.

49. *Physics,* VIII, 9, 265 a 25, after Hardie and Gaye.

50. *ibid.,* 265 b 12.

51. *De Generatione et Corruptione,* II, 10, 337 a 33, transl. H.H. Joachim, in McKeon: *Op.cit.*

52. *Physics,* IV, 14, 223 b 13, after Hardie and Gaye; see also 221 a 1.

53. *De Coelo,* I, 2, 287 a 23; after Guthrie.

54. *Physics,* IV, 14, 223 a 33; after Hardie and Gaye.

Nietzsche and the Concept of Time

D. W. DAUER

1. INTRODUCTION

Yes, I do know my origin!
Unquenchable like the flame
I consume myself and glow.
Into light turns all I touch,
Into coal all that I leave:
Flame is surely what I am!

Ja! ich weiss woher ich stamme!
Ungesättigt gleich der Flamme
Glühe und verzeh'r ich mich.
Licht wird alles, was ich fasse,
Kohle alles, was ich lasse:
Flamme bin ich sicherlich![1]*

Nietzsche often uses the flame as a symbol for various concepts; however, in this poem, he clearly identifies himself with this unquenched, unsatiated and — alas! — consuming flame. This flame, after burning with unbelievable force and splendor, turned his spirit suddenly, as Lange-Eichbaum[2] phrases it, into a completely burned out crater, a unique case in the history of medicine.

However, let us listen to his most famous work, *Thus Spake Zarathustra,* without consulting a medical diagnosis, particularly since Dr. Lange-Eichbaum is convinced that Nietzsche's mind was not impaired until December 1888; that, to the contrary, its productivity was actually enhanced to superhuman heights. His genius-like productive period fell into the years of 1881-1884 with its height in the years of 1883-1884 when he wrote with an incredible speed most of his *Zarathustra,* on which I shall concentrate in analyzing his concept of time.

Nietzsche's fate is tragic in his life as well as in his death. Schopenhauer, his spiritual teacher in his early, formative years, remained an influence from which he could never free himself entirely. Could he have been satisfied with his pro tem master's *nirvāṇa*? He could not, since that would have meant a capitulation to the school of self-negation. His will was too important to him to accept such extinction of life; thus, he had to invent something new. Hence, his idea of "Eternal Return", the idea which with the vision of "Zarathustra" in 1881 struck him, as he described it, like a thunderbolt.

Schopenhauer did not know that *nirvāṇa* had no definition in original Buddhism, and that *nirvāṇa* could mean absolute extinction as well as positive, blissful reality. Had Nietzsche known that fact and had he known the *Mahāyāna* teaching of renunciation of *nirvāṇa,* he might have given a different twist to his grandiose ideas of "Eternal Return" and "Superman."

---

*Unless otherwise indicated, the quotations in the following have been translated into English by Dorothea Watanabe Dauer from the German original identified in the Notes at the end.

2. VARIOUS TYPES OF CONCEPTS OF TIME

In order to locate Nietzsche's philosophy in the context of the Study of Time, I shall give a rough classification of the different types of ideas about time. Probably the most conspicuous distinction that divides the idea of time into two groups is that between (a) the linear and (b) the circular concept of time. We could also consider time as continuous or discrete, but this is not the issue here. Time is a sequence, or at least it involves a sequence, and in the linear idea the sequence stretches to both directions indefinitely. In the circular idea, the sequence, if extended to one direction, will after some length come back to meet the extension from the other direction.

Among the variations in the linear concept of time, there is one which has its starting point at the beginning (creation), one which has an endpoint (eschatology), and a third which has both a starting and an endpoint.

There are concepts of time in which there is no intrinsic difference in the two directions of time, and other concepts of time in which the two directions are essentially different. The former is reversible time, the latter irreversible time. The irreversible time determines the direction past-to-future, but does not determine past and future as such. The separate definition of past and future is possible only when reference is made to *now*. A moment may be compared to a point on the line – whether linear or circular. The *now* is a moment. In this sense it is the smallest constituent of time. However, the *now* is not a fixed moment. All the past moments have been the *now* and all the future moments will be the *now*. In this sense *now* encompasses all time. It is in a sense longer than any time segment.

3. TIME AND ETERNITY

Eternity is the antithesis of the concept of Time. Yet, this opposition does not mean a difference in the basic nature, as exemplified by the relation of space to time. Eternity is so to speak made of the same "stuff" as time, yet the changing aspect of time is abstracted away. Eternity is the unchanging aspect of time, hence it has to include, in some sense, the entirety of time sequence.

The concept of eternity, therefore, depends on the concept of time. The lowest, or the most primitive concept of eternity is the idea of all time. We spatialize the sequence of time as a straight line or as a circle, and we consider the entire sequence as a single object. For this purpose, the circular time is particularly convenient, because a circle is the most rudimentary way to represent infinity within finiteness. This primitive eternity will be called spatialized eternity.

There have been many philosophical thoughts that recognize some kind of an unchanging basis in the world. The Parmenides-Zeno tradition of the idea of the One that never changes or moves and can never be divided provides the basis for eternity. The world of Ideas, of Plato, exists not in time, but in eternity. When St. Augustine states that the world was created with time, the one who created the world must have been in time-less eternity. The world of nirvana, at least in the Hinayana school, is essentially a timeless eternity too. This type of concept can be classified as timeless eternity.

The Heraclitian idea of incessant change seems to mean that no concept of eternity is possible. But, this is not quite so. The law or mode by which changes take place may not change and thus it constitutes an eternal element. The Hegelian law of dialectic development, and the Bergsonian ever-changing *élan vital* have definitively an everlasting aspect. The Darwinian law of natural selection may be considered as an unchanging law governing changes. We may think of this as a reification of process. This kind of eternity may be called the eternalization of becoming, or being made out of becoming.

Finally, in connection with the concepts of the *moment* and the *now*, there is a school of thought that considers the *moment* as a representation of eternity. There is a good reason for this because the moment is not time, yet time would not exist if there were no *moment*. The concept of *moment* leads us to the idea of eternity in four ways. One is to consider each *moment* as reflecting and containing the macrocosmos. The second is to consider the moving *now* as representing eternity, since the past was *now* and the future will be *now*. The third is to consider the *moment* taking the place of the subject, making decisions for actions; it is the place of life. There is also an idea which considers the *moment* as the intersection of time and the timeless ground.[3]

4. ESCHATOLOGY

Eschatology is a concept about the end of time. The end is usually connected with some kind of an event of great magnitude. In some cases, eschatology applies as taking place individually (for individuals separately); in other cases, eschatology applies collectively (for the entire world). The idea of eschatology seems to have originated from the Zoroastrians who believed in a big conflagration that would end the world. The Judeo-Christian eschatology is believed to have been introduced under the influence of Zoroastrianism. Western thought could never free itself from eschatology of one kind or another. Marxism is no exception. In India *moksa* and *nirvāna* are cases of individual eschatology.

The eschatology takes very different forms according to the religious beliefs involved. But, one thing they have in common is that eschatology is not only the end of life on earth, but also a passage from Time to Eternity. I shall explain the Indian type of eschatology further in the following section.

5. CIRCULAR TIME AND ESCAPE FROM IT

The circular concept of time is one of the most easily conceived theoretical images of time for at least two reasons. First, practically all the changes that we observe in life are periodic. Second, the circle is the oldest device to encompass the infinite in the finite. It is true that circularity of time and circularity of the sequence of events are two different things. But by assuming a kind of principle of identity of indiscernibles, we can consider these two notions as equivalent. Two major currents of ideas of the circular time can be found in India and in Greece. Pythagoras, Empedocles, Heraclitus, Plato, among others, believed in circular time. It may look strange to place Plato and Heraclitus on the same side because Plato believed in the changeless world of ideas while Heraclitus regarded everything to be in a state of flux. However, Plato talked also about reincarnation. The circular return of the world, in the strict sense, was believed by the Pythagoreans. The Indian idea of circular time is inseparably related to the idea of chain of rebirth — whether it is interpreted as Palingenesis or as Metempsychosis according to Schopenhauer's terminology.[4] In addition to this basic cycle of rebirth of individuals, there are such times as Vishnu's periodic reincarnation.

The most characteristic feature of the Indian concept of circular time is that the Indian religions provided an escape or outlet from the cycle of rebirths which they considered intolerably painful. In Brahmanism, this outlet is called *mokṣa* which means the reunion of the individual soul with the world soul. In Buddhism, the outlet is *nirvāṇa* which means the departure from *saṃsāra*. The chain of metempsychosis ends up with an entry into *nirvāṇa*, i.e., extinction. *Nirvāṇa* is interpreted differently depending on the sects and the preachers, ranging from pure nothingness to blissful paradise. It should be emphasized that the eschatology inducing the escape from circular time applies to individuals – i.e., it proclaims the passing from time to eternity for each individual separately. Schopenhauer added to the Indian thought his idea of world *nirvāṇa*[5] – which in our language should be considered as a simultaneous eschatology of the entire universe generalized from the Buddhist eschatology for individuals.

Nietzsche's concept of circular time is different from that of his Greek and Indian predecessors in the sense that in his conception exactly the same situation must recur or return indefinitely with everything coming back to exactly the same state. He wants us to be able to withstand the indefinite number of repetitions of the same. This idea is logically untenable, since in order that one may notice that the encounter with a situation is a repeated encounter there must be something different from the last encounter informing one that it is not the same encounter; but this difference, slight as it may be, would violate the condition that everything be exactly the same. If one talks about repetition, one must remember past occurrences of the same situation. But if one remembers one more occurrence than one did the previous time, one's mental situation is no longer the same as the last time. In other words, if everything is exactly the same, no mind in the world will notice that it is a repetition. What is possible is a single ring of time. The one who could speak of repetition must be outside the cycle of time. Is Nietzsche that one? Or is it God? The Indian concept of circular time is free from this logical fallacy since it does not require the rigorous return of the same.

The other difference between the Indian concept of circular time and Nietzsche's is that the latter does not allow an escape from it. I shall discuss this problem further in the last section.

6. NIETZSCHE'S IDEA OF ETERNAL RETURN

There are two approaches to Nietzsche's idea of "Eternal Return": one from the physical sciences, the other from the standpoint of moral consideration. Let me first quote a passage from Nietzsche's *Posthumous Fragments:*

> If the universe *may* be conceived as a definite quantity of energy, and as a definite number of energy-centers, – and every other concept remains indefinite and therefore *useless,* – it follows therefrom that the universe must go through a calculable number of combinations in this great game of chance during its existence. In an infinite time, at some time or another every possible combination would have been attained; still more, it would have been attained an infinite number of times; and since between each "combination" and its next "recurrence" all other still possible combinations must have been undergone and each of these combinations must have determined the whole series of combinations in the same order, thus, a circular movement of absolutely identical series would have been proven: the universe is

thus shown to be a circular movement which has already repeated itself an infinite number of times, and which plays its game ad infinitum.[6]

Since Nietzsche refused to recognize any hidden world of noumenon, he considered the physical world not as an appearance but as the only real world. Hence, the eternal return of the physical world had to be regarded as real and unavoidable. It is important to remember that he considered the return as a rigorous return of all things in every respect.

Nietzsche, however, considers also the moral aspect of the Eternal Return and attaches an interesting meaning to it. In the chapter called "Redemption" in Part II of *Zarathustra,* he [Zarathustra] says:

> The present and the bygone upon earth – ah! my friends –
> that is my most unbearable feeling;[7] ...

> That time does not run backward, that is its [the Will's] wrath;
> "That which was" – so is called the stone, which it [the Will] cannot roll.[8]

The irreversibility of time is the cause of anger for the Will. Zarathustra once announced that the Will is "the deliverer and joybringer,"[9] but he has to admit now that the Will itself is still a prisoner.[10]

> "It was:" thus is called the Will's teethgnashing and loneliest affliction.
> Powerless towards that which has been done – it [the Will] is a malicious
> spectator of all that is past ... that it [the Will] cannot break time and time's
> greed.[11] ... This, indeed this alone is *revenge* itself: the Will's aversion to
> time, and its [the time's] *"It was."*[12]

Nietzsche has to find a solution, and at first that very solution sounds like a rationalization. Because he cannot change it, he claims that he wanted it, yet, this solution prepares the next plot.

> All "It was" is a fragment, a riddle, an agonizing chance – until the creating
> Will says to it: "but this is the way I wanted it!" – Until the creating Will
> says to it: "but thus do I want it! Thus shall I want it."[13]

Nietzsche does not appear to imply that something important is in preparation, but he has now come close to the beginning of the idea of the Eternal Return. We have seen the various changes of tenses, from "so I wanted it," to "thus do I want it," to "thus shall I want it." This means that the next time the same situation comes, he will want the same again. Thus, Zarathustra prepares those who are listening to him to the forthcoming idea of the Eternal Return without saying so at this particular time.

"But at this point in his discourse it happened that Zarathustra suddenly paused, and looked like a person who was frightened to the extreme."[14]

After three more chapters of preparation, Nietzsche introduces now the idea of the Eternal Return in the Second Chapter of Part III entitled "Vision and Enigma":

> "Look at this gateway! Dwarf!" I continued talking: "it [the gateway] has
> two faces. Here two roads come together: these no one has yet gone on to
> the end. This long lane backwards: that takes one eternity. And that long
> lane forward – that is another eternity. They contradict each other, these
> lanes; – they hit each other at their foreheads: – and it is here, at this gate-
> way where they abut upon each other. The name of the gateway is inscribed
> above: "Moment."

But he who would follow one of them further — and further and further
in great distance: do you think, dwarf, that these roads would eternally
contradict each other?" —
"Everything straight lies," murmured the dwarf contemptuously.
"All truth is crooked; time itself is a circle."
"You spirit of gravity!" said I angrily, "do not take matters too lightly!
Or I shall let you squat, where you are squatting, Lamefoot,
— and I carried you *high*! ..."[15]

Although Zarathustra admonishes the spirit of weight, embodied in the dwarf, he agrees actually with what the latter said: "Time is circular." Thus, he introduces the concept of circular time without resorting to physical theory.

Incidentally, it is interesting in this connection to note that Nietzsche was very specific about the irreversibility of time, because he says: " ... if I consider the direction (forwards or backwards) as logically indifferent, I would take hold of the head of the serpent, this very moment, and think that I am holding the tail: this pleasure I leave to you, dear Mr. Dühring."[16]

7. AMOR FATI, WILL, SUPERMAN

The moral meaning Nietzsche attached to the circular time is the essence of his entire philosophy. One should become strong enough to withstand, enjoy, and even want the eternal return. In short: "Become the one you are."[17] In order to be more precise let me re-phrase it: be such that you want to become who you are. But this seemingly "original" idea of Nietzsche had been previously conceived by his pro tem "teacher" Schopenhauer. Indeed, we read in Schopenhauer's *World as Will and Idea:*

A man ... who found satisfaction ... in life, and could calmly and deliberately
desire that his life as he had known so far, should endure for ever or repeat
itself anew, and whose love of life was so great that he willingly and gladly
accepted all the hardships and miseries to which it is exposed for the sake of
its pleasures, — such a man would stand "with firm-knit bones on the well-
rounded, enduring earth," and would have nothing to fear.[18]

By this transvaluation of values, that which would have appeared to be arbitrary and accidental events of fate becomes now what the will has wanted. The *Amor Fati*, love of fate, which otherwise would have looked like a weakling's self-deception, is transformed in Nietzsche's fancy into the virtue of a courageous man.

The Will is a free creator and therefore, by its nature, future-oriented. His Eternal Return has the effect of transforming the past into the future, thus bringing the past within the reach of the Will. In the same way fatalism is transformed into freedom as is accident into goal, nihilism into acceptance of values, negation of life into affirmation of life. Nietzsche himself was not aware of these two opposing aspects of the Eternal Return. But this transformation works backward too; future becomes past freedom becomes fatalism etc. When the return is interpreted as absence of values and goals, the theory of return looked like what Nietzsche called the "most extreme form of nihilism ... European form of Buddhism (Europäischer Buddhismus.)"[19] But Nietzsche did not want to stay in the abyss of nihilism forever. He became a conqueror of nihilism by pushing nihilism to its logical extreme. This break-through was made possible by his idea of the Will to Power, the will to create the future, the will for eternal return. The cynic may call this a deception, the sympathizer may call it an act of courage.

Nietzsche did not believe in being but only in becoming. To use Nietzsche's favorite vocabulary of the Dionysian and the Apollonian that he had actually learned from Richard Wagner,[20] Nietzsche was Dionysian rather than Apollonian. The Will to Power is the driving Dionysian force.

One way to look at the Eternal Return is to consider the Will to Power as "essence" and the Eternal Return as "existence" according to Heidegger. To this Nietzsche would probably object, because he wanted to prove his Will to be entirely different from Schopenhauer's Will, the Ding an sich. Schopenhauer's Will is insatiable, therefore an eternal lack of something, while Nietzsche's Will is fullness. The Will to Power is the Will to *More*. It is the Will to have more returns of the same. The Superman is the personification of such Will. His ethics is not "I ought" but "I will."

8. NIETZSCHE'S IDEA OF MOMENT

Nietzsche's idea of Eternal Return is widely known, but his emphasis on the concept of the *moment* is less known. Actually, he introduces the idea of the *moment* at three places in *Zarathustra* with three different meanings. I shall examine the relation of these concepts to the concept of Eternal Return in the concluding section. In this section, I shall point out the three notions of the *moment*. The concept of the *moment* in the first sense appears in the last chapter of Part I entitled "The Bestowing Virtue."

> And once again you shall have become my friends and children of our hope:
> then I will be with you for the third time, in order to celebrate the Great
> Midday with you.
> And that is the Great Midday, since the human being is standing in the middle
> of his course between animal and Superman, celebrating his path to the evening
> as his highest hope: for it is the path to a new morning.
> At such time the descender will bless himself, that he should be an over-goer,
> and the sun of his knowledge will be standing for him at Midday.
> "*Dead are all the Gods: now we want the Superman to live.*" –
> Let this be, at some future time our final will at the Great Midday.[21]

The idea of Midday is a motif that comes back again and again in various meanings, eliciting different associations. For careful analysis of the meanings in which Nietzsche uses the word Midday, I refer the reader to the most enlightening book on this subject by Karl Schlechta, *Nietzsches Grosser Mittag*. In the foregoing passage, the idea is that the great Midday is a moment of decision, the point where man goes down and the Superman rises.

The second place where the concept of the moment appears is in the second Chapter of Part III entitled "The Vision and the Enigma". In the passage quoted in Section 6 where Nietzsche introduces the idea of Eternal Return, he speaks of the Gateway as the meeting point of past and future, saying: "The name of the gateway is inscribed above: 'Moment.'"[22] This *moment* is the *now* that repeats itself indefinitely. He then questions the dwarf:

> And if everything has already existed; what do you, dwarf, think of this
> Moment. Must not this gateway too – have already existed: And are not

all things closely entangled in such a way that this Moment draws all
coming things after it? *Consequently* − − itself also?[23]

It is customary to consider the *now* eternal because the *now* covers all moments of time. But with Nietzsche the *now*-moment is, in a sense, given a character of eternity because it repeats itself. How seriously Nietzsche took the exact identity of repetitions, contrary to his earlier thoughts, can be seen in the following passage which declares the return of Midday and Superman in the chapter called "The Convalescent" in Part III of *Zarathustra*:

> I come again with this sun, with this earth, with this eagle, with this serpent
> − not to a new [life] or a better [life] or a similar life: − I come back again and
> again in all eternity to this identical and selfsame life, in the greatest and also
> in its smallest, so that I may again teach the Eternal Return of all things, − so
> that I may speak again the word of the great Midday of earth and man, to
> announce anew the Superman to men.[24]

The third place where the concept of the moment appears is toward the end of *Zarathustra* in the chapter called "Midday." Zarathustra falls asleep − with open eyes. He asks himself:

> Quiet! Quiet! Right now, has not the world now become perfect?
> What is happening to me?[25]

> − What has happened to me: Listen! Has time flown away? Do I not fall?
> Did I not fall − listen! into the well of eternity?[26]

He continues sleeping while he says finally:

> "Get up!" said Zarathustra [to himself] "you little thief, you idler!
> What, still stretching yourself, yawning, sighing, falling down into deep wells?[27]
> ... Well of eternity! you joyous, ghastly Midday-abyss! When will you drink
> my soul back into yourself?"[28]

This mental image comes very close to Brahmanism, to the idea of reabsorption of *Atman,* individual soul, into the Eternal World Soul, *Brahman.*

This Midday is the quiet noon of Pan. Already in "The Wanderer and his Shadow" we read:

> At Noon ... on a hidden forest meadow he sees the Great Pan sleeping,
> all things in nature have fallen asleep with him, an expression of eternity
> on his face − ... He does not want anything, he does not worry about
> anything, his heart stands still, only his eye lives, − it is a death with open
> eyes.[29]

It is a moment that does not move any more. It is at the same time eternity. This is a moment of perfection, complete moment. In this I see a foretaste of the satiated time of the Superman. In the chapter called "The Drunken Song," the same idea comes back in connection with Midnight instead of Midday. " − Midnight is also Midday."[30]

> Woe to me! Whither has time gone? Did I not sink into deep wells? The
> world is asleep. − ...
> The hour approaches; O man, you higher man, be cautious! ...
> *what does the deep midnight tell you?*[31]
> − Of drunken midnight-death happiness, which sings: the world is deep
> *and deeper than the day could be imagined! ...*

Gone! gone! O youth! O Midday! O Afternoon! Now evening came and
night and midnight, – ... Ah! Ah! how she sighs! how she laughs, how she
wheezes and gasps, the midnight!
How she speaks soberly, this drunken poetess! Maybe she over-drunk her
drunkenness. Has she become overawake?
Does she ruminate?
She ruminates over her sorrows in her dream, the old, deep midnight – and
still more her joy. For joy, although sorrow is deep, *joy is deeper still than
grief can be.*[32]

Now the idea of Eternal Return and Superman comes back. The Superman is happy and wants the
repetition of himself but does not want heirs:

Joy, however, wants no heirs, no children, – Joy wants itself, wants
eternity, wants recurrence, wants everything eternally like itself.[33] ...

Did you ever want once to come twice, did you ever say "I like you,
happiness! Hush! Moment!" then you wanted *everything* to come back
again.[34]

Thus, the moment – the ripe moment – wants its eternal return. Here two concepts of the *moment*
are mixed if not confused. One is the *moment* where time has stopped and the other is the *moment*
that wants to repeat itself. According to Lange-Eichbaum, Nietzsche got the idea of the timeless *moment* as a result of his illness which – as previously mentioned – gave him "superhuman" insight,
enabled him to write his *Zarathustra,* and made him immortal. Lange-Eichbaum is also convinced that
Nietzsche clearly showed schizophrenia-like symptoms. Today it is well-known that schizophrenia
causes the sensation that time has stopped. Nietzsche's experience with narcotics might have intensified this feeling of timelessness since he himself writes about the slowing-down of his time-sense as a
result of his indulging in narcotics.[35]

9. NIETZSCHE'S IDEA AND ITS PARALLELISM WITH PHYSICS

Ten years after Nietzsche became a professor at the University of Basle at the age of twenty-four, he
felt that he had wasted his time with the study of philology and reached out for new fields. He wrote:
"A deep burning thirst got hold of me: in fact, from then on I did nothing else but physiology, medicine and the natural sciences."[36] This new study was motivated by his desire to reach truth by "value-
free" scientific investigation. In the fourth book of *Joyful Wisdom* a paragraph starts with "Cheers for
Physics!":

We must be physicists in order to be creators in that sense, – whereas until
now all appreciations and ideals have been based on *ignorance* of physics, or
in the contradiction thereto. And therefore, three cheers for physics; and
still louder cheers for that which *impels* us thereto – our honesty.[37]

Seeing this strong interest in science, in particular in physics, it is surprising that about the same time
when Nietzsche was considering the problem of irreversibility and circularity of time, i.e., towards the
end of the nineteenth century, theoretical physicists were struggling with essentially the same problem
of time: irreversibility versus recurrence of physical phenomena. The most outstanding scholar in this
area of theoretical physics was the Austrian, Ludwig Boltzmann (1844-1906)* who was born in the
same year as Nietzsche.[38] I became very curious to know if there was any personal contact or corres-

---

*Professor at the Universities of Gratz, Vienna, Munich and Leipzig.

pondence between the two intellectual giants who confronted the same problem from entirely different approaches. When I was doing research at the Goethe-and-Schiller Archives at Weimar, which now keeps also all the documents by Nietzsche, I tried to locate some evidence of such contact, but was unsuccessful. The present Director of the Archives, Professor Dr. Karl-Heinz Hahn, agreed with me that there must have been some kind of bridge between scientists and philosophers of that time.

It is well-known that the names of Boscovich and Robert Mayer are often mentioned in Nietzsche's posthumous writings.[39] Boscovich is an atomic physicist and Robert Mayer is one of the discoverers of the law of conservation of energy. At the *Schlossbibliothek* in Weimar, I looked also through all the books of Nietzsche's personal library — or what is left of it — a little over 800 volumes, and I was amazed to see how many of them were in the natural sciences and, in particular, in physics. Nietzsche positively studied the "Conservation of Energy" by the physicist Balfour Stewart published in 1875.[40] In this book, both conservation of energy and dissipation or degeneration of energy were discussed. The former, conservation of energy, is the content of the so-called First Law of Thermodynamics, which is closely related to the inevitable recurrence of natural phenomena, and the latter, degeneration of energy, is the content of the so-called Second Law of Thermodynamics, which expresses the irreversibility of natural phenomena, or in technical terms, the increase of entropy with time. Furthermore, among Nietzsche's collection of books was Max Seiling's *Mainländer, ein neuer Messias,* which contained a clear description of the conflicting views of mere conservation of energy and the degeneration of energy.[41] In any event, it is safe to assume that Nietzsche knew about the First and Second Laws of Thermodynamics, i.e., the law of conservation of energy and the law of degeneration of energy and it is quite possible that he knew that there had been a controversy in physics about the irreversibility and recurrence of time. In *The Will to Power* we find the following sentence: "The principle of the conservation of energy inevitably involves eternal recurrence."[42] This is roughly correct from the point of view of physics; it may even indicate that he knew perhaps more about physics than he is given credit for.

Boltzmann's major work was his mathematical demonstration of what he called the H-Theorem (eta-theorem). His aim was to derive the Second Law of Thermodynamics from the basic laws of physics with the atomistic model of matter. He formally succeeded in this enterprise, but his argument was open to two objections: One is called "Wiederkehreinwand" or "recurrence objection" which was first raised by Zermelo.[43] It points out that the basic law from which Boltzmann started allows only recurrent phenomena; if Boltzmann claimed that he succeeded in deriving the Second Law of Thermodynamics, that would amount to deriving one-way time from circular time, which would show some logical gaps in the derivation. In fact, according to Poincaré's recurrence theorem, a physical system enclosed in a finite domain of space and under a finite amount of energy will come back to the original state. It is noteworthy that Nietzsche's argument for eternal return that we have already quoted is a perfectly acceptable description of the recurrence theorem in physics. It is inconceivable that Nietzsche thought of his argument without having a fairly good knowledge of what was going on in the physics of his time.

The second objection against Boltzmann's derivation of the H-Theorem is called "Umkehreinwand" or "reversibility objection" which was first raised by Loschmidt.[44] It is pointed out that the fundamental laws are perfectly symmetrical with respect to both directions of time and that, if Boltzmann has proved that the entropy should increase in the future, then it must be possible to prove that the entropy must have been larger in the past, contrary to the assertion of the Second Law. In Nietzsche's thought there is no assertion of fundamental reversibility of time as such, but his circular time implies implicitly reversibility of time, since if time were irreversible, it could never come back to the original state. We have seen that Nietzsche, particularly in the chapter on "Vision and Enigma," observes the essential irreversibility of time. His idea of going backward in time, in connection with the irreversibility is so foreign to any traditional thought in philosophy and literature that he may have borrowed it from some physicist's writings. In summary, Nietzsche, starting from the idea of irreversible but infinite time, finishes with the circular time which implies reversibility. Here lies his basic self-contradiction. Boltzmann, on the other hand, starting from the basic laws of physics that are reversible, somehow manages to derive an irreversible conclusion. This contradiction was the essence of Loschmidt's reversibility objection. Nietzsche's theory is open to the same objection only in the reverse sense.**

For the purpose of a later reference, I shall add a very brief description of how physicists have overcome the difficulties Boltzmann encountered. Time is reversible and recurrent, but the time required for the universe to come back to the starting state is so large that it can be considered as infinitely long. Besides, before the entropy starts to decrease again the universe has to go through the state of maximum entropy where everything will die. Therefore, as far as man's experience is concerned, the entropy will increase with time and for all practical purposes time will indefinitely continue to be irreversible.[45]***

In connection with the theorem of recurrence, physicists proved also another theorem called ergodic theorem. This theorem states that no matter in what state the (finite) Universe may be at a given instant, it will go through all other possible states in a given sequence and come back eventually to the starting state. (To be exact I should say a state very close to the starting state according to the quasi-ergodic theorem). This theorem entails the theorem of recurrence, but says much more than this theorem. If the reader re-reads the quotation made at the beginning of Section 6 concerning Nietzsche's Idea of Eternal Return, he will be astonished to see that Nietzsche uses practically the same kind of argument as that for the ergodic theorem.

10. BUDDHISM AND NIETZSCHE'S IDEAS

The strong kinship between Buddhism and Schopenhauer's philosophy is well known.[46] But few people have studied carefully the similarity between Buddhism and Nietzsche's philosophy. On the ethical level, Buddhism and Nietzsche's thoughts are diametrically opposed because Buddha preaches *par excellence* the ethics of compassion while Nietzsche opposes it. Furthermore, in Buddhism the ethical ideal is the eventual release from the cycle of reincarnation by the accumulation of good deeds through rebirths. With Nietzsche, the ethical ideal is that one's deeds should be such that one would want the repetition of the same. There is no release from the cycle.

**Purely conceptually, one-wayness and circularity are not incompatible. For instance, one can consider a circle with one direction of rotation. However, if one-wayness is understood in the sense of a one-way change of any quality or quantity of the world, they are not compatible, for the world could then not come back to the original state of affairs. In physics, the entropy increases with time and it cannot come back to the original low value. In history, any one-way development cannot bring the world back to the "less" developed stage.

***I am indebted to Professor Satosi Watanabe for the technical matters of physics described in this Section.

These very profound differences on the ethical level notwithstanding, there is a strong metaphysical similarity between Buddhism and Nietzsche. The basic reason for this similarity may be said to be deeply rooted in the fact that the concept of time in both philosophies is Heraclitian as opposed to the Eleatic.

From this basic common ground springs the denial of being in favor of becoming in both philosophies. Buddhism denied the most sacred fundamental being — Brahman or World Soul and its individuation, *Atman* or Individual Soul. Nietzsche denied everything supposedly permanent and transcendental, such as the thing in itself (das Ding an sich), substance, spirit, universal etc. Both primitive Buddhism and Nietzsche deny the existence of God. On the epistemological level, therefore, both Buddhism and Nietzsche's thoughts have some unexpected kinship with the positivistic attitude of modern sciences and phenomenological philosophies. On the existential level, Nietzsche, as well as Buddhism, emphasizes inconstancy and suffering.

Probably the most remarkable similarity between Buddhism and Nietzsche, however, may be found in the concept of the *moment.* In early writings Buddhism favored already a monadic view of the world. There was neither a world soul, nor the one and unique substance of the world. Each person reflected all other persons and each moment was at the same time eternity. This kind of idea was further developed by the *Mahāyāna* Schools of later Buddhism. There, each person has a germ of Buddha mind in him, and *samsāra* had *nirvāna* in some latent form. Particularly, the *Mahāyāna Sūtra* called in Japanese *Hokekyō* (Saddharma-Puṇḍarīka Sūtra, 1st Century A.D.) emphasized the idea of *Ichinen-Sanzen* meaning that the microcosmos and the macrocosmos are penetrated by the same unifying truth and form a single and unique existence. In another important *Mahāyāna Sūtra* called in Japanese *Kegon-kyō* (*Avataṃsaka-Sūtra,* 2nd-3rd Century A.D.) the principal tenet can be expressed by the motto: One is All, All is one. This applies not only spatially but also temporally. The temporal "one" is the *moment.*

The most important philosophical work on Buddhism written by a Japanese is *Shōbō-genzō* by Dōgen (1200-1253 A.D.). This book contains a chapter called "Uji" which may be translated as Being-Time. In this particular chapter it is stated that "Time is Being, Beings are all Time," which is singularly parallel with a major theme in Heidegger. A further interesting passage reads: "If Time is taken in its aspects as the passage from past to future, all events will appear as the *now* of Being-Time. But also if time is deprived of the passage from the Past to the Future, I have also *now* of Being-Time. Such is Being-Time."[47]

The *now* of Being-Time is the *moment.* Dōgen here considers two aspects of the *moment,* one as a moving point in the flow of time and another as the eternal *now* that is always present and does not move. Being is identified with such *Now.* Dōgen concludes the paragraph with a sentence that may be freely translated as: This moment swallows all the [past] events and ejects all the [future] events. Dōgen's concept of Being-Time is very close to Nietzsche's timeless *moment.*

Before terminating this section, I should like to mention the striking similarity and dissimilarity between Nietzsche's idea of Superman and the concept of *bodhisattva.* Nietzsche's Superman is strong enough to want any number of returns of the same in the ring of time. A *bodhisattva* is qualified to enter *nirvāna* but is compassionate, ready of his own volition to stay in the ring of reincarnation. Both are ideal types of superhuman beings who are willing to go through the circular time again and again. Yet, the ethical meanings are entirely different. The Superman wants to continue to have his pleasure and power on earth, while the *bodhisattva* is sacrificing himself by staying on this earth to help others to elevate themselves to the goal of eternity, i.e., *nirvāṇa*.

11. CONCLUSION

To summarize and draw a conclusion from this short survey of Nietzsche's concept of time: We have already mentioned the logical self-contradiction involved in the postulate of repeated returns of exactly the same; no conscious being would be capable of recognizing that it is a repetition of the same if it has exactly the same mental state as at the last visit of the same scene. But let us be a little lenient toward Nietzsche and assume either that that which returns is not exactly the same or that there is a super-mind staying outside the "universe" which is capable of recognizing the return of the universe to the same state. A more fundamental and significant contradiction is hidden in Nietzsche's concept of time. From the beginning, Nietzsche had two irreconcilable features he wanted to assign to time. One is the essentially irreversible upward aspiration of life (essentially irreversible time), and the other is the endless circular repetition of time (circular time). It is the same difficulty that Nietzsche's contemporary, the physicist Boltzmann, had to cope with.

The upward drive of life which is a Dionysian element in Nietzsche's thought is evident everywhere in his *Zarathustra*. Just to quote one of the many similar statements, we have in the second part of *Zarathustra*, "The Tarantulas": "Life strives to rise and while rising, it wants to surpass itself."[48] This upward motion culminates in the appearance of the Superman. Throughout *Zarathustra*, the dwarf of gravity plays the role of preventing the upward flight of life. Only toward the end, in "The Awakening," just before the signs of the advent of the Superman approach, do we hear about the final victory: "he already yields, he flees, the spirit of gravity, my old arch-enemy!"[49]

But can Nietzsche reconcile this essentially one-way process with the circular time? We can think of only three possible ways out. All three, however, somehow interfere with the basic premises of his philosophy. One is to consider the time before the appearance of the Superman as linear one-way time and exclude it from the circular time. After the Superman has appeared time can become circular because there is nothing else for him to become, and he wants the repetition of the same. This solution certainly destroys consistency and universality of his postulate of circular time. The entry into circular time may be characterized as a kind of eschatology because it is the happy ending which does not evolve any further.

The second possibility is to accept the postulate of circular time at its face value. In that case, life is to be considered as eternally "en route" and its only "goal" is the endless circular movement itself. If this is the case, the Superman himself after some time has to decay and disappear, and the world has to restart with beings inferior to the Superman. If we take this view, we can no longer speak only of the upward motion and we have to consider a whole period of downward motion of life as well. The whole matter of breeding the Superman must be complemented by a whole story of decline and decay of the Superman. I do not think that Nietzsche would have liked to accept that kind of picture.

The third possibility is to replace the circle of time by a helix, i.e., the one-way motion combined with a circular motion. This would imply that at the end of a cycle one would come back to a similar situation at a "higher" level. But this would contradict Nietzsche's explicit statement that exactly the same situation will repeat itself.

Thus, we come inevitably to the conclusion that his concept of time is self-contradictory, or at least incompatible with his philosophy of life. This contradiction has nothing to do with whether time is continuous or discrete, flow or succession, measurable or beyond measurement. In the case of Boltzmann's problem, the physicist had either to assume that the entire development takes place only once or else to assume that the one-wayness is only a partial truth applicable only to small sections of time separately. If Nietzsche had kept his mental health longer, I believe that he would have seen his self-contradiction, and he might have struggled to find a way out of the aporia into which he pushed himself. What would have been his solution is a question no one can answer. One might, however, hazard a guess as to what the general orientation of his solution might have been.

Nietzsche started out by denying being and believing only in becoming. But it is ironical to see that at the end of this intellectual journey he wound up by introducing something like being. In fact, once the cycle is deterministically determined, the cycle is a fixed, unchangeable indivisible object. It is no longer becoming, it is being. It is the One.

It is interesting that Nietzsche himself partially admits this inevitable reification of process. He writes: "That everything recurs, is the maximum approximation of a world of becoming to a world of being! — The height of contemplation. ... To *stamp* becoming with the character of being — this is the *highest Will to Power.*"[50] The passage from becoming to being is a passage from changing time to changeless eternity, a passage from the Dionysian to the Apollonian. Whoever tries to escape the Judeo-Christian eschatology has to introduce some kind of eschatology because there is something that is eternally there, after all. As I stated earlier, if we consider Nietzsche's "time" as consisting of linear time followed by circular time, the entry to the circular time may be considered as an eschatology. Whether this picture is appropriate, it is obvious that the time of the Superman has a timeless character because he is perfect. The timeless character is clear in the idea of Midday where everything stops. Everything is perfect, and the Superman wants the repetition of the same. If we come to this stage, time loses its need for a change; time stops, and the moment becomes eternity. Thus, Nietzsche created Eternity out of Time. What else should we call it except Eschatology? This particular eschatology finds eternity in the cycle itself. This may be one of the possible interpretations of the *Mahāyāna* adage: *Saṃsāra* is *nirvāṇa, nirvāṇa* is *saṃsāra*, i.e., in Nietzsche's poetical language the Dionysian becomes the Apollonian.

The author would like to express her thanks to the Nietzsche Archives in Weimar which allowed her to use all their facilities for this research.

NOTES

1. Karl Schlechta, ed.: *Friedrich Nietzsche* (München: Carl Hanser Verlag 1966), Vol. II, No. 62, p. 32. Hereinafter referred to as *Nietzsche-Schlechta.*

2. W. Lange-Eichbaum: *Nietzsche, Krankheit und Wirkung* (Hamburg: Verlag Anton Lettenbauer 1947), p. 15.

3. K. Barth: *Der Römerbrief* (Zürich: Evangelischer Verlag 1947).

4. A. Schopenhauer: *Parerga und Paralipomena, kleine philosophische Schriften* (Leipzig: R.A. Brockhaus 1877), p. 293.

5. D.W. Dauer: *Schopenhauer as Transmitter of Buddhist Ideas* (Berne: H. Lang and Co. 1969), pp. 29-31.

6. Giorgio Colli and Mazzino Montinari, ed.: *Friedrich Nietzsche, Nachgelassene Fragmente Anfang 1888 bis Anfang Januar 1889* (Berlin: Walter de Gruvter 1972), hereinafter referred to as *Nietzsche-Montinari*, Vol. VIII_3, p. 168.

7. *Nietzsche-Montinari*, Vol. VI_1, p. 175.

8. *Ibid.*, p. 176.

9. *Ibid.*, p. 175.

10. *Idem.*

11. *Ibid.*, pp. 175-76.

12. *Ibid.*, p. 176.

13. *Ibid.*, p. 177.

14. *Idem.*

15. *Ibid.*, p. 196.

16. *Nietzsche-Schlechta*, Vol. III, p. 703.

17. *Nietzsche-Montinari*, VI_1, p. 293.

18. A. Schopenhauer: *The World as Will and Idea*, translated by R.B. Haldane and J. Kemp (London: Trübner and Co. 1883), Vol. I, Fourth Book, p. 365-66.

19. *Nietzsche-Schlechta*, Vol. III, p. 853.

20. C. von Westernhagen: *Wagner* (Zürich/Freiburg i. Br.: Atlantic Verlag AG 1968) pp. 373-80.

21. *Nietzsche-Montinari*, Vol. VI$_1$, p. 98.

22. *Ibid.*, p. 197.

23. *Idem.*

24. *Ibid.*, p. 272.

25. *Ibid.*, p. 338.

26. *Ibid.*, p. 340.

27. *Idem.*

28. *Ibid.*, p. 341.

29. *Nietzsche-Schlechta*, Vol. I, No. 308, p. 996.

30. *Nietzsche-Montinari*, Vol. VI$_1$, p. 398.

31. *Ibid.*, p. 394-95.

32. *Ibid.*, p. 396-97.

33. *Ibid.*, p. 398.

34. *Idem.*

35. *Nietzsche- Schlechta*, Vol. III, p. 785.

36. *Nietzsche- Schlechta*, Vol. II, No. 3, p. 1120.

37. *Ibid.*, (335), p. 197.

38. L. Boltzmann: *Vorlesungen über Gastheorie*, 2 vols. (Leipzig: 1896-98).

39. *Nietzsche-Schlechta*, II, p. 577; III, 1178-79 refers to Boscovich and Robert Mayer, whereby the latter is evaluated as being nothing but a great specialist, but Nietzsche places Boscovich here next to Kopernikus.

     The Jesuit R.-J. Boscovich is an atom physicist and Robert Mayer is one of the discoverers of the law of conservation of energy, but there is nothing pertinent to my topic in these passages of Nietzsche's posthumous works.

40. Balfour Stewart: *Die Erhaltung der Energie, das Grundgesetz her heutigen Naturlehre gemein-fasslich dargestellt*, translated from the English in *Internationale Wissenschaftliche Bibliothek*, (Leipzig: 1875), Vol. 9, 1879.

41. M. Seiling: *Mainländer, ein neuer Messias* — Eine frohe Botschaft inmitten der *herrschenden Geistesverwirrung* (München: Theodor Ackermann 1888).

42. F. Nietzsche: *The Will to Power,* translated by A.M. Ludovici (New York: Russell & Russell 1964), (1063), p. 427.

43. E. Zermelo: "Ueber einen Satz der Dynamik und der mechanischen Wärmetheorie,"in *Wieder-mannsche Annalen der Physik und Chemie,* Vol. LVII, 1896.

44. J. Loschmidt: "Ueber das Warmegleichgewicht eines Systems von Körpern mit Rücksicht auf die Schwere," in *Sitzungsberichte der Akademie der Wissenschaften,* Wien, Vol. LXXIII, 1876, p. 139 and Vol. LXXV, 1877, p. 67.

45. S. Watanabe: "Time and the Probabilistic View of the World," in *The Voices of Time,* ed. J.T. Fraser (New York: G. Braziller 1966) and "Creative Time," in *The Study of Time,* eds. J.T. Fraser, F.C. Haber, G.H. Müller (Heidelberg: J. Springer 1972), p. 159.

46. D.W. Dauer, the work quoted under Note 5.

47. Dōgen: *Shōbō-Genzō,* Iwanami-Bunko Series No. 1989-1992 (Tokyo: Iwanami-Shoten 1939), Vol. 1, Chapter 11, p. 159. Quoted passages were translated for me by S. Watanabe.

48. *Nietzsche-Montinari,* Vol. VI_1, p. 126.

49. *Ibid.,* pp. 382-83.

50. *Nietzsche-Schlechta,* Vol. III, p. 895.

Temporality and Time in Hegel and Marx

W. MAYS

I. INTRODUCTION

In this paper I will deal with Hegel's account of time, its relation to the dialectic and with his views on history. I will do this by examining some of Hegel's basic texts on time and the various interpretations which have been given to them. I will also note that Hegel's account has some affinities with the analysis of time-consciousness of such existentialists as Heidegger and Kierkegaard. Finally, I will touch upon the impact on Marx's thought of Hegel's ideas on history and temporality. Although Marx did not put forward a theory of time he emphasised, like Hegel, the role of process and change in historical and social development.

Hegel's references to time in his writings are somewhat scattered[1] and are often obscurely expressed. Nevertheless, time plays a considerable role in his philosophy. For example, his conception of mind is essentially an historical one, and time is therefore an intrinsic feature of it. As Marcuse points out in discussing Hegel's work, "Mind is of its very essence affected by time, for it exists only in the temporal process of history. The forms of the mind manifest themselves in time, and the history of the world is an exposition of mind in time."[2] He continues by noting that the dialectic manifests itself in reality temporally, and that the negativity which in Hegel's *Logic*[3] determined the process of thought appears in his *Philosophy of History*[4] as "the destructive power of time."

In Hegel's philosophy process or development is expressed by him in dialectical terms. Everything in the world involves opposed and contradictory aspects, so that contradiction (or negation) may be said to be its motive force, just as it is in the case of the development of thought in his *Logic*. Since our mode of existence is made up of contradictory relations, a particular state of affairs can be unfolded only through passing through its opposite. There is a transition from an initial positive stage (thesis) through a contradictory stage (antithesis) to one of adaptation, where stability is reintroduced at a higher level (synthesis). Development in nature and history thus takes place through a series of integrations of opposites.

II. HEGEL'S 'ENCYCLOPAEDIA' ACCOUNT OF TIME

The best known account of Hegel's dialectical conception of time is to be found in the *Encyclopaedia* "Philosophy of Nature," where it appears wedged in between a discussion of space and place and motion in his chapter on mathematical mechanics. We are told that determinations within space, i.e. points, lines and planes, take place in time. Hegel conceives time as introducing diversity (or negation) into space and thus actualising it. And this presumably is what he means when he says the truth of space is time. For Hegel we are never aware of points and instants, but always of 'here-nows'. In his doctrine of the close connection between time and space, we would seem to have a forerunner of the more modern conception of space-time.[5]

Hegel defines time formally and dialectically as follows: "*I t   i s   t h e   b e i n g* which, in that it *is,* is *not,* and in that it is *not, is. I t   i s   i n t u i t e d   b e c o m i n g*".[6] Hegel is describing here in formal dialectical terms the way the present (which is) changes into the past (which is not) and how its future (which is not) becomes present (or is). It is interesting to note that an existentialist like Sartre who emphasises the important part played by time in consciousness defines consciousness dialectically as "that which is what it is not, and is not what it is." In this way Sartre brings out the changing character of consciousness through time. And as we shall see for Hegel too, there is a close relationship between spirit and time.

Hegel goes on, "Time, like space, is a *pure form* of *sensibility* or *intuition; i t   i s   t h e   i n s e n s i b l e   f a c t o r   i n   s e n s i b i l i t y.*"[7] Superficially, at least, there would seem to be some similarity here with Kant's account of time as a pure form of intuition underlying all our experiences. Nevertheless, Hegel's position does seem essentially different from that of Kent. This is clear from his statement "It is said that everything *arises* and *passes away* in time, and that if one abstracts from *everything,* that is to say from the content of time and space, then empty time and empty space will be left, i.e. time and space are posited as abstractions of externality, *a n d   r e p r e s e n t e d   a s   i f   t h e y   w e r e   f o r   t h e m s e l v e s.* But *e v e r y t h i n g   d o e s   n o t   a p p e a r   a n d   p a s s   i n   t i m e;* time itself is this *becoming,* arising and passing away."[8] And he amplifies this point when he further remarks "Time does not resemble a container in which everything is as it were borne away and swallowed up in the flow of a stream. Time is merely this abstraction of destroying. Things are in time because they are finite; they do not pass away because they are in time, but are themselves that which is temporal ... . It is therefore the process of actual things which constitutes time."[9]

For Hegel then time is to be regarded as a process of change in finite things rather than a flow of some independent stuff in which things are as it were immersed.[10] For Hegel time taken by itself and represented as being independent of things, becomes an "abstraction of externality" – the time of ordinary common sense and science, the world-time, which for Hegel is of a pure abstract and ideal nature. What Hegel seems to be saying is that Kant's conception of time as a pure form of sensibility has such an abstract character and is derivative from the more basic process of temporal change.

Hegel further brings out the abstract character of time taken as a pure concept, when he says, "Its Notion is, like all Notion, eternal."[11] In other words, the notion of time like all other notions considered in themselves is non-temporal. Earlier, he makes the same point, when discussing the essential negative character of the Notion, "The Notion, however, in its *f r e e l y   e x i s t i n g* identity with itself, as ego = ego, is in and for itself absolute negativity ... . *O n l y* that which is natural, in that it is finite, is subject to time; that which is true however, the Idea, spirit is *eternal.*"[12] As to what exactly Hegel means, when he identifies the notion as freely existing identity with itself (free from the power of time) with the identity, ego = ego, may be seen from another statement of his in the *Phenomenology of Mind*[13] where he points out that self-consciousness as thus defined is not a genuine distinction of a type in which the self is distinguished from another, "self-consciousness is only the lifeless tauto-logy Ego is Ego, I am I: since for self-consciousness the distinction does not also have the shape of *being*."[14] For self-consciousness to have the shape of being it would have to be expressed in temporal terms. The consequence of this would be that "just as formerly ultimate Reality was expressed as unity of thought and *extension*, it would here be interpreted as unity of thought and *time*."[15]

Findlay in discussing the relation between spirit and time in Hegel's *Phenomenology of Mind*, throws some light on this question, when he says "Hegel is not teaching any doctrine of the 'unreality' of Time, such as is accepted by McTaggart and Bradley, and is widely thought to be Hegelian. On the contrary he is holding that it is only by achieving self-consciousness *through* a temporal process that self-conscious Spirit can *be* at all." Time, so far from being unreal, is the very form of that creative unrest which represents Spirit as it becomes conscious of itself."[16] As Hegel himself remarks "spirit necessarily appears in time, and it appears in time so long as it does not grasp its pure notion, i.e. so long as it does not anull time."[17] And by this he means that only when one tries to conceive the self purely as a Cartesian *cogito*, i.e. as an object of thought, is it out of time. It then becomes a pure tautology (an ego = ego) different from our actual empirical experience of the self. From this point of view it may be said to be similar in principle to time taken as a pure concept, and indeed Hegel makes this identification.

If we consider pure self-consciousness in the form of the Cartesian *cogito* or the Kantian transcenden-tal ego, we then deal with a purely non-temporal conceptual activity. However, in our everyday expe-rience of ourselves as self-conscious beings in relation to things and other human beings in the world around us, the self always has a temporal character. Merleau-Ponty, for example, argues that it is diffi-cult to understand how a thinking or constituting subject is able to posit or become aware of itself in time if the 'I' is taken as the transcendental ego of Kant. But, he goes on, "If, however, the subject is identified with temporality, then self-positing ceases to be a contradiction, because it exactly expresses the essence of living time."[18] Hegel in considering the identity Ego = Ego, makes a similar point when he says that in contrast to this identity, which is what he terms an absolute pure distinction, the self-identity of the ego has to be expressed in time.[19]

Hegel's concept of the 'now' seems to depend on a dialectical relationship in which past, present and future are inextricably bound together. Thus he states "The present is, only because the past is not: the being of the now has the determination of not-being, and the not-being of its being is the future."[20] Hegel is here describing dialectically (i.e. in terms of negation) the creation and perishing so charac-teristic of the experienced temporal process. As he says, "The present makes a tremendous demand, yet as the individual present it is nothing, for even as I pronounce it, its all-excluding pretentiousness dwindles, dissolves and falls into dust."[21] Hegel is here again referring to the changing character of events from which our conception of time arises. As soon as we name the present event it has ceased to be and become past. The present for him is not simple, but is determined by the future which is not-being for it: the now which was present has become not-being and slipped into the past. This would seem to be the purport of his remark "The truth of time is that its goal is the past and not the future."[22]

Hegel also brings out the experiential (or subjective) character of the dimensions of time: past, present and future. They are, he says, "only necessary in subjective representation, in *memory* and in *fear* or *hope*."[23] They do not occur in nature "where time is *now* as separately subsistent differences." In chronological time, for example, the present hour has an ideal character and can be subdivided into minutes and seconds, and can thus be regarded as "a plurality all taken together." On the other hand, in our actual experience of time, the present, future and past play a crucial role and are intimately related to memory and expectation.

III. HEIDEGGER ON HEGEL'S CONCEPTION OF TIME

Heidegger has given an interesting discussion of Hegel's views on time, within the context of his own 'ecstatico-ontological' conception of time.[24] Since for Heidegger time has a creative immediacy about it, the classical moments of time [past, present and future] are described as 'ecstases' and are defined in terms of our actual experience and not as points along a spatialised time-dimension. Heidegger, who gives a priority to the future over the present, describes the future in terms of our expectations and projects, and the past by our memories: the present is the experiential phase in which our expectations fall into the past to become our memories. Thus, these three 'ecstases', or experiential temporal phases, replace the common-sense dimensions of time, past, present and future.[25] Heidegger contrasts his view of time with the way time is ordinarily understood in the form of a pure sequence of nows, without beginning or end, in which the ecstatic character of primordial temporality, or as Heidegger also terms it, "the moment of vision"[26] (*Augenblick*) has been lost.

Heidegger points out that when we give a philosophical interpretation of time it has to be in terms of some system. In the physics of Aristotle, for example, in the context of an ontology of nature, the ordinary way of understanding time received its first interpretation: 'time', 'location' and 'movement' stand together.[27] In his "Philosophy of Nature," Heidegger goes on, time for Hegel is intuited becoming, which Heidegger identifies with the transition that shows itself in the sequence of nows. He believes that Hegel accepts the ordinary everyday conception of time in terms of the 'now' intuited as something present to hand, though ideally.

Since for Hegel "spirit falls into time," Heidegger next turns to Hegel's attempt to establish a distinction between time and spirit. He points out that the essence of spirit for Hegel is the concept and by this, Heidegger notes, Hegel means not a universal form or genus, but the grasping of a differentiation[28] or a negation, as the elements thus distinguished negate each other. As for Hegel the movement of spirit is through the negation of contradictory aspects (as is also the case with time) Hegel can, Heidegger argues, then define the essence of spirit formally as 'the negation of the negation'. And in this way he "gives a logically formalized Interpretation of Descartes' *cogito me cogitare rem*'."[29] By showing that despite the independence of both spirit and time they yet possess a similar formal structure, Hegel, says Heidegger, believes that he has demonstrated how spirit can be actualised in time. Nevertheless, Heidegger concludes that Hegel's account casts little light on the source of time and leaves unexamined the way in which it is essentially constituted. On the other hand, his own existentialist analysis of *Dasein* (or human existence) starts with immediate human experience itself, and time regarded as being derivative from this experience.

However, Derrida evaluating Heidegger's discussion, notes certain resemblances between Hegel's and Heidegger's positions on time.[30] He points out that Hegel's assertion that time is the same principle as the I = I of pure self-consciousness, resembles Heidegger's own statement in *Kant and the Problem of Metaphysics* "Time and the 'I -Think' no longer face one another as incompatible and heterogenous; they are the same."[31] Similarly, for Hegel as for Heidegger "the whole does not arise and move *in time (in der Zeit)*, but time is itself this *becoming*, this upsurge and movement."[32] Derrida sums this up by saying that "it would be too hasty and simplifying to say that the Hegelian concept of time is borrowed from a 'physics' or from a 'philosophy of nature', and that it passes essentially unmodified into a 'philosophy of spirit' or into a 'philosophy of history.'"[33]

IV. TIME IN THE "JENA LOGIC" AND "REALPHILOSOPHIE"

A very different sort of interpretation of Hegel's conception of time is found in the work of Koyré,[34] perhaps better known through his *Etudes galiléennes* as an interpreter of Galileo. Starting from Hegel's early work on time in the *Jena Logic* and the *Jena Realphilosophie*,[35] he claims that Hegel, far from putting forward the ordinary conception of time with its emphasis on the present, gives priority to the future. In these early writings, Hegel, Koyré says, is not concerned with analysing the abstract spatial concept of time, such as it is presented in physics, in Newtonian or Kantian time. Hegel is rather concerned with time itself, with the spiritual reality of time.

It is therefore difficult to reconcile Heidegger's description of Hegel's account of time as the most radical formulation of the traditional concept of time – a mere string of nows – and a paraphrase of Aristotle's essay on time,[36] with Koyre's interpretation. As we shall see there would seem to be a substantial similarity between Hegel's view and that of Heidegger. In the *Jena Logic*, for example, Hegel specifically says that the future is the reality of the now, that the now transforms itself through the future into the contrary of itself. It is only the whole cycle which is real time through the now and the future become past.[37] Although Heidegger does quote from the *Jena Logic*, he seems largely to quote those passages which confirm his own interpretation: that it is the present rather than the future which is central for Hegel.

It is true that Hegel's discussion of time in the *Jena Logic* appears in a physical context. Nevertheless, it resembles Heidegger's own doctrine of the primacy of the future except that whereas the latter emphasises the existential side of the process, Hegel tends to describe it in dialectical terms (i.e. the negation of the negation). The difference between them seems largely bound up with their respective terminologies. According to Koyré, Hegel makes a crucial distinction in his *Jena Logic* between the *now* and the *present*: the *now* refers to a moment of time in its full dialectical character, and resembles Heidegger's "moment of vision", whereas the present refers to the common-sense notion of an ideal moment of time. Koyré further notes that this distinction is absent in Hegel's later writings.

In commenting on the *Jena Logic*, Koyre says, Hegel does not start from the analysis of the notion of an instant, a punctual non-temporal limit between the past and the future, but from the concrete apprehension we have of it – time as it constitutes itself in us starting from the now. The Hegelian *now*, he goes on, is a directed instant, not toward the present, but toward the future: it is the future which develops in the now, and which in some way is before the past. Koyré's point is that Hegel is here describing the flux of concrete spiritual life and is not concerned with an abstract image of becoming.[38] Koyré also brings out the primordial importance of the notions of history and time for Hegel. Spirit (or mind) for him has a historicity, and time a dialectical character and it is only because spirit is temporal and time dialectical that a dialectic of spirit is possible.

In connection with the *Encyclopaedia* account of time, on which Heidegger largely bases his interpretation, Koyré remarks that we find very little in this work relating to the supremacy of the future over the past: a phrase about Chronos, another on the necessity of separating eternity and time, as well as the order in which Hegel enumerates the moments or dimensions of time: now, future, past.[39] He makes the point that in reading Hegel's *Encyclosaedia* discussion of time "one can only with difficulty take note of the phenomenological substructure by which Hegel has arrived at his account."[40] On this question Hyppolite remarks that Hegel's early work (i.e. his writings on Abraham or on love) has a much more concrete existential character than appears in his later writings in which Hegel "intends to *conceptualize* life, whereas earlier he was content to describe it."[41] What Heidegger seems to have done is to have interpreted Hegel's text not so much in the context of Hegel's own work, but in that of Aristotle.

Koyré considers that it is this emphasis on the priority of the future over the past, that constitutes Hegel's great originality: the now, the future, the past are coordinated, imply each other and are not juxtaposed in indifferent equivalence as are the three moments of space. Hegel's time is one which negates itself, throwing itself into the future before falling back into the past, and is essentially the principle of the creation of novelty. For Hegel, Koyré continues, it is we who project ourselves, in negating our present and making of it a past. And it is we who in our memory take up again and relive this dead and accomplished past.

At this juncture it is worthwhile to point out that although in his account of time, Kierkegaard is critical of Hegel's conception of time as being abstract and eternal, one may also find certain similarities between their views on time. Kierkegaard attacks the view of time which makes the present the only reality, the past a mere 'no-longer-now' and the future a 'not-yet-now'. Experienced time is for him a unity in which the modalities are distinguished by reference to human purposes and decisions. His account of time is that of a past held in memory and a future that is projected in our decisions. In Kierkegaard's existential time the three modes are given in the unity of immediate experience.[42]

V. KOJÈVE ON TIME AND THE CONCEPT IN HEGEL

Koyré's discussion of Hegel's account of time was taken up by Kojève in his commentary on the *Phenomenology of Mind*. Kojève was a Russian Marxist with existentialist leanings, living in France. His lectures on Hegel's *Phenomenology of Mind* (1933-39)[43] at the *École practique des Hautes Études* in Paris, much influenced philosophical thinking about Hegel in France. As Spiegelberg puts it "In these lectures he simply asserted that Hegel's phenomenology was "phenomenological description (in Husserl's sense of the word)" dealing with man as an existential phenomenon like Heidegger's phenomenology. Such an interpretation becomes possible because to Kojève the Hegelian method, in contrast to the reality which it tried to explore, was "by no means dialectical: it is purely contemplative and descriptive, i.e. phenomenological in the Husserlian sense of the term."[44]

Hence, in discussing Kojève's interpretation of Hegel one must make clear his bias, if it is a bias. Voeglin, for example, thinks of Kojève's work as "a break-through in the understanding of Hegel."[45] In any case, Kojève does give a valuable insight into Hegel's treatment of time, relating it to Hegel's thought in both the *Phenomenology of Mind* and the *Philosophy of History*. Although Kojève's account may have existentialist overtones, one cannot overlook that existentialism has Hegelian roots, even though Kierkegaard's philosophy was a reaction to that of Hegel. Koyré, who had at one time worked with Husserl at Göttingen, pointed out that the method of Hegel is above all phenomenological, and that although he then suppressed it like a useless scaffolding, such analyses, nevertheless, underlay his constructions.[46]

Kojève notes that Hegel identifies the Concept with Time and that this point was made by Hegel in the Preface to the *Phenomenology of Mind*, when he says "As for *Time*, it is the empirically existing Concept itself,"[47] and this is repeated word for word in Chapter VIII. Kojève remarks that in previous philosophies concepts had been thought to be outside time. On such a view time had nothing to do with the Concept (which was identified with eternity), with absolute knowledge, *epistēmē* or truth, nor with man to the extent that he exemplifies the empirical existence of knowledge in the temporal world.

According to Kojève, those philosophers who do not identify the Concept and Time cannot give an account of history — and this was Hegel's principal aim. Hegel's philosophy, he goes on, describes the existence of Man, who sees that he lives in a world in which he knows that he is a free and historical individual. The time Hegel identifies with the Concept is, Kojève argues, historical time — the time in which human history unfolds. In the *Phenomenology of Mind*, for example, Hegel says "that Nature is Space, whereas Time *is* History."[48] This would seem to mean, according to Kojève, that there is no natural cosmic time: there is time only to the extent that there is history.

In the time considered by pre-Hegelian philosophy Kojève goes on, the movement went from the past toward the future by way of the present. It may be, Kojève remarks, that the time in which the present takes primacy is cosmic or physical time, whereas biological time could be characterised by the primacy of the past. It could be, he argues, that cosmic and biological forms of time exist only in relation to man — that is in relation to historical time.[49] In Hegelian time, on the other hand, Kojève points out, the movement is engendered in the Future and goes toward the Present by way of the Past; and it is this, he claims, which gives human, that is, historical time its specific structure.

Kojève next considers what he terms the phenomenological, or rather the anthropological, projection of Hegel's metaphysical analysis of time, which is to be found in his *Phenomenology of Mind.*[50] Hegel's discussion of time, he tells us, is worked out there in terms of desire. For Hegel once self-consciousness has been achieved, it enters into relationship with things in the wrold and finds itself in a state of desire. Man desires the objects around him and appropriates them in order to satisfy his basic needs. Since man is also a social being, self-consciousness attains its satisfaction only in another self-consciousness. In Chapter IV of the *Phenomenology of Mind,* Kojève continues,[51] Hegel shows that the desire that is directed toward another person is the desire for recognition — that is the desire to achieve prestige in the eyes of the other; and this by generating the Master-Slave relationship (and the resulting alienation) engenders history. A distinctive feature of the historical process is what Kojève terms an "historic moment"[52] in which the present is ordered in terms of the future. He gives as examples of such historic moments, the declaration of a war or the signing of a peace treaty. Each such moment realises a project for the future, and may be said to mark out the shape of things to come. It will be seen that on this conception of history, the historical process is not a fortuitous succession of occurrences but one in which man's projects for the future play an important role.

In evaluating Koyré's and Kojève's interpretations of Hegel's conception of time, one must note that they were writing after Heidegger's *Being and Time* had appeared, that they were acquainted with this work and had both phenomenological and existentialist interests. Nevertheless, there seems some evidence that the early work of Hegel employed phenomenological description rather than conceptual formulation. One must, however, remember that even the account of time in the *Jena Logic* quoted by Koyré has already a somewhat formalised dialectical character. It certainly appears from both Koyré's and Kojève's interpretations that there is a marked and significant resemblance between Heidegger's and Hegel's conceptions of time. Insofar as Hegel's account is historical in character, it has a closer connection with human experience than Heidegger acknowledges. Much of what nowadays passes as the existential analysis of time seems then to have been foreshadowed by Hegel. One cannot, of course, rule out the fact that some of Hegel's doctrines influenced Heidegger, and were consequently reflected in the latter's work. Indeed Kojève asserts that the philosophical anthropology of *Being and Time* "adds, fundamentally, nothing new to the anthropology of the *Phenomenology* (which, by the way, would probably never have been understood if Heidegger had not published his book)."[53]

VI. HEGEL AND THE PHILOSOPHY OF HISTORY

Before looking at the way temporal themes appear in Marx's philosophy, which is strongly oriented to socio-historic questions, we need to give some account of Hegel's philosophy of history, since Hegel's views on history and Marx's are closely related — even though Marx does say that he stood Hegel's account on its feet. As we have already seen it is through history that time enters into Hegel's philosophy, and one may say the same thing about Marx's.

In his *Philosophy of History* Hegel puts forward the view that in order that we may understand history, we need to make use of general categories, such as the idea of progress. Men's actions proceed from their needs, interests and characters, but as they follow them out, individuals through their actions promote the progress of mind (i.e. the universal) and advance freedom. In the historical process there are periods of regression, alternating with periods of advance. Nevertheless, there is an overall progress to self-consciousness of freedom which exhibits itself in man's ability to grasp the universal and make it a reality. Historical individuals, such as Caesar and Napoleon, have anticipated the necessary stages of progress which the world was to take.

To a cursory observer history may not exhibit the sort of progress that Hegel says that it does. When he looks around him he may observe a less cheerful picture: wars, famines, natural disasters — calamity heaped upon calamity. Hegel tries to rationalise this state of affairs by introducing the concept of the *cunning of reason,* which, as it were, directs the passions to work for it: individuals perish, but the 'Idea' prevails. In Hegel's approach to history the rational (in this case, the idea of progress) permeates the temporal process of historical change. Even when the historical individual believes that he is alone responsible for his actions, he is really playing out a role assigned to him by reason.

VII. MARX'S PHILOSOPHY, HISTORY AND TEMPORALITY

As we have already seen there is no independent discussion of time in Marx's work. Unlike Hegel, he was not interested in discussing abstract problems relating to the philosophy of time, but rather in such practical questions as the analysis of the capitalist system for the purpose of bringing about a socialist society in which there would be a more equitable distribution of wealth. Philosophy for Marx was concerned neither with general questions relating to the nature of the universe, nor with the language or meaning categories we use, but with specific probelms having an ethical, social and political relevance. The purpose of philosophy for him was to enable us to obtain knowledge of those social tendencies which would most effectively enable us to change our society: and this is summed up in the phrase "the unity of theory and practice.".

How then, one might ask, is it possible to discuss Marx's views on time and temporality if he never overtly discussed such concepts himself? Although Marx is not essentially interested in abstract discussions about time, he would nevertheless regard this concept as a derivative abstraction from the more concrete phenomenon of historical change. As Marx said "But *nature* too, taken abstractly, for itself — nature fixed in isolation from man — is *nothing* for man."[54] Insofar as the historical process essentially involves change, time at least in its concrete exemplifications in human and social affairs plays an essential role in Marx's thought.

Marx was, as Hook points out, particularly sensitive to the place and importance of time. He quotes Marx as saying "Ideas and categories are not more eternal than the relations they express. *They are historical and transitory products.*"[55] Marx is here referring to social relations and categories which have for him a history in time. Marx's emphasis on historical development is evident in his account of the class struggle and the productive economic process which he regards as exemplifying the Hegelian dialectic. However, as we shall see, the dialectic for Marx is intrinsic to history and not — as it is with Hegel — a logical formula applicable to history.

Hegel, as Hook also notes, distinguishes between time and eternity, which serves "to justify the contention that the *logical* development of categories is outside the temporal process; only our knowledge of that development is temporal."[56] For Marx, on the other hand, one cannot distinguish the development of categories from the historical facts in which they manifest themselves. Thus, categorial conflicts are first experienced in our social existence as concrete conflicts of interest among individual men living and working together in society. When they appear in our verbal formulations as logical contradictions, they only have an abstract derivative form. Marx believes that everything which is given in our culture — be it the value of a commodity, a moral code or an art form — is essentially a social product. Philosophical categories too are not passive Platonic universals but have a history in time, i.e. they have a genetic character.

Marx contends that Hegel instead of taking as his starting point the activities of human beings pursuing their everyday aims, tried to show that the facts of history progressively realise a pre-established rational order. History for him, as we have observed, manifests the *cunning of reason,* so that although historical facts may give the impression of "the play of the contingent and unforeseen" nothing in history is without significance, if we only have the wit to make it out. Like Hegel, Marx believed that the order of historical events is more than a confused record of chance occurrences; but unlike Hegel, he would deny that the order was a rigidly predetermined one. Although he argued that the direction of history comes from the productive economic forces within history itself, he believed that the individual has himself a part to play in its making: that he could influence the course of history, at least by accelerating or slowing down its advance.

Whereas for Hegel history (or rather reason) uses man as a means to carry out its ends, for Marx history is nothing but the activity of man in pursuit of his ends. Marx explains historical activity by the concrete needs of people and the conditions out of which they arise. In *Capital*, however, Marx does not seem to be interested in the individual man, but rather in the behaviour of groups or classes and in the individual only insofar as he is a member of a group, so that for him, the social may seem to be prior to the individual.[57] But if we go back to his early writings we see that for Marx, the individual is identified with the social. Man for him is a *species being*,[58] he is conscious of sharing a common nature with others. It is therefore futile on Marx's view to try to distinguish the individual from the social group.

Central to the Marxist dialectic is the concept of reciprocal action or interaction between parts or elements of a given social or historical system, and this is the ultimate source of its change and development. As Marx puts it, civilisation is based on conflict, "No antagonism, no progress; this is the law that civilization has followed up to our days."[59] A dialectical situation may be said to occur when two successive historical phases are opposed to each other so that the result of their interaction exhibits something qualitatively new, preserving some of the structural elements of both phases and eliminating others. An example of this relationship may be seen in the rise and fall of social groups or classes, as when, for example, feudalism was followed by capitalism.

It is in this sense that we are to take Marx's view that his dialectical method is the direct opposite of Hegel's. There is a continuous and explicit reference to empirical reality in his account. It is not, as in the case of Hegel, a universal procedure applicable to any subject-matter. The concepts used by Marx are avowedly abstractions and not unchanging universals, and reflect historical and transient states of affairs. The dialectic for Marx then is a purely historical and social concept and applies exclusively to social phenomena and historical events.

The Marxian dialectic has been criticised by philosophers and others who argue that they can find no such triad of thesis, antithesis and synthesis in natural and historical processes. It is, however, Engel's extension of the dialectic into the field of natural phenomena (in his *Dialectics of Nature*) which is perhaps more open to criticism than Marx's application of it to history. Nevertheless, Popper, particularly with Hegel and Marx in mind, has forcibly argued against there being any determinate order in history, and this belief he dubs *historicism*. As he says, "the historicist does not recognize that it is we who select and order the facts of history, but he believes that 'history itself', or the 'history of mankind', determines, by its inherent laws, ourselves, our problems, our future, and even our point of view."[60]

When Popper talks about the 'we' who select and order historical facts, he seems to assume that history has no intrinsic order in itself, and that the order we find in it is due to ourselves. In marked contrast to this seemingly conventionalist approach to history, Popper takes as the touchstone of science an objective rationalist theory of truth. Marx, however, defined the truth of a theory pragmatically. If a theory can be made to serve some useful purpose either in the field of technological invention or social construction, then it is true — since it works in practice.[61] And he might give the same justification for his application of the dialectic to socio-historical phenomena.

Despite Popper's criticisms of historicism and the dialectic, which he takes as a special case of historicism, there are certainly elements in the social process which lend themselves to a rational description, although perhaps not in terms of the formalised dialectical triad. The dialectic in its origin in Socratic-Platonic thought, represented the way in which rational discourse develops. Whatever one might say about the contingency of history, one cannot deny that among other things history has to concern itself with individuals who not only feel and act, but also think and plan for the future, and this feature of their experience must to some extent exhibit itself in the warp and woof of history.

Socio-historical processes are by their very nature dynamic — manifesting processes of equilibrium occurring in time. The theory of the trade cycle, for example, is an illustration of such a movement in the economic field, a theory which owes much to Marx's conception of the dialectical nature of the economic process. Further, historical development is not linear, but seems to exhibit phases of progression and regression. This may be traced to the element of conflict one finds between men in society. Hegel has epitomised this opposition in the Master-Slave relation, and it is also to be found in Darwin's notion of the struggle for existence and his doctrine of the survival of the fittest. Marx himself thought that human need, which he interpreted in economic terms — in terms of our struggle for existence in a hostile world — is the driving force behind our actions and behind the rationalisations we produce to justify them. Although we can explain history rationally, it is, however, based on need and not on reason, and in this way once again Marx's thought differs from Hegel's: "Need — 'the practical expression of necessity' as Marx calls it — brings human beings to consciousness, to class-consciousness, to revolution."[62]

Thus, time may be said to enter into Marx's thought through the dialectic of social change. What may seem to be absent from Marx's account is the subjective element which both Koyré and Kojève claim to find in Hegel's account of time, where man's projects can influence the historic process. But this factor is not entirely lacking in Marx, at least in his early writings. One must take account of the Utopian element in Marx's thought, which assumes that man is not simply at the mercy of extra-human forces, but that he can, by planning ahead, exercise some measure of control over his natural and social environment. In keeping with the practical function assigned by Marx to philosophy, he claims that instead of its being, as in the past, an apologetic for a social system or a form of quietism, it now becomes an instrument of social change. To quote his own words "Until now philosophers have only *interpreted* the world differently: the point is to change it."

Time would therefore be for Marx in its origin, as is the case with other categories, a social product and one derivative from the process of social change itself. The shift from Hegel to Marx, as far as time is concerned, is from a subjective dialectical conception of time to an objective one. In both cases the future — taken almost in a prophetic sense — seems to have primacy over the present, giving a direction to the temporal process. It is interesting that Kojève who was a Marxist himself, characterised the Hegelian conception of time as follows: "The Time that Hegel has in view, then, is human or historical Time: it is the Time of conscious and voluntary action which realizes in the *present* a Project for the future, which Project is formed on the basis of knowledge of the *past*."[63] If we substitute 'social' for 'conscious' in the above passage, we would get a pretty good approximation to what Marx would have said about the structure of social time, if he had got round to describing it.

NOTES

1. Hegel discussed the nature of time in some of his early writings, particularly the *Jena Logic* and *Jena Realphilosophie.* These are not so well known as his condensed and somewhat cryptic discussions of time in his *Encyclopaedia* "Philosophy of Nature." Hegel's references to time in his *Phenomenology of Mind* are somewhat fragmentary and are developed in relation to history in his *Philosophy of History.*

2. Herbert Marcuse: *Reason and Revolution* (New York 1941), p. 224.

3. G.W.F. Hegel: *Science of Logic,* tr. by W.H. Johnson and L.G. Struthers, 2 vols. (London and New York 1929).

4. G.W.F. Hegel: *The Philosophy of History,* tr. by J.B. Sibree, with a new introd. by C.J. Friedrich (New York 1956).

5. Cf. *Hegel's Philosophy of Nature,* Vol. I, II and III, ed., tr. by M.J. Petry (London and New York 1970), Vol. I, p. 229.

6. *Ibid.,* pp. 229-30.

7. *Ibid.,* p. 230.

8. *Ibid.*

9. *Ibid.,* p. 231

10. For example, for Newton "Absolute true and mathematical time of itself and from its nature flows equably without relation to anything external." I. Newton: *Mathematical Principles,* tr. by A. Motte, ed. by F. Cajori (Berkeley 1934), p. 6. Such a conception of time would for Hegel be an abstraction from the changing pattern of things.

11. *Hegel's Philosophy of Nature,* p. 232.

12. *Ibid.,* p. 231.

13. G.W.F. Hegel: *The Phenomenology of Mind,* Vols. I, II, tr. by J.B. Ballie (London and New York 1910).

14. *Ibid.,* Vol. I, p. 166.

15. *Ibid.,* Vol. II, p. 816.

16. J.N. Findlay: *Hegel: A Re-Examination* (London 1958), p. 146.

17. Hegel, *Phenomenology of Mind,* Vol. II, p. 813.

18.   Maurice Merleau-Ponty: *Phenomenology of Perception* (London 1962), p. 425.

19.   Hegel, *Phenomenology of Mind,* cf. Vol. II, p. 816.

20.   *Hegel's Philosophy of Nature,* p. 235.

21.   *Ibid.,* p. 231.

22.   *Ibid.,* p. 240.

23.   *Ibid,,* p. 233.

24   Martin Heidegger: *Being and Time,* tr. by John Macquarie and Edward Robinson (Oxford 1967), especially sect. 82, pp. 480-6.

25.   Cf. *Ibid.,* p. 377. Heidegger brings out the difference between the *ecstases* of temporality and the common sense dimensions of time as follows: "Correspondingly, the future as ecstatically understood — the datable and significant 'then' — does not coincide with the ordinary conception of the 'future' in the sense of a pure "now" which has not yet come along but is only coming along. And the concept of the past in the sense of the pure "now" which has passed away, is just as far from coinciding with the ecstatical "having-been" — the datable and significant 'on a former occasion'. The "now" is not pregnant with the "not-yet-now", but the Present arises from the future in the primordial ecstatical unity of the temporalizing of temporality." (p. 479)

26.   Heidegger believes that Kierkegaard has seen with the most penetration the *existentiell* phenomenon of the moment of vision, but he does not think that he has been successful in interpreting it existentially: "He clings to the ordinary conception of time, and defines the "moment of vision' with the help of 'now' and 'eternity' ... . If, however, such a moment gets experienced in an existentiell manner, then a more primordial temporality has been presupposed, although existentially it has not been made explicit." (p. 497, iii. (H. 338).

27.   *Ibid.,* p. 481.

28.   *Ibid.,* cf. p. 484.

29.   *Ibid.,* p. 484.

30.   Jacques Derrida: "Ουδία Γραμμή": A Note to a Footnote in *Being and Time,*" *Phenomenology in Perspective,* ed. F.J. Smith (The Hague 1970). pp. 54-93.

31.   *Ibid.,* p. 68.

32.   *Ibid.*

33.   *Ibid.,* p. 69 n. 14. In this footnote to *Being and Time,* which as Derrida remarks is the longest in the work (p. 500 xxx (H. 432-33), Heidegger contends, with examples, that the section on time in the "Philosophy of Nature" in the *Jena Logic* (System of the Sun) in which Hegel defines time in an exposition of the concept of movement, and in which the *Encyclopaedia's* analysis of time has been developed in all its essential parts, is but a paraphrase of Aristotle's essay on time (in *Physics IV*).

34. A. Koyré: "'Hegel à Jéna': A propos de publications récentes," *Revue d'histoire et de philosophie religieuse de Strasbourg,* No. 5 (1935), pp. 420-58.

35. G.W.F. Hegel: *Jenenser Logik, Metaphysik und Naturphilosophie,* Sämtliche Werke, Bd. XVIII ed. G. Lasson (Leipzig 1925).

  G.W.F. Hegel: *Jenenser Realphilosophie,* Vorlesung manuscripte von 1803/4 und 1805/6, ed. Johannes Hoffmeister (Werke, Bde XIX-XX), (Leipzig 1932).

36. The phrase "Aristotle's essay on time" is that of Heidegger.

37. This occurs in a passage from the *Jena Logic* quoted by Koyré who gives a number of extracts from both the *Jena Logic* and the *Jena Realphilosophie* to prove his point that the future has a priority over the present in Hegel's conception of time.

> The present anulls itself in such a way that it is rather the future which engenders itself (becomes) in it: it is itself this future: ... . In fact, there is neither present nor future, but only this mutual relationship between the two, equally negative each in relation to the other ... . The difference between the two is reduced in the quiescence of the past. The now has its non-being in itself and becomes for itself immediately another, but this other, the future in which the present transforms itself, is immediately the other of itself, for it is now present ... a now which is engendered from the present by the future, a now in which the future and the present are equally anulled and absorbed ... . The past is this time turned upon itself which has absorbed in itself the former two dimensions ... it is only the whole cycle which is real time which through the now and the future becomes past. *(Jenenser Logik,* p. 202: Koyré, pp. 439-41).

38. Koyré, "Hegel à Jéna," p. 446.

39. *Ibid.,* cf. p. 456.

40. *Ibid.,* p. 436.

41. Jean Hyppolite: *Studies on Marx and Hegel,* tr. John O'Neill (London 1969), p. 4.

42. Cf. J. Heywood Thomas: "Kierkegaard's View of Time," *Journal of the British Society for Phenomenology,* Vol. 4, No. 1, pp. 33-40.

  Koyré, "Hegel à Jéna," p. 431 n. 3, tries to show that one can trace Hegel's view as to the primacy of the future back to his early theological writings: "The first category of 'historical' consciousness is not memory; it is the expectation, the foretelling, the promise. The first example of this consciousness is that of Abraham which Hegel analyses in his *Ecrits théologiques* (see p. 371 sq.). And perhaps might one not say that it is by the comparative analysis of the essentially ahistorical mentality of classical antiquity and of the essentially historical mentality of the Hebrew bible, which has revealed to Hegel the dialectical importance of time and the specific historicity of the spirit."

43. Alexander Kojève: *Introduction to the Reading of Hegel,* ed. Alan Bloom (New York 1969).

44.   Herbert Spiegelberg: *The Phenomenological Movement*, Vols I, II (The Hague 1960), Vol. II, pp. 414-15.

45.   Eric Voegelin: "On Hegel — A Study in Sorcery," *The Study of Time*, eds. J.T. Fraser, F.C. Haber, G.H. Müller (Heidelberg 1972), p. 448 n. 18.

46.   Koyré, "Hegel à Jéna," cf. p. 436.

47.   *Introduction to the Reading of Hegel*, p. 132.

48.   *Ibid.*, p. 133.

49.   *Ibid.*, p. 134 n. 21.

50.   *Ibid.*, p. 134.

51.   *Ibid.*, cf. p. 135.

52.   *Ibid.*, p. 136 n. 25.

53.   *Ibid.*, p. 259 n. 41.

54.   Karl Marx: *The Economic and Philosophic Manuscripts of 1844*, ed. Dirk J. Struik (New York 1964), p. 191.

55.   Sidney Hook: *From Hegel to Marx* (Michigan 1962), p. 33.

56.   *Ibid.*, p. 33. This view does not seem to be all that far removed from such modern accounts of time given by writers such as Grünbaum, who give the topology of time a purely formal but basic character, except that Grünbaum would deal with 'before' and 'after' relations, rather than with formal dialectical ones. But Hegel would not, however, dismiss as Grunbaum does the character of the passage of time as illusory. For Hegel actual becoming is of the very essence of time, whereas the ideal concept of time is derived of actual being for him.

57.   *Ibid.*, cf. p. 39.
        Although Marx was the first to state that history is the activity of men in pursuit of their varied ends, this insight seems to have become blurred in some of his later work. In *Capital*, for example, he conceives history as largely determined by the economic productive forces. It is this which has given rise to the conception of Marx as a rigorous historical determinist.

58.   Karl Marx, *Economic and Philosophic Manuscripts*, cf. p. 159.

59.   Karl Marx: *Poverty of Philosophy* (London 1941), p. 53.

60.   K. Popper: *The Open Society and its Enemies*, Vols 1, 2 (London 1962), Vol. 2, p. 269.

61.   It is Francis Bacon rather than Galileo who is Marx's mentor as far as scientific method is concerned. The function of science is to improve the lot of man rather than to satisfy what Marx would consider an artificially abstract set of truth criteria.

62.   *From Hegel to Marx*, p. 39.

63.   *Introduction to the Reading of Hegel*, p. 136.

On Historical Time in the Works of Leibniz

W. VOISÉ

Being one of humanity's most universal minds, Leibniz was also a historian and creator of very original conceptions concerning historical time, although he never presented them in a particular work.

As a representative also of scholarly historiography, always in search of old documents reaching back to the sources of the problems with which he happened to be concerned, Leibniz considered his sources to be the more trustworthy and reliable the older they were. He often touched upon this subject in his works, writing for instance in *New Essays Concerning Human Understanding* that in every effort tending to the reconstruction of the past an important thing is to know the beginnings (*connoistre les origines*), that is, to go back to the time when the event occurred which particularly interests us. Only in that way did he expect, like other historians, to achieve simultaneously two goals: to hit upon the real truth, and to confound sceptics who had questioned the likelihood of finding it.

On the other hand, the conception of "series temporum", which is frequently found in Leibniz's works, denoted a factor favouring the discovery of hidden truth. To one of his friends he wrote in March 1700: "denn accurata series temporum gibt den Dingen das beste Licht." He was also well aware that arranging events exclusively in time does not disclose their true sense, as chronological order should by no means be an essential aim, although it presents the most precise form of a historical elaboration of the material.

The chroniclers' *annalistic* conception of history presented an exactly chronological order which did not, however, mean that single historical facts were isolated by them, since chroniclers often associated them, convinced that every detail may have its meaning. Therefore, Leibniz stressed not only the important role of certain places in space, but also the crucial meaning of certain moments in time. For instance, he drew attention to the fact that the battle of Poitiers, won in 732, could have been lost several years earlier or later. So — wrote Leibniz in his letter to Peter the Great — future generations can have a right to blame their statesmen for having sometimes overlooked the right moment (*moment favorable*).

These remarks are evidence of complications connected with the apparently simple conception of annalistic time in historians' works. Fontenelle wrote about this in his "Praise" of Leibniz, showing that his interest in the beginnings of events (the origins of nations, languages, customs and learning) was linked to the tendency for there to be connecting events in the course of time, which give the possibility of some sort of prediction of the future (*prophéties*).

This last remark leads us to the heart of the matter, just as that way of connecting the past, the present and the future, reminds us that Leibniz influenced the development of historical thinking, being above all the creator of a general conception of the law of continuity (*lex continuitatis*).

Many authors have written about the connection between these ideas of Leibniz and the birth of modern conceptions of historical thought. Such authors, however, have tended to discuss only certain aspects of that problem, or to have treated it very generally, often shifting moreover the centre of gravity of influence of Leibniz's thinking to the decline of the Age of Enlightenment. Tending to consider the question as a whole and desiring to connect it more closely with Leibniz's ideas on the methodology of the historical sciences, we should begin by stating that the Leibnizian conception of history began against the background of his deliberations concerning a number of scientific disciplines. We can today reconstruct the process of formation of Leibniz's conception of historical time, observing attentively not only the great philosopher's metaphysics and epistemology, but also his studies on mathematics, physics, biology, law and even religion.

Leibniz had already, in his first "academician" writings, introduced the idea of continuity, and soon after, in connection with his mathematical investigations (mainly the account of infinity) and studies on the philosophy of nature, he attended to it more closely. Well read in the writings of Aristotle and the neo-scholastics (especially the Spanish ones such as Suarez, Soncinas and Pereira), he made use of their ideas. In the pages of Leibniz's early writings we may discern the influence of his reading among the books of lawyers of the 16th century. And so in *Questiones philosophicae ex jure collectae* (1664) he considered the idea of continuity with reference to the works of Francis Hotman. The list of authors who had inspired Leibniz in this range can be further multiplied, by mentioning here the names of contemporary writers such as Thomas Anglus and also ancient authors, with the book of J.Ch. Magnenus *Democritus reviviscens sive de atomis* (1646) playing here an outstanding role. The most important thing, however, is that in his early writings Leibniz had already connected the conception of continuity with other basic conceptions of his system, which later became the corner stone of his cardinal methodological and historical ideas.

Investigating the idea of substance, Leibniz concluded that all substances are subjected to unceasing change, that is, that the principle of change is linked with substance itself. No wonder therefore that in his letter of November 1671 to Arnauld he said among other things that no body in Nature is continually maintained in a state of rest. Somewhat later, while discussing problems of dynamics, he considered more deeply the principle of continuity. He did so in correspondence as well as in his writings for publication. In the second part of *Specimen Dynamicum,* he was opposed to the interpretation of Nature formulated by Malebranche, saying that the law of continuity is the basic principle of an initial and general order in Nature, being infinite from the very beginning. It was precisely on this basis of the law of continuity that Leibniz critically considered the problem of the origin of the World and of life on Earth. He concluded, namely, that in the light of that law, the statement that the creation of the World and of all its inhabitants was accomplished by a single act is not immune from criticism. He further stated that neither did the World appear all at once, nor was the animal world created at one time. In his opinion, Nature had been formed by a slow, gradual, continual process, rather than by a single, mighty, creative act.

In the Introduction and First Book of *New Essays Concerning Human Understanding,* a polemic against Locke, he said that nothing in Nature occurs violently, that everything is formed by degrees, and that a basic principle of Science is that "Nature never leaps " thus affirming the validity of the law of continuity. "C'est une de mes grandes maximes," wrote Leibniz, "et des plus vérifiées, que la nature ne fait jamais de sauts: ce que j'appelais la Loi de Continuité."

Leibniz had a broad conception of Nature, treating it as a union of both worlds: the world of men and that of things. When Rémond put questions concerning metempsychosis and resurrection, Leibniz answered that such conceptions are opposed to the order of Nature, which knows no leaps. (The thesis "natura non facit saltus" has become a well-known saying of Leibniz.) He argued that every change occurs in accordance with the principle of gradual transformation. Bodies – he wrote – are in a state of continual, incessant change, similar to the course of a running river, while birth and death are relative phenomena consisting in an unbroken passage from one state to another. In his *Considérations sur les Principes de la Vie* (1705) he claimed that death is no destruction of life, and neither is birth its beginning, as both these events are only the continuous transformation of life to death, and vice-versa. Here the influence of the ancient atomists, chiefly Democritus, can easily be traced, for the latter treated death not as a single event, but as a phenomenon occurring gradually, and stated that death (or rather dying) is the result of continual changes continuing even beyond the point at which we consider a man to be dead. (He mentioned examples of hair and fingernails growing after death.) More important, however, than tracing these distant influences is the fact that Leibniz – conceiving similar phenomena to occur in Nature and Society, and starting from analogies existing between a unit and a human community – not only conquered Cartesian dualism of spiritual and material phenomena, but also made a significant contribution to later views on the history of mankind, treated as an infinite sequence of changes occurring in the course of time according to the general law of continuity.

This was no less a bold step forward in regard to the Middle Ages than to the Renaissance, for the belief that historical cycles (the golden or silver age, etc.) marked out the rhythm of history was then still vivid and widely diffused. "Les hommes du moyen-âge n'avaient pas conscience des modifications successives que le temps apporte avec lui dans les choses humaines" was a remark made by G. Monod in his article entitled "Du progrès des études historiques en France depuis le XVI-e siècle " published in the first number of *Revue Historique* (Paris 1876). The same thought occurred also to many later historians, even though genetic views on social and political transformations had appeared before Leibniz, but always without any general theory which would allow the comparison of disparate materials. It was Bodin who risked a first attempt in his work *Methodus ad facilem historiarum cognitionem* (1565) and later came De La Popelinière, with his book entitled *Histoire des histoires* (1599). Neither of these authors yet knew the deeper methodological construction which would allow a synthetic conception of an entire historical process. Naudé, in his *La Bibliographie politique*, published in Paris in 1642, remarked that Popelinière, instead of a methodical review of problems, had left only a catalogue of historians. This was perhaps too severe a judgment, yet it proved that the requirements writers had to meet were growing apace.

In order to understand the meaning of the Leibnizian conception of historical time, it should be remembered that in his classical work, speaking of the methodology of history, Bernheim distinguished three states in the development of historical learning: the referring (when an author simply refers to certain events), the pragmatic (when an author endeavours to edify the reader morally) and the genetic, in which history becomes a science in the proper sense of the word, resting as it then does on general principles of the evolution of mankind. Genetic history (called by Bernheim also "entwickelnde Geschichte") rests on several principal bases, among which two are leading elements: a feeling of community comprising all mankind and its history, and the idea of continuity in the fate of man.

It will be easy to find both of these leading motives of modern historiography in many of Leibniz's works. Admitting the existence of an infinite number of individual substances (monads) ruled by the uniform law of a pre-established harmony, he among others, claimed that all facts jointly forming reality are closely interconnected, and therefore mutually dependent. So — in contradistinction to metaphysics — science examines phenomena in which monads act on each other to form a mutually dependent whole, although they work individually. At the same time, all the phenomena range themselves into sequences, being ruled by the well-known law of continuity. Leibniz as a philosopher thus established theoretical premises that became a basis for the methodology of Leibniz the historian. He maintained that the acquisition of certainty in historical research demands that the greatest possible number of facts be collected and ranged in an unbroken sequence, in a way that would make their present state appear simultaneously as the result of the previous stage and the starting point of the following one. The activity of Leibniz as an indefatiguable researcher of historical sources, collecting them as he did throughout his busy life, was bound up in this way with his historiographical ideas. He formulated them many a time in his writings and letters. His letters to Arnauld merit special attention, chiefly one bearing the date of June 1686. In the Arnauld letter, Leibniz argues that every state, be it present, past or future, is in itself a whole and a continuity. Here he used both terms: the Latin "continuum" and the Greek "syneches". In the same letter we find the idea, formulated by Leibniz and repeated in his "Apokastasis" and "Monadology", that the present communicates with the future. The idea, frequently repeated by others, became the basis of modern thought on the subject of man and society, and its importance in this domain was conceived in the first half of the 19th century by the "father of positivism", August Comte, in the following words:

> The attitude of dynamic sociology is expressed in understanding every successive social state as a result of the preceding one and as a force activating the following state, according to the dictum of Leibniz: the present is gravid with the future.

For any study of the genesis of a historical outlook on society, the prospective outlook (turning to the future) is not the only important thing. We should also remember the retrospective point of view (viewing the past), treating the present as the subject of scientific research. The saying of Leibniz to be found in the introduction of his *New Essays* is for this reason worth quoting here. While explaining the meaning of small observations in the development of the human mind, he emphasized: "the present is full of the future and weighted by the past." This very "weighting by the past", as a characteristic feature of the present, allows us to understand why Leibniz treated so many problems from a historical point of view, and endeavoured to present them in the context of bonds linking the past, the present and the future.

In contradistinction to Descartes, who rejected the idea that history is a source of creative inspiration, and to Bossuet, who being himself a historian later arrived at a similar conclusion, Leibniz characterized the attitude of the enemies of history as a narrow-minded one (*la petitesse d'esprit*). The past was for him an archive of experiences that could bring profit to the present and the future; he also judged the writings of Utopian authors to be of considerable value. As the French writer Gabriel Naudé had done many years earlier, Leibniz considered it unjust to place the Utopias of More and Campanella "inter chimaeras", believing as he did that they contained proposals for a better organized world. In such an attitude we may detect the silent assumption that even in cases where such proposals seem nonsensical, we may always improve upon them by virtue of our own experience, or that of others, which is to say that the mistakes of our predecessors can be a useful lesson to all of us. Man's wisdom, multiplied by his knowledge of records of the past, can contribute to the improvement of the future fate of mankind. As Grua had indicated, Leibniz never ceased to ponder on the possibilities of improving the World, and eliminated in turn various hypotheses concerning the course of mankind's history. He therefore rejected the idea of a gradual degeneration of the World, as well as that of its incessant improvement, and finally the conception of the immutability of all human institutions. Admitting that there were unlimited opportunities for their development in the future, Leibniz predicted a gradual process of continual perfectioning of the World, which would come about despite those momentary disturbances which were likely to occur. The past, the present and the future, become in this way an inseparable whole, the knowledge of which cannot be a matter of indifference to anyone who desires to make the future more perfect. In his *Discourse on Metaphysics,* Leibniz says:

> The past is our inheritance from previous generations, whose imperfections are not our fault, and we may only acquiesce in it in accordance with God's will. But as to the future, we ought not to be quietists, standing with folded arms ridiculously waiting to see what God will do, in conformity with the sophism which the ancients called the lazy reason. We must rather act with the presumptive will of God, so far as we are able to know it, trying with all our might to contribute to the general welfare and particularly to the ornament and perfection of that which concerns us.

Risking the translation of Leibniz's segments of historical time into grammatical tenses, his conception may be presented as follows: he conceived the past as an imperfect past tense, and the present as a past tense that is still occurring, or present perfect, which means that he did not separate these forms of time so sharply as did many of his contemporaries (first and foremost, Descartes). He treated the future, however, in a certain sense as "futurum gnomicum", which means that conclusions were to be drawn as well out of the past as out of the present, allowing us to form more consciously the future fate of mankind. That kind of three-dimensional conception seems to be the most characteristic feature of the Leibnizian idea of historical time.

It is finally time to summarize briefly the material assembled here, and to present it in categories of chronosophy. We are here much impressed by the conciliatory character of Leibniz's mind, which allows him to reconcile opposite tendencies. Thus, when reading in *New Essays* that "space as well as time have their reality from God only" ("l'espace comme le temps n'ont leur réalité que de Dieu"), it may be assumed that there is reason to qualify his conception as a providential ordering of historical events. Matters are, however, different: in his correspondence with Samuel Clarke, Leibniz determines space as "an order of co-existence" and applies another formula to time – "existent order". And responding to Bayle's criticism, he writes in 1702 that time "is the order of inconstant possibilities," which seems to suggest that he also treats historical time as a real relation. We see a similar sort of reasoning in his 5th letter to Clarke, where, discussing the subject of the "anticipated creation of the World", Leibniz remarks that "if that creation does not change anything in the visible World, it means nothing real."

Leibniz, therefore in some measure, leaves in suspense the providential ordering of events, placing next to it a mathematical ordering, and stressing moreover the role of chronological order. This is clearly said in the letter he addressed to Thévenot in 1691: "Once having arrived at series, you have a guarantee of truth and the affair is won, just as when you find series in numbers" ("Lorsqu'on est venu ad series on a garans de la vérité et le procès est gagné, comme lorsqu'on trouve des séries in numeris").

The quantitative (= numerical) order of events is not, however, so important here, as we know with what care Leibniz considered the qualitative priority of certain of them, that decided a specific development of a given historical process and allowed no alternative. The genesis of these events he regarded as having been, in a certain sense, determined by their antecedents, while these in turn were to determine the further course of the following stages of history. Neither single facts, nor their segments, nor historical events, are therefore isolated fragments of a process of development forming a historical continuum. For this reason, every period was for Leibniz equally unique, just as every human being is a unique individual. (He wrote of this in his "Apokastasis panton".) This resulted, moreover, from indiscernible principles, according to which it was claimed that two things can never be absolutely alike, and that indiscernibles are indeed identical.

The chronological order answers as well to the law of continuity as to the conviction of historical finality, obvious in the progressive improvement of mankind, about which he wrote in a letter addressed to Vota, of April 4, 1703: "All is going on in a wonderful order by stages marking an advance towards the infinite." It is truly only an earthly order, but that he believed it necessary to treat this from a cosmic point of view is evident from a letter written to Princess Sophie, dated November 4, 1696, where it is said that owing to the discovery of Copernicus we are able to rise up to the Sun "with the eyes of our spirit", and admire from there the "wonderful order" established owing to the "perpetual life of souls", causing a "continual advance and development of the Universe."

Two matters raised in this very context – eternity and duration – require clarification. Following the old philosophical tradition, Leibniz rejected the popular conception of eternity as a temporal existence with neither beginning nor end, and emphasized that eternity is an unalterable and infinite presence. Also opposed to common sense, and close to the neoplatonic tradition, Leibniz sharply distinguished between the two ideas, duration and time. From his letter to Princess Sophie (October 31, 1705), and from Chapter 12 of the second part of his *New Essays,* it can be assumed that Leibniz distinguished in the course of historical time a number of events that do indeed co-exist simultaneously – which, however, does not imply that their succession is identical with historical time, since the duration of these particular events does not result in the formation of a regular sequence.

For historical time is not only a regular succession of events, but also their certain and well ordered continuity, which coincides moreover with the determination of time in general; in his letter of December 1700 addressed to de Volder, he wrote that time is "continuus ordo existende successive."

Another consequence of the law of continuity is the determinism resulting from a causal connection of past, present and future, or the "concatenation" of these three parts of historical time. Leibniz wrote about it, for example, in his "Reflexions on the work of Hobbes": "Everything occurs by virtue of determined reasons, the knowledge of which – should we have had it – would let us understand both why it happened and why it did not take a different course." We know, however, that this reflection was to Leibniz no hindrance in opening up a large field of human activity by consciously planning what was to be done in the future. He was intent here not only on transcendental finality, but also (and perhaps still more) on immanent finality, resulting from monads attaining perfection as subjects of a historical drama, leading to the emergence of new forms of human co-existence. In such a way did Leibniz solve the eternal problem of conflict between free will and the lawfulness of historical processes.

Were we, however, to maintain that Leibniz connected "a progressivistic conception of history with the faith in an infinite progress", it would be necessary to qualify the idea further. In contradistinction to Vico, Leibniz did not represent a cyclical theory of the sort we encounter in the *Scienza Nuova* (1725) which was published after Leibniz's death; neither did Leibniz manage to produce such a scheme as that in which historical changes are presented in diagrams, containing, as did Bossuet's lines, religious and lay elements, Christian and Jewish, ancient and modern ones. This was achieved by Voltaire, who in his *Essai sur les moeurs et l'esprit des nations* (1757), a hundred years before Darwin, gave the first alternative to a providential model of human history.

We cannot speak without qualification about the conception of linear progress in Leibniz's works, since we know that he indicated examples of interruptions impeding the progress of improvement. His was rather a conception of sinusoidal progress, which probably resulted from his methodological principle of epistemology, expressed in the letter already mentioned which he wrote to Thévenot. There, after the words "Naturam cognosci per analogiam" ("I learned to know Nature by analogies"), he added: "I wish above all to be able to find, in what I call 'series', an order, a progression, this being the result of several analogies or comparisons" ("et surtout je voudrais qu'on put trouver en ce que j'appelle series, un ordre, une progression, qui est le résultat de plusieurs analogies ou comparaisons"). This is an example of what was perhaps the first attempt at a modern and conscious unification of the world of things and the human world, resting on the new commonly exploited integration of methods such as are applied in two branches of learning, natural science and the science of man, branches which in Leibniz's time suffered a period of widening separation.

In his letter to an unknown correspondent, dated December 16, 1707, Leibniz stressed the importance of the idea of continuity in all scientific disciplines, and complained that his age did not favour its reception ("le siècle n'est point fait pour les recevoir"). If, today, we conceive history not only as a science indicating the genealogy of the present, but also pointing out ways leading to a better future, then we are very close to the conception of historical time pleaded for by Gottfried Wilhelm Leibniz.

BIBLIOGRAPHY

Bernheim, E.: *Lehrbuch der Historischen Methode und de Geschichtsphilosophie.* Leipzig 1903.

Davillé, L.: *Leibniz historien.* Paris 1909.

Eliade, M.: *The Myth of the Eternal Return.* New York 1954.

Ettlinger, M.: *Leibniz als Geschichtsphilosoph.* München 1921.

Grua, G.: *Jurisprudence universelle et théodicée selon Leibniz.* Paris 1953.

Jalabert, J.: *La théorie leibnizienne de la substance et ses rapports avec la notion de temps.* Paris 1946.

Leibniz, G.W.: *Philosophical Papers and Letters.* L.E. Loemker, Ed. Dordrecht 1969.

——————. *Das Problem der Kontinuität.* P. Schneider and O. Saame, Eds. Mainz 1970.

Sticker, B.: "Leibniz Beitrag zur Theorie der Erde." *Sudhofs Archiv* (Bd. 51), 1967.

Thompson, J.W.: *A History of Historical Writing.* New York 1942.

Uferman, K.: *Untersuchungen über das Gesetz der Kontinuität bei Leibniz.* Berlin 1927.

The author is deeply grateful to Dr. Albert Heinekamp of Hanover for his critical evaluation of this text, and to Dr. J.D. North of Oxford for certain stylistic improvements.

IV. LITERATURE

Four Phases of Time and Literary Modernism

R.J. QUINONES

"Isn't time the essential dynamic of the West?" This was the rhetorical question put to me by a Ruma-
nian sociologist in New York City a few years back. Having just completed a book whose subject was
that very thesis, I could not disagree. And indeed it does seem that whenever there is a new stirring of
this dynamic, time as a concept becomes extremely active in literature. In Jacqueline de Romilly's
Time in Greek Tragedy, the Sorbonne classicist argues that time and a sense of history emerged together and were integral to the great age of Greek tragedy.[1] This book, incidentally, should be read in
order to counter the hasty assumption that the Greek sense of time was uniformly cyclical. My own
The Renaissance Discovery of Time shows the many ways that time entered quite specifically into the
re-awakening and quickening of life among the European countries in the Renaissance, and retrospectively indicated how many of the Renaissance sources for their inspired addresses to time were from
Roman literature.[2] One notices the gap of the Middle Ages; this is not because I deplore that period
of our cultural history – quite the contrary is true (although it might represent my ignorance of it) –
but rather my scholarly belief, which has not yet been countered, that in the Middle Ages, this dynamic lay fallow and, as a consequence, the concept of time as we have later come to regard it was
largely non-existent. And whenever in public gatherings I express this belief, medievalists immediately become angry, as if I am shortchanging their period. When asked, however, to cite some instances
similar to those I show in the Renaissance, and even when given several days advance notice they
normally can muster only two or three relatively minor and even disputable utterances. From the Renaissance, however, if one were called to such a presentation, within a few minutes one could cite
some thirty major works where the sense of time is a central, vital and dynamic concept. When we
come to Victorian society, a society on the move if there ever were, time was again something of an
obsession, as Jerome Buckley's *The Triumph of Time* attests.[3] Time has been, and should be, treated
as a major theme of Western literature; I can be more specific and refer to it as an *indicator-theme*,
one that clearly points to and is even instrumental in the surges and sags of Western society.

Time is especially significant then in the cultural movement of the first half of the twentieth century,
that period of literary modernism when so many of the values that emerged in the Renaissance and
were codified in the nineteenth century were challenged. As a general indication of the sort of cultural difference I am marking, it is instructive to be reminded that nineteenth-century society grasped
at Darwin's evolutionary theories, and relatively ignored the second law of thermodynamics, formulated seven years before Darwin's own results were published. But at the end of the century, and in
the years before World War I, in Europe as well as in America, so-called degradationist theories began
to prevail, indicating that bourgeois society was losing its faith as well as its energy, perhaps even, its

very will-to-be. In this atmosphere, it was natural that a new sense of time would dominate the literature most responsive to modern currents, the literature of that intellectual generation born in the 70's and the 80's, who came into maturity before World War I, found their experiments and published such masterpieces as Proust's *A la Recherche de Temps Perdu,* Mann's *The Magic Mountain,* Eliot's *The Waste Land* and Joyce's *Ulysses.*

Such works that I mention coming all together as a release that the end of the war brought about, of course, do not of themselves constitute the canon of these masters of modernism. Their early works, *Buddenbrooks, The Dead,* "Prufrock," must be considered, as well as their final achievements, *Finnegan's Wake, Four Quartets,* and *Joseph and his Brothers.* What is remarkable about these modernists is the role that sheer development has in their literary careers — it is a special mark of their genius. Consequently, their work is not all of one piece, although it might be of one direction. A literary scholar and historian is pained when one aspect of a writer's work is singled out and made typical, ignoring its complex evolution. My belief is that in their developments all of these writers underwent similar evolutions and that these searchings and processes of growth can be summarized under four temporal phases. The first one expresses the collapse of historical values, the atmosphere of an age of bourgeois decline and anti-democratic thought from which our writers necessarily had to retrieve themselves, the second phase, and perhaps the most critical in its determining function, shows the development of the complex central consciousness — it is perhaps here that we have modernism's most essential contribution to our sense of time; phase three reveals their concerns with the recovery of the self and larger mythic associations; and phase four, registered in all of their works, returns them, and as it were completes the cycle, to a sense of historical reality. Mainly for reasons of time and because of the critical problems presented, this report will concentrate on phases one and two and their interrelations.

The first phase of time and modernism, the collapse of historical values, is directly connected with the Renaissance, that period of our cultural past in which these values emerged. In his relation to the goods of life, medieval man suffered a radical alienation. His ideal image is that of a pilgrim, assured of his destination, but paying only passing attention to the things along the way. In the Renaissance man began to feel more secure with, and invest more value in, those earthly things that promised continuity: the *polis,* marriage, family, and children and also fame. While this is an obvious simplification, and while these commitments were not free from tensions and religious compunctions, still it is safe to say that in the Renaissance man acquired historical existence, and that his historical existence assumed value. But these possibilities of extending his existence were not automatic; they were based on fundamental recognitions. And one key recognition was that of man's precarious existence in time. In my book I called this the argument of time. Scrupulous surveillance of time became the occupation of the new men of the Renaissance.[4] This placed much more responsibility on the individual; it was he who had to open his eyes to the dangers in life, and to manage his resources diligently, if he were to make full use of his talents. This code was one that joined early capitalist and humanist alike. In the Renaissance time was a great discovery, with metaphysical and cosmic reverberations, but it also led to practical decisions and mundane consequences, such as scheduling one's daily activities, that inevitable *emploi du temps.* These mundane deductions and thrilling illuminations were part of a new order of life.[5] Hence, that sleepy Gargantua when aroused by the new humanistic program of study became a vigorous young man diligently exploiting all available time so that he did not lose one hour of the day. A vision of possibility, closely allied with the realizations of time, stirs from slumber energies that had long lay dormant. Another Renaissance prince, the young Hal, in Shakespeare's second series of history plays, also must learn the value of time; to redeem time for him means to come into full possession of his historical identity. In this time-world the father plays a pivotal role, and the generational link between fathers and children becomes crucial in Shakespeare's political and historical ethic. Shakespeare's history plays are not merely historical in subject, they are historical in value. Such historical identity that Hal assumes carries with it much limitation and loss of freedom, but the positive results more than compensate. In becoming his father's son and true successor, Hal with a rare show of exhilaration embraces the codes of continuity:

> The tide of blood in me
> Hath proudly flowed in vanity till now.
> Now doth it turn and ebb back to the sea,
> Where it shall mingle with the state of floods
> And flow henceforth in formal majesty.   (2 *Henry IV*   5.2. 129-133)

When we come to the first stirrings of modernism, this code had lost its fervor; it has become rigidified. Nietzsche himself portentously declared that in process there can be no salvation. Rather than multiply instances I think we can at first succinctly illustrate what happened in the nineteenth century to these historical values by juxtaposing Thomas Mann's *Buddenbrooks* with some famous Renaissance texts. In this novel written at the juncture of the nineteenth and twentieth centuries, the third generation of Buddenbrooks enter adulthood by sacrificing their personal lives. Unlike the two Renaissance cases mentioned earlier, these sacrifices are not attended by any redeeming compensations. Tonia surrenders a youthful summer romance in order to marry according to her parents' wishes (as it turns out they have been deceived by the prospective suitor). Her father's letter should be used as a forbidding example of the sclerosis that had penetrated a code formerly so full of individual choice and lively response; in fact, it would not be too unfair to contrast it with Gargantua's famous letter to Pantagruel, where the father's insistence on familial continuity and the higher, but also more practical uses of learning, is an inspiration for the eager son. The Herr Consul writes:

> My child we are not born for that which, with our own short-sighted vision,
> we reckon to be our own small personal happiness. We are not free, separate,
> independent entities, but like links in a chain, and we would not by any means
> be what we are without those who went before us and showed us the way, by
> following the strait and narrow path, not looking to right or to left. Your path,
> it seems to me, has lain all these weeks sharply marked out for you, and you
> would not be my daughter, nor the granddaughter of your grandfather who
> rests in God, if you really have it in your heart, alone, willfully and light-
> headedly to choose your own unregulated path.[6]

In the nineteenth century, as Jerome Buckley suggests, time did indeed triumph, but it brought with it such an emotional backlash as to lose all sense of personal fulfillment. Such constructions, only latent in the Renaissance, or even in the time of his grandfather, become suffocating for Tonia's brother, Thomas. Like many men in business and other endeavors, he started out thinking he had imagination and all he had was energy. As he encounters the inevitable patterns of sameness, which incidentally, the familial context, with its rituals of repetition, reinforces, as he achieves the limit of his attainment, with no new world to conquer, Thomas could very well serve as the portrait of the end of Faustian man. Summarily put, the values of history have failed to satisfy, have failed to answer the question that young Hans Castorp in *The Magic Mountain* felt, but did not formulate: "To what end all of this endeavor?" Thomas did not have the healthy, eighteenth-century irreverence of his grandfather, nor the restoration religious piety of his father: "He ended by finding in *evolution* the answer to all his questions about eternity and immortality." A mistranslation in this English passage provides us with a clue to the sort of values under discussion. Mann did not write that Thomas tried to find an answer in *evolution*, but rather in historical values: "hatte er sich die Fragen der Ewigkeit and Unsterblichkeit historisch beantwortet." What he means by *historisch* is explained in the next clause, "dass er in seinem Vorfahren gelebt habe und in seinem Nachfahren leben werde (He said to himself that he had lived in his forbears and would live on in his descendants)."[7] Familial continuity, for Mann as it did for the bourgeois society of the nineteenth century, clearly provided the flesh-and-blood meaning of historical value. And the same was true for Shakespeare's royal history plays. These two efforts deserve to confront one another over the centuries. If Shakespeare's histories were historical in theme as well as in subject, Mann's first novel, indeed one that sets the stage for his later works as well as those of modernism, is unhistorical in subject as well as value. It shows not merely the decline of generations, but the decline of the generational ideal. Shakespeare's historical plays center around the time-conscious father, who saves and is saved by his son. Mann's novel has been called "a reckoning with the world of the fathers," where redemption can only lie outside its processes.

The guiding purpose of Thomas' life had been his "sense of family, his patrician self-consciousness, his ancestor-worship," but in the shadow of death these values proved insufficient; not bearing a philosophy, they provided him "with not a single hour of calm or readiness for the end." It is at this time that he makes his momentous, but tragically belated discovery of Schopenhauer. He is thrilled by the passionate freedom of the discourse on the indestructibility of the soul, and for once feels liberated from the responsibilities that had so weighted him down: "Have I hoped to live on in my son? In a personality yet more feeble, flickering, and timorous than my own? Blind, childish folly. What can my son do for me — what need have I of a son? Where shall I be when I am dead? ... I shall be in all those who had ever, do ever, or ever shall say "I" — especially, however, in all those who say it most fully, potently, and gladly!" For the first time in his adult life he has broken with the bonds of family line, that now seem accidental, and reaches out to exercise free passionate choice. Wherever there is a child growing up who fully embraces and represents life, that child is his son. The presence of death revealed the insufficiency of his former code; the same presence now opens up the possibilities of broader and freer affiliations. Mann's narration then proceeds to link his rejection of historical values with Schopenhauer's own thoughts about time and space, and for our purposes brings the novel *Buddenbrooks* and its issues within the main stream of modern temporal speculations:

> The deceptive perceptions of space, time and history, the preoccupation
> with a glorious historical continuity of life in the person of his own
> descendants, the dread of some future dissolution and decomposition, —
> all this his spirit now put aside. He was no longer prevented from gras-
> ping eternity. Nothing began, nothing left off. There was only an end-
> less present ...[8]

For once, Thomas breaks out of the middle channel of historical continuity and experiences a freedom that is both cosmic and psychic, and in so doing recapitulates what John Henry Raleigh suggests in "The English Novel and Three Kinds of Time" to be the pattern of modernism as a whole.[9]

Different from its Renaissance origins, this nineteenth-century code of historical continuity has become a mausoleum harboring no life. This is more than Mann's perception; it is the property of an entire intellectual generation, and forms one of the key concepts in Ortega y Gasset's *The Revolt of the Masses*. Ortega attacks the notion that the nineteenth-century was a period of plenitude, and, in so doing, provides one of the best glosses on this crucial episode of *Buddenbrooks*. Thomas' collapse comes after he had achieved the utmost of his possibilities as the head of his firm. As Ortega writes of the so-called fullness of the past century, "When a period has satisfied its desires, its ideals, this means that it desires nothing more; that the wells of desire have been dired up. That is to say, our famous plenitude is in reality a coming to an end." Indeed, Ortega regards the nineteenth-century notion of progress as a gloomy faith. "it meant that tomorrow was to be in all essentials similar to today, that progress consisted merely in advancing, for all time to be, along a road identical to the one already under our feet." As the Schopenhauerian episode reveals Thomas Buddenbrooks did have other desires but these were submerged and unacknowledged, and, in any event constituted a thorough repudiation of his life's commitments. These late-revealed desires were for freedom, the sense of which Ortega describes as escape. When we abandon the narrow limits that circumstances have imposed on us, "we enjoy a delightful impression of having escaped from a hermetically sealed enclosure, of having regained freedom, of coming out once again under the stars into the world of reality, the world of the profound, the terrible, the unforeseeable, the inexhaustible, where everything is possible, the best and the worst."[10] These passages from the Spanish philosopher joined to the crucial moments in *Buddenbrooks* help us to see the preliminary movements of modernism as being based on the personal need to retrieve oneself from an ethical code that smelled of decay. Historical continuity, process, even progress had become a kind of static logic that formed a block to real freedom. Modernists were then not altogether the fragmented victims of a dynamic past, the fallers-off from a wholeness that their fathers had possessed. Rather they were searchers for new values, unsatisfied by the apparent dead-end of nineteenth-century beliefs.

Of course, the leader in all these attacks on the preeminence of historical values was a man from the preceding generation, Nietzsche, the first real philosopher of modernism. I refer to the Nietzsche of the 70's and early 80's, the Nietzsche of the great and creative negative critiques of bourgeois society and values. I shall take a passage from *The Birth of Tragedy,* and treat it as key because it completes the network of assumptions common to time and historical values. Protesting against the middle class triumph of time that since the Renaissance tended to view all things *sub specie temporis,* Nietzsche advocated a new tragic realism, and a return to myth, the vision of man *sub specie aeternitatis.* In the Schopenhauerian episode of *Buddenbrooks* the connection is made clear between the sense of time as linear sequence and the bourgeois notion of familial continuity. It is left for Nietzsche to add to this complex of temporal attitudes by linking bourgeois values and the linear assumptions of causality necessary to science. In praising the new men of genius (he means Kant and Schopenhauer) who broke this chain of logic, Nietzsche is aware that he is destroying the underpinnings of an entire culture: "Whereas the current optimism treated the universe as knowable, in the presumption of eternal truths, and space, time and causality as absolute and universally valid laws, Kant showed how these supposed laws serve only to raise appearance ... to the status of true reality: in the words of Schopenhauer, binding the dreamer even faster in sleep."[11] In attacking these assumptions of causality, linear sequence and succession, one is attacking the very arteries of nineteenth-century society: science, bourgeois values, and the joint product of technology and continuity, the faith in progress. These speculations stemming from German nineteenth-century thought, and of which Thomas Mann is only the first and most immediate heir, form the ideational content of modernism, and the complex of temporal attitudes understood in phase one of this study.

As if indicating its position at the frontiers of modern thought, a new and changed conception of time is very early introduced in Mann's fiction as a clue to pending alterations in bourgeois character. Gustave von Aschenbach, taking flight from his controlled existence as Thomas Buddenbrooks was unable to do, encounters with the spatial removal an altered conception of time. In fact, contemplating the measureless sea, his own capacity to compute time has markedly diminished: "the time-sense falters and grows dim."[12] This summary alteration then seems to open the door to a host of shadowy subliminal creatures who, begging their release, crowd his consciousness. In that great *Zeitroman,* as Thomas Mann himself has termed *The Magic Mountain,*[13] Hans Castorp almost upon arrival begins speculating on other possible conceptions of time. One cannot say "objective" time, as measured by the clock, he contradicts his more stable cousin, Joachim. Later on in his protracted stay, this first philosophical questioning about time is flung back at Hans by Joachim, "Und mit der Zeit hattest du es gleich am ersten Tage zu tun (From the very beginning you began bothering about time),"[14] a phrase which might very well stand as the motto of our discourse. Indeed at the very beginning of modernism the concern was with time. But as *Buddenbrooks* showed — and as was apparent throughout my studies of Renaissance time — given the nature of time in the West as a sort of bellwether, when one encounters time one encounters a host of related issues, such as the values of history and of children, of continuity and of parenthood. Thomas Mann explains why this should be so when he writes in *The Magic Mountain* that "the perception of time is so closely bound up with consciousness of life that one may not be weakened without the other suffering a sensible impairment."[15] Consequently, an alteration of the sense of time and a weakening of the values of life concomitant with that sense of time occur together. Often quoted is Stephen Dedalus' cry that history is a nightmare from which he is trying to awake.[16] He desires a spiritual father more commensurate with the freedom of his speculative self; his real father is too much part of local limitations and of the dismal history of dissolution and decay — national as well as familial. The mother in *Women in Love,* madly yet prophetically exclaims, "Pray for yourselves to God for there's no help for you from your parents."[17] In Proust this concern constitutes what could be called a sub-theme, for the fact is that in his enormous novel every parent who entrusts himself or herself to the continuity of his children in a conventional sense is rudely disappointed. In *The Waste Land,* part of whose biographical background was the recent death of Eliot's father, the grief of the son and the hope of resurrection find no modern corollary. Ariel's lines from *The Tempest,* "Full fathom five thy father lies ... " are broken, and left incomplete, as if an emotional circuit had been interrupted.

Fragmentariness and a lack of connection in the artistic work itself is a further product of this disruption of historical values. Sequence suffers violent upheaval in the plot and line of narrative as well as in the family. Virginia Woolf considers narrative skill to be unimportant, even cheap; she admits her preference for situations rather than story. *The Waste Land* is precisely that: a series of situations, vignettes of modern life, with no immediate plot line, story or "character to sepak of." And indeed the flight from the plot involves, in Arnold Hauser's phrase, the flight from the hero,[18] so that T.S. Eliot could say in his notes to *The Waste Land* that Tiresias, while "not indeed a character," is still the poem's "most important personage." We come full circle with a phrase from Lionel Trilling's recent and admirable, *Sincerity and Authenticity,* where he discusses the falling off of narrative in relation to the fall of the family: "Traditionally the family has been a narrative institution, with counsel to give and a tale to tell of how things began, including the child himself."[19] The collapse of historical values has consequently great implications for the ways in which literature is presented. The importance of the book *Buddenbrooks,* with which we opened this discussion, can be summarized by saying it is the last great narrative novel that traces the decline of those very values that made narration possible.

Before we proceed to that critically important second phase of modernism, I mean the development of the complex central consciousness, which clearly grows out of the collapse of historical values, I would like to suggest two related forces that furthered the demise of those values. From within even the deadest of shells there is needed some strong force to emerge before it can be pushed aside. One such force in modernism is eroticism. Thomas Mann, in a later essay, regards the message Schopenhauer held for Thomas Buddenbrooks as being essentially erotic in nature. The force that is driving the super-controlled Gustave von Aschenbach to his destruction is unconsciously erotic, and Hans Castorp seems to be following his own inner need when he dismisses his rational counsellor Settembrini in order to return the pencil to Clavdia Chauchat. As Nietzsche had said, in modern times all values must be brought before the bar of Dionysus. To refer to Nietzsche is to suggest, with *The Birth of Tragedy* in mind, that the force allied with eroticism is primitivism. And the great text, or series of texts, that exerted so fundamental an influence in this regard is Sir James Frazer's *The Golden Bough.*[20] The literary impact of this study was far different from Darwin's work; rather than suggesting evolutionary formulae, it suggested a co-temporality; it was in Eliot's words, "a revelation of the vanished mind of which our mind is a continuation."[21] It inspired authors to look for the interpenetrations of past and present, and became a storehouse of mythic suggestion. As such it should be discussed in the third phase. Almost invariably, however, the experience of the primitive, if even in a mythic sense, is bound up with some sense of terror or struggle, some emphasis on blood and that unregenerate part of human experience that constitutes a challenge to the vision of nineteenth-century liberal humanism. This latter is the vision of that Mediterranean figure, Settembrini, so easily eclipsed and so enduring, the voice of the concept of time that emerged from the Italian cities of the Renaissance, of humanism and of liberal politics and economics, in short, a figure summarizing, in Mann's essayistic and intellectual *roman,* the very continuities of Renaissance and nineteenth-century thought that it is the function of this paper to present Yet, his message does seem anachronistic when challenged by Naphta, that son of the *schlacht*-house and the exponent of terror, that voice of reactionary radicalism that was heard so often in the early decades of the twentieth century. Dangerously lost in the snow, and even more precariously tossed between the arguments of these antagonists, Hans Castorp, like others of his nation before him, returns to the world of the Greeks in order to reconcile these antinomies. In this sense the classic Mediterranean world has always represented the nostalgia of the West. He dreams of a beautiful, sunlit world and ideal community where human form dominates unseemly process. But as he penetrates the veil of this world, he encounters the irrational even among the Greeks. At an inner temple he pictures two old hags, their mouths running with blood, devouring children. Classical myth is made more sombre by an awareness of primitive myth. Any faith in a humanistic view is starkly challenged by this vision of an unregenerate brutishness in experience. The scene is made contemporary, the continuities of past and present re-established, when the hags speak in the gutteral dialect of Castorp's native Hamburg.[22]

Throughout modernism we find evidence of the impact of this kind of experience. Almost literally combining *The Golden Bough* with Nietzsche is Yeats's remarkable version of the "nativity" in the opening lines of "Two Songs from a Play":

> I saw a staring Virgin stand
> Where holy Dionysus died,
> And tear the heart out of his side,
> And lay the heart upon her hand
> And bear the beating heart away ...[23]

Eliot sees the same recurrence in Christian experience: "In the juvescence of the year / Came Christ the tiger."[24] Indeed, for our purposes perhaps the most interesting figure in his accessibility to primitive experience is T.S. Eliot, a poet who normally, because of his declared allegiances to royalism, classicism and anglo-catholicism, one would not suspect to be so inclined, but who on the pages of *Criterion* supported Sir James Frazer as F.H. Bradley's successor to the Order of Merit. Essentially we can say in Eliot the primitive, in its challenge to the securities of nineteenth-century humanism, serves to reinforce the religious. Christ is a God whom we need to devour:

> The dripping blood our only drink,
> The bloody flesh our only food:
> In spite of which we like to think
> That we are sound, substantial flesh and blood —
> Again, in spite of that, we call this Friday good.[25]

The primitive terror, like the sea and the river, is all around us and is within us. Conventional humanism based on faiths in human perfectibility and progress breaks apart on these hard rocks of permanent human experience.

As Nietzsche's older colleague, Jacob Burckhardt, was fond of saying in his study of the Renaissance, "As usual our first witness is Dante," so, in relation to modernism, the same can be said of Nietzsche himself; that is, except when we come to the second phase of modernism, the development of the complex central consciousness. Normally the first philosopher of modernism, Nietzsche's role is here altered to that of an antagonist. In all other phases Nietzsche is prophetic; he is even partially formative in the second phase by virtue of his strenuous assault on the unconditional in the middle class ethic; he found it too unresponsive to the facts of life and the needs of man. But when he attacks the relativism of modern thought, its cosmopolitanism, the hybridization of modern society, the malaise and anxiety of modernity then Nietzsche is at odds with that creative second generation of modern writers. He finally thought complexity could be overcome by an act of the will, but it was their particular genius to abide within complexity, and to register diversity with patience and reflective intelligence.

Ecumenism is the hallmark of their social, racial, geographic backgrounds as it is of their works: Eliot an American in England; Auden an Englishman in America; D.H. Lawrence globe-trotting; Joyce in Paris, Trieste, Zurich, and back to Paris; Thomas Mann pursued by the fortunes of war from Zurich to the California coast, and back to Switzerland. The list goes on, and the litany of modern cities traversed or inhabited sounds like a passage from *The Waste Land*. While the travels of Proust seem confined to the narrow triangle, fictionally represented as Paris, Combray and Balbec, socially he is the son of a professional family risen to the level of the aristocracy, and on the one side of his family he is Jewish — a fact that adds another dimension and perspective to his work. The bourgeois and the artist, the closed German North and the open Catholic South represent great dichotomies in Mann's work. Such dilemmas lead Joyce to the greatest wanderer of all time, Ulysses, who informs the activities of a "jew-greek" in Catholic Dublin. Ecumenism as part of their personal histories naturally feeds the broader visions of their works, rescuing modernism from the largely national preoccupations of nineteenth-century fiction. As the disruption of historical values leads to an interpenetration of past and present, a psychic and historical co-temporality preparatory to myth, so too it leads to spatial fluidity, an international mixture that unsettles any national identification. The break-up of the world of the fathers implies a loss of the fatherland's commanding position. "Todo por la patria" is not the voice of modernism. In English literature the result was particularly salutary, and, to George Orwell's way of thinking, had a civilizing effect. Orwell wrote of Joseph Conrad that he was one of those non-English-born authors "who in the present century civilized English literature and brought it back into contact with Europe, from which it had almost been severed for a hundred years."[26] Conrad said that all Europe went into the making of Kurtz, and he meant it — spiritually as well as geographically. Thomas Mann could have said the same for his *Magic Mountain*: all Europe from the Caucasus to Spain (and even Mexico) extended its influence on the growing German soul of young Hans. Switzerland figures centrally as a receiving spot, a fluid environment for the modern soul in search of itself. As Hans, himself, somewhat innocently remarks, "The atmosphere up here is so international." Such dislocation is necessary in space as well as in time: in *The Waste Land*, Eliot's polyglot first section "Burial of the Dead" is international in setting.

This appreciation of a catholicity of background and of a wide range of experience receives its embodiment in the central characters of each of the classic works of modernism. In Joyce it became something of a critical principle of selection, determining his choice of the hero suited to overlay his modern epic. If the criterion is the variety of relationships into which one enters, then Ulysses is a more fitting hero than Faust or Christ or Hamlet. Ulysses was father, son, husband, lover, adventurer and home-seeker, the first gentleman of Europe and the inventor of the tank. But beyond the mere breadth of experience, Leopold Bloom — and Ulysses to a certain extent — joins with the other major characters, Marcel in the *Recherche*, Hans Castorp, Tiresias in *The Waste Land* in that they are all reflective, passive, selfless and tolerant witnesses. Their complex central consciousnesses constitute one of the major achievements of modernism, an achievement that is epoch-making in several ways. In these characters and their new modes of being and perceiving, we were given the form of the intellect for the next sixty years of our culture, the form of twentieth-century man. A new type emerged — "ein neuer Menschentypus" — in whom to this day we continue to see ourselves.[27]

This new type of man represents a psychic corollary to the radical alteration of historical values suggested in phase one. If the lengthened shadow of man is history, then man strides vigorously toward his future. He is assured in his ego and assertive in his will; he has "character" and can serve as a "character". The validity of these personal qualities emerged in the Renaissance, and formed the nucleus of values that grew out of man's newly-awakened sense of time. They are based on a fundamental recognition of life's precariousness. But if this true perception is made, man can marshall his resources and effect the critical choices necessary to his safety and success. The new sense of time saw life as struggle, and its later manifestations were called variously a Protestant ethic or, in America, the Puritan work-ethic. One can gauge then the extent and importance of time when a change in its perception effects so radical a change in basic character values. We must not make the mistake of J.F. Revel who, not finding in Proust the linear, destinal time of a Balzac, believed therefore that time was not important to his thought.

It is no accident that some of the best writing on modernism is devoted to Proust, and to what I have called the complex central consciousness. Ortega y Gasset gave another name to this new awareness when in 1916 he called it *perspectivismo*; he quite early saw the significance of modernism when he allied it with Einsteinian physics as being both repudiations of the Newtonian world-view of an absolute space and time. This view, according to Ortega, fails to take account of the fact that since there can be no absolute beholder there can be no absolute time; instead we have changing perspectives. Consequently, perspective itself must enter into the picture of reality, "La perspectiva es uno de los componentes de la realidad."[28] Following Orgega, two of the major early critics of modernism, Ernst Robert Curtius and Edmund Wilson, make similar assertions. To take only Wilson, in his *Axel's Castle*, the first large attempt to explain modernism to an older generation of naturalists, he writes of Proust (and he goes on to say the same thing about Joyce) that "he has recreated the world of the novel from the point of view of relativity: he has supplied for the first time in literature an equivalent on the full scale for the new theory of modern physics."[29]

Not only in the central consciousness, but in the total register of sensibility we find a new complexity at work. A single channel of experience had been exploded and man bears witness to what Ortega called "the profound, the terrible, the unforeseeable ... the best and the worst." A whole new range of material is thus able to be presented in modern art, forcing an alteration of what had formerly been thought to constitute poetic material. The image of the poet seems to be brought closer to ordinary experience. He is no Aeolian harp, no prophetic voice of any Cult, no competitor with the religious. As the bourgeois ethic lost "character," so literary aesthetics abandoned "personality." The new role of the poet is that of a medium, or an amalgamator of disparate experiences. Rather than a creator he is a compositor, and the experiences he brings together are not uniform, but as diverse as reading Spinoza, falling in love, the smell of cooking, or the noise of a typewriter[30] — these are the bits and pieces that go to make up the mosaic of modern art. The modern line is a hard-line, down-to-earth, sexually frank or ironic and also scientific. It regards with suspicion and discomfort any posturizing of the personality, any romantic egoism. The poet must not look into his heart and write, he must consult his "cerebral cortex, his nervous system and the digestive tracts." One is reminded of the Walpurgis-Nacht episode of *The Magic Mountain,* where Hans's scientific lexicon does little to detract from his ardor. In short poetry must be desentimentalized. This is as large an attack on nineteenth-century values as was that on temporal continuities in phase one. However much they might express disenchantment, Romantic poets in their imagery suggested an essential unity of experience. Goethe declared that "human nature knows itself one with the world." Mankind can be sure that the outer world is an "answering counterpoint to the sensations of its own inner world." For Wordsworth the "beauteous evening, calm and free" is like unto a "nun breathless with adoration." The human, the divine and the natural interpenetrate to suggest a sentiment of being in which man can find a spiritual peace. Such possibilities are precluded in modernism, where those spheres of existence are kept distinct — confirming the theories of T.E. Hulme. The human is not spiritualized, nor is the natural anthropomorphized. Contrasting with Wordsworth's sonnet are the opening lines of *Prufrock,* where "the evening is spread out against the sky, / Like a patient etherised upon a table." While this might seem far from the discussion of time in phase one, it is part of that very transformation, deriving from it, and applying the same rebelliousness to matters of consciousness and poetic theory.

Throughout his work, particularly in the essay, "Humanism and the Religious Attitude,"[31] T.E. Hulme makes plain that the period from the Renaissance to the nineteenth century was a single period, that it was bound together by common assumptions, assumptions so common that they were unchallenged. These unquestioned assumptions he calls pseudo-categories: they were unconsciously satisfying personal preferences that became laws. Although Hulme does not discuss it as such, one of the pseudo-categories that emerged from the Renaissance and triumphed in the nineteenth century was the sense of time and its attendant values of causality, continuity, and progress. Like other values it seemed absolute until subject to philosophical and historical inquiry, and only then did it assume its rightful status of relativity. By placing it in historical perspective we can trace the growth of this idea of time, see the reasons for its emergence as well as for its triumph, we can see the social as well as personal needs that it satisfied. By similar historical inquiry we can see why it failed, why it failed on the same level that it had triumphed, the level of personal need and satisfaction.

The confines of this paper do not permit any full discussion of the reasons why the older sense of time and its values were — if only temporarily — discarded. One explanation, and only one of several has to deal with the greatest paradox of time, namely, that the essential dynamic of the West can, by its very triumph produce a condition that is far from dynamic, that is in fact routinized and standardized. The ever-lasting soup of *The Magic Mountain* or the taking of toast and tea in *Prufrock* has a way of reducing life to nothingness, and in essence reproducing a kind of alienation that prevailed for theological reasons in the middle ages. Historical continuity can lead to a kind of sameness, stressed in the ritual repetitions of *Buddenbrooks*. Progress then is only gloomy, and science itself contributes to a "nightmare of determinism." In this sense then modernism, far from being a remedievalization of the West, is actually a movement intent on saving things from a process of standardization that reduced all to nothingness. In its attempt to salvage distinctiveness from nothingness, modernism continues its Renaissance heritage. But in order for this to be done a complex process had to be initiated whereby modernists were obliged to repudiate and attack the obsolescence of that very ethic. The first phase of modernism was then, one of removal and retrieval, in T.E. Hulme's phrase, "the closing of all the roads." By a strange reversal, however, this interdiction of the future can lead, as Ortega and Eliot have attested, to new openings of experience. The development of the complex central consciousness cultivated multiperspectives, and, freed from any channelising unity, was able to regard objects in all the freshness and freedom of their separate identities. If we were to terminate there, however, the picture would be incomplete. All of the modernists were masters of architectural design; they tried to see wholes and patterns of development; they welcomed associations with grander figures from the past and mythic interpenetrations of past and present. But this line of thought would bring us to the third stage of time and modernism.

Suffice it to say that each of these phases, while separate in itself, derives from the preceding and prepares what is to follow. Nevertheless, the second phase is critical and determining because those writers or those works that seem to proceed from the collapse of historical values to the mythic without the intercession of the second phase are doomed to failure. Ortega explains this by his concept of "the height of the times."[32] There is an intellectual level from which one cannot recede, but only pass through if one is to transcend it. The level of our times, according to Ortega, involves liberal democracy based on scientific knowledge. The second phase of modernism here described would be the literary equivalent of the historical level of the times. It does not seek to redress complexity by any act of the will, by any pact with the diabolic. "Teach me to sit still," is the prayer of this second phase as it is of Eliot in *Ash Wednesday*. This does not mean that grace is not hoped for, nor the illumination of some epiphany, nor the recognitions of mythic enhancement, or the blessings of *moments bienheureux*. But the point is that for all of these great writers, when they move from the everyday, and their accustomed rounds of natural experience and reason, to these far greater experiences, they do so involuntarily. In fact, for Proust, the authenticity of a great moment is manifest in its involuntariness, by the fact that we did not force it, or compel it. Hence, the importance of the virtues developed in their works of passivity, reflectiveness, selflessness and tolerance. This may have been because in the twentieth century the voices of temptation were calls to action:

> The temptation to entrust oneself to a sect which solves all problems with a
> single formula, whose power of suggestion imposed solidarity, and which
> ocstracized everything which would not fit in and submit — this temptation
> was so great that, with many people, fascism hardly had to employ force when
> the time came for it to spread through the countries of old European culture ...[33]

I close with this quotation from Erich Auerbach in order to support my contention that in the finest writers of modernism, those who underwent the natural phases of its development, an inherent contradiction exists (whatever the temptations and even flirtations) between it and the willed and simple solutions so appealing to fascism. This conclusion brings us to an even more final and fitting one; as it was in the Renaissance, so it is in the twentieth century: to address its sense of time is to speak to the profoundest issues of a culture and of a society.

NOTES

1. (Ithaca, N.Y.: Cornell University Press 1968), p. 11.

2. (Cambridge, Mass.: Harvard University Press 1972).

3. (Cambridge, Mass.: Harvard University Press 1966).

4. In addition to my study and its supporting bibliography, here is a recent, forceful statement by Jacques LeGoff: "Perhaps the most important way the urban bourgeoisie spread its culture was the revolution it effected in the mental categories of medieval man. The most spectacular of these revolutions, without a doubt, was the one that concerned the concept and measurement of time." "The Town as an Agent of Civilization," *The Fontana Economic History of Europe: The Middle Ages*, ed. Carlo Cipolla (London: Collins 1972), p. 86.

5. This double face of time can be called, on the one side, predictive, and on the other, innovative. "The sheer need for a more precise control of predictions thus might have helped to give birth to the notion of time as such." (Quoted from the abstract of Masanao Toda's paper at this conference of the International Society for the Study of Time; the sentence did not occur in the actual delivery.) The other side is given by R. Schlegel, "In a human culture in which there is active development of new knowledge and new ways of living there is perhaps concomitantly an emphasis on history and the role of time." "Time and Entropy," *Time in Science and Philosophy*, ed. Jiri Zeman (Prague: Academia 1971), p. 29.

6. *Buddenbrooks*, Part III, chapter 10, translated by H.T. Lowe-Porter.

7. Part X, chapter 5.

8. Part X, chapter 5.

9. *Time, Place and Idea: Essays on the Novel* (Carbondale, Ill.: Southern Illinois University Press 1968), p. 43.

10. *Revolt of the Masses* [trans. anonymous], (New York: Norton 1932), pp. 32-33.

11. In *The Birth of Tragedy and the Genealogy of Morals*, trans. Francis Golffing (Garden City, N.Y.: Doubleday - Anchor 1956), p. 111.

12. In *"Death in Venice" and Seven Other Stories*, trans. H.T. Lowe-Porter (New York: Vintage 1957), p. 18.

13. In the essay, "The making of *The Magic Mountain*," rpt. as an appendix to *The Magic Mountain*, trans. H.T. Lowe-Porter (New York: Vintage 1969).

14. Chapter VI, "A New-Comer."

15. Chapter IV, "Excursus on the Sense of Time."

16.   *Ulysses* (New York: The Modern Library 1934), p. 35.

17.   Viking Press reprint 1960, p. 327.

18.   *The Social History of Art* (New York: Vintage n.d.), IV, 238.

19.   (Cambridge, Mass.: Harvard University Press 1972), p. 139.

20.   See John B. Vickery: *The Literary Impact of "The Golden Bough"* (Princeton, N.J.: Princeton University Press 1973).

21.   "London Letter," *The Dial*, Sept. 1921. Quoted by A. Walton Litz: *"The Waste Land*: Fifty Years After," in *Eliot in His Time*, ed. Litz (Princeton, N.J.: Princeton University Press 1973).

22.   *The Magic Mountain*, chapter VI, "Snow."

23.   See also the poem, "Parnell's Funeral" from *A Full Moon in March*, in *The Collected Poems* (New York: Macmillan 1951), p. 275.

24.   From "Gerontion," in *The Complete Poems and Plays* (New York: Harcourt, Brace 1952), p. 21.

25.   From "East Coker," in *Four Quartets, ibid.*, p. 128.

26.   *The Collected Essays, Journalism and Letters of George Orwell*, edited by Sonia Orwell and Ian Angus (New York: Harcourt, Brace and World 1968), IV, 489-90.

27.   The phrase is taken from E.R. Curtius: "T.S. Eliot: 1", in *Kritische Essays zur europaischen Literatur*, second edition (Bern: Francke 1950), p. 326.

28.   "Perspective is one of the component parts of reality," from "The Doctrine of the Point of View," *The Modern Theme*, trans. James Cleugh (New York: Harper Torchbook 1961), p. 90.

29.   *Axel's Castle* (New York: Scribner's 1931), p. 189. For Joyce see pp. 221-22: Eliot himself translated Charles Mauron's "On Reading Einstein," for *The Criterion*, X (1930), 27-8. See the valuable article by C.A. Patrides: "The Renascence of the Renaissance: T.S. Eliot and the Pattern of Time," *Michigan Quarterly Review*, XII (1973), pp. 172-96.

30.   From Eliot's famous essay, "The Metaphysical Poets," in *Selected Essays* (New York: Harcourt, Brace 1950), p. 247.

31.   In *Speculations*, ed. Herbert Read (New York: Harcourt, Brace 1924), p. 1, from which the quotation from Goethe is taken, p. 54.

32.   *The Revolt of the Masses*, p. 94.

33.   *Mimesis*, trans. Willard Trask (Garden City, N.Y.: Doubleday - Anchor 1953), p. 486.

V. MUSIC

The Structure of Time in Music: Traditional and Contemporary Ramifications and Consequences

G. ROCHBERG

I

If we say that "time is an essential feature of the universe,"[1] presumably we mean not only that time is built into the very fabric of the physical reality of the universe but that it also exists apart from man. But, since man is himself part of the fabric of that same physical reality, there is ultimately no way to separate his consciousness from it. He is inextricably bound up in it; his perception and consciousness of the universe are mental reflections and refractions of it. Because temporal order in music is an ordering or *rhythmatization* of something other than itself – sound vibrations which have their own interacting laws of autonomous physical structure and artistic musical combination, music provides us with a cosmic metaphor in which we may examine more closely and directly the properties of duration, continuity, and the direction of events, as well as their relation to repetition and recurrence of event patterns and their connections with the functions of perception and memory.

Music, of all the arts of man, enjoys that peculiar position of intermediary link or interface between the physical universe and ourselves because music endows physical vibrations, i.e., objective, measurable properties of sound, with human form. The special feature of that human form is the creation of temporal order without which sound qua sound cannot be raised to the level of music. Seen in this way, music becomes that unique aesthetic form which arises spontaneously in every culture from the action of an autonomous, deeply intuitive sense of time (implanted in man's mental makeup) on a fundamental reality of the physical universe. Man creates in music a palpable, perceptual metaphor of the universe in a passionately colored succession of expressive events which become a transformation of the raw data of physical reality in and of themselves. Man's rhythmic energies, rooted in his physiological and neural constitution, are inextricably bound up with his ability to create time in music.

II

Even the most cursory examination of the history and theory of Western art music must reveal the curious fact that the study of the structure of time in music – or to use the simpler terminology of the musician, the study of rhythm, tempo and motion – has received scant attention compared with the conscious care and concern lavished on problems of sound in the form of pitch and its possible combinations in the construction of melody, harmony and polyphony. One would have thought that musicians, who became literally *artificers of time* practicing the temporal art *par excellence*, would certainly have concerned themselves consciously and systematically with the devices and practices by means of which they created and regulated the flow of time through sounding forms. Yet this has not been the case, generally speaking, as the history of music of the past three centuries, particularly, shows. Yet, over this same period of 300 years, composers developed the articulations of temporal structure to remarkable degrees of subtlety and imaginative skill in matters affecting phrase lengths, rhythmic combinations and the architectonics of musical form and structural proportions. Those 19th century composers who came after Beethoven especially accomplished brilliant feats of temporal ordering, became master artificers of time; but their achievements were neither matched nor accompanied by rational/theoretical considerations of the problems they faced or the solutions they arrived at.

One must assume, therefore, that with two notable exceptions, the approach to the handling of time in music has been and remains largely intuitive rather than rational. However, these special cases exhibit all the characteristics of a rationally ordered approach to time in music but from radically different directions and for totally different reasons. They are the periods in the history of music when the intuitive necessarily had to give way to the systematization of rhythmic modes in order for music to develop into an art form in the first place and to extricate itself from what appeared to be a dead-end in the second place by arbitrary, rationally determined systems or principles of order. I am referring, first, to the medieval period, extending from about the middle of the 13th century when the establishment of fixed rhythmic modes, based on what musicologists claim to be garbled versions of Greek poetic foot lengths, resolved problems of notating two or more simultaneous parts to the more subtle and flexible mensural notation of the 14th century and beyond; and second, to the 20th century, when, after the crucial break with tonal practices, developed from the end of the Renaissance through the 19th century, composers once again took up a more self-conscious, rational-minded approach to their problems in order to find new ways of creating and controlling time in music.

It is of fundamental significance to the ideas I wish to advance in this paper that, except for these two not insignificant periods, the first of which gave birth to the basic formulations of the very language of Western music, the second of which ushered in a radical break with that same language at the point of its maximum development and maturation, composers essentially behaved intuitively toward the handling of time, accepting as either convention or tradition or both, the stylistic practices and devices of rhythmic motion of their particular day. Admittedly, these practices and devices changed as the styles of music themselves changed — but by no means radically or abruptly. It would appear then, that especially between 1600 and 1900, the three centuries which saw music evolve toward an increasing command, subtlety and flexibility of melodic, harmonic and polyphonic expression and the concomitant expansion of temporal structures, climaxing in the 70-80 minute symphonies of Bruckner and Mahler and the two to three hour (and sometimes longer) operas of Wagner and Verdi, composers struggled with and overcame problems of giving adequate size and shape to their increasingly intense and expressive intentions. As far as theory was concerned, however, it lagged behind in the academic doldrums of the same old harmonic/polyphonic considerations. The development of harmony itself, however, cannot be divorced from the development of temporal structure; and, not surprisingly, in a symbiotic relationship such as exists between pitch and rhythm, what happens to the one must necessarily affect the other. As it turns out, the theory of music of the Classical and Romantic eras remains woefully deficient in dealing with the conditions flowing out of this symbiotic relation. The failure to deal with this very problem arises from the fact that music theory is obsessed with the specificities of pitch relations. The vast majority of theorists treat such relations as static ones unrelated to temporal motion. A further difficulty arises from the fact that temporal motion, though entirely specific as to notation, becomes meaningful only as aggregate time structures which become increasingly resistant to analysis by the means traditionally at theorists' disposal.

It seems logical to infer, then, that when a musical language is being consciously formulated or a new one consciously invented, rational involvement with all aspects of the craft and syntax of that potential language becomes mandatory. From the fact that an intuitive grasp of that craft and syntax is slow to develop and mature it also follows that composition itself is slowed down. Donald Tovey has remarked: "It is very seldom that a composer or writer can simultaneously concentrate his energies on the working out of a new language and on the construction of big designs."[2] In the case of early medieval music, composers had first to find a clear means by which to fix and establish the relationship between two or more parts. Otherwise, it would not have been possible to regulate the flow of harmony of two or more layers of pitch motion. When such regulation was suspended or became nonexistent, centuries later, as in certain varieties of 20th century music, disorder and chaos resulted. But medieval composers were trying to create order by combating potential disorder, not making disorder part of their aesthetic arsenal. The suspension or non-existence of regulated flow, resulting in either chaos or a new-won freedom (depending on who is perceiving the phenomenon and from what point of view), became only one aspect of 20th century music as composers, releasing themselves from traditional melodic/harmonic flow, sought out new means of pitch organization and motion and were thus led into the necessity of investigating new possibilities of creating time in music. To what extent they have succeeded or failed is a major concern of my paper.

While it is undeniable that the medieval and contemporary periods share a rational approach to musical time, it needs to be pointed out that there is a vast difference between the two rational approaches. On the one hand, we see the first application of rhythmic modes to pitch in the medieval period, and on the other, the application of rationally based rhythmic and non-rhythmic schemata to new pitch possibilities in recent contemporary developments. It is not known for certain how pitch and rhythm finally came together in the dawning of polyphonic music. Donald Grout points out, "It was obviously necessary to find some way of showing rhythm and duration by the shapes of notes themselves. Strange to say, a complete solution to this difficulty was not achieved until after the middle of the 13th century, when a simple set of signs was evolved for the notation of rhythm. In the meantime, 11th and 12th century composers and singers of polyphonic music devised a system in which the prevailing rhythmic pattern of a melody called the 'rhythmic mode', was specified by certain conventional combinations of notes and note-groups."[3] It can be reasonably conjectured, therefore, that the use of rhythmic modes helped to establish a musical practice and style of composition and performance which arose, if not spontaneously, certainly intuitively. Grout makes clear that the rational organization of rhythmic pattern followed after at least 200 years or more of slow emergence and development. This separates the rationalist character of the medieval period definitively and distinctly from that of the 20th century; the medieval being distinguished by slow, intuitive root growth, the contemporary by an arbitrary rationalism born in a cataclysmic upheaval whose effects are still present.

III

Having gone to some lengths to establish the notion of a basic intuitive approach to the creation and regulation of time in music, I am now faced with the necessity of trying to locate, i.e., to discover, the source of this intuitive capacity and to demonstrate the link between the rhythmic forces and energies bound up in this fundamental capacity and the flow of sound events as it emerges from the depths of the mental/perceptual structure of consciousness. As an absolutely unavoidable corollary, I am also faced with the necessity of trying to make clear how it happens that an intuitive temporal sense, minimally guided by the rational via a practical, working system of conventional notation for marking off beats and groups of beats controlling the flow of events articulated in phrases and their accumulation into larger and larger groups eventuating in precise and clearly ordered macro-structures, produces immediately graspable, clearly perceptible body-rhythms or pulsations; while, in the case of rationally based contemporary systems and/or schemata which suppress or circumvent the physiologically-neurologically based intuition of time, we observe the introduction of a spectrum of perceptual difficulty and lack of graspability ranging from the merely vague or uncertain to the disordered and chaotic, not only affecting the perception of beats and/or groups of beats – if indeed they are present at all – but also stamping the macro-structural flow of events with the same characteristics.

It must be noted here we are dealing with different languages of pitch combination and construction. Directly associated with the intuitive temporal sense is the diatonically based language of key-centered tonality and its melodic/harmonic constituents which emerged from a long process of development and maturation. Directly associated with the contemporary rational temporal sense, the product of a cultural break that occurred in the first decade of this century, is the chromatically and/or microtonally based language of non-key centered atonality and its range of arbitrarily systematized pitch formulations, commonly known as 12-tone music or serialism, to unsystematized pitch happenings resulting from what can only be called "game plans" based on chance or choice, otherwise known as aleatoric music – happenings which may or may not include noise elements or noise resultants. The application of electronics to either side of this atonal (or non-tonal) spectrum of composition alters in no way the nature of the perceptual results I have outlined.

I wish to introduce here the criteria by which the conditions I have described in a general way may be recognized. Since I have made clear that, in my view, the repertoire of traditional music derives primarily from the exercise of the intuitive sense of time and that, in the case of the so called advanced or experimental music of the 20th century, the intuitive is usurped by efforts to establish a more rationally based system of time controls, these criteria will necessarily reveal positive and negative faces with respect to the conditions of aural perception depending on which way they are turned.

These criteria flow from the kinds of relationships which connect music with *identity*[4] and, in turn, *identity* with memory. I am defining "identity" here as perceptual impression or gestalt, an organic totality, which can be remembered in all its constituent details as well as overall design; i.e., committed to memory, or recognized on repetition as essentially the same gestalt previously experienced whether with large or small variations in the mode of projection, a common experience in the performance of music where the same work is subject to variant interpretations affecting tempo and/or dynamics, etc.

The properties of identity[5] are the profile or shape of motivic or melodic ideas, the direction of harmonic progression and repetition or recurrence of melodic and/or harmonic ideas in the identical form as first stated, or variant structures retaining essential features of the initial statement or statements. These properties refer primarily to the shape and form given to pitch through the action of time articulated as rhythmic periodicities. Such periodicities constitute the particular ways in which pulsation of beats and their organization into larger and larger groups affect phrase lengths, and ultimately, the extension and accumulation of phrasal, sectional and movemental structure. In every case memory is potentially activated by the properties of identity. The degree to which these properties are well-defined and artistically well-wrought will affect their relationship to memory. Conceivably, one can move from a condition of absolute identity to vague or uncertain identity to absolute non-identity with their corresponding degrees of absolute memorability to difficult memorability to non-memorability. Thus, we can see that music as a creation of temporally ordered events must achieve clarity, precision and definition on all levels pertaining to the articulation of its structural flow, if it is to possess identity and arrive at memorability.

IV

I should like now to trace out or to suggest the location of the source of the intuitive capacity for creating temporal order in music and also to attempt to show in what way or ways the criteria of identity and memory are related to that source. In order to do so, I intend to develop a series of speculations and conjectures suggested to me by a reading of John von Neumann's book, *The Computer and the Brain.*[6]

In the language of computer technology as described by von Neumann, "serial" and "parallel" serve to define successive and simultaneous operations. These two operations or functions are, not surprisingly (since the brain is the model for the computer), comparable to similar operations or functions of the central nervous system, with the important difference, however, that "large and efficient natural automata are likely to be highly *parallel,* while large and efficient artificial automata will tend to be less so, and rather *serial.*"[7] If parallel operations are multiple serial operations occurring simultaneously, we not only have our first real clue to the location of the source of the intuitive time sense, we also have a potential paradigm or model for music itself.

It takes little effort to translate the serial and parallel functions of the CNS into the characteristic functions of pitch combination and temporal flow. They match up perfectly and with extraordinary ease. Under the serial function which governs successive events on a single plane, we can subsume both melody and its metric/rhythmic structure. Under the parallel function which governs simultaneously successive events on multiple planes, we can subsume harmony as a complex of multi-voiced melodic/rhythmic structures. Combinations of serial and parallel functions also arise when, for example, we accompany a melody with harmonic patterns organized on the basis of a different use and disposition of the same metric forces which guide the melody, or when a harmonic progression in three or more parts moves along the same time axis, or when we pit different tempos or speeds against each other, etc.. We grasp the essence of a serial or parallel structure directly and intuitively whether in the act of creation, in the act of performance, or in the act of perception. As far as the serial/parallel functions — printed as it were on the CNS — are concerned, it makes no difference which of these acts we engage in; they are all equally available to us through music. It is conceivable therefore, that because the logic of melodic structure based on the metric/rhythmic positioning of pitch is absolutely essential on at least one clearly perceivable level in order for our nervous system to be able to organize a continuous stream of parallel functions related to that melody, the *sine qua non* of musical organization is the serial mode which establishes successive relations and connections between the individually synchronous aspects of parallel structures occurring simultaneously. In other words, the serial function is absolutely basic and primary in the organization of temporal flow — even for parallel structures if music is to have clarity of design and clarity of direction.[8]

Not only is the CNS the probable source of the intuitive temporal sense, it is also the probable source of memory functions. Von Neumann admits that "the presence of a memory — or, not improbably, of several memories — within the nervous system is a matter of surmise and postulation, but one that all our experience with artificial computing automata suggests and confirms." He considers "all physical assertions about the nature, embodiments and location of this sub-assembly, or sub-assemblies, are equally hypothetical."[9] Not only do we not know "where in the physically viewed nervous system a memory resides; we do not know whether it is a separate organ or a collection of specific parts of other already known organs, etc.."[10] Von Neumann suggests that the memory function "may well be residing in a system of specific nerves"[11] or "have something to do with the genetic mechanism of the body."[12] As he affirms, "the only thing we know is that it must be a rather large-capacity memory, and that it is hard to see how a complicated automaton like the human nervous system could do without one."[13] Among the possible embodiments of memory functions von Neumann describes are: nerve cell connections which relate to variability of stimulation criteria and involve past history of individual cells or connections (axons) with other cells; genetic memory in which "chromosomes and their state effect, and to a certain extent determine, the functioning of the entire system";[14] self-perpetuating traits of chemical composition of certain areas of the body which may be possible memory elements; and nerve cell systems "which stimulate each other in various possible cyclical ways."[15] Obviously, musical structures in themselves cannot physically constitute memory functions as such. If they could, music could remember itself. Music cannot remember itself physically in the way the CNS can, but there are ways in which a memory analog may be said to function in music. In the same way that a memory bank is installed in a computer, the composer installs a kind of memory bank in music through various processes of repetition, recurrence, consistency of motivic and rhythmic elements, variant forms of association, etc.. In almost every piece of music since the time of Gregorian chant codification to at least 1950, we find structural devices and patterns whose fundamental purpose is self-perpetuation. These forms of musical self-perpetuation, like von Neumann's genetic memory-bearing chromosomes or self-identifying, chemically stable memory elements in certain areas of the body, are absolutely essential to the establishment of the identity of a work, whether on implicit or explicit levels — and, as I believe, actually on both simultaneously. Lacking such identity, a work could hardly be said to exist as an organic entity, which is to say, it cannot be described as art. At the very least, a work of art is an identifiable organic structure. Whether it is a good or bad work has to do, I believe, with its profile of identity. If the delineation of that profile is very high, we will undoubtedly discover self-perpetuation operating on every significant level. If it is deficient and low, we will undoubtedly find that the urge toward self-perpetuation is either minimally present, or poorly organized or both. In determining whether a piece of music is "good", "mediocre", or "poor", these are considerations which must be included.

What are the devices for structural self-perpetuation at the disposal of the composer? They fall essentially into two groups: one which operates at the level of pitch and/or rhythmic motives, where analogically speaking, primal cellular identity must be located through a process of organic saturation; the other which operates at the level of large-scale form which grows out of the process of organic saturation and achieves complex, cumulative shapes and gestures which must also emerge as perceivable identities. This process which occurs through the self-perpetuation of musical cellular structures takes place on both levels simultaneously through repetition, recurrence and variant associations which, then, literally saturate the serial/parallel functions and provide the whole with an organic sense of identity.

Conceivably, one can also apply the notions of redundancy and information to the process of achieving identity through self-perpetuating functions. For example, a high level of redundancy and a low one of information are essential to the memorability of a musical work. Proceeding from this level through stages approaching zero redundancy, we can see diminishing degrees of identity and therefore memorability. Tentatively, it is possible to conclude therefore, that where serial and/or parallel functions in music are structured along lines of self-perpetuating forms of identity which exist at higher levels of redundancy, they will correspond at some level (or levels) to the memory function of the CNS. Where this correspondence shades into areas of decreasing profiles of identity and decreasing levels of redundancy, the potential relationships to memory begin to attenuate. And should the serial/ parallel functions of music not be structured on the principle of self-perpetuating forms of identity and possess both too high a level of information and too low a level of redundancy, then it can be said that the resultant structure, whatever its nature, will probably deviate significantly from the memory functions of the CNS.

The application of these speculative, conjectural probabilities to traditional or traditionally conceived music reveals an almost complete correspondence with every significant factor and level of this discussion. A similar application of these same probabilities to experimental or advanced music of the 20th century, particularly music since 1950, reveals difficulties all along the potential lines of correspondence. This is not the place to enter into value judgments of contemporary music, but I feel it can be said with some justice that the difficulties contemporary music (of the 12 -tone/serialism persuasion as well as that of aleatory) have encountered, always granting the possibilities of exceptions, are due not to the dynamics and problems of acculturation (as apologists frequently claim is the case), but to a far more basic and crucial problem which arises from the fact that such aesthetic attitudes and compositional procedures — however they may be rationalized or justified in the light of present world cultural conditions — do not have their roots in the functions of the CNS and therefore have failed to enter into an easy, natural relation and correspondence with the CNS.[16]

Memory is the way we, as human beings, know and can be assured of the existence of time. While everything that happens occurs in time, not everything that occurs in time is or can be remembered. Music is not only the creation and concretization of time, it is also the creation and concretization of memorable temporal events. The bond between musical events and memory is deep and thoroughgoing and can only be found in the central nervous system. That bond is both the creative and perceptual source of this profound relationship. In its own way, music provides a durational model of past, present and future through the experience of events which not only return but whose return is anticipated. Only that music escapes the net of time which does not or cannot recall itself or develop anticipation of recall.

V

At the root of musical time — its creation and regulation — are two unmistakable conditions which determine its essentially goal-oriented nature: *directionality*, which is produced by the relations between pulsations and periodicities which move in only one way — toward the goal of structurally organic completion of the rhythmic forces and energies released in the music; and *causality*, which may be seen not only in the ways in which pitch elements affect each other through the response of melodic consequents to melodic antecedents, the striving of harmonic progressions toward points of cadence and rest and new beginnings or, for that matter, endings and the relations polyphony creates through the uses of contrapuntal devices among two or more independent moving parts; but also in ways which are equally evident in the tendency of metric/rhythmic forces to accumulate and drive to climactic points in phrases, sections and movements, and thereby release new momentum or fall away to make approaches to cadential points of temporary rest from which new beginnings may arise. The goal-oriented structure of time in traditional music starts from the smallest units of pulsation and, gradually through a multiplication of the effects of accretion and cause-and-effect relationships, builds a temporal organism which not only affects us physically by linking itself to our bodily physiology, but also by engaging us emotionally and mentally in the actions that time produces on and through pitch complexes. Comparing traditional music — which he calls "teleological" — to avant garde music, Leonard B. Meyer reminds us that, despite obvious differences in structure and pattern as exists in the music of, say, Bach or Haydn, Wagner or Bartok, "such music is perceived as having a purposeful direction and goal ... . But the music of the avant-garde directs us toward no points of culmination — establishes no goals toward which to move ... . It is simply *there*. And this is the way it is supposed to be. Such directionless, unkinetic art, whether carefully contrived or created by chance, I shall call *anti-teleological* ...."[17] There are numberless statements by avant-garde composers which corroborate Meyer's view of the situation — but not his critical attitude. One such statement will suffice: "The music has a static character. It goes in no particular direction. There is no necessary concern with time as a measure of distance from a point in the past to a point in the future ... . It is not a question of getting anywhere, or making progress, or having come from anywhere in particular ... ."[18] Meyer sums up his position toward anti-teleological music in these terms: " ... underlying this new aesthetic is a conception of man and the universe which is almost the opposite of the view that has dominated Western thought since its beginnings."[19] The juxtaposition of the two views could not be more clear. On the one hand, teleology, exercising the role of purposeful, directed tendencies moving along an axis of continuity and internally meaningful relations between cause and effect which serve to corroborate the unidimensional flow of events, becomes the tacit premise on which traditional music built up its power and effectiveness; on the other hand, anti-teleology, exercising the role of aborting, frustrating or circumventing direction and continuity, purpose and internal causal relationships is established as the avowed premise on which aleatoric music rests. Once again the presence of the intuitive or the rational may be clearly traced out in these opposite aesthetics and implied conceptions of man and his relation to the universe. It is interesting to note that avant-garde composers are much drawn to philosophies which prefer *stasis* to *kinesis*, and call on the implications of quantum mechanics, which revealed the potentially indeterminate nature of the micro-world, for support of their aesthetic of macro-cosmic indeterminacy.

If it were simply a matter of having to choose between the teleological and the anti-teleological view, personal preference could decide the case. However, it is not that simple, for the very reason that the anti-teleological approach to composition has brought out and emphasized an aspect which broadens my whole enquiry into the nature of the structure of time in music and music itself as an artistic representation of the modes by which man knows the universe. I refer to the curious phenomenon of *music as space*.

If the temporal nature of music has received scant attention from composers and theorists, the spatial nature of music has received even less — probably because the idea that music could represent aspects of spatiality did not occur to anyone until after the premises of traditional music came into serious question and were replaced by a different set largely adhering to anti-teleological views. While the case for time in music, difficult as it is, is susceptible to various approaches, including the one I have been making, the case for a spatial dimension in music is refractory in the extreme. Nevertheless, in view of the conscious efforts to suspend teleology in contemporary composition, it is imperative to understand the nature of the results which have emerged from this negative attitude toward the structure of time in music.

We must ask: what happens to music if continuity and directionality, pulsation and periodicity are suppressed or subverted or circumvented in favor of discontinuity and non-directionality, and the suspension of pulsation and periodicity? The answer is two-fold: first, the regulated flow of time is replaced by mere duration which now assumes a passive, non-dynamic, non-kinetic character; and second, musical sound loses its connection with melodic/harmonic motion and is, therefore, transformed into potential mass or density of vibratory forms which exist in and for themselves. Since pitch is that aspect of music whose source is the physical universe and cannot move by itself, but depends entirely on being moved by the action of the intuitive time sense, i.e., on being rhythmatized, pitch assumes new properties and presumably, a new independence. When structured time acts on tone, motion or the illusion of motion is created. When structured time is divorced from tone, the illusion of space is created.

Unfortunately, the limitations I have necessarily imposed on my discussion prevent an inquiry into the questions my last remarks raise. Whether motion in music is real or illusory, whether space in music is real or illusory are beyond the scope of this paper. However, for the sake of penetrating the problem of spatiality in music which the new aesthetics has raised, I am making the necessary assumption that the results of this new view of music have approached a condition of stasis, and, therefore, spatiality by removing themselves from traditional usages of temporal structure.

The most significant effect of these radical departures from tradition is discerned in the nature of the musical discourse itself. Traditional processes of thought had assumed tonality as the basis of melodic/harmonic organization and periodicity as the regulators of metric organization and rhythmic flow. When these fundamental direction-producing forces were replaced by freely chromatic atonality and an essentially non-periodic rhythmic structure, a musical discourse based on readily predictable continuity was no longer possible. With the advent of atonality, melodic motion, lacking the organizing factor of a tonal center, became almost completely disjunct. As a result, melodic shape or contour tended to become less predictable, its direction less certain, its topological formation showing the strain of distortion and stretching. Because of this attenuation of shape, musical discourse, though it might be precisely ordered through the new procedures and mechanics of the 12 -tone method of serialism, stood in danger of perceptual disorder. With the advent of total serialization, this tendency was reinforced. While the stream of events in totally serialized works may be continuous in the sense that sound is always in motion, the discourse seems to have lost its sense of direction, resulting in a kind of unplanned indeterminacy in which the position and motion of any given pitch element become matters of considerable perceptual uncertainty. Planned indeterminacy in the guise of aleatory, on the other hand, seeks to be utterly discontinuous and perceptually unpredictable in order, according to the pseudo-philosophy of its adherents, to preserve the spontaneity of the pure moment divorced from any past or any future, the absolute freedom and freshness of the living situation. In this kind of music, sounds emerge from silence only to sink back again. This view is considered to approximate the real conditions of "life" as human beings experience it. It is unequivocally described as something which "is always starting and stopping, rising and falling ... so listening to this music one takes as a springboard the first sound that comes along: the first something springs up into nothing and out of that nothing arises the next something, etc. ... like an alternating current."[21]

Thus we see that the new autonomy granted the sound material is largely due to its liberation from the process of establishing traditionally causal connections between melodic phrase shapes and the harmonic progressions which support them, all of which, in turn, derive their energy and movement from structured time. This liberation permits sound to create its own context apart from the forces of periodic time, a reversal of the traditional procedure. The widely held view of sounds as objective, concrete, quantitative entities in themselves is revealed in the terminology which has grown up around it: densities, vertical pitch aggregates, sound objects, sound structures, etc.. The non-periodic motion of these new sound constructs requires that it not interfere with their projection but rather enhance it; i.e., the duration assigned them must give them full and ample opportunity to reveal their purely vibratory characteristics. Inevitably, the suppression of pulsation and periodicity radically affects the perception of time in music. One of its more obvious results is to slow down the passage of events, sometimes to the point of near immobility; and even when volleys of rapid projections of sound tend to increase the speed of the passage of events, the perceptual sense of the motion remains essentially non-dynamic. It follows, then, that musical discourse in the traditional sense is out of the question and is replaced by a new phenomenon characterized by discontinuity and unpredictability.

The approach to spatiality, however, was taken consciously and directly by Edgard Varèse. One might almost say that he invented spatial music. Certainly he saw more clearly than anyone else the possibilities in a situation which appeared confused and confusing. Though it does not give up the identifiable motive, Varèse's music relies heavily on densities or aggregates organized by the imaginative control of timbre primarily. These densities gather themselves up into asymmetric spatial configurations by a process in which entry and attack determine rhythmic position and motion. Varèse confirms this when he says of one of his works, *Intégrales,* that it

> ... was conceived as a spatial projection ...... constructed according to certain acoustical principles which had not existed previously, but which I knew, could be realized and made use of ... . While in our musical system we divide quantities whose values are fixed, in the realization which I sought, the values would be continuously changed in relation to a common factor. In other words, it would be like a series of variations, the changes resulting from slight alterations in the form of a function or the transformation of one function into another. In order to make my meaning clear — for the eye is quicker and more disciplined than the ear — let us transfer this conception into the visual field and consider the shifting projection of a geometric figure on a ground with the figure and ground both moving in space, but each with its own individual speeds and varied according to position and rotation. The immediate form of the projection is determined by the relative position between the figure and the ground at this moment. But in permitting the figure and ground to have their own movement one is capable of representing with the projection a highly complex image seemingly unpredictable. In addition, these qualities can be further developed by letting the forms of the geometric figures vary as well as the speeds ... .[22]

The release of sound from tonal functions accomplished by atonality, the gradual suppression of beat and pulsation which parallels the evolution of atonality and its total systematization, the resultant emergence of the unpredictable and the discontinuous — these are the paths which led to the overthrow of a long-dominant temporal structure and the spatialization of music where the sound substance is formed as the primary object of projection and perception, its motion entirely secondary and contingent on the emerging structure of the sounding forms themselves. The motion of this music — it must be emphasized that we are dealing with "musics" which cover a wide range of points of departure and compositional procedures — tends either toward a state of equilibrium or toward erratic gestures and rapidly changing shapes and speeds. The temporal aspect, long dominant, must give way to the exigencies of producing these new sound forms.

I raise once again the analogy of music to the CNS with respect to the serial function which we saw operating in my sketch of the structure of traditional music. It is apparent that, if the serial function which operates along the time axis to produce direction and continuity has no place — or at best an uncertain or diminished place — in the forms of 20th century music I have been describing, we are left with only the shell of parallel function and the question of its operative or non-operative relation to a spatially oriented music. It could be argued that the remaining shell of parallel function of music is precisely that which permits many tones to sound together in some kind of durational succession and still remain comprehensible — and, if not comprehensible, then certainly perceivable, though not always with the means of precise identification. Theoretically, I would agree that this is so; and if it is, then spatiality in music has some potential ground in the CNS. But unfortunately, the factor of memory, which we cannot turn on and off at will, enters into all human experience; and, particularly in music, where we all seek meaning of one kind or another, when we are presented with a mere succession of sonorous sensations which seem to possess no apparent connections to each other either in pitch or in time, it would appear that even in musical approaches to spatiality and states of stasis we continue to seek identity, consistency and redundancy because it is an unconscious, deeply intuitive urge or complex of urges arising out of the structure of the CNS which makes these demands of us. Perhaps this is why Varèse's music succeeds where other music of the same or similar tendencies fails. Paradoxically, on closer examination, it turns out that while his intentions are indeed contemporary in the anti-teleological sense, Varèse remains within the orbit of tradition precisely because he asserts the power of the serial function by precise and careful control of the motions of his geometric figure-ground images and by aligning his use of the parallel function with the serial, albeit in a new and personal manner. He also maintains the same ratio as traditional music does between a high level of redundancy and a low one of information. In his special way he is as economical and sparing of pitch motive and harmony as Beethoven — and like Beethoven, he was unafraid to repeat his ideas.

The tendency of almost all other music which is structured through the sound object — whether it is the result of organization or random occurrence — is toward a high level of information and a low one of redundancy. Generally speaking, then, the parallel function, confined primarily to the building of sound events and structures, appears to become vitiated, if not actually helpless, unless it is directly allied in a meaningful way with the serial. Music cannot exist in a half or partial state. It cannot, in the last analysis, be transformed into merely successive, vibratory forms which accidentally or on purpose approximate color fields or luminous objects hanging in space, or quasi-stochastic gaseous cloud formations having mass and density. The process of hearing cannot be converted into the process of seeing. The only real possibility for the inclusion in music of such limited constructs is as a device to produce perceptual dissonance, disorder or imbalance coupled with their opposites. And in truth, this is what is beginning to happen in music as we approach the end of our century. Composers are engaged in the act of attempting to reconcile the old and the new, seeking musically meaningful embodiments of the traditional handling of structured time and the contemporary tendency toward spatializing sound elements.

VI

When Whitrow says, "mind, as manifested in consciousness, exists only in time: it is purely a 'process' and not a 'thing',"[23] one immediately recognizes music as a manifestation of that mental process — an objectified manifestation through which one can more readily read back into the contents and structure of the process than by confronting consciousness itself or its effects head on.

It strikes me as a reasonable certainty that the conception and realization of a musical work is, in its special way, a complete manifestation of mental process, a projection of fundamental aspects of consciousness including time and memory; and that the physical projection and performance of such a work activates, to the highest degree possible, both the contents and structure of that process as well as the structure, shapes and gestures of the sound material of that work through which the process is revealed and may be perceived.

Like spoken languages and mathematics, music is a deep expression of the central nervous system,[24] perhaps even deeper than the others — although I see no way, and frankly, no need to try to prove this. Just as languages and logics are rooted in the CNS, music's roots can be traced to the same source. But between music and languages and logics is a vast difference in the mode of manifestation: music is not simply a patterned structure of notated symbols or self-generated sounds and vocables possessing potential meaning and communicability; it arises from the action of the mental process on the entire range of audible vibrations, from the interaction of human biological reality and external physical reality and is made manifest by the specific ways in which that process makes use of the physical substance of vibratory phenomena. By the interplay and interpenetration of mind and sound — and it must be remembered the mind not only perceives sounds coming from outside itself but can also imagine them apart from physical stimuli — music achieves a special form of unification which offers us an immediate and concrete way to study the phenomenon of time and its relation to memory and provides us with one of the major symbols of human culture in which we recognize man as the microcosm in whose nature we may trace out the lineaments of the macrocosm. Time is truly perceived through the ear as space is truly perceived through the eye; and both major organs of perception are the openings onto the living universe of which we, in turn, are physical/psychic manifestations and by which we may learn to know it and ourselves.

NOTES

1.   G.J. Whitrow:*The Natural Philosophy of Time* (London: Thomas Nelson & Sons 1961), p. 313.

2.   Donald Francis Tovey: *Beethoven* (London: Oxford University Press, Humphrey Milford, 1944), p. 51.

3.   Donald Jay Grout: *A History of Western Music* (New York: W.W. Norton 1960), p. 74.

4.   I am indebted to John Michon for the suggestion that, in the sense I intend, "identity" may be considered as a *perceptual invariant* thus reinforcing my discussion of redundancy. (See pp. 16-17).

5.   It must be remembered that aural identity must, in every case, corroborate visual identity, i.e., the notated representation of the music. Not surprisingly, this is not necessarily the case with aleatoric music; i.e., no necessary relationship exists between the sound and the appearance of the "score" which, often enough, may be a graphic design or a set of verbal instructions, etc..

6.   (New Haven and London: Yale University Press 1958). I am basing this portion of my paper on von Neumann's ideas not because they may or may not be scientifically "correct", i.e., valid in the sense of current knowledge, but because they open the way to examine the musical problem under consideration from a direction which I believe to be fruitful and rich with possibilities.

7.   *Ibid.,* p. 51.

8.   One cannot imagine the serial function of temporal flow requiring conscious rational thought beyond the initial act of deciding the precise pattern of that flow. Once established, its continuation is guaranteed by the physiological continuum which controls bodily movement. To think otherwise would be to raise such natural acts as walking, trotting, running to a self-conscious level. We all know that these physical acts of motion and movement are best accomplished in a smoothly articulated rhythm — although I grant that they may be consciously regulated. But once set, they go, so to speak, by themselves. This explains, perhaps, why once the tempo of a musical performance is established, the basic pulse can be taken for granted and performers are free to concentrate their mental energies on the rigors and nuances of performance problems. One need only observe athletes, circus performers, dancers — all of whom use their bodies at levels of trained performance far beyond the ordinary — to recognize that rhythmic, physiological functions are at work at levels below that of the conscious and rational. Much in the same way, it is possible to think of the musician expressing bodily rhythmic functions translated, however, from the plane of physical,spatial motion to musical, temporal motion. In his peculiar way the musician is also an athlete whose natural, intuitive sense of rhythm arises out of his body, to transform itself into the creation and projection of time through sound.

9.   *Op. cit.,* p. 60-1.

10.   *Op. cit.*, p. 61.

11.   *Ibid.*

12.   *Ibid.*

13.   *Ibid.*

14.   *Op. cit.*, p. 65.

15.   *Op. cit.*, p. 66.

16.   It would appear that their case will ultimately be decided on the well-known physiological principle of "rejection of foreign bodies."

17.   Leonard B. Meyer: "The End of the Renaissance," *Music, The Arts and Ideas* (Chicago: University of Chicago Press 1967), p. 72.

18.   Christian Wolfe quoted in John Cage: *Silence* (Middletown, Conn.: Wesleyan University Press 1961), p. 54.

19.   *Op. cit.*, p. 72.

20.   It could be argued that this condition of stasis approaches the state of "non-temporality" and, therefore, reinforces the illusion of spatiality. The difficulty here is that the literature of traditional music is replete with instances of music (e.g., Beethoven's 9th Symphony *Adagio*) which approach stasis not through suppression of pulsation and periodicity but through *slowness of tempo or motion.* In the latter instance, direction and causality still operate, leaving unaltered the essential bases for temporality.

21.   John Cage, *op. cit.*, p. 135.

22.   My translation from the original French, published in a record review by Istvan Anhalt, *The Canadian Music Journal* (Winter 1961): 34-5.

23.   *Op. cit.*, p. 113.

24.   For a provocative and whimsical corroboration of this view, I refer the reader to the chapter, "The Music of *this* Sphere," in Lewis Thomas' *The Lives of a Cell* (New York, Viking Press 1974).

VI. PHILOSOPHY

Human Temporality

H.L. Dreyfus

I. INTRODUCTION

The most sophisticated account of human temporality so far developed is set forth by Heidegger in *Being and Time*. Heidegger has formed his conception by combining Kierkegaard's and Husserl's. He has used Kierkegaard's study of eternity in time as the inspiration for his notion of authentic temporality, and Husserl's phenomenology of internal time-consciousness for his description of inauthentic temporality. He has not, however, just combined these ideas eclectically; he has added his own account of a primordial temporality, which can misinterpret itself so as to appear to itself as dispersed in the way described by Husserl, or can reintegrate itself into the temporal structure discovered by Kierkegaard.

This account of Heidegger's sources, though systematic and satisfying, is unfortunately overly simple, for, from Heidegger's point of view, both Kierkegaard's and Husserl's theories of temporality are impure, inadequate, and incomplete. First, each view still retains a residue of the tradition. Second, and more serious, neither view relates human being to temporality closely enough: Kierkegaard fails to show why the self necessarily posits itself as temporal; Husserl fails to show why consciousness inevitably constitutes itself in time. Finally, neither philosopher works out the relation of human temporality to physical time. Heidegger proposes to complete the work of his predecessors by arguing that Dasein [human Being] *is* temporality, i.e., that the structure of a human being as a self-interpreting entity necessarily has the dimensions past, present and future. Furthermore, he claims to show that the tendency of human beings to misinterpret their own temporality is the source of the mistaken belief in the primacy of physical time.

II. KIERKEGAARD'S ACCOUNT OF TEMPORALITY

Kierkegaard poses the problem we will be following, by distinguishing between time and the temporal. Kierkegaard defines time as an infinite succession of moments: an ordering with respect to before and after. He then notes that "When time is correctly defined as infinite succession, it seems plausible to define it also as the present, the past and the future."[1] But, he warns, this effort to derive the three temporal dimensions from time conceived as a succession of before and after is self-defeating. In fact, according to Kierkegaard, when we try to conceive time abstractly (i.e. apart from human experience), we cannot even make sense of the present.

(i)   If we try to think of the succession of moments from within the series, it seems that although the past is gone, and the future does not yet exist, there is a perpetually changing present moment, But since there is no basic unit in terms of which to find a "foothold",[2] the present too vanishes as infinitely brief, so that this line of thought cannot make sense even of the present.

> If one would now employ the instant to define time, and let the instant indicate the purely abstract exclusion of the past and the future, and by the same token of the present also, then the instant precisely is not the present, for that which in purely abstract thinking lies between the past and the future has no existence at all.[3]

(ii)   If, on the other hand, we detach ourselves from our current "location" in time and think of time from the perspective of eternity — as Kierkegaard says, spacially — the distinctions among past, present and future also disappear. Past, present and future are all laid out before us at once, all simultaneously. This way, too, the present vanishes, and although the formal ordering of the series remains, the notion of succession as passage disappears.

> For thought, the eternal is the present as an annulled succession ... in the eternal there is not to be found any division of the past and the future, because the present is posited as the annulled succession.[4]

According to Kierkegaard, the ancient Greeks were stuck with this puzzle and could only conceive of the relation of time and eternity as an incomprehensible opposition reflected in the opposition of body and soul. Kierkegaard, like Hegel, holds that when two contradictory concepts cannot be reconciled, the opposition can only be overcome when the opposed factors are seen to be abstract aspects of some third thing. The opposition of body and soul is overcome with the introduction of the Christian notion of spirit. What third term, Kierkegaard asks, unites time and eternity?

> Where is the third term? And if there be no third term, there is really no synthesis; for a synthesis of that which is a contradiction cannot be completed as a synthesis without a third term.[5]

What unites time and eternity, according to Kierkegaard, is the instant: "If time and eternity are to touch one another, it must be in time — and with this we have reached the instant [Øjeblikket]."[6] This new notion of the instant, unlike the traditional concept of the instant as that which excludes the past and future, refers to any occasion in human life when one makes a decisive choice, a commitment which gives a definitive form to one's future and a retro-active meaning to one's past. In this sense, it possesses a sort of eternity.

> The instant is that ambiguous moment in which time and eternity touch one another, thereby positing *the temporal* ... Only now does that division we talked about acquire significance: the present, the past, and the future.[7]

This human temporality which is posited in the instant of commitment may seem evanescent and subjective compared with the objective succession of natural time, but, Kierkegaard holds, it has its own special importance.

> ... the temporal seems even more imperfect, and the instant still more insignificant, then the apparently secure persistence of nature in time. And yet it is

exactly the converse, for nature's security is due to the fact that time has
no significance for it. Only in the instant does history begin.[8]

History "begins" in the sense that only when an individual has an experience of a privileged instant of commitment does he transcend natural time and get beyond what Kierkegaard calls the sensuous. Only when an individual makes a choice in which he takes responsibility for what he has been, and dedicates the rest of his life, can he hold together in one unified biography past, present and future. In this instant, one grasps the temporal and the eternal (the immutable) together, and gives to each of the temporal dimensions its unique and full significance. Neither the Greeks nor the Hebrews achieved this synthesis.

For the Greeks, who had no notion of a decisive instant, the past was the only important dimension. The past is fixed and known; the future only repeats the past. The Greeks identified eternity, as the fixed and necessary, with the past. "Here the significance of the Platonic recollection is evident. The Greek eternity lies behind, as the past into which one enters only backwards ... . If there is no instant, then the eternal appears to be behind, like the past."[9]

For the Hebrews, the future is *going* to be decisive: the significance of everything will be explained with the coming of the Messiah. But the fullness of time has not yet come. So the eternal significance of events in time is put off to the future limit. "If the instant is posited, but merely as a *discrimen*, then the future is the eternal."[10] " ... The future is the incognito in which the eternal, as incommensurable for time, would nevertheless maintain its relations with time."[11]

Only the Christian view of the cultural instant, when God becomes man, and the personal instant, when a believer commits himself to the God-Man, does justice to both past and future.

> The concept around which everything turns in Christianity, the concept
> which makes all things new, is the fullness of time, is the instant as eter-
> nity, and yet this eternity is at once the future and the past.[12]

For Kierkegaard, in the instant of total commitment to Christ (conversion), the believer takes up and reinterprets his whole past (atonement) and prepares to repeat this commitment for his whole future (redemption). Or, as Kierkegaard puts it, "Only in existing do I become eternal."[13]

III. HEIDEGGER'S ACCOUNT OF AUTHENTIC TEMPORALITY

Heidegger takes over and secularizes Kierkegaard's insight. According to Heidegger, Dasein is related to its past by its indebtedness (*Schuld*) to its culture for possibilities of action it did not create (this indebtedness Heidegger calls thrownness), and since it does not create its own possibilities, Dasein is related to its future as always having the possibility of having no more possibilities (this Heidegger calls Being-unto-Death). In the instant of insight (*Augenblick*)[14] Dasein takes itself over as thrown-unto-death and so unifies its temporal field.

In living out this insight Dasein gives full significance to the past and future, which it *is*. Authentic Dasein lives in a way that expresses its indebtedness and its possibility of having no more possibilities. Heidegger calls this twofold attitude *repetition* and *anticipation*. Moreover, authentic Dasein resolves to express in all its activities the insight it has achieved concerning tis own structure. As Heidegger puts it: "Authentic resoluteness ... resolves to keep repeating itself."[15] "To the anticipation which goes with resoluteness, there belongs a Present ... That Present which is held in authentic temporality and which thus is authentic itself, we call the 'instant of insight'."[16] The elements that constitute the temporal structure of authenticity, then, are repetition, the instant, and anticipation, held together in resoluteness.

Now we can see how Heidegger has adapted the secularized Kierkegaard. Conversion as the instant of commitment in Kierkegaard corresponds to Heidegger's notion of the instant of insight in which Dasein owns up to its own structure. Atonement corresponds to Heidegger's notion of repetition, in which a person takes over his indebtedness. Redemption corresponds to Heidegger's notion of anticipation, in which a person understands himself in terms of his ultimate future. (See Table 1).

	past	*present*	*future*
Kierkegaard	atonement	conversion or instant	redemption
Heidegger	repetition	instant of insight	anticipation

Table 1: Structures of authentic temporality in Kierkegaard and Heidegger

Heidegger partially acknowledges his debt to Kierkegaard. In an intriguing and obscure footnote, he says that Kierkegaard's existentiell – autobiographical – account of the instant and time is sound but that Kierkegaard lacks the radically new concepts that would be needed for an existential – i.e., philosophical – account of temporality.

> S. Kierkegaard is probably the one who has seen the *existentiell* phenomenon
> of the instant of insight with the most penetration; but this does not signify
> that he has been correspondingly successful in interpreting it existentially. He
> clings to the ordinary conception of time, and defines the "instant" with the
> help of "now" and "eternity".[17]

Kierkegaard, it would seem, articulates his insight in the traditional concepts of time and eternity. Heidegger claims that in order to give a coherent account of the experience of temporality, Kierkegaard would have to adopt Heidegger's view.

> When Kierkegaard speaks of "temporality", what he has in mind is man's
> "Being-in-time". Time as within-time-ness knows only the "now"; it never
> knows an instant of insight. If, however, such a moment gets experienced
> in an existentiell manner, then a more primordial temporality has been pre-
> supposed, although existentially it has not been made explicit.[18]

Is Heidegger right? Does Kierkegaard hold that man lives *in* time, a succession of nows, and that in the instant man relates to some eternity outside time? Or does Kierkegaard see that neither the succession of nows nor eternity can be understood until the two are synthesized? For the answer we must try to interpret a very difficult passage in *The Concept of Dread*.

> The synthesis of the soulish and the bodily is to be posited by spirit, but the spirit is the eternal, and therefore this is accomplished only when the spirit posits at the same time along with this the second synthesis of the eternal and the temporal.[19]

If we read this passage in the light of the one already quoted affirming that "the instant is that ambiguous moment in which time and eternity touch one another thereby positing the temporal", it might seem that the spirit, or the eternal, stands apart from time, and then somehow combines with time to produce the temporal. In that case Heidegger's criticism would be justified. But before we write off Kierkegaard as unable to conceptualize his religious experience, we must turn to Kierkegaard's definition of spirit in his most religious work, *The Sickness Unto Death*.

> Man is spirit. But what is spirit? Spirit is the self. But what is the self? The self is a relation which relates itself to its own self.[20]

This seems clear enough until it is followed by the explanation that:

> Man is a synthesis of the infinite, and the finite, of the temporal and the eternal ... A synthesis is a relation between two factors. So regarded, man is not yet a self.[21]

What this means is that man is not just some combination or synthesis of factors, but is the stand that combination takes on itself. Moreover, not just any stand will do. To deny the eternal aspect like the man who tries to live in the moment, or, on the other hand, to claim that man is essentially an eternal soul stuck with a temporal body, like Plato, is still to fail to be a self. To be "a self in equilibrium"[22] one must affirm both factors as fully significant and fully interdependent, which one does only when one develops an infinite commitment to something or someone finite, and thus achieves eternity in time. In this highest stage of self-interpretation, man is truly spirit or truly a human self, "a relation which relates itself to its own self, and in relating itself to its own self relates itself to another."[23] Only in this highest stage, which Kierkegaard calls the religious sphere, can the relation of the "now" and the "eternal" be consistently lived, and presumably, only in terms of this religious stage can these concepts be coherently understood.

In this light, what we should stress in reading the passage from *The Concept of Dread* is that spirit (the eternal) comes into being along with temporality. This means that for the religious individual — and the existential thinker reflecting on the religious life — spirit, the self, the instant, the temporal, and the eternal are all understood together. Up to this point each is understood incompletely — abstractly as Kierkegaard would say.

> By the Greeks time and eternity alike were conceived abstractly, since the Greeks lacked the concept of the temporal owing to the fact that they lacked the concept of spirit.[24]

If this interpretation is correct, Kierkegaard does not make the mistake Heidegger accuses him of: he would agree with Heidegger that one cannot coherently understand human temporality in terms of some combination of antecedently intelligible (Greek) notions of the now and eternity. He does, however, seem to suppose that the time of nature as pure succession makes sense whether or not spirit posits the temporal. For the argument that time too is an abstraction from temporality we must turn to Heidegger's argument in *Being and Time*. But first we must, for an intermediate step, stop at Husserl.

IV. HUSSERL'S PHENOMENOLOGICAL DESCRIPTION OF TIME-CONSCIOUSNESS

Husserl's contribution is a phenomenology of the experience of time, or more precisely, of the experience of everyday temporal objects. As Husserl points out, we do not, when experiencing an object in time, experience a succession of "pure nows", or we would never experience succession at all. Rather, the "living present" contains anticipations of further experiences, which he calls "protentions", and reverberations of experiences just slipping away, which he calls "retentions".

> The immanent contents [of consciousness] are what they are only so far
> as during their "actual" duration they refer ahead to something futural and
> back to something past ... In each primal phase which primordially consti-
> tutes the immanent content we have retentions of the preceding and pro-
> tentions of the coming phases of precisely this content.[25]

Husserl develops this view by analyzing the experience of listening to a melody. If I am familiar with a melody the sound of each note depends on the surrounding notes. I do not hear merely a C sharp, for example, but a note with a quality which would be altered if it appeared in a different melody. In this way, the future and past of the melody are involved in the experience of any given note.

But this cannot be the whole story, since if it were, I would hear the melody all at once. We still need to account for the experience of hearing the melody one note at a time. In Husserl's language, the protentions are *filled* one by one, and then become retentions as new protentions are filled. This requires a second principle: not only does my present experience imply the other notes of the melody, but at each moment a filled note is fading away but still reverberating in my consciousness.

> When a new note sounds, the one just preceding it does not disappear
> without a trace; otherwise, we should be incapable of observing the
> relations between the notes which follow one another.[26]

Thus, retentions as filled are not explicit judgments about what I remember nor are they memories in which I represent to myself the contents of the past. Rather, they are the way the contents of the past are directly present to me.

> If we call perception the act in which all "origination" lies, which
> constitutes originarily, then primary remembrance is perception.
> For only in primary remembrance do we see what is past; only in
> it is the past constituted, i.e., not in a representative but in a pre-
> sentative way.[27]

When I am hearing the fourth note (say) of a familiar melody, the third note is slipping away and the fifth is looming up. When I am listening to the fifth note, the fourth will be fading out, and the retention of note four that I have when I am listening to note five includes the retention of note three that I had as I was listening to note four. Thus, the awareness I have of successive fillings is made possible by a nesting of previous retentions in the present one. Likewise, when I have a protention of note 5, I have through it progressively dimmer protentions of notes 6 and 7 and 8, etc.. When note 6 arrives, I have retentions of notes 5, 4, 3, etc., and protentions of notes 7 and 8, etc.. (See Table 2).

P1(P2(P3(P4(P5))))
[P1]P2(P3(P4(P5)))
[[P1]P2]P3(P4(P5))
[[[P1]P2]P3]P4(P5)
[[[[P1]P2]P3]P4]P5

Table 2: Imbedded Structure of Time-Consciousness

As the melody goes on, the sequence of notes to come is unpacked, and the sequence of past notes becomes compressed until finally I lose them. On Husserl's model this limitation on the mediated presentation of the past, which is neither directly present nor represented, resembles more a perceptual failure to discriminate fine details than a failure of memory.

> Of the interval that has expired we say that we are conscious of it in retentions, specifically, that we are conscious of those parts or phases of the duration, not sharply to be differentiated, which lie closest to the actual now-point with diminishing clarity, while those parts lying further back in the past are wholly unclear; we are conscious of them only as empty ... As the temporal object moves into the past, it is drawn together on itself and thereby also becomes obscure.[28]

Husserl follows this analysis of the role of protentions and retentions in the experience of temporal objects, by a study of the way remembering and re-remembering constitutes repeatable objects and orders them in an objective time series. This, in turn, leads to the difficult question of the relation of consciousness to the objective temporal series its activities have organized.

Husserl's characterization of conscious as the "living present" here becomes treacherous, for, as we have seen, Husserl's important discovery is that the now-phase is only an abstract aspect of a complicated structure of protentions passing into retentions. Only in objective time does the present, understood as a pure now, have a privileged actuality. His terminology, however, seems to give a priority to objective time. Moreover, Husserl likes to speak of the "living present" as a "primordial flux", thus raising the question of the time in which this flux flows.

Husserl is aware of his dilemma. He sees he has been led to speak of consciousness as if it were in time:

> But is not the flux a succession? Does it not, therefore, have a now, an actual phase, and a continuity of pasts of which we are conscious in retentions? We can only say that this flux is something which we name in conformity with what is constituted, but it is nothing temporally "Objective". It is absolute subjectivity and has the absolute properties of something to be denoted metaphorically as "flux", as a point of actuality, primal source-point, that from which springs the "now", and so on. In the lived experience of actuality, we have the primal source-point and a continuity of moments of reverberation. For all this, names are lacking.[29]

But Husserl also sees that the primal source-point itself cannot be *in* time: one cannot locate a pure source-point — one cannot say when it arrives — since the temporal dimensions are constituted *in* the primal flowing source-point. So Husserl has to say the source-point is not temporal at all: "Subjective time is constituted in absolute, timeless consciousness."[30]

Kierkegaard would say that Husserl gets into these difficulties because his phenomenological method leads him to take a detached attitude which excludes the instant of commitment. He thus falls back into the dilemmas posed by an abstract analysis of time in terms of now and eternity. Heidegger would add that in spite of the sophistication of his analysis. Husserl never succeeds in showing the necessary relation between human being and temporality and so ends up treating consciousness in a traditional way, i.e., not as temporality but as a thing which either is in time or has time in it. Heidegger cautions:

> We must, in our investigation, make ourselves familiar beforehand with the primordial phenomenon of temporality, so that in terms of this we may cast light on the necessity, the source and the reason for the domination of the way [time] is ordinarily understood.[31]

V. HEIDEGGER'S ACCOUNT OF INAUTHENTIC TEMPORALITY

For Heidegger, as we have seen, Dasein is neither in time nor timeless but, as thrown-projection, human Being is fundamentally temporal. Another, more basic way to put the same point is to say that for Heidegger, as for Kierkegaard, human being is a being that must take a stand on itself. Heidegger calls this stand "existential understanding". Dasein is constantly *ahead* of itself making projects on the basis of its sense of what it *already* is as revealed in its moods. Thus, Dasein has a threefold asymmetrical structure simply by virtue of being a self-interpreting given. It is this structure which Heidegger calls primordial temporality, and which he says is neither in time nor atemporal.

> While the "ahead" includes the notion of a "before", neither the 'before' in the 'ahead' nor the 'already' is to be taken in terms of the way time is ordinarily understood; this has been automatically ruled out by what has been said above. With this 'before' we do not have in mind 'in advance of something' in the sense of 'not yet now — but later'; the 'already' is just as far from signifying 'no longer now — but earlier'. If the expressions 'before' and 'already' were to have a time-oriented signification such as *this* (and they can have this signification too), then to say that [Dasein] has temporality would be   to say that it is something which is 'earlier' and 'later', 'not yet' and 'no longer'. [Dasein] would then be conceived as an entity which occurs and runs its course 'in time'.[32]

In trying to understand this claim, it is important to note that while specific events in the past are before events in the present (or after them, depending on where the speaker situates himself in objective time), the past as thrownness is neither before nor after the present. Dasein just is a thrown-projection. Past, present and future are interrelated dimensions of the field of disclosure of self-interpreting beings. Another way to put this is that the field of disclosure does not flow (it is not timeless, either), but events succeed one another in this field.

Now Dasein, as self-interpreting, also has to interpret its temporality. "Temporality has different possibilities and different ways of temporalizing itself."[33] The most common way is the temporality of everyday practical activity. Heidegger calls this the undifferentiated mode of Dasein's being, and the time which it discloses is called world time. Heidegger means by this the temporality of the world of human society, not the time of the physical universe. World time consists in the public self-interpretation in which there is a time to wake up, dinner time, etc., and the time of nature is experienced as daytime, nighttime, etc.. Dasein is for the most part absorbed in this public time, which is the same for everyone and which has no place for the self-interpretation of individual human beings.

Individuality emerges only when one of the cogs in this social system experiences an anxiety attack. This is an awareness that there is no special place for me in this public system of roles and goals, and that as far as the public is concerned, I could equally well not exist. This could be the instant of insight we have already discussed, in which Dasein interprets itself resolutely as indebted-being-unto-death. But it is more frequently the occasion for Dasein to fall — to flee back into the public world, bent upon covering up and misinterpreting its authentic temporal structure. In that case Dasein, as primordial temporality, temporalizes itself in a way which Heidegger calls inauthentic.

The inauthentic temporal dimensions are awaiting (future), making present (present) and forgetting (past). These dimensions characterize the way the lives of most people, most of the time, are temporalized. First, Heidegger introduces the notion of awaiting.

> Dasein does not come towards itself primarily in its ownmost non-relational potentiality-for Being, but it *awaits this* concernfully *in terms of that which yields or denies the object of its concern* ...The inauthentic future has the character of *awaiting* .[34]

We thus relate ourselves to the future (which includes our demise) as though it were something distinct from the present, something that has not yet come.

Correlatively, Dasein interprets the present, as Husserl interpreted the living present, as a now-point rather than an instant of insight embracing the whole structure of a life.

> In contradistinction to the instant of insight as the authentic Present, we call the inauthentic Present "making *present*".[35]

Likewise, forgetting is the inauthentic mode of having been.

> If *Being*-as-having-been is authentic, we call it *"repetition"*. But when one projects oneself inauthentically towards those possibilities which have been drawn from the object of concern in making it present, this is possible only because Dasein has *forgotten* itself in its ownmost *thrown* potentiality-for-Being.[36]

Finally, Heidegger gives a convenient summary of authentic and inauthentic temporality. (See Table 3)

> Understanding is grounded primarily in the future (whether in anticipation or in awaiting). Moods temporalize themselves primarily in having been (whether in repetition or in having forgotten). Falling has its temporal roots primarily in the present (whether in making-present or in the instant).[37]

	past	*present*	*future*
inauthentic	forgetting	making present	awaiting
authentic	repetition	instant of insight	anticipation

Table 3: Authentic and Inauthentic Temporality

In the inauthentic mode of self-interpretation, the dimensions of time are held apart, so that Dasein's thrownness and mortality seem to be in the distant past and future, remote from what Dasein is now, while Dasein busies itself in what is present in order to forget its anxiety. Its life is identified with concrete, unrelated projects which fill the time between the present and what is awaited, and which, when completed, are gradually forgotten. This parade of projects corresponds to Husserl's succession of protentions and retentions (see Table 4). The more superficial Dasein's life, the more short-range and easily forgotten are its projects, so that at the limit inauthentic temporality approaches a series of unconnected presents — pure now — lacking even Husserlian protentions and retentions.

	past	*present*	*future*
Husserl	retention	now phase	protention
Heidegger	forgetting	making present	awaiting

Table 4: Structures of Inauthentic Temporality in Husserl and Heidegger

VI. HEIDEGGER'S DERIVATION OF SUCCESSION

Heidegger, we have seen, criticized Kierkegaard for taking over a traditional account of the now and of eternity. We have also seen that this criticism is unjustified, in that Kierkegaard understands the eternal and the now as abstract and superficial forms of human temporality. But Kierkegaard *did* suppose that natural time, as infinite succession, makes sense apart from man's temporality. Heidegger set out to show that

> Time as ordinarily understood does indeed represent a genuine phenomenon,
> but one which is derivative. It arises from inauthentic temporality.[38]

To carry out this "derivation" Heidegger first gives a detailed analysis of how natural clocks are used as we set up dates in world time. This enables busy Dasein to organize the many projects that make up the field of significance in which everyday Dasein is absorbed. Then Heidegger shows how the superficial interpretation of time, into which inauthentic Dasein flees to avoid experiencing the instant of insight, makes possible a detached attitude which abstracts even from everyday practical concerns. This scientific attitude abstracts itself from the concrete practical situations in which dinnertime and nighttime have meaning, in order to study the relations of pure qualities at pure moments. These moments then appear as a series of pure nows from which even awaiting and forgetting have been eliminated.

> In the ordinary interpretations of time as a sequence of "nows", both data-
> bility and significance are *missing* ... The "nows" get shorn of these relations,
> as it were; and, as thus shorn, they simply range themselves along after one
> another so as to make up the succession.[39]

Ontology then comes along and makes a final move: claiming that these context-free moments are ultimately real. This view, explicit in Descartes, who thinks of time as an infinite succession of durationless moments, is the one Kierkegaard seems to sanction. That this detached view is untenable, however, is suggested by the difficulties into which Husserl is driven when he tries to think of a timeless yet flowing present source-point as alone actual.

Heidegger's analysis definitely breaks with traditional ontology on just this point:

> Not only must the now-time be oriented primarily by temporality in order of possible interpretation, but it temporalizes itself only in the inauthentic temporality of Dasein; so if one has regard for the way the now-time is derived from temporality, one is justified in considering temporality as the *time which is primordial.*[40]

VII. CONCLUSION

We have now seen that Kierkegaard and Heidegger agree that the trouble with the traditional view (and Husserl's) is that it illegitimately abstracts the self from time. As Kierkegaard first saw, the self can be unified and fulfilled only by total commitment to something temporal. In Heidegger, this becomes the view that when Dasein tries to deny its finitude, its life is dispersed into a series of moments which approach mere succession, but that when Dasein accepts its finitude as thrown-unto-death, it surmounts being-within-time, not by getting out of time, but by reintegrating its own temporal structure.

NOTES

1. Sören Kierkegaard: *The Concept of Dread* (Princeton, N.J.: Princeton University Press 1957), p. 76.

2. *Ibid.*, p. 76.

3. *Ibid.*, p. 78.

4. *Ibid.*, p. 77

5. *Ibid.*, p. 76.

6. *Ibid.*, p. 78.

7. *Ibid.*, p. 80.

8. *Ibid.*, p. 79, 80.

9. *Ibid.*, p. 80.

10. *Ibid.*, p. 81.

11. *Ibid.*, p. 80.

12. *Ibid.*, p. 81.

13. Sören Kierkegaard: *Concluding Unscientific Postscript* (Princeton, N.J.: Princeton University Press 1941), p. 508.

14. The German word *Augenblick*, used by Heidegger, is translated by Robinson and Macquarrie as "moment of vision". However, since *Augenblick* is the same word as the Danish *Øjeblikket* used by Kierkegaard, it seems more appropriate to translate it by the same English work, viz., the instant. As a compromise I prefer to speak of the instant of insight and I have changed the English translation accordingly.

15. Martin Heidegger: *Being and Time* (New York: Harper and Row 1962), p. 355.

16. *Ibid.*, p. 387.

17. *Ibid.*, p. 497.

18. *Ibid.*

19. *Op. cit.*, p. 81.

20. Sören Kierkegaard: *The Sickness unto Death* (Princeton, N.J.: Princeton University Press 1964) p. 146.

21. *Ibid.*

22. *Ibid.,* p. 147.

23. *Ibid.,* p. 146.

24. *Op. cit.,* p. 79.

25. Edmund Husserl: *The Phenomenology of Internal Time-Consciousness* (Bloomington, Ind.: Indiana University Press 1966), p. 110.

26. *Ibid.,* p. 30.

27. *Ibid.,* p. 64.

28. *Ibid.,* p. 46, 47.

29. *Ibid.,* p. 100.

30. *Ibid.,* p. 150.

31. *Being and Time,* p. 352.

32. *Ibid.,* p. 375.

33. *Ibid.,* p. 351.

34. *Ibid.,* p. 386.

35. *Ibid.,* p. 388.

36. *Ibid.*

37. *Ibid.,* p. 401. [To simplify my exposition I've translated *Befindlichkeit* as mood although a more literal translation would be "where-you're-at-ness".]

38. *Ibid.,* p. 374.

39. *Ibid.,* p. 474.

40. *Ibid.,* p. 479.

Structures of the 'Living Present': Husserl and Proust

J. HUERTAS-JOURDA

The mediate aim of this essay is to make good two related claims: that phenomenology in Husserl's sense is adequate to the task of describing the most concrete structures of subjectivity; and that the universal may be apprehended directly through the concrete.

With the first claim, I have in mind to counter what I call the fallacy of the supposed concreteness of ontology. Those committing this fallacy suppose mistakenly that they can escape the abstractive or universalizing effect of language simply by describing solely concrete and particular entities. It is hoped that by limiting oneself to the description of such "real" objects one reaches for,however fleeting a moment,an apodictically founded, ontologically prior, hence scientifically grounded layer, upon which the confrontation with truth takes place. The fact that such a description itself remains "abstract" — in the sense in which Husserl's work has been accused of being "abstract" by those who refuse to follow him in his "transcendental turn" — thanks to the universalizing (or "abstractive") power of language, is minimized or relegated to the level of an existential feature inherent in all speech and partially overcome through "thought", or "hermeneutics", or "l'engagement", and the like. Husserl fought this fallacy as psychologism, historicism and anthropologism, in divers successive essays. Rather than trust to a nebulous existential emotion, he advocated the careful use of the abstractive qualities of language against themselves, as it were. By insisting upon the careful description, not only of the concrete objectivities under scrutiny, but also of the intentional operations required to focus upon such concrete "matters of fact", Husserl found the only means with which to avoid "abstraction" and to return "to the things themselves". Indeed, anyone can perform such intentional operations ("noeses", in Husserl's terminology), and thereby focus upon (or "constitute") the "meaning" aimed at ("noema") in the operation, no matter how concrete the "objective referent" now perceivable *through* the text, though not *in* it, and *in spite* of the abstractive force of the language of description. Whether one focuses upon an "abstract" or a "concrete" "object" (objective referent of some kind) depends upon the operation performed. Abstraction, in Husserl's view, is an intentional operation, a structure of awareness, to be enacted by someone,and different from the operation required to focus upon some concrete entity of some kind.

Hence, on this view, all those who claim that the abstract may *not* be apprehended *through* the concrete, can be shown to commit an elementary fallacy. Thus, Proust's meticulous rendering of the

---

*My warmest thanks to those who have helped me with their critical comments in earlier stages of this essay, and most particularly to Mr. Harry Reeder. The final state, however, is my sole responsibility.

unique and temporally moored particularizing features of a live-subjectivity (the narrator's) can be shown to be "exemplary" in Husserl's sense, and made to yield "abstract" (universal, or ideal; in Husserl's terminology, "eidetic") structures, when "asked" (that is to say, when one exercises the peculiar mode of observation required to focus upon such "abstractions"). This mode of observation Husserl called the *Wesensschau* or intuition of essences (essences, or "eide", are ideal and universal structures). It is a particular intentional operation, in the sense mentioned above, and consists in recognizing, after appropriate testing has allowed the emending of contingent elements, the necessary features universal to all the members of the group under scrutiny. The phenomenological method as advocated by Husserl culminates in the exercise of the *Wesensschau*.

By showing where, and in what way, eidetic features of subjectivity as recognized by Husserl and whatever is visible of such features in Proust's novel, when Proust's narrative is looked upon as describing an "exemplary" subjectivity, corroborate or contradict each other, I hope to make good both of the claims stated initially. If, along the way, Proust helps throw a more familiar light on Husserl's distant "abstractions", the independent corroboration thus provided should give every serious philosopher pause when he is tempted by his own brand of rigour to misappreciate other varieties.

After giving some *prima facie* grounds for a comparison of Proust's studies with Husserl's, the demonstration windingly proceeds from selected features in Proust's novel to the corresponding structures in Husserl's analyses of subjectivity. I attempt to show, first, that Proust's novel "takes place" in the "living 'now'" of the narrator as he attends the matinée at the prince de Guermantes, and that the very structures Husserl has recognised in "a" "living 'now'" are instantiated in the novel. Thus, I analyze the time at which the one action the narrator actually performs in the novel takes place, and then turn to Husserl's corresponding descriptions which I summarize. Using Husserl's discoveries (retention, protention, the primary perceptual upsurge of the real in the "living 'now'"), I proceed to show concrete instances of the very same structures in the novel. The careful descriptions of such events as the narrator's encounter with the ravages of age on the guests of the prince's matinée or his introduction to Mlle de Saint-Loup, allow me to use the text of the novel without resort to any interpretation.

A preliminary case thus presented, I turn to Husserl's description of the apperception of the unity of the flux of lived-experiences, through that flux, by the consciousness concerned, and I summarize it. This present phase of the study culminates then in the attempt to demonstrate that the instances of involuntary memory so meticulously recorded by Proust's narrator exhibit the essential features Husserl had recognized. The lesson to be drawn from this and the previous analyses is then summarized, and some of its results are outlined by way of a conclusion which is also an invitation to the Husserlian phenomenologist to mine the rich material available to him in Proust's work. To go beyond this would be to stretch further an already extended essay. Let it be considered complete insofar as it does provide a refutation of the two fallacies stated initially. This presented, philosophy in the narrow sense may be deemed to be satisfied. Whatever wider horizons and incidental finds, philosophy, in the wider sense (as phenomenology and *askesis*, all in one), may encounter in the immediate topics considered in accomplishing the more restricted task, let each allow himself to pursue. This essay properly seen is but a beginning contributed, in spite of its blemishes, for the use of those who will want to go beyond its limits.

I. SOME PRELINIMARY GROUNDS FOR A COMPARISON BETWEEN HUSSERL AND PROUST

1.

The most widely known probes into subjectivity which this century has seen all began under the aegis of classical causal theory and all followed the pattern of tracing effect to specific causes inherited from a primordially mechanical model of the universe. This is true most particularly of behavioural psychology, and *mutatis mutandis*, of psychoanalysis as well. As against this approach, another one has made its sway which purports to be strictly descriptive of formal patterns rather than causal ones. Phenomenological description, especially as advocated and practised by Husserl, eschews the metaphysics of causality in preference for purely descriptive moves. However, while the studies of Heidegger, Sartre and Merleau-Ponty are widely known and translated, the bulk of the work on the topic by the founder of the Phenomenological School himself, Edmund Husserl, is even now difficult to consult. It is for the most part unedited and has surfaced only here and there in critical essays based upon it like Gerd Brand's[1], Alwin Diemer's[2] or Klaus Held's[3], to name a few, the most important for us being the last one whose work concentrates upon manuscripts of Husserl's last period, namely, the studies on the "Living Present". Little wonder, therefore, that the two most searching probes into the phenomenology of subjectivity have never been the object of a thorough comparative study. Proust whose relationship to Bergson has been the subject of much debate in spite of his own explicit distanciation from the thought of the French master,[4] has been seriously linked to the existentialists,[5] but the fruitfulness of a comparison between his work and that of the later Husserl has been mentioned favorably, to the best of my knowledge, only once (by Pascal Fieschi)[6] and never seriously explored.[7] But without letting this influence us unduly, let us reconnoitre the grounds for further comparison between Proust and Husserl, and in doing so, sketch, however scantily, what results such a comparison might yield.

2.

It may, at first blush, seem strange that one should refer in one and the same breath to both Proust and Husserl, let alone refer to both as to fellow phenomenologists similarly engaged in a pure phenomenology of subjectivity. Proust, the novelist of the decadent Parisian Grand Monde at the turn of the century, paired with the retiring scholar of Freiburg! But distances are misleading and a closer look may reveal strong affinities. The narrator, in *La Recherche du Temps Perdu*, in speaking about the novel he projects, asserts that he wants to express in it transcendent truths, claims that he wishes to arrive at the "essence of things".[8] Proust himself had learned, we are told in other sources, from his first teacher of philosophy and a major influence on his thought, Alphonse Darlu, to reject a too facile idealism and to begin with the objects of the senses, with, one might perhaps paraphrase, "the things themselves".[9] Admittedly, these may be only superficial similarities; it is in the practice of these ideals that we shall see whether a stronger case can be made. The results of Proust's investigations, if they turn out to be phenomenological at all, could then be expected to reach structures similar to the ones Husserl reached by dint of having also inner subjectivity as their subject matter. Let us, therefore, without further delay turn to the texts themselves and see what a critical, but unbiased analysis, may yield.

II. THE TEMPORAL UNITY OF THE SINGLE ACTION REPORTED IN THE NOVEL

3.

The first result, not unexpectedly, is to find that, in *La Recherche du Temps Perdu* we are very far indeed from the "abstraction" of the language of Husserl. Obviously, the "generality" the narrator talks about,[10] must be other than the generality of the language of eidetic description, (i.e. of the description of essential structures of experience; cf. above, p. 3). For we are confronted from the very first line of the novel with a singular first person pronoun, a very concrete ego, a well-defined, finely particularized "me" or individual personality who is seemingly talking to us, the readers, now, and in confidence. Whatever generality is to be observed there, will have to be reached through this *concrete* presence. The move to a transcendental description is not explicitly made for us, rather, a concrete appearance makes itself manifest, right from the start. Perhaps this concrete presence in its very uniqueness will exemplify — instantiate — general (transcendental or eidetic) truths. This remains to be seen and is not, *ab initio*; perceptible. All we have to begin with is a double concreteness, a double particularity: that of the speaker and that of the moment of his speaking. The speaker as is immediately obvious from the first phrase: "Longtemps, je me suis couché de bonne heure ... ", is a live human being, a man, possibly talking to us, possibly talking to someone else or to himself, whose speech the *novel purports to reproduce*. The moment of his speaking is a rather more complex thing to ascertain.[11]

It could be in our present, our present as reader-hearers, here and now reading (hearing) the story the narrator has to tell. And indeed, in a manner of speaking, it is in our present that this speaking is done at least at one level of concrete lived experience, a fact that we, as phenomenologists, might well bear in mind. Within the novel, however, the time of this speaking must be a certain moment in the narrator's life, and we, as "readers-listeners" (if indeed he is actually talking to us), would have to enter in this narrative in some way. Proust was too clever to involve himself and us in such a charade, and, although the speaking does take place at a particular moment in the narrator's life, we are in effect eavesdropping on the narrator talking to himself. In this case also, the speaking takes place in a "present" moment. It is a moment well-circumscribed within the narrative, however, one the duration of which is clearly demarcated for us by specific literary devices. The most direct of these is used when, in the library of the prince de Guermantes, while awaiting admission to the matinée to which he has been invited but has come somewhat late, the narrator refers to his own reflections as occuring "en ce moment".[12] This, in effect, is the *présent* which we have been expecting ever since the *passé composé* of the first sentence of the book. The *passé composé* is related to the present, not because (as R. Shattuck would have it,[13] who is the only one to have noted the meticulous use Proust makes of this tense in some specific cases for a specific effect), the action denoted continues to the present (indeed, this is more the usual function of the *imparfait*), but because *the action of speaking about* the action denoted is taking place *right now*. The *passé composé* is a conversational tense, not a "literary" one. It has directness and bespeaks actual encounter. In that manner it is related to present speaking. Hence, when the narrator speaks about the reflections he is making in the library of the prince de Guermantes, he is not referring only to the immediate reflections suggested by the re-occurrence of involuntary memory in the incident of the paving stones or of the spoon hitting a glass. He is referring also to the reflections we have heard from the beginning. What we have heard from the beginning are the reflections suggested by both incidents and not simply those immediately summarized then.[14]

The singling out of this moment in the life of the narrator as the one at which the novel's only real action takes place — namely, the narration itself — is very carefully marked once again by Proust, if only with a characteristic covering of tracks, by the introduction at its close of another tense, indeed, another mode: the *conditionnel présent*. With it, many pages and incidents after the passage we have just referred to, Proust signifies that the moment of reflection is coming to an end, as is the novel, while the narrator is still at the matineé of the prince de Guermantes, shortly after he has met Mlle de St. Loup.

The *conditionnel présent* expresses a future less-vivid, an optative. Proust uses it here advisedly in order to confer a certain opacity to the narrator himself, the opacity of his own future which may or may not bring the completion of the projected novel. Indeed, the narrator himself is aware of this as is shown in the very passage in which the *conditionnel présent* is made to forecast the close of the "moment" of narration: "Quand je retournerais tout à l'heure par les Champs Elysées, qui me disait que je ne serais pas frappé par le même mal que ma grand mère ... ".[15]

The economy of method used here by Proust and the parallel with the introductory ploy are very striking. The *conditionnel présent* is also a present-based tense. Although less direct and more literary than the more conversational *futur* — particularly where strongly held projects are concerned — it serves nonetheless to describe the probable future as seen from a particular present. Here it is used to greatest effect both to help terminate the narration and to give it its future projections as visible only from the moment of narration, and not as they will have occurred when they are realized.

Notice, then, both corresponding shifts from the more expected usages. They enable Proust, first, in the beginning, to mark with the single use of the conversational *passé composé* that the action of speaking *is* to have the directness of presence. Thus, Proust prepares the way for the temporal location of the novel's single *real* action within the narrative itself. And, second, at the end, with the use of an unexpected and less-vivid form, Proust preserves the same directness to the same single action by preventing the slackening of belief that the use of the *futur*, with its self-assurance, would have caused. Were the narrator actually speaking to a friend or to himself, he might, given the present strength of his literary avocation recovered, express himself in the *futur*. In the novel this would indeed appear too strong — besides not being in character — all the more so that it would tempt the reader into believing that he is reading the narrator's novel, something he has no right to assert. Thus reinforced, the *indicatif futur* would have force of fact. The very ties with the "live" present that the *conditionnel présent*, as less-vivid, preserves (continuously reminding us not only that the action it denotes is to come, but that it may not come at all), would be all but severed. The unity of the single action of the novel would be shattered.

A close study of the tense structure used to depict the novel's single action shows therefore that this action takes place in a limited present moment lasting whatever a *matineé mondaine* may last, and that this present moment (leaving aside for now the question of its actual extension) has a "thickness" of sorts. It has a past depth starting from the moment the narration begins, and a future height extending into the conditional projections. Note that I am speaking here only of the time of the act of narration, and not of the other times of the novel which are themselves specified in different ways throughout the narrative, and which can be related more or less directly to various anchor points: the narrator himself, Swann, Proust, social and political history, or the reader, to name a few.

III. HUSSERL, AND THE STRUCTURES OF THE "LIVING 'NOW'"

4.

The thickness of the time of the act of narration is one which students of Husserl's works will immediately recognize. The depth of the past narration, from the opening sentence on, can be seen to be "retentively" held in immediate presence throughout the incidents occurring from the moment the narrator enters the library of the prince to, and including, the moment he meets Mlle de St. Loup. In a similar way, all the projections concerning the writing of the novel, all of them expressed in the *conditionnel présent,* can be seen as "protentions" presently held by the narrator as he reminisces and as he meets with the same shock the guests of the *matinée,* now transformed by age, and Mlle de St. Loup. "Retention, protention" are technical Husserlian terms which need some further explanation. They were used by Husserl in his *Lectures on the Phenomenology of Inner Time-Consciousness* to characterize the specific temporal thickness of the present as immediately observable when, for example, listening to a single note in pedal point, or to a melody.[16] Proust, as we have shown, marks a similar thickness governing a similar present. By "retention" is meant the presence in the living "now", of the "just past", as, for example, the sense of the first words of this sentence linger, as it proceeds, and light the meaning of even these words here with the very special twist of sense that the first alone contribute to the last now read. Thus, some thing is experienced "retentively" when its perdurance in the "now" is experienced as a temporally, internally related network of "just past" persistence in the "living", *present* present. The table on which I am writing, this sentence, are examples of this internal present persistence, insofar as they are presently perceived, and, as lived experiences of perceptions, retained in the present perceiving of them. Similarly, the living present is structurally defined towards the oncoming flow of lived experiences. The "period" which ended the previous sentence, structures my expectations in a different manner than does this comma, in that nothing more is expected to complete the sense of the previous sentence, whereas something more was expected after the comma, and disappointment would have occurred as a live experience in the flow if no other words had followed. Protentively, the paper I write on is perceived as perduring, and hence, my living present possesses as an internal structure, the unitary expectation of the paper's perdurance all the while I am "busy with" the paper. Likewise, the expectations which guide the narrator's thoughts as he shapes them, while he ruminates over his future in the salon of the Guermantes, after meeting Mlle de St. Loup, open his living present in a particular way towards the oncoming flow of his lived experiences, and this particular way involves the gestation of a novel, and perhaps the actual writing of it, *si Dieu lui prête vie.*

This, then, is the thickness of the living present as seen by both Husserl and Proust. Immediate objections arise, not the least of which is the fact that, where Husserl's retentional-protentional thickness extends in the examples he chose over the attention span required by a tone or a melody, and perhaps a little beyond; Proust's, on the other hand, purports to extend over at least a few hours, if not, as we shall see in a moment, over a lifetime. Still, given these discrepancies, it can be conceded that the single action of the novel takes place in an extended present, a present thick with retentional and protentional auras.

5.

Such a present, as Husserl described it, *originates* in a "living 'now'",[17] that is to say, a "now" from which the retentionally held tone recedes and towards which the protentionally-held future moves. This "now" involves "primary givenness, primary perception" rather than "remembrance" or "presentification".[18] The "now" borning and beheld a-borning, presents itself as occurring for *the first time* in some, at least, of its features. It involves a bedrock of perceptual contact with something announcing itself as "new", as other than what the "stream of lived mental experiences" *already* contains,[19] even if only with "newer" phases of the same aspects of what it already holds. Thus, as I sit in this room writing, although I pause now and again to ponder over what I have written so far, the now within which I review — presently — in my mind what I have already done, contains as well in primary givenness the ever changing aspects of the room — visual as well as auditory and tactile, not to mention olfactory and gustatory. Thus, noises occur which do not become the theme of my present focus of attention, though they remain on the "horizon" of it, as do the hardness of my chair, the smell of food cooking and the accompanying awakening of my tastebuds.

The living "now", therefore, is continuously anchored in perceptual primacy, though this primacy need not be the focus of my immediate attention and may remain strictly "horizontal" to it. By "horizontal" is meant here the temporal analogue to the visual phenomenon according to which the limit of my field of vision is tacitly (non-thematically) "understood" (to use a vernacular expression) to promise the continuation of what it does not encompass at the moment of perception; that is to say, it is experienced as holding the possibility of focus on what is not yet focal, but only liminally held by it, in any one instance. Similarly, the temporal field holds "horizontal" promises which hover on the outer limits of a present focus, as the protentive possibility of focus, "understood", or "comprised" in the living "now", but not thematically beheld in its present focal plane. Thus primary perception may recede into an horizonal background, while a flow of memories, verbally recalled or re-lived involuntarily, may flood the focal plane of my "living 'now'" in a continuing thematic upsurge, complete with its own retentional and protentional auras and with the now new index "again", or "for the nth time", (if what it contains is something "I cannot get out of my head", like a persistently recurring melody). But, fundamentally and inescapably, the "living 'now'" *is* the upsurge of primary givenness as primary perception, whether the objects of primary perception occupy the focal plane or fuse into the horizonal background, as they do whenever memories verbally recalled, or involuntary memories, imaginations or the apprehension of eidetic objects intrude. All these "happenings" surging in my "living 'now'" fuse at first retentionally, as they occur, later as sedimentations of my past into what Husserl chose to call a "stream of lived experiences", ever receding into the past, ever upsurging from its originary present, with its ever changing protentions towards an ever opaque future.

As the "origin" of this stream — in point of description, rather than in any causal sense — the "living 'now'" is, in effect, the origin of "time" as I experience it[20] and the anchor point upon which stands the individualized, concrete ego I am. We have seen so far how Proust's novel exhibits the thickness of the living present and presents structures similar to the ones Husserl recognized it to have. Does the living present of the narrator exhibit as well the originary primary perception which Husserl found to be the source of the "living 'now'"? If it did, the motives for undertaking the comparison we have launched would be greatly strengthened and we would be one step closer to showing in what way Proust's work provides Husserl's with an unexpected independent corroboration and contributes as well a wealth of further data to the phenomenology of subjectivity. Before we can turn back to Proust's text for an answer to this question, it will be well to analyze more closely the relevant fundamental characteristics Husserl found in the consciousness of internal time. It will suffice to the task of the present paper that I limit myself to what Husserl described in the *Lessons on the Phenomenology of Inner Time Consciousness,* [21] although a more extended treatment of what Proust has shown would require that we take into account not only Husserl's later works on time-consciousness, but some of my own researches as well.

The guiding theme for these *Lessons* is apparently a description of the operation of constitution of temporal and a-temporal objectivities of whatever sort — real, imaginary, ideal — by any concrete awareness whatever. These descriptions were undertaken in answer to questions left somewhat in the dark in the *VIth Logical Investigation*[22] and they provide such an answer up to a point. Given the descriptive schema "intentional operation", or "noesis", focus of such an operation (or "meaning", or "noema") which I explained briefly at the beginning of this essay, the constitution of temporal and a-temporal objectivities can be accounted for in terms of operational structures and corresponding foci. The *Lessons,* therefore, culminate in a flat "mapping" of the noetic structures required for the constitution of temporal and a-temporal *noemata*. If they stopped there, the lacuna left in the *VIth Logical Investigation* would not have been filled. For in the constitution of *temporal* objectivities the constituting agent is itself essentially implicated, as against its essential distance from a-temporal objectivities. The agent itself *is* a temporal objectivity. The action of constitution — the action of performance of an intentional operation — *is* a temporal objectivity. Hence, behind the "flat" mapping involving operational structures and their foci, when these concern temporal objectivities, lurks the contour-map of the acting agent, that is to say, the phenomenology of *performing* consciousness. Such a phenomenology is different from the previous one — usually called "static" — in that it consists in the recognition within consciousness of the different "regions", occurring at different (deeper or higher) levels, within which the distinct types of activities which a consciousness performs are seen to "take place". Thus, some "activities" are seen to be "prior" (in the "genetic" sense that other activities appear as complication of these) and said to occur at a deeper or more elementary level. A new, "genetic" phenomenology fleshes out the previous "static" one with the full weight of performance and presence. This new phenomenology has its birthplace in the *Lessons* which contain the first formal presentation of some of its most fundamental features, such as the recognition within awareness of various "regions" characterized according to their relationship to a *primal upsurge,* as we have seen. This is an upsurge of lived mental experiences. They flow as in a stream within which all other constitutive activities take place. The "absolute, temporally constitutive flux of consciousness"[23] is the most fundamental region of the live awareness. It originates in an "all together", an "all-at-once" of "primal sensations" which embraces all in the total "now",[24] as we have seen briefly. The peculiarities of this "total now" and of the so-called flux at this originary point are that neither can be described as changing, or genuinely flowing, or evolving with respect to form, although, as a later move, contents may be selected which "have a place" in a succession and recede in the flow with respect to the self-same upsurging "primal now".

With the recognition of the "primal now" we have reached the absolute originary source of enactment of a particular consciousness. To go up from this source-point back to the unities of meaning constituted by a concrete ego actively concerned involves the most intricate phenomenological descriptions. The source-point is the level of the barest "live-awareness", a mere sentient presence recipient of a flow it endures as it perdures in numb passivity. Reactive presence, active involvement with selected aspects of this flow is already a higher level. Higher still is the level at which some of these aspects are numbly lived as its own by a live awareness which is as yet neither reflexibly conscious nor unitarily aware of itself as ego. Above are the levels of active ego recognition and involvement, levels at which a Platonic *doxa* may be constituted and beyond which the *epistemic* layer of a-temporal *eide* and universal truths beckons.[25] Husserl, characteristically, takes over the Platonic terminology but with modifications of his own which we need not consider here.

In the *Lessons*, a gap is left between this primal region and the level of egoic intentionalities, which Husserl tried to fill in his later studies on *Passive Synthesis*[26] and to the filling of which I see my own researches as contributing. For the present state of the analysis of Proust's novel, the descriptions of *Lessons* may be deemed sufficient which do not as yet distinguish clearly the finely separated layers of conscious life subtending active ego involvement at intellectual (that is to say, *doxic* and *epistemic*) levels. These higher levels are treated as simply contiguous to the level of primal sensation and up-surge, and the genesis of doxic and epistemic elements from this primal level is treated as if it were intentional in the sense of acts of consciousness as described in the *Vth Logical Investigation.*[27] That this cannot be so, that below a certain point no active ego involvement takes place though a "con-scious life" occurs, we need not go into any more explicitly here. For the time being, the vocabulary of the *Lessons* is sufficient to describe in details adequate to our purpose what the *Lessons* call the "living 'now'" of primal sensation, as if intentional involvement were indeed simply contiguous to it.

Concentrating now upon this primal level we see that neither simultaneity, nor flow, nor indeed any "temporal" adjective can be predicated of it. All these predications occur on the basis of the configu-rations the primal now exhibits *now*. And indeed primal sensations can exhibit a wide range of confi-gurations some of which appear in present immediacy as "recessed-receding" with respect to some others which cannot be called simultaneous with them, but which are best characterized as "in the same plane", to use a geometrical metaphor (a later usage of Husserl's to denote the primal now is "Praesenzfeld" or "field of presence"; this warrants the analogy I am making here).[28] As in antiphony simultaneous sounds appear to come from different points in space "all-at-once", so in primal sensa-tion some elements appear to come from different points in time "all-at-once". This paradoxical pre-sence of the past in the mode of "recessed-receding" in the primal now is the modal difference upon which higher conscious activities such as retention are enacted. But it would be phenomenologically wrong to call this difference in mode, at this level of appearing, a "temporal" difference. Antiphonic sounds are all heard in the same place, which is why some "sound" more distant than others *from that plane* and without any higher interpretative move. Such an interpretative move can be the immediate manner of attending to what is heard, it will be constituted on the specific difference of the immediate hearing in the immediate plane of the sounds as they gave themselves "all together – all-at-once", albeit in different modes. A similar example could be provided by perspectival drawing (in which all the lines are in the same plane all-at-once, but some give themselves as receding with respect to others even though they appear all together in the same plane).[29]

Exhibiting, therefore, the "perspectival" modalities I have just been describing, a "primal now" may also be further complicated by interpretative intentionalities which, in the a-temporality of the actual now, set in sharper relief these very perspectives, as when one "sees" the drawing perspectively, or "hears" the sound antiphonally – i.e. when one makes thematic (one pays attention to, one focuses one's attention upon) these perceptual differences, (when one makes these differences the object of an intentional act of some kind).

Husserl recognizes two such interpretative intentionalities: a "transverse" intentionality and a "longi-tudinal" one.[30] These terms have been chosen by him with reference to the governing metaphor he uses to characterize the perceptual depth of an actual now, namely, the "stream of lived mental ex-periences". In that depth viewed as a stream, "transverse" is the dimension along which unitary appea-rance is given to perceptual elements "whenever" they may "appear" on the depth dimension (one might say that the "transverse" intentional object is secant to the depth dimension);[31] "longitudinal" is the side of such a perceptual unit parallel to the depth dimension. Thus, for example, "if I adapt myself to the 'longitudinal intentionality' and to what is self-constituting in it, then I turn my reflective regard from the sound (which has endured for such and such a period) to what is new in the primal sensation at a point in the before-all-at-once and to what is retained "conjointly" therewith following a continuous series. What is retained is past consciousness in its series of phases (first of all, its preceding phase). Then, in the constant flowing-forth of consciousness, I grasp the retained series of expired consciousness with the boundary-point of the actual primal sensation and the continuous shoving-back of this series with the fresh onset of retentions and primal sensations."[32]

In the *transverse* intentionality, on the other hand, "if I orient myself on a sound, I enter attentively into 'transverse intentionality' (always experiencing unity in primal sensation as sensation of the actual tonal now, in retentional modification as primary remembrances of the series of tonal points which have expired and in the flux of retentional modifications of primal sensations and retentions already on hand); then the enduring sound is present there, ever widening in its duration."[33]

To be sure, Husserl later abandoned the use he made here of the term "intentionality" when he described the same syntheses, the better to mark the fact that prior to any active intentional involvement a "passive fusion" takes place, but we do not need, at this point, to go into such descriptive refinement. As a matter of fact, the comparison is more easily made between the "living 'now'" as "primal upsurge" described by Husserl then and corresponding structures in Proust if we keep the macroscopic lenses on for a little while longer. The question now is, are there in the novel, in the moment we have isolated as the present of the act of narration, instances of antiphonal structures appearing in the single plane of encounter (the single plane of a "living 'now'" complete with its "primal perception") similar to the ones which Husserl has so carefully – and abstractly – described, and traceable both to similar "intentionalities" and a similar upsurge of "primary givenness"?

IV. AGE AND THE GUESTS OF THE PRINCE DE GUERMANTES

6.

Bearing in mind the "stereoscopic" intentionalities Husserl has characterized, we can immediately recognize them in a number of concrete examples Proust describes with his usual care. Starting with the encounter with M. de Charlus on the way to the matinée of the princesse de Guermantes (formerly Mme Verdurin),[34] and continuing with the famous "coup de théâtre" experienced by the narrator upon entering the main salon,[35] we are led by the narrator from surprise to surprise as we discover, with him, in the appearance of people we cannot at first recognize, the distant traits which make them who they are, buried within the present traits with which time has made them other. On the one hand, "the prince still has, as he receives, this good-natured look of a fairy tale prince [the narrator] had discovered him to have the first time (they) met";[36] on the other hand, this particular look has achieved an immeasurable "temporal"depth with respect to the white beard and the dragging feet which the present encounter brings. In the duc de Chatellerault only "a tiny sparkle in the glance" has remained "the same" and enables the narrator to recognize the young man he had once met at Mme de Villeparisis.[37] In each case, the pattern is the same: the present encounter contains in antiphonal distance juxtaposed elements some of which bespeak temporal depth with respect to others, or, better said, upon some of which the longitudinal intentionality enacted reaches into a very distant past (appears as originating from a perspectively very distant point). The element of surprise is essential here to guarantee that what we are confronted with is genuinely a perceptual datum. Both the surprise upon meeting someone unexpectedly and the accompanying shock of recognition serve to emphasize that what is taking place is a direct encounter in immediate upsurge, that what we are witnessing is in each case a living "now" in the sense in which Husserl meant this in the *Lessons*. On the other hand, there is nothing intrinsically surprising in this living now exhibiting elements on the basis of which temporal distance may be constituted, thanks to the stereoscopic effect of the joint intentionalities with which the living now is apprehended. Note that this is not a matter of memory in any usual sense. The narrator in the cases mentioned does not *first* meet a stranger, *then* in memory run through a rogue's gallery of possible identifications *finally* to settle on the most plausible one. There is never a doubt that *this* is M. de Charlus, *this* le prince de Guermantes, *this* M. d'Argencourt, etc. ... But *this* is each, *changed*; that is to say, each in unities which transverse intentionality constitutes at a distance from the corresponding present unities it also constitutes. Longitudinal intentionality apprehends this retentional distance as all the greater, that the originary unity announces itself *immediately* as placed farther away in the stream of lived experiences the narrator is. "Announces itself immediately", that is: *presents itself in this way*, not "is remembered this way", for what is described is not remembrance but immediate apprehension. As against this, there are, of course, instances in which the narrator *has* to have recourse to memory, and these instances are very carefully recorded by Proust — the case of Bloch is one such. Bloch has changed his name, his appearance and even his voice, having preserved from yesteryear only a peculiarity of pronunciation.[38] Gilberte as a fat lady is also hard to recognize and the narrator does, in her case, try out identifications suggested by memory; in this instance, he mistakes her momentarily for her mother, the former Mme Swann, only to be set aright by Gilberte who tells him: "You thought I was my mother ...".[39]

Finally, there are cases in which even though immediate recognition is warranted and correct, the differences evidenced in direct perception are too great and the narrator finds rather that a stranger has taken the place of the person he once knew, though this stranger be the same person (for example, the case of the former school friend with the darting and laughing blue eyes whose glance has become fixed but who has preserved the same laughter).[40]

7.

There are more examples of this sort, but I think this is enough to demonstrate that in those instances, what, in his own way, Proust characterized, was the primal upsurge of a living "now" exhibiting the different antiphonal resonances Husserl had described. Only, Proust did this in the concreteness of specific instances, rather than in the full generality of a technical language. To the possible objection that instead of *one* living "now" we have a series, the answer must be that this is a pseudo-problem: I can only have a series *now*, that is, from the absolute perspective of the now in which these other instances appear in antiphonal distinction. It may also be objected that, whereas Husserl uses throughout the *Lessons* the metaphor of the "flux" of consciousness, of the "stream" of lived mental experiences, I have consistently referred to a *plane* or to an a-temporal, absolute "now" within which specific points exhibit different "perspectival" locations. To this I can only answer that I find the metaphor of the *stream* dangerous and misleading, and point to the text in which Husserl warns against taking it too literally.[41] The analogy with antiphony seems to me to portray more faithfully what he described than does the "stream", even though phenomena such as the "receding" of a present instance, first in retention, then in the past, are more easily portrayed by a flow than by the rather static antiphonic distance. Both the "flowing stream" and the "static antiphonic distance" capture some elements in the living "now"; neither accounts for all. The challenge is not to let either blind us to the elements represented only by the other — a most difficult task. Finally, it may be objected that in the encounters described, no genuine example of retention or protention has been singled out for observation. This objection stems from a misunderstanding of the nature of the living "now". It is not from elements in the living "now" that retentions and protentions are "later" constituted. Retentions and protentions are immediate *modes of apprehending* elements in the living "now". Thus, the unity of presence of the prince de Guermantes, or of Gilberte, throughout the conversation is a retentional unity which is constituted in the flow by both "intentionalities" working together. The "expectations" the narrator has with respect to their appearance and behaviour are also so constituted and maintained in the living "now"; even as it transforms itself in the varying antiphonal index of its contents. This, in effect, is what "protention" means, as Husserl tirelessly points out.[42] This *is* the thickness of the living "now", or of what Husserl will later term the living present.

V. OF MONADS, TEMPORALITY AND ART: THE PROUSTIAN VIEW

8.

Granted, then, that the structure of the one action Proust's novel describes is that of a "living present" in Husserl's sense, we should now be able to recognize other features of such a living present, more particularly the "temporalizing" and the personalizing which such a present generates. At the root of all phenomenological reflexion (as we have briefly seen) is a consciousness, an absolute concrete subjectivity, not even personalized or polarized about an ego but from which such an ego is generated. This monadic[43] "live awareness" *with* its stream of lived mental experiences forms the fundamental originary source from which all other objectivities are constituted. This is not to be understood in a *causal* fashion. We are not re-introducing here ontological or metaphysical presuppositions. Whatever the ontological dimensions may be which best describe the live-awareness, its "stream" or "antiphonal plane", the objectivities it finds in it, and the relationships between these three descriptive "regions", such ontological dimensions would be out of place in a purely phenomenological description. Hence, we cannot presume to characterize in any causal model, or any other model, for that matter, what is described here; all we can do is recognize the mode of appearance of these different elements, and let ontology become a later task.

Perhaps no longer unexpectedly, Proust, through the narrator, very meticulously renders visible in direct concrete experience the "temporalizing" which a live-awareness is and in so doing goes beyond (or perhaps "beneath" might be a better term) the recognition of antiphonal distance we have just analyzed. That is to say, in so doing, he makes visible the unique locus of such temporalizing, namely the narrator himself *qua* live awareness.

The first example of this we shall take up is the culminating moment of the novel and one of the loveliest "pictures" in the cathedral it purports to be. From it as from a rose window above the main door radiate all the themes which the faithful upon leaving the edifice should bear with them. I am speaking, of course, of the meeting with Mlle de St. Loup. As Proust is at great pains to show, she more than any other person, summarizes the full path of the narrator's life, the unification of all the paths of Combray, the path which went by way of Swann's property as well as the path which went by way of the Guermantes'. Moreover, in the flesh, she embodies both (a last minute addition to the text underlines this intent on the part of Proust).[44] Thus, the narrator says: "Time, colorless and ungraspable, ..., had materialized itself in her."[45] But the reader asks as Proust wants him to: "Whose time?" Certainly not hers. Her interests are elsewhere and for her, as the brief sketch of her future shows,[46] other things are more important. We are unavoidably referred back to the narrator himself who, in the immediacy of a living present *sees* in her not simply a pretty face, but the depths of his own temporal dimensions. Her face is that of Robert de St. Loup, and it resonates throughout the narrator's life in immediate retentional depth. Her nose ends in a manner characteristic of her mother's and her grandmother's, and there, too, echoes are immediately awakened which reverberate throughout the narrator's own history, not intellectually, at first, but rather in immediate perception: this is who she is *for him*. She is *his* temporal thickness embodied, beyond her years, in ways she will never *really* know.

To be sure, had not the narrator just been involved in the kind of reflections he has been doing ever since the uneven paving stones in the courtyard of the Guermantes' triggered yet one more experience of "involuntary memory" (as Proust calls it),[47] Mlle de St. Loup might not have been perceived by him immediately and unreflectively in this way. She might have been yet one more "jeune fille en fleur", a surrogate Albertine — which is what he was looking for when he asked Gilberte to introduce him to "young girls", a delicate task at best, an indiscrete request in any case, which Gilberte very deftly neutralized by introducing him to her own daughter.[48]

But the narrator *had* been busy with the presence of his own past, not simply in verbal remembering, not even in the usual kind of vague, half-fantasying, half-re-living remembrance but in what Husserl terms primal memory, in genuine retention. Hence for him, Mlle de St. Loup appeared as the other living embodiment of the past he *was* in its totality and in its full, live presence to him in retention. Which, of course, she is but only for him. Proust, in this way, marks the unbridgeable distance between the otherness of the other, no matter how closely associated with our life he may be, and the monad each of us is. Mlle de St. Loup, for herself, is altogether irreducible to what she is for the narrator. As a cipher on the walls of the narrator's monad she is representative of a temporality, a subjectivity which she, in the hermeticity of her own monad, is not at all. Thus, the very cipher she is, in herself, throws the narrator back upon himself and upon his own absolute isolation in a paradoxical solipsism. Equally paradoxically, the very opacity, the very uniqueness she is, the fact that her appearance in the living "now" of the narrator is not reducible to his experience of her but announces itself as irreducible to it, all this breaks the solipsistic circle. The living "now" is confrontation, upsurge, not auto-creation. The mode of appearance of what is genuinely given in the living "now" is this opacity against which protentions extend their fishing nets, filling them only some of the time.

The narrator, at the time of the encounter with Mlle de St. Loup, has learned that the other cannot be completely and fully absorbed — or *known* — and enjoys the new lighting her presence throws upon his own presence, rather than try to capture in her the tangibility he had been unable to find in himself, that is, until his reflections in the library of the prince. In fact, what *is* changed about him is that, although he finds time to have materialized in her, he no longer harbours the illusion that this materialization is better than the one it might achieve through him, through his work, more particularly through the work of recording its trace both within and without him in a work of art of his own creation. For the narrator has learned to see how he himself as "live-awareness" structures protentively what he encounters. The corollary to the opacity of the other is that that part of the other which is visible to us is visible *in fact* under the lighting *we* provide. The other, in its aspect for us is in part our creation. We, as monads, have each an "index of refraction" all our own which accounts for our opacity to others, but accounts as well for the slight distortion between what we expect from others — or see in them — and what is really there. In concrete encounters we never have the eidetic transparency of essences giving all at once its object in ideal purity.[49] Rather, we propose, and the object disposes, answering as best it can our expectations. Thus, for a large part we create what we experience, though for a large part only, since the resiliency of the object in any concrete encounter determines the range of fulfillment possible. The meeting with Mlle de St. Loup marks this distance. As a surrogate Albertine, she is more and she is less. As the other upon whose tangibility the narrator once would have placed all his hopes of happiness she is no longer sufficient. The real other (as we shall see more explicitly further on) is the *me* in the very temporal extension she materializes so well and so unsatisfactorily, now that the narrator realizes that the *me, in its otherness,* is the world as he has lived it, *his* lived-world, with *his* resonance, diflected by *his* index of refraction, thick with the sediments of *his* creative protentions, that is to say, thick with *his* memories when they are held in the full depth of *primal memory,* in the full depth of retention. For this "his-ness" bespeaks distance: the narrator *is* his past, but he is more, his past is *his* past protentions as well. That is why *his* materialization of *his* own past, and Mlle de St. Loup's materialization of *his* past are not congruent. Leaving aside Mlle de St. Loup's own past and judging her strictly as a representation of the narrator's temporal distance (since she is portrayed by the narrator as representing *his* temporal distance), we may indeed look upon her solely as a possible symbol for his temporal distance, as the narrator briefly allows. As "embodiment" of this elapsed time, that is immediate objective, bodily, or perceptual presence (in primary perception) of this time, she may seem to be able to lend *her* tangibility to its continued presence, as marble lends its tangibility to the sculpture. And, indeed, as we shall see, the temptation to find a surcease from time in another's tangibility is one of the most fundamental reflexes of the narrator's, manifesting itself as early in his life as the "drame du coucher". But, as "carrier" for this "symbolic meaning" she has an adequacy that is limited by her own protentions which lead her to discard in her own life one of the major axes in the narrator's, namely, his "social climbing", or his "social awareness". Her own protentions do not involve at all social climbing; hence, even as a present materialization of *his* temporal thickness, she is not adequate since she is lacking the direction of his present, its "polarities", its *index of refraction.*

If, lacking this, she were to be considered an adequate image of the materialization of time he experienced in his past, this would mean that the narrator would be reducible to his past as a completed — and, at the end of his life, absolutely completed — chain of events and as nothing else. His protentions and the *index of refraction* through which they *filter* (I am using this word as it is used in photography) and assume the shape they do in the "living present" would be left out of consideration altogether. But this would do an intolerable violence to the phenomenological data as it is presented for us to see. The link between past and protentions is the particular "filter" the narrator has been and still is, through which the upsurge of his lived experiences take their relief and color. If one were to talk in a Sartrian vocabulary, what I call "filter" here would be identifiable to the "project", this most fundamental relationship to Being, each "pour-soi" is, which can be apprehended only by means of "existential psychoanalysis". But in adopting Sartre's ontological vocabulary, one forces upon the myriad colorings this fundamental project may take in its actualization, the oversimplified diadic structure based upon the distinction between Being-in-itself and Being-for-itself. Thus, the "project" would be reduced to two possible forms: absorption in Being, or distanciation from Being and the eventual diag-

nostic would be rather uninteresting, if it were to remain this spare. Even Sartre has to resort to analyses placed erroneously by him at a higher level than this one and having to do with contextual descriptions through which the enacted project-structure can be apprehended. Proust, on the other hand,
shows concretely that the enacted filter is not simply reducible to a diadic structure; the world of the
narrator, as I plan to show in more detail later, has *three* main polarities visible already in the "drame
du coucher": the pole best represented by Swann, the one represented by the other in that scene, and
the one represented by the grandmother whose "presence" by way of the books she had planned to
give the narrator is nonetheless as important as the other two.

Thus, not only is a life completed in death not simply the sum of one's actions as Sartre says, but the
project is more than what is visible through one's actions; it is a complex "filter" with, possibly, a
larger number of polarities than the Sartrian two. This "filter" it is which sets in relief or places in the
background the features immediately apperceived in the "living present" through what Husserl called
"passive synthesis". Only on the basis of this "three dimensional picture" the "filter" produces passively — does the project manifest itself. The project is therefore a "later" move occurring on the face
of the more fundamental passive syntheses. What the passive syntheses "described" in the novel are,
we shall see later in more detail; suffice here to say that, against the *me in its otherness* which is my
personalized lived-world as I have lived it, there is the *me in its ownness* which is the "filter" I have
just mentioned. To describe this filter one must make perceptible its index of refraction — the *index
of refraction* the narrator *is* in his protentions and has been in the past (and *is* in *his* past, just as Mlle
de St. Loup is *hers* in *her* protentions and in *her* past). Hence, no surrogate embodiment from out of
one's lived world will do as an adequate representative, and Mlle de St. Loup, who comes closest to
such, nonetheless fails ineluctably in this. Only the gaze of the narrator can light the rose window she
is with the brilliance of meaning and find back, through her but not in her, the sacrificial focal point
upon which his own life has played out its rituals and its passion. To enable another's gaze to plunge,
like the sun's rays, through the diflecting panes of stained glass and focus upon and animate once
again the stilled passion of the narrator, one would have to re-create not only the colors of the glass,
but also their *index of refraction.*

The challenge to be met by the work of art, and by the work of art alone, is to preserve this very index
of refraction and to make it perceptible to others in their monad. What the other as the object of love
achieves only through opacity, hiddenness, betrayal or lack, namely, autonomous presence, the work
of art alone can present in total possession, for the work of art alone can transmit the direct experience of the other through the index of refraction the artist *is* and hence restitute to *both* the artist
and the other their monadic tangibility.[50] This is why the truly great painter paints only one picture,
though through many canvasses, and the great novelist writes only one novel, though through many
titles, as Proust makes the narrator remark.[51]

The Proustian lesson is here very clear and far richer than Husserl's less wide-ranging account. Still,
after briefly recalling the points at which Husserl's account coincides with Proust's, let us follow Husserl to see whether the eidetic structures he recognizes give us the skeleton of yet one more Proustian
experience: involuntary memory.

VI. OF MONADS AND THE RETENTIVE UNITY OF THE FLUX OF LIVED EXPERIENCES: THE HUSSERLIAN VIEW

9.

The dual constitution of the other and of the me, that is to say, of the sphere of ownness of the concrete ego, are the major themes of the *IVth* and *Vth -Cartesian Meditation* of Husserl, in which he borrows the term "monad"[52] from Leibniz. The constitution of the other as alter-ego is taken up by Husserl as the key to the constitution of intersubjectivity. The problems of interpretation of this text are enormous and it is known that Husserl himself was dissatisfied with it. However, if one bears in mind that its object is an eidetic description of the constitution of the concrete ego, that is to say, a description of the universal structures required by transcendental subjectivity for the meaning "concrete ego" to be focused upon, then the text becomes somewhat less tangled. A concrete ego can then be seen to be anchored upon a subjective pole which organizes in a stream of lived mental experiences some of these experiences into a "sphere of ownness"[53] — that is to say, calls them its own, recognizes them as its own. For this subjective pole, the world is a lived-world, the world of its experiences. As an oriented, active consciousness, this subjective pole has passively synthesized the unity of its own flux of experiences in a primal self-apprehension similar to but not identical with what Sartre was to call the pre-reflexive cogito.

Husserl's own descriptions of the genesis of the concrete ego, in the *Cartesian Meditations*[54] and elsewhere do not go beyond what were perforce very abstract terms — when compared to the concreteness of the genesis of an actual living concrete ego, personalized in its social and historical origins and in its own history. Nowhere does Husserl pick up the challenge of capturing and representing the unique index of refraction through which such an ego lives the world, although his recognition of the fact that the stream of lived mental experience individualizes,[55] in its generality, provides the outline of the frame within which a unique personal experience takes place. To recapture and represent such an experience means must be found which will render immediately present — presentify — in immediate retention, or primal memory, enough of a stream of lived experiences for the manner (the tone) in which this particular concrete ego lives the wrold to become visible. It is the way in which the world resonates in me and through me that makes me who I am. It is this particular resonance which distinguishes the sphere of ownness of this particular *me* from every other and makes my constitution of the other within my monad a self-constitution as well. Whether representing this is at all possible is the question. Husserl answers it affirmatively in the following way: he asks in the *Lessons*: "Can I find and lay hold of at a glance the entire retentional consciousness of the past flow of consciousness, this retentional consciousness being enclosed in a before-all-at-once?" And he describes what needs to be done: "Obviously, the necessary process is this: I must first grasp the before-all-at-once, which is retentionally modified; indeed, it is what it is only in flux. Now, the flux, so far as it modifies this before-all-at-once, is intentionally in coincidence with itself. This constitutes unity in the flux and the one and identical element maintains a constant mode of being shoved back. An ever new element is joined on in front only to immerge again immediately as its momentary nexus. During this process, the regard can remain fixed on the momentary all-at-once which sinks down, but the constitution of the retentional unity reaches out beyond this and adds to the ever-new. The regard can be directed thereon in the process and it is always consciousness in flux as constituted unity."[56]

Perhaps Husserl did not have in mind quite the extent of presentification to which Proust makes his narrative pretend. No law of essence can be eidetically apprehended, however, which would prevent the kind of total recall the narrator makes retentionally present ot himself in the library of the prince de Guermantes, only contingent limitations due to the psychological differences and particularities each one of us possesses. Those are *causal*-limitations, not *formal* ones, as when we say that we are unable *actually* to count past a certain number, we are not describing a formal limit to the iterability of the operation of addition, but we are describing limitations due to our own psychological make-up, essential to this make-up though they might be. Hence, there is no *a priori* formal limit to what can retentively be held in presentification.

As Husserl writes: "The flux of the immanent, temporally constitutive consciousness not only *is*, but is so remarkably and yet so intelligibly constituted that a self-appearance of the flux necessarily subsists in it, and hence the flux itself must necessarily be comprehensible in the flowing. The self-appearance of the flux does not require a second flux, but *qua* phenomenon it is constituted in itself."[57] Thus, the flux both subsists in itself and can appear to itself, i.e. constituted instances may be recalled representified, but, better still, the full extent of the flow from the present upsurge to some "beginning phase" may in principle be the object of such a retention. However, one must not forget that the primal now within which I consciously re-presentify such a past stream is still an upsurging primal now whose properly perceptual contents, while not thematic, are nonetheless perceived, though not in a consciously apprehending act. (This is why Husserl resorts to terminology such as "primal intentionality", "passive synthesis", "primary receptivity", "apperception", when referring to the contents of the primal now, of the living present). Thus, in the retentional apprehension of the self-appearance of the flux, in the re-presentification of the immediately past phases of the flux, "the constituting and the constituted coincide, yet naturally they cannot coincide in every respect. The phases of the flux of consciousness in which phases of the same flux of consciousness are phenomenally constituted cannot be identical with these constituted phases, and they are not. What is caused to appear in the momentary actualities (momentan Aktuellen) of the flux of consciousness is the past phase of the flux of consciousness in the series ot retentional moments of this flux."[58]

Husserl is very clear in distinguishing "retention" from "reflection" and in emphasizing the a-voluntary aspect of retention. "Retention itself is not an act of looking back which makes thematic an object of the phase which has expired. Because I have the phase which has expired in hand, I live through (durchlebe) the one actually present, take it — thanks to retention — "in addition to" and am directed to what is coming (in a protention)."[59] And again: "Retention is itself not an 'act' (i.e. an immanent unity of duration constituted in a series of retentional phases) but a momentary consciousness of the phase which has expired and, at the same time, a foundation for the retentional consciousness of the next phase. Since each phase is retentionally cognizant of the preceding one, it encloses in itself, in a chain of mediate intentions, the entire series of retentions which has expired."[60]

Is it possible for some occurrence in the primal upsurge of the living now to trigger a kind of double exposure within which a past phase of the lived flux might be presentified together with the non-thematic apperception of the primal now of sensation? There is no law of essence against it. The mechanics of this or that occurrence are psychological laws having to do with this rather than that objective entity and do not concern us here. The retention may be one of a past instance presentified rather than remembered — i.e. re-lived — or the retention may be a present primal fusion of a "something" originally given, *or an overlapping of both*; these are the modes of appearing of retentions as eidetically given; they are, therefore, co-possible modes. Husserl does not seem to have described the third one. Proust, on the other hand, most certainly did, together with Chateaubriand and Baudelaire whom he mentions as his precursors in this regard. Husserl, at any rate, does not rule out expressly this third mode; in fact, in the passages quoted, he leaves room for it. Perhaps, since at this stage in his investigations he was interested in distinguishing between rather different operations of appearance such as retention, on the one hand, reflection, on the other, he did not deem it appropriate to go into all the modalities of either at that point.

Proust's interest, as we have seen, was the very presentification of enough of a stream of lived mental experiences to render visible what I have called the "index of refraction" with which the concrete ego generated in that stream experiences – lives – the world. As we shall see, as against the fairly commonplace (as Germaine Brée points out)[61] so-called psychological laws with which the narrator continuously regales himself (and us), the transcendental truths for which Proust was reaching through the history of the narrator, were the laws of the genesis of just such an index of refraction, of just such a concrete ego. For Proust, therefore, the mode of retention which could serve him best was the third one, the one through which the lived temporal thickness of the flux would be itself presentified in immediate retentional givenness.

VII. INVOLUNTARY MEMORY AS THE PROUSTIAN ANALOGUE TO PHENOMENOLOGICAL SUSPENSION

10.

This brings us to the second example (in this case, a series of examples actually) of the manner in which Proust makes visible in what way the narrator himself *qua* live awareness is the unique focus of temporalization: namely, to *la mémoire involontaire, la mémoire proustienne*. The description of those instances are justly famous, from the bite in the *madeleine* while drinking a cup of tea out of which all of the childhood world of Combray resuscitates,[62] to the uneven paving stones in the courtyard of the Guermantes',[63] the ringing of the spoon against a glass,[64] and the feel of the damask napkin.[65] In each case, the primal upsurge of sensations brings in the living "now" an encounter with an antiphonal immediacy that has all the qualities of a photographic double exposure and is suffused with a feeling of happiness and relief. This is so fleeting that the first time this happens, in the case of the *madeleine*, the narrator perceives only the feeling of happiness and the sensation of something trying to make itself manifest. He then bites again into the *madeleine*, finds the sensation still there but diminishing, puts off for a while further direct stimulation by other bites, and, later, eventually succeeds in allowing the world of his childhood to surface. The method is to allow himself to become open to whatever it is that tried to manifest itself in whatever way it wished to do so at the prompting of his senses. The narrator's explanation for this is strictly causal; it is the direct sensation of the taste of the *madeleine* which, by virtue of being intimately connected with a particular set of past events, "caused" the experience of these events to surface in memory. It did not cause them to appear with the vague fadedness of something recalled, rather it caused them to re-appear with the full intensity of presentness, so much so that, in being thus re-lived, although they bear the index "re-lived" they are almost successful in pushing out of even horizonal presence the immediate sensory promptings of the actual living "now". They are *almost* successful, but not completely successful, as the narrator stresses for the primacy of the sensory upsurge the living "now" is, is not to be supplanted by any other, and the index "re-lived" maintains in antiphonal distance what is thus "presentified", to use Husserl's term.[66]

What is thus brought back "involuntarily", and lived "again" with full retentional and protentional intensity, is an instance lived "already" by the narrator, and it is brought back with exactly the features it had as lived by the narrator the first time. The immediate certainty, one is tempted to say, the apodictic evidence with which these events presentify themselves is stressed by Proust time and again, and is contrasted by him with the unreliability of memory [taking this term in its usual sense].[67]

The immediate perceptual differences between the two consists in the fact that what involuntary memory brings back occurs at the prompting of a sensory trigger and has the power of almost totally supplanting a particular retentional and protentional thickness by another. Only the continuing presence in primal upsurge of contradictory sensory promptings, horizonally or non-thematically, prevents the presentified "thickness" from being "all there is" and gives it the index of "remembrance". The narrator is immediately aware that what has happened is the recapturing of a part of his own lived experiences from his past, and that this recapturing, by virtue of being "the same", albeit with the index "again", demonstrates the iterability, under specific conditions, of the lived, and hence its potential a-temporality, if by time we understand a unique flow of events "which never come back." Time, in a sense, is thereby "recaptured", however fleetingly, in the very uniqueness of the concrete manner in which it was lived. A happiness comes from this the sources of which will have to be investigated further. Suffice it to say here with the narrator that it involves being unconcerned with death, delighting in the a-temporality achieved in a thickness of presence which goes far beyond the usual retentional-protentional living "now", even though it does so by a jump back into another retentional-protentional "now" which is now past, rather than extending back to this past now in an unbroken retentional connection. In spite of the "jump", one might even say *through* it, the presence of the narrator to himself becomes thematic in the apodicticity of a *cogito*. (Pascal Fieschi, in his very rich article on "Le temps retrouvé" mentions just such a "cogito").[68]

The narrator apprehends *himself* in *his* "temporality" and uniqueness. This, as against the *"pure"* cogitos attempted by Descartes and successfully carried through by Husserl in *Ideen I,* is a *concrete cogito* one which is concerned with the apodicticity of concrete presence, rather than with the apodicticity of *a priori* or formal presence in *every* act of knowing, with which both Descartes and Husserl concerned themselves. Whereas, one might say Descartes and Husserl used the cogito to demonstrate apodictically the formal foundation of knowledge, Proust, in involuntary memory, discovers a cogito which demonstrates apodictically the occasional, or actual foundation of knowledge: i.e., makes manifest to itself the unique unity of a particular stream of lived experiences as it exists in the living present. And he does this, as Husserl had stipulated it would have to be done, in that stream itself, marking clearly as well the slight "décalage" Husserl was careful to note between the flux as the primal upsurge continues to generate it, and the past phase as presentified, a "décalage" which results in the unsuccessful "fight" between the phase "presentified" and the continuing upsurge in originary givenness of sensory promptings in "horizonal" presence.

The immediate benefit to be derived from the presentification of such past instances is that, although one lives them "lengthwise" as it were, in their full protentional and retentional dimensions, they can become the object of detached descriptions. Their objects were the foci of intentional involvements which now themselves become manifest in being "re-lived" under the mode in which involuntary me-mory presents them. In other words, *what* is being re-lived, is re-lived *passively*, in what amounts to a detachment from the activity with which it was lived the first time, and because of this, this activity may now itself become thematic.[69] The very passivity with which the whole episode is re-lived amounts to a "suspension" of involvement. Thus, the contemplation which involuntary memory per-mits amounts to a phenomenological contemplation in the best Husserlian sense of the term. The nar-rator may thus become aware not only of *what* interested him at the time re-lived, but also of the *modalities* of that interest, and of their *intentional structures*. These can now be described by him, phenomenologically in what amounts then to a "concrete" — as against a "pure" — phenomenology. And indeed, that is what the narrator does, rather tirelessly, throughout his long monologue. With this, one may say that the case for a comparison between Husserl and Proust has been made.

VIII. BY WAY OF CONCLUSION, A PRELIMINARY WIDENING OF HORIZONS BRIEFLY SKETCHED

11.

Let us pause a moment and review what has been analyzed. We have seen that the novel takes place in a living present, the living present of the narrator, which exhibits the features Husserl recognizes: pri-mary perception, a stream of lived experiences, individuated and individuating, retentional and pro-tentional thickness, the constituted unity of that stream (as lived unreflexively and as apperceived in the "concrete cogito" of involuntary memory), a sphere of ownness, a concrete ego or directed con-sciousness and, passively synthesing both these latter "unities of meaning" along with many others ac-cording to the foci of its "interests" (*its index of refraction*), a conscious life, pre-doxic and pre-egoic, and generative of the individuated "lifeworld" or "lived-world" within which all "higher" "unities of meaning" arise.

We have seen as well the Proustian refutation of solipsism and have all too briefly recognized the con-crete solution of the problem of intersubjectivity one might work out on the basis of Proust's descrip-tions. That this is not a *de-jure* solution but merely a *de-facto* one need not bother us at this point. What is more intriguing is that it falls within the phenomenology of knowledge described in Husserl's *Logi-cal Investigations* and can be shown to involve the classical structures act of signification, expression, intended meaning, objective referent, act of fulfillment, fulfillment, partial or non-fulfillment based upon the coincidence or lack thereof of the ideal structures of the intended meaning and of those of the objective referent. The resiliency of the objective referent, its resistance to the meanings the narrator attempts to foist upon it are among the most tirelessly illustrated "findings" of the narrator's. Examples of this are his discovery of the inadequacy of names and his realization that the truth about any one person can never be completely apprehended in any fully self-legitimizing apperception, be-cause anticipations are always betrayed (la Berma, Venice, the Guermantes), or because no one is ever reducible to any one facet or group of facets of his personality, that is, not even to those facets direct-ly observed (Charlus and Jupien, Saint Loup and his medal, not to mention Albertine and her friends), since "facets" may well be a function of the foci of interest of the observer, and of the modes of apprehension they allow and prescribe.

Thus, for Proust as for Husserl, the "lived world", that is to say, the world as lived, as polarized by an oriented consciousness, is the unsurpassable actual foundation of all objectivities. All of Husserl's work was devoted to the progressive uncovering within the world as he lived it of the universal structures present in any lived-world whatever, apodictically, and to the fully legitimized (fully "grounded") phenomenological description of these structures by means of a language and a method that were meant to guarantee the preservation of universality through expression and decipherment — both. All of Proust's work was devoted to the capturing and expressing as faithfully as possible the ideal uniqueness of the world as he lived it — which is not to be confused with some realist rendition of the same. What universal structures there are, were to pierce through and be made visible by the faithfulness of the concrete description.

They are not to be found in the Balzacian generalities with which the narrator draws the lesson of his experiences, however apt these may seem in the light of one's own practice of the world. Rather than these "laws" (based upon the usual kind of statistical extrapolations over a sample group deemed representative, and corroborated in the same way), the "transcendental truths" illustrated by the novel are "formal" truths, apodictically evident to any consciousness whatever because they are exemplary of what a consciousness is and does.[70] The method chosen is the exhaustive description of the example, not the description of a phenomenological "residue" as in Husserl. The "reduction" will take place when the reader, by transposing for the situations described by Proust, situations he himself had experienced, will sift, in a progressive testing, what cannot be detached from individual circumstances from what is independent of any such circumstances. In so doing, the reader will perform what amounts to imaginary variations upon the concrete example Proust so faithfully represents (this faithfulness understood in the light of what we said earlier about the work of art). The method we have followed so far illustrates the kind of procedure that is meant here although we have left out any explicit reference to personal situations. Instead, we have had direct recourse to Husserl's descriptions and have endeavored to show how they fit Proust's own as a substructure.

The recourse to Husserl, which may have seemed at first rather far-fetched, began to yield results the moment *the static substructure* of the novel began to emerge. No one need be surprised if the elements of this substructure exhibited underneath their concrete representation (use of the *passé composé*, immediate antiphony apperceptively grasped, retentional and protentional thickness, double exposure effect of primally perceived data, flow of lived experiences and apperceptive unity of this flow) correspond to elements described with full transcendental and eidetic universality by Husserl. Both Husserl and Proust, concentrated their efforts on the same object of investigation: "this wonder of wonder", as Husserl termed it, "subjectivity". Both, each in his own way, applied the same fundamental canons of scientific scrupulousness and clinical precision, albeit one did so with an eye to the transcendental, the universal, and the other with an eye to the concrete, the unique.

The result of our investigation so far is to force us to dismiss the two corresponding misinterpretations we mentioned initially. First, Husserl's "abstraction" does *not* require a particularizing *ontological* "foundation" in the activity of any particular or actual *Dasein* or *Pour-soi* for its ultimate legitimation. Its legitimation comes from its universality and its essential independence from any particular instantiation. This independence is the primary distinguishing feature of the "abstractive" regard: the *Wesensschau*. Its operation performable by anyone, the *Wesensschau* in principle and *a priori* has for ideal focus the essence of the objectivity so observed. Whether the objectivity observed in any one case yields this or that essence neither legitimizes nor disconfirms the Wesensschau as operation *a priori* possible; it only legitimizes or disconfirms the particular instance as a genuine, partially proper or totally inadequate instance of the essence 'x' (the "eidos" 'x'). Given that the subject of Proust's novel is a particular subjectivity, the applicability of Husserl's findings in his study of subjectivity is an *a priori* possibility stemming from the law of essence governing the general theory of phenomenological description. To find such features as Husserl had recognized clearly described, albeit in completely different terms, by someone whose methodological conceits differed markedly from Husserl's save for the unconditional obedience to "the things themselves" in their appearance, is to find an independent corroboration of the primacy of transcendental phenomenology as advocated by Husserl. Were the activity as founding as it has been claimed, were the particularizing circumstances of the actual *Dasein* or *Pour-soi* as unsurpassable as has been claimed by Heidegger or Sartre, then the Proustian descriptions would bear only the most distant similarity to those of Husserl, and the most trivial, and not the precise congruence on key features we have observed so far.

And, indeed, the view of Proust which makes him the meticulous recorder of "la durée concrète" and nothing else, eventually has to accuse him of gratuitous triviality, for the life thus recorded in its uniqueness remains under this view as gratuitously unique and as useless as a pebble of sand on a beach. A sterile and devastating nihilism emerges over and against Proust's own pronouncements which are simply set aside as a late-blooming intellectualism. Admittedly, this is an extreme view vitiated at the outset by the desire to demonstrate Proust's purported debt to Bergson, over and against the former's protests deemed at worst disingenuous, at best warranted only by the worst features of his novel. But I think that in this case the author knew what he was about better than the commentator.[71] *Proust was not mesmerized by the flow of time, he was mesmerized by the permanence of specific polarities in spite of the flow of time.* The great "themes" of the novel illustrate this, and among these, most particularly, the snobbery of the narrator, his unblushing social climbing, and the drama of jealousy, of the elusiveness of the object of love even in direct presence. A third theme, that of artistic creation, constitutes the third most important permanent polarity orienting the apparent diversity of the flow of the narrator's life. Together, these three structural elements are present in the "drame du coucher", and one cannot dismiss as an afterthought the remark Proust places in the mouth of the narrator at the end of his re-living this particularly harrowing experience, to the effect that his whole life was only the continuing resonance of this one moment.

As a string vibrates and provides the continuing tone it does by virtue of its length and the strength of the stroke that set it in motion, so does the narrator's fate unfold itself as it does within fixed boundaries. Swann, the mother and the grandmother are the major representatives of the poles defining the field in tension within which the avatars of the narrator's life will create the disturbances with which he will mesmerize himself and us. Swann, the prestigious and mysterious "mondain" whose visits shatter the security of the young Marcel, will long mesmerize Marcel the adult with the glitter of salons and lure him to search for a token of tangibility and permanence in the evanescent gilt of ephemeral society successes. Swann possesses as well the added attraction, largely unconscious, of having lived the most shattering passion of his life for a woman he could not trust, that is to say, a woman who could never bring him an unperturbed quietude, but would bring even within the most fulfilling moments, the enervating spices of otherness and betrayal, of elusiveness.

Thus, the very man who shatters the boy's refuge by causing his mother not to come and kiss him goodnight, this man has known as well what it means to consent to a solace granted under duress. He has experienced what it means to be consoled by a reluctant kiss, given half-heartedly as a ransom to the weakness of the receiver, because it is the simplest way to get rid of the importunate petitioner and while the mind is elsewhere, with the other, whose shadow poisons and exacerbates the moments thus wrested from the unwilling giver. In the case of the boy, as the narrator, with the detachment of involuntary memory, clearly perceives, the "other" was a conglomerate: his father, his own mental health, the educational principles of his grandmother, his mother's own better judgment, Françoise. In the case of the man as in the case of Swann the "other" will be all the possible others; one instance will illustrate this: the narrator falls in love irrevocably with Albertine only when he discovers that he can be jealous of her, when, on the train returning from la Raspelière, he becomes aware that her relationships with women may be more intimate than simple friendship.[72]

Swann and the mother, "society" and "the reluctant lover", both those poles are indissolubly united and complete one another. Both concern the search for tangibility and solace outside the self in the other, the other as refuge, as source of order and legitimacy. As against these, alone and openly flouted, stands the desolate and tragic figure of the grandmother, walking out into the garden in the storm, seeking the perdurance of the self in the unique manifestation of the self, in a work of art, in its fore-defeated but nonetheless longer perduring uniqueness. Thus, in the very moment of defeat and surrender, when the narrator as a boy receives as a poisoned reward for his own self-abandonment and self-betrayal (his search for tangibility and peace outside himself in the seemingly more sturdy and perduring "world" represented by his mother) the grudging and ashamed embrace of his mother, his grandmother urges him still to find within himself the true source of strength and solace, though she does so unwittingly, by way of the books she had been meaning to give him and which his mother reads aloud to him to put an end to his tears.

These three poles represent, grossly schematized the concrete manifestation of the "filter", of the "index of refraction" the narrator is in his monad. The work of art must, over and against the continuous "loss" of time, make manifest this perdurance in its uniqueness and in its creative force. Far from being the intuitive immersion in the "flowing", the "moving", a certain bergsonism would have us believe, the work of art is a freezing, a hardening for preservation and perdurance of the unchanging filter through which the unique "point of view on the world" each of us is, to borrow a phrase from Merleau-Ponty, lights the world as he lives it. This "lighting" of the world is creative; the preservation of its index of refraction is what artistic creation does at its lowest rung of achievement. Thus, lived-world and unique point of view on the world, originary source, are the two facets of the creative upsurge a living presence is, and, when they issue in hard and fast expressions beyond the now of their surging, they achieve the a-temporal resiliency conferred by art. But this, in itself, is not sufficient to explicate the "deliverance from time" which the narrator experiences in his living present, for a fleeting instant, in the privileged moments and through involuntary memory. In order to do this, we would have to go beneath the layers of consciousness we have already explored, into the pre-doxic layer within which is elaborated, in passive synthesis, the index of refraction the doxic configurations of which we have just outlined. Here again, Husserl, rather than Bergson shall be our guide, and a-temporal structures rather than inchoate durée shall be our reward. This, however, requires a study at least equal in length to the present one in which I would have to expatiate fully on the results of my own researches in the phenomenology of the living present. Suffice it here to say in conclusion that the analyses involve the three major "themes" we have just outlined and the phenomenological genesis of the "objects" by which each is illustrated. Permanence and flux as antithetical but mutually referent aspects of reality figure prominently in these genetic descriptions, as do the notions of proto-symbols, proto-symbolic valence, and proto-symbolic valence carriers, all of which would require formal introduction. In the light of this, then, perhaps it will not be taken amiss if I consider the aim of the present essay fulfilled on the basis of the results so far obtained, and, having made the case for a husserlian approach to *La Recherche du Temps Perdu*, leave for a later work the further exploration of the territory thus opened to the strict phenomenologist.

NOTES

1. Brand, Gerd: *Welt, Ich, und Zeit,* Nach unveröffentlichten Manuscripten Edmund Husserls, Martinus Nijhoff, Den Haag 1955.

2. Diemer, Alwin: *Edmund Husserl,* Versuch einer systematischen Darstellung seiner Phänomenologie, 2te, verbesserte Auflage,Monographien zur Philosophischen Forschung, Band XV, Verlag Anton Hain, Meisenheim am Glan 1965.

3. Held, Klaus: *Lebendige Gegenwart,* die Frage nach der Seinsweise des transzendentalen Ich bei Edmund Husserl, entwickelt am Leitfaden der Zeiproblematik, Phaenomenologica, Martinus Nijhoff, Den Haag 1966.

4. The famous remark in the interview of Proust by Elie-Joseph Bois in *Le Temps* of 12 November 1913, p. 2. This interview is reproduced in Robert Dreyfus: *Souvenirs sur Marcel Proust,* Grasset, Paris, 1926, pp. 287-92. Here is the relevant passage: "It will not be solely the case that the same characters will reappear under different aspects throughout this work, as (they do) in some of Balzac's cycles, but (also that) in one (and the same) character, certain deep, almost unconscious impressions (will reappear).

    From this point of view, my book might be an attempt at a sequence of 'novels of the Unconscious': I would have no shame in saying 'bergsonian novels', if I believed this, for, in every period it happens that literature attempts to relate itself — after the fact, naturally — to the dominating philosophy. But this would not be exact, for my work is dominated by the distinction between involuntary memory and voluntary memory, a distinction which not only does not appear in the philosophy of M. Bergson, but is even contradicted by it." (pp. 288-9)

    As Robert Dreyfus judiciously remarks: "This affirmation of Marcel Proust's seems debatable." (p. 289, note); but this is not the place to enter that debate. For an incisive treatment, cf. Françoise Fabre-Luce de Gruson, Bergson et Proust, in *Entretiens sur Marcel Proust,* Mouton, Paris -La Haye 1966, pp. 234-46.

    N.B. All translations unless otherwise stated are mine.

5. Newman, Pauline: *Marcel Proust et l'Existentialisme,* Nouvelles Editions Latines, Paris 1952. Let us note in passing that in a discussion of Elstir's "impressionism" the author briefly refers to husserlian phenomenology: "Still, in spite of his conceptions of involuntary memory, Proust's work, by announcing already the radical return to subjectivity, manifests a tendency towards phenomenology. It is surprising to observe how two points of view so radically opposed may end up on the same terrain. Husserlian phenomenology, taken up by Heidegger and Sartre, may be compared to a description of the immediate data of consciousness, from a certain point of view." (p. 139) In the present essay I shall try to show that both Husserl and Proust *begin on the same terrain*: the most radical kernel of a concrete subjectivity, the "living present", and that they proceed in parallel ways though with *diametrically distant means.*

6. Fieschi, Pascal: "Le Temps perdu est retrouvé," in *Proust, Collection Génies et Réalitiés,* Hachette Paris 1965: "The dimensions of this study do not allow us to show the deep analogies between proustian "eternity" and Husserl's "living present" which seems to us closer than all the others to the *nunc stans* of Boethius." (p. 260)

7.	The obvious differences between Husserl's methodological conceits and Proust's chosen mode of expression have discouraged beforehand further comparison of their results. Cf., for example, Florival, Ghislaine: *Le Désir chez Proust*, à lá recherche du sens, Neuwelaerts, Paris-Louvain 1971: "It is quote evident that we shan't find a parallel with Proust in Husserl. But we shall be happier in seeking either in the perspective of Hegel, or, in a completely different direction, in the line and following the method of Merleau-Ponty." (p. 14) To be sure, the author notes Merleau-Ponty's acknowledgement of his debt to Husserl for his method but to no further effect.

8.	Proust, Marcel: *A la Recherche du Temps Perdu*, 3 vols., Bibliothèque de la Pléiade, Gallimard, Paris 1954, I use "transcendent truths" as an adequate heading under which to include such disparate items as "extra-temporal being" (III, 871), "real but not actual, ideal but not abstract" ... "impressions" (III, 873), "the permanent essence '" of things" (*id.*), "our true self" (*id.*) Cf. also: "Only the impression, ... , is a criterion of truth, and because of that it alone deserves to be apprehended by the mind (l'esprit) ...". (III, 880); which is to be read in the light of: " ... one must attempt to interpret sensations as the sign of so many laws and ideas ..." (III, 879). The novel must "restore generality" to "the love of an Albertine" or of a Gilberte "and give this love, the understanding of this love, to all, to the universal mind (l'esprit universel) and not to this (woman) or that one in whom this (self) or this other (self) we have been successively would wish to fuse." (III, 897) Whether the "extra-temporal self" bears any deeper resemblance to Husserl's "absolute, timeless consciousness" (Husserl, Edmund: *Zur Phänomenologie des inneren Zeitbewusstseins, 1893-1917, Husserliana*, Band X, Martinus Nijhoff, Den Haag 1966, p. 112[464] − the pagination in brackets is that of the original edition, as given in the margins in this one; the translation used here is that of James Churchill: *The Phenomenology of Internal Time-Consciousness*, Indiana University Press, Bloomington, 1964) remains to be seen, as well as whether − and in what way − the curious proustian tandem "impression-essence of things" resembles the complex husserlian experiential datum "lived-experience − enacted noesis and corresponding noema."

9.	Alphonse Darlu (1849-1921), like Franz Brentano, had the genius to awaken lasting philosophic avocations, so striking were his ambitions for philosophy, the rigour of his methods and the seriousness of his intent. As in Brentano's case, the roster of those he influenced and whose thought he formed and matured is most impressive: Léon Brunschvicg, Elie Halévy, Xavier Léon, L. Couturat, to name the philosophers. Others, like Marcel Proust, turned to literature. Except for one or two circumstantial speeches and some contributions to the *Revue de Métaphysique et de Morale* which a "team" of former students of his asked him to help found and placed under his spiritual tutelage in 1893, his "presence" is preserved rather in the works of his students; the axes of his thought must be rediscovered through its traces in theirs. André Ferré summarizes his "doctrine" thus: "His poistion in the attempt to know and to explicate the world and man is that of a realist; but the first reality he encounters in his investigation of facts and their causes is that of the mind (l'esprit), of the activities and demands of his own mind. His position is therefore wholly spiritualist, and shuns accounting for (what is) superior in terms of (what is) inferior. As against this, his spirituality is purely secular. It does not take its source in the adherence to some revealed supra-human truth. Both (his positivism) and (his spiritualism) are founded upon an observation of things which sets aside all pre-conceived systems." (Ferré, André: *Les Années de Collège de Marcel Proust*, Gallimard, Paris 1959, p.221) For a longer treatment in which what Ferré terms "spiritualisme" is labelled "idéalisme" (p.84) and the priority of Darlu's influence on Proust over Bergson's is clearly shown both in terms of chronology and of contents (pp. 80-1), cf. the excellent study of H. Bonnet: *Alphonse Darlu, maître de philosophie de Marcel Proust*, A.G. Nizet, Paris 1961. Among other passages from Darlu's contributions to the *Revue*, H. Bonnet quotes the following which has the inspired imprint of direct oral delivery and may be an echo from Darlu's lectures: "Every time we reflect upon things [*les choses*; this has in French the force of "the things themselves" although it is not customary to translate this construction this way; the echo with Husserl's independently proclaimed justly famous admonition: "Zu den Sachen selbst!" is staggering.], our thought has its centre in our individual consciousness (that is the *cartesian cogito*); and it applies to objects which hold to one another and are the pieces and the parts of the universe." (p. 50) It is tempting to try to see in the "spiritualist/idealist – realist" positions of Darlu, theses not too far removed from the realist "mentalism" of Brentano's psychology which was also directed against the too narrow understanding of empiricism of the positivists, and a too nebulous idealism. The following admonition recorded by Proust in *Jean Santeuil* and in all likelihood attributable to Darlu (as H. Bonnet agrues, who reports it; *op. cit.*, p.60) seems to point with the force and concreteness of a *koan* to the same "guiding principle" as Brentano's "Experience is my only teacher": "Doubtless you have experienced, just like everyone, the noble voluptuousness caused by certain perfumes. Do try to represent it to us; it will be a thousand times more interesting."

10.   Cf. note 8, above; also Proust, *op. cit.*, III, 904-5.

11.   At least initially.

12.   *Id.*, III, 873.

13.   *Ibid.*; cf. also 1044 and 1046. (cf. below, note 14, for the quote). The *passé simple* and the *imparfait*, both narrative tenses used throughout, are also used to describe this moment, this "aujourd'hui-même" in which the novel's single action takes place, to be sure, but that is the only way available to Proust not to destroy the credibility of the situation, as a present tense might through its constantly renewed demand that we suspend disbelief totally. By allowing us to acknowledge that we are reading a novel, Proust enables us to believe that we hear a confession and thus maintains the directness he wanted to achieve throughout. The reader, indeed, "reads himself" in the work. For a shrewd discussion of the tense structure of the opening sentence of the novel and its relation to the rest of the work, cf. Roger Shattuck: *Proust's Binoculars, a study of memory, time and recognition in A la Recherche du Temps Perdu*, Random House, New York 1963, pp. 80 sqq.

14.   "And it was this notion of the time evaporated, of the spent years not separated from us which I now intended to put so strongly in relief, because at this very moment, in the hotel of the prince de Guermantes, the noise of my parents' footsteps accompanying Mr. Swann to the door, this tinkling of the little bell, rebounding, ferrous, irrepressible, shrieky and fresh which announced that at last Mr. Swann had left and that my mother would come up, I heard them again, I heard them themselves, they situated for all that so far into the past." (*id.*, III,1046 the passage continues and develops into both a proustian *cogito* and a claim to total "retentional" presence of the narrator's past).

15.   *Id.*, III, 1038. To be sure, the *conditionnel présent* is used also to talk about the projected novel; however, the text referred to here involves the "real" termination of the one "real" action in the novel.

16.   Husserl, E., *op. cit.*, cf. for example Sec. 11 et sqq. for "retention", Sec. 16 for "primary expectation" (retention and protention), and Sec. 24 for "primordial protention". Cf also the excellent article by John Brough, "The Emergence of an Absolute Consciousness in Husserl's Early Writings on Time-consciousness," in *Man and World*, vol. 5, No. 3, August 1972, pp. 298-326; esp. Sec. 2, Time-consciousness as originary consciousness in a three-fold sense, pp. 300-3.

17.   Husserl, E.: *op. cit.*, p. 55[413]. Because the "living 'now'" is a phenomenological datum and not viewed as an object of the real world, it is fundamentally different from Bergson's "lived duration" or "lived time" and owes but little, if anything, to the French philosopher's prior discoveries. For a fuller treatment of Bergson's relevance to this study, cf. below, note 68.

18.   *Id.*, pp. 33[393] sqq.; also Sec. 16, pp. 38[397] sqq.

19.   *Id.*, p. 76[431].

20.   *Id.*, p. 29[390].

21.   As put together for purposes of eventual publication by Edith Stein and published by Martin Heidegger in 1928 from texts dating as far back as 1893 and not later than 1910; the canonical edition of this text was prepared by Rudolf Boehm for the volume of *Husserliana* already mentioned in which are included some of the texts on which the published version was based.

22.   Namely, the "self-constitution" of universal objects in "universal intuitions", that is to say, the age-old problem of the actual apprehension by an agent limited in time and place of an object by essence independent of time and place. The outline of an answer is given in the *VIth Logical Investigation* (Husserl, E.: *Logische Untersuchungen*, Max Niemeyer Verlag, Halle a.d.S. 1922, Band II, Teil II, Sec. 52, pp. 161 sqq.) and within this outline the studies of the *Lessons on Internal Time-consciousness* provide the more concrete answer. (Husserl, E., *op. cit., Husserliana*, X, Sec. 45, pp. 96[448] sqq.).

23.   *Id.*, p. 73[428].

24.   *Id.*, Sec. 36, p. 74[429]; Sec., 38, pp. 76[431] sqq.

25.   I summarize here the path followed in the studies published as *Erfahrung und Urteil*, F. Meiner Verlag, Hamburg 1972.

26.   These are the studies on which the works of G. Brand, A. Diemer and K. Held, mentioned earlier (cf. notes 1, 2 and 3, above) shed a precious light. For Husserl's own texts on the matter, aside from consulting the holdings of the Husserl-Archives at Louvain, cf. for example among the published works: Sec. 28 of the *IVth Cartesian Meditation*, (*Husserliana* I, pp. 111 sqq.), Sec. 51 of the *Vth Cartesian Meditation*, (*id.*, pp. 141 sqq.).

27.   Husserl, E.: *Logische Untersuchungen*, Band II, Teil I, chap. II, pp. 363 sqq..

28.   Husserl, E.: "Analysen zur passive Synthesis," *Husserliana*, XI, p. 160; cf. also Klaus Held, *op. cit.*, p. 26.

29.   Counter-arguments having to do with the fact that experience in judging perspective may be culturally derived would interest only the causal modalities of the recognition of perspectives and not its formal requirements: if the plane does not exhibit the right kind of lines no cultural experience will enable anyone to find perspectives in it.

30.   Husserl, E.: "Zur Phanomenologie des inneren Zeitbewusstseins," *Husserliana*, X, pp. 81[435] and 82[435].

31. Cf. on this John Brough, *op. cit.*, pp. 318 sqq.. For what I call, following James Churchill (*op, cit.*, note 8, above), "transverse", Brough uses "vertical", Husserl's term in "Querintentionalität"; for what I call following Churchill "longitudinal", Brough uses "horizontal", Husserl's term is "Längsintentionalität". I reserve in my own practice the term "vertical" for a different but related use.

32. Husserl, E, *op. cit.*, Sec. 39, p. 82[435]; the translation used here is that of James Churchill, *op. cit.*, p. 108.

33. *Id.*.

34. Proust, M., *op. cit.*, III, 857.

35. *Id.*, III, 920.

36. *Id.*.

37. *Id.*, III, 921.

38. *Id.*, III, 953.

39. *Id.*, III, 980.

40. *Id.*, III, 941.

41. Husserl, E.: "Zur Phänomenologie des inneren Zeitbewussteins," *Husserliana*, X, Sec. 36, pp. 63[429].

42. Cf., for example, *id.*, Sec. 40, p. 83[437], and also the relevant part of the commentary on the famous diagram of time, Sec. 43, p. 94[446].

43. I introduce informally this term as a purely descriptive one warranted by Proust's text. A more formal treatment occurs below in the discussion of Husserl's handling of similar problems (cf. part VI).

44. Proust, M., *op. cit.*, III, pp. 1031-2; cf. also the note p. 1032.

45. *Id.*, 1031.

46. *Id.*, 1028-9.

47. Cf. note 4 above; for a clear distinction between "involuntary" and "voluntary" memory in the text, cf. *id.*, 873.

48. *Id.*, 988, 989, 1028.

49. Proust writes concerning the vision of Balbec involuntarily remembered: "And it was not only these colors I was enjoying, but a whole instant of my life which brought them up, which doubtless had been aspiration toward them, though some feeling of tiredness or sadness had prevented me from enjoying them at Balbec, and which now, rid of what there is that is imperfect in external perception, pure and disembodied, filled me with happiness." (*id.*, 869).

50. "Through art only can we get out of ourselves, know what another sees of this universe which is not the same as ours, and the landscapes of which would have remained as unknown to us as those there may be on the moon. Thanks to art, instead of seeing only one world, our own, we see it multiply, and, for as many original artists as there are, as many worlds have we at our disposal, more different the ones from the others than those rolling in infinity; and, many centuries after the fire from which it emanated was extinguished, be it called Rembrandt or Vermeer, their special ray of light is still sent to us by them." (*id.*, 895-6).

51. *Id.*, 376-7.

52. Husserl, E.: "Cartesianische Meditationen," *Husserliana*, I, Sec. 33, p. 102 (IVth Meditation)

53. *Id.*, Vth Meditation, Sec. 44, p. 125.

54. Cf. in particular the IVth Cartesian Meditation and more especially Sec. 32 entitled "The Ego as Substrate of Habitualities," (*id.*, 100-1; Dorion Cairns Translation, pp. 66-7, *Cartesian Meditations*, M. Nijhoff, The Hague 1960).

55. Husserl, E.: "Ideen zu einer reinen Phänomenologie und phänomenologischen Philosophie, erstes Buch, *Husserliana*, III, Sec. 83, 203[167].

56. Husserl, E.: "Zur Phänomenologie des inneren Zeitbewusstseins, *Husserliana*, X, Sec. 39, pp. 82-3[436]; Churchill, 108-9.

57. *Id.*.

58. *Id.*, 83[437]; Churchill, 109-10.

59. *Id.*, Beilage IX, 118[472]; Churchill, 161.

60. *Id.*.

61. Brée, Germaine: *Du Temps Perdu au Temps Retrouvé*, introduction à l'oeuvre de Marcel Proust Société d'Editions "Les Belles Lettres", Paris 1950, pp. 260-1.

62. Proust, M., *op. cit.*, I, 45 sqq..

63. *Id.*, III, 866 sqq..

64. *Id.*, 868.

65. *Id.*; this list of "moments bienheureux ou privilégiés" is by no means complete; cf. on this R. Shattuck, *op. cit.*, chapters II and III, and especially the table, pp. 70-4.

66. Proust, M., *op. cit.*, III, 866-80; most particularly 872-3 and 874-5.

67. *Id.*, 873.

68. Fieschi, Pascal, *op. cit.*.

69.   A particularly compact example of this was used by Leo Spitzer as the basis for an analysis which results in the emphasis of the "intentional" aspect made visible thanks to the "detachment", the "purification" involuntary memory allows. I am referring to the analysis of the sentence "Ce nom de Gilberte ... plumet bleu de son chapeau." (Proust, *op. cit.*, I, 394-5 reprinted in *Les Critiques de notre Temps et Proust*, Garnier, Paris 1971, pp. 40-3.

70.   These, if I may anticipate on a sequel to this study, are of two kinds: first, they consist in the "intra-monadic" "extra-temporal" "essences of things" through the contemplation of which the narrator recaptures his own unique "presence" in extra-temporal stillness; one such essence is visible through the unique destiny of the narrator which acquires exemplary force; thus, by suitable phenomenological procedures, one may apprehend the laws of essence of the "eidos" "human consciousness" (concrete ego), and the readers, thus, do indeed become "readers of themselves" (*id.*, 1033), in ways I plan to show elsewhere in more detail. Second, the philosophical and phenomenological structure of the "extra-temporal being" thus revealed in the concrete stillness of a-temporality, once it is described in what I take to be husserlian terms, shows itself to be identical with the concrete awareness of what I call "the point of equilibrium, the Archimedean point beneath all proto-symbolizations, and from which both fundamental ontifyings are equally possible: the point at which philosophy in the pregnant sense is concretely presented as a real, lived possibility." But these brief indications are all that can be mentioned here.

71.   This is indeed the thesis advanced by Floris Delattre, a most partisan advocate of the "strong bergsonian influence" thesis, who does go to the extremes summarized here; (Delattre, F. "Bergson et Proust, accords et dissonnances," in *Les Etudes Bergsoniennes*, vol. 1, pp. 7-127, Albin-Michel, Paris 1948).

While it was part of my original intent to show in what way and within what limits Proust might have been said, as he himself suggested, to have written a "bergsonian novel" (cf. note 3 above), the actual development of this project has reduced the intrinsic interest of this question in relation to the whole to such an extent that a lengthy footnote may be deemed adequate. This is because most of the elements required in answering have already been touched upon in other contexts, where the properly "proustian" treatment of temporality and self have been considered.

As the Editors of the volume on *Proust* in the collection *Génies et Réalités* (*op. cit.*, cf. note 6 above) summarize it, the most accepted current scholarly opinions on this topic run as follows: "Knowledge of the mind by the mind (l'esprit), *La Recherche* reveals between Proust and Bergson an intellectual affinity. Henri Bergson, Professor at the College de France, and through his marriage with Mlle Neuberger, having become Proust's cousin, will have on *La Recherche du Temps Perdu* a primordial influence: Proust, however, does not apply bergsonism, he lives it, feels it, personally rediscovers it. The involuntary associations which surrender the key to a past buried in memory, the wanderings of intuition in the unexplored fields of the unconscious, the perpetual evolution of personality in duration, the inadequacies of intelligence alone with respect to the task of understanding and grasping life, the sovereignty of art, unique reality of the world: as many themes common to both bergsonism and the proustian view." (*op. cit.*, p. 40, caption accompanying a photograph of Bergson). Using this nuanced statement as a fair rendition of the present state of this question, let us take each of its particulars, one at a time.

First, the "two memories". They are indeed clearly recognized by Bergson as has been noted by commentators from F. Delattre to F. Fabre-Luce de Gruson (cf. note 4, above). But the more nuanced bergsonian account involves two involuntary memories: *la mémoire-habitude,* largely unconscious and immediately present in everyday performances, and the more "proustian" *mémoire-image* which can be triggered by sensation and does involve total and dated — temporally placed in the stream of lived experiences — recall (cf. on this Bergson, H.: "Matière et Mémoire," Chap. II and III, in *Oeuvres*, Edition du Centenaire, P.U.F., Paris 1963, pp. 223[81] - 316[198]. Moreover, the bergsonian distinction corresponds only in part to the Proustian account which deals solely with two different modes of approach to the *mémoire-image*: "Impressional" or sensory trigger, and voluntary recall. Furthermore, as we have seen, what the *mémoire-image* surrenders under the trigger of impression is different from what it lets appear under the prodding of the will, and it is not merely the fragility of a fleeting moment of past duration, rather to the contrary it is "the essence of things," an "extra-temporal being" iterable and unchangeable in its purity and in its fixity. We recall thus not solely the located remembrance but also the manner of involvement which generated it, that is to say, the total, unique, spiritual event now *frozen* in its a-temporal iterability, ready to become the object of a unique description, the "content" of a work of art. In the Proustian view intelligence is not the only faculty which generates fixed and unchangeable entities; life does too: that is what life does, in fact, and only that, and art, insofar as it preserves the traces of its passing by changing its sedimented alluvions into the sandstone within which its fossilprints may endure, art, like intelligence, far from rendering duration visible, in its flow, rather immobilizes that flow in its then temporary now a-temporal patterns.

Needless to say, such fundamental, if seemingly minute differences, will reverberate throughout the other points of contiguity listed above in such a way as to make its overall coloring "non-bergsonian". True, intuition wanders in the fields of the unconscious in both cases, but what it finds there for Proust is the permanence of poles of tension (as I go on briefly to show in the text) which organize from one moment to the next each fleeting apprehension, and, in each such apprehension, the static uniqueness of its one manifestation within the monadic walls of the narrator's presence, as the tangibility of this presence. The unconscious is *static* and is best characterized as a lens with a particular index of refraction all its own: personality *does not* evolve in duration. The celebrated proustian *selves* are the changing manifestations in the changing social and intellectual context of the selfsame unconscious configuration which predures unchanged. That is why the little bell still rings in a genuinely eternal present in the library of the prince de Guermantes as it did in the garden at Combray. The ringing bell is one of the permanent elements that contribute to the personality of the narrator. He *is the* one hearing *that* bell, *that* way, still, and hearing it thus, the one who must wander from glittering salon to deceitful lover to his writing table to art galleries to salon to deceitful lover. The only "evolution" in or through duration are the different weights each of these three poles receives in the sequence of encounters (occasions) life is; and, when caught in the momentaneity of their appearances, each of the "selves" thus manifested in the different encounters seems irreducible to the preceding ones by virtue of the irreconcilability of its contingent components. One has to view their filigree to distinguish the warp under the weave, much as Cl. Levi-Strauss does in his analyses of myths. Intelligence is inadequate to this task, this understanding and grasping of life, not because life is an evolutionary process too evanescent to be apprehended by the freezing glare of the intellect, but, more traditionally because the intellect cannot make a science of the unique and must operate inductively over a number of samples. Thus, the narrator himself, who is intelligent, presents us with statistical generalities, "laws" of behaviour, observed, in Balzacian fashion, to be valid severally throughout society. This single outstanding characteristic of his argues categorically against mistaking Proust's novel for that of the narrator and would alone be sufficient to decide the case, had not Proust, whenever he could not avoid it, referred to the narrator's monologue as a "récit", thus marking clearly his distances.

The proper understanding and the grasping of life comes through another "freezing glare" — art as we have seen. Art preserves, in *stilled* perdurance, a unique, *monadic* world; and "there are as many universes as there are artists", says Proust. What the artist restores and gives us is this frozen index of refraction, this *lens* he is, permanently. Through this we may see *his* world, not duration, and *his* world, though unique, is not the one world the unique reality of which is mentioned above, but *a* unique world among many, one of which is ours for the creating were we to freeze it and not simply endure it, Swann-like, in mere participation. Art is sovereign, not because it alone permits the direct intuition of *durée,* but because it alone permits the preservation — the creation — of the unique. And it does so by concentrating not on what flows and disappears — as social lions and lovers do — but on the sedimented traces of that flow within — and thus without — one's temporality, on the declivities of the a-temporal rock which gave the flow its evanescent features, and were the fixed boundaries of its seemingly erratic course.

In this manner one can make a case for the non-bergsonism of Proust's "bergsonian novels": the bergsonism of Proust reveals itself to be merely superficial. But the deeper causes of this surface effect are in the intellectual antecedents to both Bergson and Proust, from which both derived the starting point of their enquiry. Taking the larger perspective one can see that Husserl, by way of Brentano, takes the sources of his philosophizing from sets of problems similar to those agitating his French contemporaries. All five — if we include Darlu — gave original solutions to problems which seem to have common parameters, though each gave differing import to their various elements. This, rather than a matter of inlfuence, is what the relationship of Proust to Bergson reduces to.

72. To be sure, the love affair with Albertine has already known two stages in which this possibility has entered only subliminally at best; however, the narrator's own remark concerning his long dating negative reaction to Andrée's presence near Albertine — a reaction going back to the very beginning of their acquaintance — reveals both the actual content of the perennial worry and one of the unconscious elements in what made Albertine attractive to him (Proust, *op. cit,* II, 1115 sqq.). The pattern illustrated here is present in the mother, as I have mentioned, appears in the relationship with Gilberte where it causes the narrator to break off the relationship, is present in a different way in both Mme de Guermantes and Mme de Stermaria where, at its least virulent, it is simply a sense of "mystery" emanating from the beloved. (*id.,* II, 363).

Temporal Passage and Spatial Metaphor

N. LAWRENCE

The subject of time is so multifaceted that the papers we read to one another are extremely diverse. Each of us must take only one small area and explicate a few of its features. The casual observer might wonder, "Are these pundits really talking about the same thing?" I believe we are, and that part of our task is to explore ways of drawing together the diversity of interests.

Before getting to the limited subject of temporal passage and spatial metaphor, I shall begin by making a brief general presentation of *four* major metaphors for time, illustrated by a few expressions. I am not sure these four comprise the whole range of available metaphors, nor indeed that there are just these four or even just four. It is essential to the general context of this paper, however, to insist that the spatial metaphor is only one metaphor, that it has important but restricted uses, and that if we are to create a net strong enough to hold the enormous burden of the idea of time we shall require a whole set of metaphors which refer to one another reciprocally. What I mean by the reciprocity of metaphors is that any one of them, pressed sufficiently, will lead us to the others. The need for inter-dependent metaphors holds with other very pervasive notions such as the notion of "love" or that of "life." Consider, for example, the difference between the way the microbiologist speaks of the "secret of life" and the way the marriage counselor uses the same term. Or again, think of what the social re-former means by the "quality of life."

In general, ubiquitous notions cannot be captured by the vocabulary of a single discipline, since a discipline is usually defined by the inclusion of some restricted data and the exclusion of others. The general task, then, far beyond the scope of this paper, is the coordination of these vocabularies, rather than trying to show how one limited vocabulary can "really" embrace all the others.

It is important to observe that I am using "metaphor" in a somewhat extended sense. I wish to avoid the ironic problem of whether the metaphors are literally metaphors or only figuratively so. "Literal" itself is multiply definable, and the unexplained use of it is often a swift begging of the question. The metaphors I shall describe might perhaps better be called "pro-metaphors," "master metaphors," or "mental metaphors." I extend the idea of metaphor beyond instances of linguistic usage, but I do not disengage it from such instances generally. The relationship between them and the instances in which they are employed is reflexive; they inform such instances and can be derived by distillation from them. Thus to say, "After a space of two minutes or so he spoke," both employs and mentions "space." The expression "a short time off" does not mention space, yet space at least partly governs its mea-ning. One further digression might be added here before we return to what I shall call "master" meta-phors. It is always a moot point whether one of these master metaphors, or any discipline especially dependent upon one of them "owns" any particular word, i.e. has the right to say, "Our usage is literal; all others are metaphorical." Consider "way," for example: "The law penalizes the discharge of refuse in the public way." But what about a "way of life" or the "way" to do x ? We may assign these ambiguities, or broadened meanings, when we find them so used in other European languages as well – *via, Weg,* etc. – to linguistic usage or even useful appropriation. But when we discover the same breadth of usage in the Japanese word, *michi,* we may suspect that the word *way,* as an idea, rather than as a linguistic entity, may be in the public domain and is not the property of a would-be literalist. Nonetheless, I shall at the outset use the term "master metaphor," since "mental metaphor" raises the troubled problem of whether mental entities are discovered or created, and whether they are explicitly present to attention or are only implicitly so (as Polanyi has suggested).

There seem to me to be four master metaphors for time. I shall mention them briefly and then go on to the collision between the idea of temporal passage with one of these metaphors: time represented as space. Even in this restricted subject it will be necessary to be more economical than comprehensive. I shall concentrate on two major complaints against the notion of temporal passage: (i) that it does not make sense and moreover is supernumerary; (ii) that it is merely subjective and has nothing to do with objective reality.

On the theme that the "passage" of time makes no sense and adds nothing to our understanding, I shall examine a paper written nearly a quarter of a century ago by the Harvard philosopher, Donald C. Williams, called "The Myth of Passage."[1] On the theme that the passage of time is purely subjective, not belonging to the objective world, I shall examine an essay by Hermann Weyl, written nearly a half century ago and first published as a contribution to Oldenbourg's *Handbuch der Philosophie,* "Space and Time, the Transcendental External World."[2] I have chosen these two essays because of their clarity and their substantiality, as well as the fact that they have been long in the public domain. My rejoinder to each of these themes cannot be comprehensive enough to satisfy all their supporters, I suspect. But where these themes take other forms, I believe the rejoinder can as well.

First, let us look at the four metaphors.

1.   Time represented as number is commonly seen in dating. We say: "This is the year 1975. The time is 10:15, here at the 74th meridian."With suitable spatial reference points we can convert the 74th meridian into a relative dating. We strike "74th meridian" and replace it with "4 hours and 56 minutes west of Greenwich."

2.   Time represented as space. We speak of the "distant past or future," a "long" time ago, or say that a war "covered a great expanse of time." Recently, no Watergate apologia is complete without the repeated use of the provisionary "at this point in time."

3.   Thirdly, there is time conceived as force or commodity. Where one regards himself as active he will think of time as submitting to commodity use. Someone may say he "spent a lot of time" in Afghanistan, for example. But where he thinks of himself or something else as relatively passive, he metaphors time as force. We say, "Time overtook me and I couldn't complete the task" or of weathered rocks that they are "time worn."

4.   Finally, there is time conceived as telos. By "telos" I mean what Aristotle meant. Telos does not mean "conscious intent or purpose," of course. Purpose is a special and restricted form of the telic. "Telos" may simply refer to a functional analysis. Roots of plants go down to

get water. The heart is for pumping blood, the dappled surface of the toad is for camouflage. Crudely speaking, then, a telic order is one whose phases or structure show organization to an end. Time as telos appears in such metaphors as "Time will tell," "His time had come," and so on.

With this background let us turn to the idea of temporal passage.

In adult consciousness we find that the sense of temporal passage underlies, in immediate awareness, everything else we try to make of time. Notice that I say "adult consciousness." The work of Piaget and others tends to show that the developing of a child's ideas of time parallels the developing comprehension of himself and the natural world. Similar developments occur in the transition from primitive cultures to more advanced ones. The roots of our ideas of time are thus themselves temporal.

No one denies, of course, that there is some basic, wide-spread notion of the passing of time. What is objectionable is that the notion of passage does not seem to bear scrutiny. What does the time pass? How fast does it pass? These are instructive questions to ask, but if they are thought to render the idea of the passage of time paradoxical, then they have been asked in too much innocence.

Donald Williams in the essay, "The Myth of Passage," makes the complaint clear. He says:

> Now, the most remarkable feature of all this is that while the modes of speech and thought which enshrine the idea of passage are universal and perhaps ineradicable, the instant one thinks about them one feels uneasy, and the most laborious effort cannot construct an intelligible theory which admits the literal truth of any of them. The obvious and notorious fault of the idea, as we have now localized it, is this. Motion is already defined and explained in the dimensional manifold as consisting of the presence of the same individual in different places at different times. It consists of bends or quirks in the world line, or the space-time worm, which is the four-dimensioned totality of the individual's existence. This is motion in space, if you like; but we can readily define a corresponding "motion in time." It comes out as nothing more dramatic than an exact equivalent: "motion in time" consists of being at different times in different places.
>
> True motion then is motion at once in time and space. Nothing can "move" in time alone any more than in space alone, and time itself cannot "move" any more than space itself. [3]

From these considerations, Williams believes, it follows that the passage of time leads to an infinite regress of times, each of them required by the motion of the previous one. [4] Williams concludes his complaint by saying that the motion of time is either contradictory of superfluous.

But the distressing problems here seem to be a bit contrived, and certainly many intelligent laymen and even "candid philosophers" (to borrow a term from Williams) are not so troubled as Williams expects them to be. Williams equates motion and passage with one another and I do not demur, although it might lead to misunderstanding. However, temporal passage, or temporal motion if you like, (Williams equates motion and passage) may occur without any special reference to space or spatiality at all. For example, the seriality of music performed is a temporal order to which spatial questions do not apply. The illustration from music should not revive the hackneyed opposition between art and science. Let that opposition seep away with the seepy snows of yesteryear. In the sciences, as well, both practical and theoretical, the sheer act of counting, whether audibly or not, invokes a temporal order and adapts to it. As the French psychologist, Guyau, says, "Hearing only locates stimuli very vaguely in space, but it locates them with admirable precision in time." His compatriot, Paul Fraisse, quoting this passage, goes one step further: "Thus hearing," he says, "is the main organ through which we perceive change; it is considered as the 'time sense,' just as sight is that of space."[5]

Both of these observations emphasize the non-spatial character of temporal apprehension, but unhappily they overemphasize the process of hearing. Temporal passage surely does not *depend* upon hearing in order to be apprehended. Persons totally deaf are not thereby devoid of a sense of time, any more than those who are totally blind are devoid of the sense of space. Nonetheless, the passage of time through which auditory data appear, warns us that — no matter how rich the data of vision may be and how much of our consciousness is spatial — there is no ground for supposing that pure spatial representation adequately comprehends temporal data, or — more precisely — the temporality of experience.

It is much simpler and more natural to say that the sense of temporal passage is integral to the whole of sensibility rather than residing in some special sense or organ. A less atomistic approach to the sense of temporal passage shows, indeed, that it is well developed, in rough proportion to how firmly consciousness grasps its own nature. Inadequacies in the time-sense seem to correspond to defects in self-awareness. In very young children, for example, the self is sufficiently non-integral so that its memory is perforated by discontinuities, and prior to a certain age, there is no memory of a personal past at all. Piaget points out in this connection that the non-focal character of early self-awareness is responsible for a lack of infant memories. He rejects the idea that a set of repressions — as Freud supposes — has obliterated the past.[6]

Corresponding to underdevelopment of self-awareness in children, there are diseases of ego-integration in middle life and the evaporating of self-consciousness of senility in old age. The common complement of these maladies is distortion or defect in the time sense.[7] Memory being what it is — the link between self and world — it is not a great surprise to find that where the ego is embryonic, diseased, or fading away, there are corresponding abnormalities in the time sense. Yet the point is worth emphasizing. Tweedledum and Tweedledee battles have been fought on whether or not there is a time sense, where the issue was whether there is a precise organ or set of organs for apprehending temporality, as if the ego were somehow the sum of its bodily parts, rather than a whole that expresses itself through them.

The search for a time sense is not the subject of this paper, and is less popular than it used to be, but it rests on an assumption of atomistic psychology, dating back to Locke, which is very pertinent to our paper, namely that what is external to us is either a thing such as we touch, or a property such as we smell, or maybe either or both such as we see. This tendency to reify time leads men to try to find the organ proper to it, and that leads them to say that if it passes, it must be doing so in space and time.

To return to our questions about passage, the idea of temporal passage need not be troubled by questions of spatial motion. Indeed, it would seem that from the point of view of systematic clarity, the physical movement which the term "passage" often invokes, would necessarily *presuppose* a temporal passage as a baseline, accompanied by a spatial variable for the moving object, where values are set into one-to-one correspondence with the temporal variable. Again, the point is obvious. Temporal passage must be the foundation of any kind of passage and cannot be replaced by one of its hybrid forms. We might as well or better try to replace it by the "movement" in a symphony, a revolutionary political "movement," or a "motion" to adjourn. It should be evident that these motions and movements are as much movement as is movement in space and that to speak of them as "mere metaphors" or as "figures of speech" simply begs the question.

What we actually have is a basic intuition of change. This change can be experientially focused in a purely temporal series in at least two commonplace ways. (i) The series may be imposed upon us as we listen to a rhythm, the flow of words, the sound of music. (ii) Or we may enter the time stream actively as in the case of counting or playing music. Counting, of course, contributes nothing to the passage of time, but that passage is required in order to measure time. The need to measure is useful in showing temporal passage as a spatial spread, that is, time without passage. It also shows that our active engagement in the time stream runs into a kind of objectivity which is all too often ignored. When I count, I engage in an activity which acknowledges the independence and non-personality of what I count from my personal resolve to do the counting. As I calculate cricket chirps, for example, against the ticking of my watch, both the order and the frequency of the two series have not only been actively sought by me, but also passively accepted. To the extent that my data are honest and sharable, I can only record what has occurred. At the risk once again of laboring the obvious, it should be observed that scientific objectivity goes no further than demanding that what one observer claims is so can be verified by another. Is the verification less real, less objective than what it verifies? In truth such objectivity is really a very reliable kind of intersubjectivity. The reasons are simple: Primary observers, whether as scientists, musicians, or what have you, are subjects. Thus to label the passage of time as merely a subjective impression has little meaning. All impressions on consciousness are subjective. What is at stake is: Are they privately subjective, personally subjective, individually subjective, exclusively subjective?

If I record thirty-two chirps per minute for some species of cricket and you do also, at the same temperature and season, then we have hopes of having transcended the peculiarity of some particular covey of crickets or of ourselves. We then have a rule for cricket chirps. This rule is about a temporal series formulated ordinally, based on the passage of time, and validatable in a nice objective future passage of time. It is a statement about the world viewed scientifically. To try to translate the objective observation, its objective statement which presupposes and requires the passage of time, and its objective verification into the language of subjectivity would be absurd. That would be to invoke a myth of non-passage, and to add superfluously the concept of subjectivity.

To put the matter briefly, any claim about reality must be based on an experience which is through and through temporal. For it to have meaning, furthermore, is for it to be applicable and verifiable in further experience, again through and through temporal.

We may, of course, fasten our attention on concepts in the domain of *ideas*, that is — not on the *presentation* of reality but on its ordered *representation*. This is in fact what the spatial metaphor does in a limited way. To represent time requires that we somehow deprive it of its original passage. That is, it must hold still so that we can share our comprehension of it. One way to do this is to show it as a line with an arrow point that faintly represents our awareness of its irreversible ordinality. This is a kind of paralysis in which the animal is immobile but still alive. Or we may wish to preserve and embalm time entirely, eliminating its arrow points and focusing on static temporal locations and the termini of temporal durations. At this higher level of abstraction we can play largely verbal games like "Is time reversible?" "Is time extension?" and so on. All answers to these questions must begin with "As represented, yes." Time presented is quite another matter.

We may then ask, "Why represent time spatially? Why deform or reform what is given as passage?" The answer is likely to be, "To get rid of the subjectivity, to present an objective account." So let us examine objectivity a bit more, since it is commonly invoked without reflection in discussions of time.

I have just undertaken to show that objectivity in one science, field biology, may be very different from objectivity in another, say physics concentrating on the location of particles, things, and events. In much science and in the world of common sense, objectivity is rather more like the cricket chirps. This is an objectivity of that which is independent of my volition and is both logically and temporally prior to the objectivity of that which is independent of my knowing. That is, it is given to me first in the temporal order itself, and it is also the logical underpinning of any ideas I have about reality.

Beginning with Parmenides who – on purely logical grounds – flushed *change* down the drain and thus eliminated *time* as well, through Leibniz who thought spatio-temporality was a confused human way of exhibiting logical relations, to Hermann Weyl, whose lonely consciousness crawls up the life-line of its isolated body, there has been a solemn agreement that objectivity and unchangingness mutually require one another. Indeed, there has been a tendency to identify them with one another. But again this seems to be objectivity-in-review, unchangingness in representation. Objectivity in presentation is quite another thing. The most surprising thing about Hermann Weyl's famous figure of speech is its innocence, for the metaphysical muddles in it lie very close to the surface, yet we know that Weyl had at least read some philosophy. He tells us that in the objective world nothing happens, it simply is; but to the gaze of my consciousness as it crawls up the life-line of my body, spatial images reveal sections of the world which change in time.

First, we should not say of Weyl's remarks, "Oh, this is mere metaphor." The question is, what is the metaphor a metaphor *of*? In Weyl's remark the expression "crawl" may be a metaphor, but a metaphor for what? It is at least a metaphor for a change and likely one for an effort as well. The latter problem, namely of the domain of the volitional as an index of objectivity, I have already briefly touched upon, but let us lay it aside and consider the problem of objectivity merely in relation to observation and description. In the objective world nothing happens, says Weyl. Now I look at his picture. Here is a Minkowki world-line. Weyl calls it a life-line, but I think he merely means by this the world line which temporally coincides with the life of my body. Indeed, he shortly speaks of the body of the ego as a physical object. Now this body is really a time-worm of one temporal axis fleshed out in three spatial dimesnions, laid out in great curves. In ordinary parlance it changes – that is, it moves, but in the happenless world all the motion is reduced to a set of variables taken in a serial order which is the direction of "up," in the Minkowski diagram. Let us call this change$_1$, the change that can be reconstructed in the happenless world. But now there is the subjective change, the gaze of my consciousness. This subjective change is invisible. Let us call it change$_2$, spaceless change; it does not appear on the Minkowski diagram. I think this is the domain of counting and of hearing music, etc., both of which essentially require the passage of time. It is said to be purely subjective.

I now wish to introduce a myth, a very dull myth but pertinent. Suppose Weyl and I agree on what he says. We are mentally holding hands. We are looking at two worlds, one happenless and objective, the other happening, in passage, and subjective. What I should like to know is where we are standing as we look at these two worlds? Certainly not in just one of them, by definition. My inclination would be to say that we have transcended them both, objectifying them both, in the sense of rendering them objects of consciousness. I think we are now — he and I — standing on ground where we agree that you can view the world as temporally frozen, or as temporally writhing and jumping. But this new standpoint from which we objectify the two alternative ways of thinking of the world — is it to be thought of also as passageless and happenless? I think not.

However, I do not think that Weyl would like this rejoinder at all. He would find it either nonsense or a way of smuggling in passage and thus begging the question. I suspect he would say that we are in both, or can be, and that we need no third world, only a clear view of the two we have got. But if that is the case, then the distinction between the two worlds is a distinction in theory only, since both are given to us, and we should lay off trying to bifurcate reality, save as a kind of convenient myth. Myths about reality are far different from myths that hold that a claimed aspect of reality is really a myth.

We must continue with myth very shortly, but Weyl does not complain that passage is a myth, rather that it is merely subjective. To him, then, the answer is, "The distinction between objective and subjective must be abandoned. A purely subjective world, confined to a single subject, could never be shared. A purely objective world could never be found or constructed, for who could find it or construct it without leaving his subjective stamp upon it? Objectivity and subjectivity are giant abstractions grounded in the same reality. Only in the mind of the theoretician does there exist a world without passage. Only in the mind of the theoretician does there exist a subjective world defined as utterly distinct from the objective world and yet mysteriously connected with it. Descartes, for all his difficulties, had a helpful God to bridge the two worlds. Weyl confines himself to equations.

Finally, let us look at the troubled topic of myth. It is difficult, of course, to tell what authors mean by myth when they use the term without telling us what they mean by it. I shall assume that a myth is supposed to embody a truth in terms which are not literally true but are easy to comprehend, and have a kind of homely usefulness, therefore. Williams goes a step beyond this. According to him, the myth of passage is superfluous. It adds nothing. There is passage but it adds nothing. "Adding" is, of course, itself a relational metaphor whose use here indicates that everything is taken care of by the Minkowski picture. (Weyl would only say that it adds nothing objective.)

What this charge of superfluity comes to is that the *presentation* of time as passing will not work when we try to add it to the spatial *representations* of time. But the reason it will not work is that you have already made up your mind to stick to time represented spatially or at least as a fixed dimension. We thus have the tautological outcome of a resolve. If my model for time is time not passing but already past, and I project this model into future time as having the same fixity of structure and as being homogeneous with past time, then it is virtually a logical consequence of this resolve that persons and consciousness will be treated as subjective and incidentally that the conception of "now" will also be extra diagrammatic, extra representational, and so on. The discovery that we do not need the idea of the "now," that subjectivity belongs to another world, and that passage adds nothing are not discoveries about the world at all, but rather discoveries about the system of representation. Ironically, one could hardly imagine a more subjective, or at least arbitrary, claim. The Cartesian problem here is still the major one, and ignoring it will not make it go away. How does one account for the interaction of what is subjective and what is objective? For the world is no mere object of observation and description. It is a world of action and interaction, of which mere perception, observation, and description are but a small part.

To re-phrase the last point in terms of persons, rather than objectively: I am not just an observer, merely gazing "à la Weyl," and then rationally reconstructing time past out of time passing, as I subjectively notice images of it unrolling before me. I am also, and perhaps primarily, agent. Observation and description are some of my activities and not my only sources of knowledge. All of these presuppose, however, the reality of the passage of time. No restricted portion of these activities can tell me something about reality which renders the rest of my knowledge merely subjective, mythical, or superfluous. The proper rejoinder to the charge that the idea of the passage of time is a myth is to ask about the myth of non-passage, tracing the idea of non-passage from the representation where it was born, and to question the illegitimate substitution of that idea for the real passage that aroused our curiosity in the first place. And to the charge that the idea of passage adds nothing we should say that its omission subtracts something. When we omit it, we cannot explain the immediate data of volition and action in which the world and consciousness are intimately connected and in which they are only theoretically and in retrospect separated.

ADDENDUM

The preceding paper includes topics which Professor Park treats in his paper, as well as some appearing in his essay in the first volume of the proceedings of this society.[8] We have both altered our papers after a very profitable series of discussions. Professor Park has incorporated his additional observations and some changes in his text. Mine largely are presented in this appendix.

My paper is not sufficiently candid about the four metaphors for temporal representation. I believe that each leads on to another and that ultimately none is *selbständig*. What does not appear is that I believe that time as telos is a kind of *primum inter pares*. My reason for so thinking (and this is far from a well-developed argument) is that whether I choose one metaphor or another, or several at once itself depends upon what function that choice performs. The metaphors are neither arbitrary nor independent, when sufficiently analyzed, but they are ruled, in their effectiveness, by the ends which they serve.

I believe that Professor Park's first paper wanted to isolate the subject of physics and show that the notion of passage is not required for time as it appears in physical theory. Further, he uses "myth" to refer to a figurative way of identifying something that is true, rather than as a name for an illusion which we should clear from our minds. His second paper takes a somewhat more pragmatic approach, allowing that current physical theory may well prove later to rest on metaphorical expression, but urging us to accept it now as if it were not. If I understand him rightly, it is one thing to admit that the ideal of literal language can only be approached asymptotically; it is another to abandon the ideal and settle for a set of interlocking metaphors.

Needless to say, this later formulation is the outcome of my juxtaposing our papers after we had given them, and I must take responsibility for the formulation. The continuing issue between us, then, concerns the nature of the language about time. I have marked doubts about the possibility of a literal language of general scope – that is, a scope sufficient to deal with cross-disciplinary topics like "time," "love," "energy," etc.. At the same time, there is nothing arbitrary about the heuristic value of an ideal of literalness *within* a given special discipline, nor is the ideal dispensable. Indeed, the major advance in the formulation of basic physical theory from the time of the Greek physicists to the present has been along just such lines.

The latter part of Professor Park's paper shows that "now" and "event" also do not belong to the language of physics, and concludes with an exposition of how time can be regarded as having a directional grain from order to disorder. It seems to me important to stress that the notions of "now" and "then" and the "passage" of "events" belong not merely to fuzzy, commonplace experience, but are part of physics as a whole, since physics arises from observations and experiments and depends upon that same world of experience, however selectively controlled, for both meaning and verification. It would seem, then, that if we venture beyond the formulaic part of physical theory – or if we identify that body of theory with its formulas – we must employ "now" (and both kinds of "then") and "events," to distinguish physics from mathematics, and in this Professor Park concurs.

NOTES

1.   Donald C. Williams: "The Myth of Passage," *The Philosophy of Time,* ed. Richard M. Gale (Garden City, N.Y.: Doubleday 1967; Anchor Books A573), pp. 98-116.

2.   Hermann Weyl: "Space and Time, the Transcendental External World," *Philosophy of Mathematics and Natural Science* (New York: Atheneum 1963), pp. 95-137.

3.   Williams, pp. 104-5.

4.   *Ibid.,* p. 106.

5.   Paul Fraisse: *The Psychology of Time* (New York: Harper & Row 1963), pp. 82-3.

6.   Jean Piaget: *Play, Dreams and Imitation in Childhood* (New York: Norton 1962; Norton Library 1971), pp. 187-8. See also pp. 189-92. The French title is *La Formation du Symbole.* What I have called the "non-focal character of early awareness" is an interpretation of what Piaget speaks of as the absence of "mental images, interiorised language, and the beginnings of conceptual intelligence." He then shortly describes this phase as one of "interiorised intelligence, [which] only gives rise to organised memory when speech and the system of concepts exist."

7.   See, for example, Fraisse, pp. 181-210 passim, pp. 246-8, and elsewhere.

8.   The latter essay, "The Myth of the Passage of Time," bears a title close to that of Donald Williams' study, "The Myth of Passage"; I chose to examine Williams' essay, since it is generally philosophical in its scope and is nearly twenty-five years old.

Time: Being or Consciousness Alone? — A Realist View

M. MATSUMOTO

PREFACE

Experience of matter can be described in the context of time and space, whereas, some people say, experience of mind may be described according to time only. Accordingly, though time and space together are regarded as objective forms, one may have a propensity for treating time alone as a particular form of the subjective consciousness. For space is indeed referred to the self-evidence of being, while time is thought to belong rather to the self-evidence of our own consciousness. According to my opinion, however, even the spatial description is indispensable for the state of "mind". For instance, the contents of our consciousness can be described only in terms of the localized phases of their images. Contrary to Kant, who regarded time and space together as forms of the outer intuition (i.e. as conditions of sensation), and time alone as form of the inner sense, I have a firm intention to assert them both as two forms of objective being because it is the being itself that can be the ultimate object of any of our cognitive powers – sensation, understanding and reason. Time and space are not forms proper to a particular "being" such as conscious existence like ours; they are also, nay, above all, two objective forms of being in general that transcends all such limited existences. It follows that these forms themselves, once abstracted in our mind, must, first of all, be valid for material beings; after that, i.e. derivatively and analogically, they may also be valid for mind-beings. The main aim of this paper is "dialectically" to elucidate that fact on the subject of time in particular.

1. TIME IS ROOTED IN THE CONSCIOUSNESS OF BEING

Experience is nothing but the evidence of beings; we can by no means doubt the fact of their existence. That is the significance of experience. In other words, it is only when we become conscious of a being, intending ourselves thereto, that our experience begins. Here, we must ascertain we have only the "consciousness of something" (because a being is but something) but *not yet* the "consciousness of consciousness", that is, the "self-consciousness". Neglecting this clear order, René Descartes, by his maxim *cogito ergo sum*, posited the "consciousness of consciousness" (*particular* consciousness of the self) in the place legitimate for the "consciousness of something". He took the former for "pure consciousness" and started his philosophizing therefrom. But the fact that the self is immediately given to the self, was not, in reality, his *patent* discovery. As father of modern philosophy, he used a well-known formula "know thyself" – the goal of Socratic philosophy – as his starting point. On the contrary, the traditional ontology since Aristotle traced the natural way ($\mu \acute{\epsilon} \theta o \delta o$ s): from the first experience of something, to the last experience of the self. We will explain this lineage of the history of ideas in the following paragraphs.

Descartes, reducing the absolutely undoubtable evidence to the *conscious* act — "I think", meant to found the philosophical evidence upon the consciousness of the self. But far before such a subjectivist foundation appeared, the traditional ontology had already established, as we have seen, the evidence of being. This ontology also began with the very moment where no doubt could possibly enter in. "Being is what is known first (*ens est id quod primo intelligitur*)".[1] Here metaphysics or primary philosophy, well coincides with Descartes' primary contention. Thomas Aquinas writes: "Whosoever may know being does not necessarily know his knowing consciousness. Yet without the knowing consciousness, one can know nothing (*nec quicumque intelligit ens, intelligit intellectum agentem; et tamen sine intellectu agente homo nihil potest intelligere*)".[2] Here he clearly establishes that the being is directly given to the consciousness, for which the evidence of the "self-consciousness" is not always necessary. Yet, where there is no existence of consciousness, the evidence of being is impossible because there is no substratum to which this being could be present. In other words, only if the consciousness exists even without the act of being conscious of itself, the evidence of being can be *sufficiently* established.

Whether ignorant or not, Descartes identified the consciousness in general with a particular self-consciousness. This confusion came to cause the birth of a new type of the "consciousness of things other than the self", which, thereby, became merely a mental phenomenon, incidental to the "self-consciousness". This prepared the way for European Idealism, from Kant and Post-Kantian idealists to Husserl and Heidegger. But this lineage of idea, far before Descartes, can be traced back to Avicenna's "Man in air", the Augustinian theory of mind, Plotinus, Gnosticism, Orphism, and finally reaches Brahman-Atman in the philosophy of the Upanishads.[3] I regard therein a great line of subjectivism, or of Identity-Philosophy, which consists in the absolute Self or in something that leads to this notion, excluding any relative existences.[4]

Time connects before and after. Cartesian time concept implies a self-consciousness which remembers "myself of yesterday" and expects "myself of tomorrow". In this sense, memory of the past is contained within the present of consciousness. The same is true of the present expectation of the future. Thus, both past and future are connected in the self-consciousness, as "the present of past", "the present of future" in "the present of present", to use Augustine's division. The time that connects past and future in the present is nothing but a form of self-identity that says "myself of yesterday" is "myself of today", "myself of today" is "myself of tomorrow". This self-consciousness became the root for the Identity-Principle, the fundamental principle of logic. The root for the identity of all experienced things was the apperception of a transcendental self. The foregoing way of thinking, successively inherited from Augustine to Descartes, from Kant to German philosophers, seeking the identity in the "consciousness of consciousness", expanded it into the "consciousness-in-general of being." I wish to call this idealistic tendency the identity-philosophical construction.

A realist cannot possibly agree that the self-consciousness is the a priori basis for the identity of all things. For, as we have remarked, consciousness consists first in the "consciousness of something". And only when this "something" is, in fact, a consciousness, does it become a self-consciousness. In an a priori, *sublimated* self-consciousness, "myself of yesterday" is identified with "myself of today", and so the perception of time is completely incorporated in the present structure of consciousness. However, the existence of "myself" on realist view is merely one of the many things that exist in the time which is itself one of the categories of the reality. Here a real ego stands side by side with other real things. Consequently, consciousness as a being is understood and measured in time, and not time in consciousness, as Augustine claims. When my consciousness can identify "myself of yesterday" with "myself of today" and "myself of today" with "myself of tomorrow", it rests upon more than two functions of memory and expectation. It must also presuppose the identity of a "thing remembered" with a "thing known at present", and that of a "thing perceived now" with a "thing expected". It is not "myself", as Kant realized, but one and the same "thing" which is remembered, perceived and expected that becomes the only clue to establish the self-consciousness. For consciousness must be consciousness of being before it is that of the self.

In order for us to confirm this reality, suffice it to take the example of a person suffering from amnesia. He is unable to connect "himself of yesterday" with "himself of today". Not to say, if a person cannot suppose the identity of a thing being expected tomorrow with a thing being recognized today, he naturally becomes very anxious since his own existence needs external reference. If this thing, the identity of which he doubts, be his own body, it would be the threat of death. However, the time-structure of self-consciousness solidly tying "myself of yesterday" with "myself of today" comes from the perception of the time-structure of everyday things: that is, it comes from our conviction that a thing could subsist how ever it changes. In other words, it presupposes the identity of the thing underlying any phenomenal transformation of that thing. Just as the evidence of the self is a particular case of the evidence of being, this identity-structure of consciousness is nothing but one example of the said identity of being as such.

That underlying identity of being is nothing but its *substance,* which, Aristotle says, is identical to itself and, thus, liable to cause various phenomenal changes.[5] This being itself whose evidence always produces in us a "consciousness of something" belongs to no other than the category of substance. Now, this substance can be of a nature endowed with consciousness. In this case, it can be conscious of its own nature when it becomes *pour soi* (für sich, for itself). For instance, when I become *pour soi,* I shall find myself being *conscious* of myself, i.e. I shall discover my own existence identical with myself. One sees thereby the emergence of the "consciousness of the self". This is what is meant by the Heideggerian existentialist term "existence".

2. THE REDUCTION OF THE EVIDENCE OF BEING TO THE "THING-IN-ITSELF"

From the aforementioned fact, clearly there is a time which connects before and after in terms of the evidence of being, and a time in terms of the evidence of consciousness. I shall discuss the relation between these two kinds of time. This is also an examination of two kinds of evidence — that of being and that of consciousness. The evidence of being is univocal, but that of consciousness has two aspects, that is, it is a part of the evidence of being, but also possesses its own proper self-evidence. From the relation between this particular evidence of consciousness and that of being in general arise these three questions:

(i)   How far can the evidence of being be interfered with or be prevented by the evidence of, so to say, *arbitrary* consciousness?

(ii)   How far can the evidence of being conserve thereby its independent identity?

(iii)   What are the conditions under which "time" of the evidence of being is recorded independent-
          ly of "time" belonging to the consciousness alone?

After all, those are the problems of objectivity: how far the evidence of being can be independent of
the evidence of consciousness. It is to determine the epistemological possibility of reducing the evidence of being to the "thing-in-itself".

It is incontestable that the consciousness cannot help being given that being because of its complete
passiveness, so that it cannot but witness the evidence of being: this is a *transparent* consciousness,
whose whole content is permeated by the light of its object. On the other hand, when the consciousness is conscious of itself, it has actively chosen one out of all possible objects, so that it acquires the
reflection of a luminous evidence of itself that fills the agent's whole content. Accordingly, while the
primary evidence is involuntary and, so to say, *heteronomous,* the secondary evidence is voluntary and
autonomous. How does the latter work upon and interfere with the former? How far can the former
assure its own evidence from the latter's evidence? What we must be careful of here is the fact that, although the primary evidence of being must be distinct from the secondary evidence of consciousness,
the consciousness itself which establishes the evidence of being and the consciousness of the self are
not at all distinct in their substratum. Every consciousness is first of all *a being capable of being conscious.* Consequently, "being" in the context of "consciousness of a being" can be, either voluntarily
or arbitrarily, replaced by "consciousness". From that, it is necessary to discern the evidence of the
independent being from that of the *arbitrary* consciousness. This epistemological manipulation – we
call it the objectification of the "self-evidence of being".

Now, if the level of self-consciousness, though it is a still vague and tiny consciousness of the self, is
realized, it implies necessarily that a *being* has already been given to the "consciousness of consciousness". Then, this stage requires that the evidence of being as such is testified anew in terms of the self-
consciousness. Though its own evidence is indeed a display of the reflexive consciousness, still the
evidence of the being included therein is beyond its reach. Therefore, we are not at all able to modify
what was in us, namely, the evidence of being, no matter whether we ourselves understand that evidence or not, confirm it or not, observe it or not. For, as we have already seen, being is directly given
to consciousness before the latter is given to itself, and it follows that nothing could now prevent us
from supposing that being as preceding any consciousness in existence. Thus, we define the objectivity as being definite in itself and independent of any consciousness no matter whether the latter is at
work or not.

How is such an idea of philosophical objectivity identified in the field of the empirical science? In
classical mechanics, the observer occupies the place of a knower such as God, who is outside the system and who influences neither the system nor its content, so that there arises no problem other than
avoidable errors of observations. In Einstein's relativity-theory, the observer's position and motion
inside the system to be observed contributes to the observed results. Yet, one can perfectly trace such
an influence according to the law given by the theory itself. In general, it is admitted that the objectivity stands firm where all the actions of the observer and their results are included within the systematic theory, that is, where all the interferences by an observer can be described and controlled according to the law of the system. However, in quantum-mechanics, interferences by observation in the
microscopic object cannot be below the scale of the same microscopic object under quantum law;
hence, although it is on the level of the complementary relation of uncertainty, the results of observa-

tion rest still uncertain. This is fundamentally different from the *subjective* error which can be, in principle, reduced indefinitely toward zero. Instead, we have an objective uncertainty founded upon the laws of quantum-physics. If one of the terms of the complementary observable quantity is fixed, the uncertainty of the other term spreads infinitely. Such confusion brought about by observation seems too colossal to be under a certain law.

Nevertheless, the definite description of the other complementary observable quantity compensates for this fact. Thus, by placing even the interferences, which are uncontrollable according to a certain law, within the framework of this uncertain relation, these interferences can be recorded as an objective *state* by Schrödinger's wave-function ψ which, in this case, is interpreted as a probability law.

Probability law is also useful for the macroscopic world as we see in statistical mechanics. When we make various observations in trial, we consider that, if the frequency of trials does reach a sufficiently large number, the sum of statistical alculations gradually reaches a definite value according to the law of large numbers, so that the subsequent trials do not seriously affect this uniformity and the recorded effect becomes more and more confirmed. A similar procedure is found in the objectification of the identity of the evidence of being. There, a large number of people are obliged to recognize such an objective identity only by means of induction, that is, by the fact that their acts are unable to affect the last core of being's evidence because these acts' *arbitrariness*, or freedom, is limited, in some degree, with respect to the being given to them. Therefore, the objectivity of the macroscopic world can be guaranteed by and only by the statistical procedure of induction. Indeed, we find it relies on probability, but this probability has no basis in any subjectivity, and so far it is an objective probability on which lies our representation of the world.[6]

Now I assert that the similar procedure of statistical description is valid for the system of quantum-mechanics. The microscopic world is observable only by the *macroscopically* significant value of any fluctuating indicator of the observation-machine equipped with macroscopic materials. For, when the results of observation are piled up to an uncertain value, this functional value, limiting itself to each time representing an uncertainty of the micro-object, still preserves the aforesaid statistical macroscopic objectivity. For it must be only within the macroscopic reach, in which we do not necessarily take into account the possible interference of any observer, that we are able to lawfully describe microscopic values. By representing a *state*, either the wave-function ψ or the uncertainty relation preserves in itself a macroscopic character of our objects. They are merely complementary elements that constitute a *state* of an *independent* object. An object is necessarily a macroscopic entity.

In Newtonian dynamics, time as represented by "t" is reversible, and also in Einstein's theory, the "t" can be replaced by "– t", if one acknowledges movement beyond light-speed. Yet, in Schrödinger's wave function, time is not reversible in so much as it describes voluntary reactions of micro-scaled objects caused by micro-scaled effects of observation. For there a kind of irreversible time is presupposed like the increase of entropy which is a macroscopic statistical effect of the second law of thermodynamics. Axiomatically, entropy may decrease and time may also be reversible. Yet, as empirically and statistically in the sense of the aforesaid objectivity, entropy always increases, so is time irreversible as a whole. In a partial system, however, entropy may decrease and thus time may be reversible. From what I have said, I conclude that time as a whole is irreversible since it is a form (a category) of the evidence of being under the condition that the being is a definite and closed wholeness, whereas entropy may decrease and thus time might become reversible in a partial evidence, i.e. evidence of consciousness which is but a part of the being.

3. AN ARISTOTELIAN'S CONCEPT OF TIME

Kant holds that the evidence of consciousness wraps the evidence of being. Thus, he flatters himself that he has given a "Copernican revolution" to Aristotelian ontology, which states that the evidence of being wraps the evidence of consciousness. But as I have said before, consciousness can be nothing but "consciousness of something" and so, its evidence is derived from the being and not from the consciousness proper. It is only when this being is limited to being a consciousness, in order for it to be "consciousness of consciousness", that the evidence of consciousness is born. Therefore, the so-called revolution must miscarry.

Now let us examine this Kantian way of thinking from the well-known Scholastic formula: "The received is received according to the mode of the recipient (*receptum recipitur per modum recipientis*)".[7] The evidence of being consists in the fact that the being is directly given to consciousness. The "received being" is, of course, received according to the form of the "receiving consciousness". Therefore, Kant regarded the form of being as able to be replaced by that of consciousness. But if "the received being" be deprived of its own form, that being could not be said to be received. Its material "content", together with its "form", must be received and become a content of consciousness. The Scholastic theory of abstraction explains the procedure by which the universal proper to the being becomes a content of the consciousness. Both Descartes and Kant, however, ignored that theory and saw the essence of cognition exclusively in the a priori form of consciousness. But clearly this is far from the starting point in which the being is given at first. For both of them, what becomes the content of cognition is nothing but the putting of the mark of consciousness upon such an *insignificant* being. Behind the "Copernican revolution" there is concealed a certain misunderstanding of the Scholastic formula.

However, as mentioned before, we must be careful not to forget that in order for us to establish objectivity the insistence on the evidence of being alone is not sufficient; we must, besides, manipulate the following procedure. Just as the fluctuation through trial and error becomes fixed because of the law of large numbers, we must keep confirming inductively the subsistence of the "thing-in-itself" whose evidence of being appears to us more and more independent of the arbitrariness of the evidence of consciousness. Thus, the objectivity can be established only gradually in its relation to the evidence of consciousness through the evidence of being. This is an a posteriori, epistemological *canonic*[8] of our objectivity. In this sense of objectivity we assert also that the irreversible succession of time is determined by the increase of entropy which itself possesses statistical, macroscopic objectivity.[9]

It is interesting to compare such a conclusion with some of Aristotle's ideas of time. The following is a philosopher's fancy that starts with various hypotheses of contemporary physics. Aristotle defines time as the number of movement concerning before and after.[10] Movement is an accident of a certain thing, and therefore its quality changes according to the nature of that thing – corporeal or spiritual. Yet, as long as they move, both body and soul possess their proper "time". When he talks of time as "numerable number", the time-number is individual and discrete, so that, as it is numbered, it increases. The modern way of thinking maintains that the increase of entropy is the increase of number of motion of the system including observers, and makes this number the standard for determining the direction of time. Aristotelian theory metrically coincides with such a contemporary concept.

Moreover, according to Aristotle, just as space "wraps" things, time "wraps" things: in other words, just as things are "within" space, they are "within" time. This expression "wrap" or "within" is his special idiom. He himself explains that when he says "a thing is within number", this number is the number of the thing. It means that number belongs to the thing just as space belongs to the thing when one says "a thing is within space." Now movement, number, time and space are no more than accidents of substance. Therefore, there exist different spaces, different times according to the different kinds of substance. In this sense, time belonging to the evidence of being can differ from time of the evidence of consciousness. In passages[11] treating the everyday vocabulary of time, Aristotle explains $\dot{\epsilon} \xi \alpha \acute{\iota} \phi \nu \eta s$ ("suddenly") as a word to be used for "a thing that slips out of a former situation with such a speed that it has not been perceived." He hints here an objective time pertaining to a being that far transcends the subjective time pertaining to the consciousness.

He goes further and says that the "nows" (time-units) are indivisible and individual. This had been already presupposed when he defined time as number of movement concerning before and after, for number is a discrete quantity that always increases by counting. Accordingly, we can see in such a concept that time is a successive series of indivisible units of various phenomena within the framework of substance, while space is a co-existence of indivisible units of various phenomena within the framework of substance. These images of time and space become interesting when we make them correspond to another pair of Aristotle's categories,[12] action ($\pi o \iota \epsilon \hat{\iota} \nu$) and passion ($\pi \acute{\alpha} \sigma \chi \epsilon \iota \nu$), if the former could be called, by modern terminology, momentum and the latter, energy. In quantum theory, as momentum and energy are quantified and individualized, so it is possible to coordinate momentum with an indivisible time-unit such as "width of synchronism",[13] and energy with an indivisible space-unit such as "electronic radius".[14] Then, through the uncertainty relation, the canonical conjugate of time becomes energy, and that of space momentum. Thus, just as the principle of complementarity has unified those complementary concepts in the concepts of "state", the Aristotelian categories of time and space, action and passion can be unified in the category of state ($\kappa \epsilon \iota \sigma \theta \alpha \iota$).[15] Physical hypothesis has always an objective value; it is never incompatible with metaphysics. When Aristotle was attacked by modern physics, it was not because his philosophy was metaphysical, but only because it was very obstructive to the tendency of scientific hypothesis of that time. The tendency of science is, in truth, philosophical. That is why modern empirical science since Galilei had for its background sometimes Platonic metaphysics, sometimes atomistic and mechanistic cosmology. I think that the present-day situation of physics requires a new examination of Aristotelian metaphysics.

POST-SCRIPTUM

If the movement whose number is time is perfectly cyclic, time will be eonized (perpetuated), as seen in Aristotle's celestial time. A circle is invariant under space-reflection, so that the direction of motion is reversed. So there is no way for us to know about the definite direction of time in this circumstance. Yet, in our empirical world, as in Aristotle's sublunary world, time is irreversible and successive in one direction, just as the increase of entropy is. From the same view-point of empirical and statistical objectivity, space-reflection seems to be generally impossible. For it is most probable that, in parallel with the one-direction of the empirical time, space might have the one-way bend or distortion.

Our general knowledge which is provided only by experiences of this existing world leads us to a knowledge that there is no essence that completely wraps and rules existence in our incomplete circumstances. From this point of view, our empirical world is but a system of an aggregate which seems to be closed whole to our experience, but, when examined closely, is incomplete and never able to determine its own term. This world is merely a part of a certain wholeness — open to the existence whose priority over essence must be acknowledged finally.

NOTES

1. Thomas Aquinas: *Quaestiones Disputatae de Veritate*, Quaestio I, Articulus 1; J. Iturrioz: *Metaphysica Generalis* (Madrid: Philosophaie Scholasticae Summa I, 1957), p. 512, n 58.

2. Thomas Aquinas: *Quaestiones Disputatae de Veritate*, Quaestio I, Articulus 1, ad 3.

3. Avicenna, Muslim philosopher of the thirteenth century, is regarded as a forerunner of Descartes by his metaphor of "Man in air" (*Avicenna's De Anima*, ed. by F. Rahman, London 1959, p. 16). One of us, he says, could be conscious of his own existence, even if he were in air having nothing to perceive by the perceptive organs of his body.

   As for Augustine's theory of mind, see *Confessiones*, Lib. XI, cap. 14-20; *Soliloquia* II. 1, n 1; *De Vera Religione* 39, n 73; *De Trinitate*, X. 10, n 14 et 11, n 26.

   As for Plotinus, see *Enneades, Plotini Opera* ed. P. Henry and H. -R. Schwyzer, III,4(5); III, 8 (5,6); III, 9 (3); V, 1 (7).

   As for Gnosticism, see C. Andresen: *Die Gnosis* (Bd. I, *Zeugnisse der Kirchenvater)*, Stuttgart 1969.

   As for Orphism, see Rohde: *Psyche*, Tübingen 1921.

   As for Brahman-Atman, see Bṛhad-Aranyaka: *Upaniṣad* I-4-10; II-5-19; IV-4-5 and *Chandogya* VI-8. See also next note.

4. We may compare this subjectivism with the monotheistic tradition of Hebrew thought. Here the "Absolute" is the ultimate "Other" that transcends the world of relative beings. Joined with the objectivism of Hellenism, this thought regards the world as objective being and makes the "Absolute" its original and final cause. Both the existence and meaning of the world exist relatively to the "Absolute", not of themselves. This outer world, founded and created by the "Absolute Other", becomes even a measure for the unstable and arbitrary, relative ego. This is the origin of Western Objectivistic realism. Here is a dialógue between the absolute "Thou" and the relative "I" and a contract or covenant between God and man. Faith is the loyalty to this contract. Christianity arises from this cosmology.

   According to Brahmanism, however, many "small egos" attach themselves so much to the outer world to substantialize it. Thus, they confine themselves to the cycle of metempsychosis. This theory teaches that one must awaken to the fundamental absolute "Ego" by enlightenment, in order to go back and to be united with the big "Ego". This enlightenment is a monologue, the return to the true "Ego". It requires a kind of *ecdysis*, as a preparation for contemplative introversion.

5. Aristotle: *De Generatione et Corruptione*, A 4, 319b 10.

6. To the possible criticism that places the problem of induction and probability within the scope of "inter-subjectivity", I answer with this passage that this problem is deeply rooted in the domain of the "objectivity". "Subjectivity" or "inter-subjectivity" can only be discussed within the reach of the "evidence of consciousness", and never of the "evidence of being" which

the induction relies upon.

7.   Thomas Aquinas: *Summa Theologica* pars I. Q. 12, art. 4; *Quaestiones Quodlibetales* 7, 1. 1; cf. Centro di studi filosofici di Gallarate, *Enciclopedia Filosofica,* Venezia – Roma 1957, Vol. III, p. 1803; IV, p. 1862.

8.   In the sense of the Epicurean epistemology.

9.   We have obtained the definition of time from thermo-dynamics, which again presupposes time because physical description consists in the succession of time. That seems to be a "vicious circle". But I think this is a normal circular definition which permits us to define again the increase of entropy by the irreversible time. For any circular definition is possible and sometimes unavoidable between two "accidents", and I call it "correlative definition" in contrast with the "absolute definition" that must be given to the "substance" only.

10.   Aristotle: *Physica,* Δ 11, 219b1. As to his whole discussion about time, see *ibid.,* Δ 10 - 14, Z 1 - 3.

11.   *Ibid.,* Δ 13, 222a10 - b16. For example, $\pi o \tau \acute{e}$ has contrary meanings – "not yet", "formerly" – according to the context that determines it in relation to the "now".

12.   Aristotelian categories are historically well-known. They are ten: substance, quality, quantity, relation, time, space, action, passion, state, possession. I summarize them into four principal groups: (1) substance, (2) attribute (which includes quality, quantity and relation), (3) accident (which includes time, space, action, passion and state), (4) possession that means value-beings such as truth and falsity, goodness and badness. The terms in the same group show some correlative characters, either between a pair (e.g. quality and quantity, time and space, action and passion) or among them all. Cf. my book, *Sonzai no Ronrigaku Kenkyu (The Logic of Being),* Tokyo 1944.

13. 10^{-27} secs.

14. 10^{-13} cm.

15.   These ideas are based on a new interpretation of Aristotle that I presented, *op.cit.,* pp. 384-85 and notes (5) and (7).

Time and Ethics: How Is Morality Possible?

C.M. SHEROVER

It has frequently been remarked that Immanuel Kant brought the concept of time into the forefront of philosophic discussion; that much of our preoccupation with time stems from his work. But it is too often forgotten that he had carefully and painstakingly restricted the dimensions of time to the cognitive functioning of the human understanding, that he had denied time applicability to the human self in its exercise of that moral freedom which he regarded as the secured foundation of moral reason.

That he bifurcated the human self and its experience between the temporal and the nontemporal raised serious questions concerning the unity of experience and of Kant's new Critical philosophy. That he denied temporality to moral reason effectively attacked the foundations of the moral philosophy and its experiential preëminence he was concerned to ground. That he felt it necessary to propound such perplexities points to the decisive import of the particular concept of time which he inherited and, in turn, passed on to us.

PART ONE

Concerned to validate the new science, Kant argued that our knowledge of the objects constituting natural phenomena — our perceptual experiences and theoretical reason's interpretive understanding of them — are completely structured in temporal terms. Our sense experiences and the principles in terms of which they are to be understood are both grounded, he argued, in the pervasiveness of the form of time which takes in every act of cognitive consciousness. This is to say that what he regarded as the root concepts of scientific explanation — quantifiability, substantiality, causality, interaction, necessary connection — were all ostensibly grounded in the form of time which structures human cognitive experience.[1]

But Kant was even more concerned to secure the foundations of morality. Doing so necessitated defending the possibility of moral responsibility, individual conscience, the autonomy of the self, and a normative ethic. Each of these constituents of morality is, itself, grounded in the moral freedom of practical reason. The importance Kant attached to moral reason is but indicated by the fact that he believed it to be here and here alone, not in the exercise of theoretical or scientific reason, that one could rise above the confines of phenomenality and catch a glimpse into noumenal or ultimate reality itself.

Ironically, he found it necessary to sequester this whole area of practical reason from the domain of time. Having demonstrated the pervasiveness of time in the functioning of cognitive reason, which had hitherto been associated with the timeless, he found himself impelled to deny the temporality of moral conscience and practical reason in order to save the integrity of the autonomous self and the essential freedom which grounded them. Although the painstaking dissection of theoretical reason was explicitly undertaken to safeguard faith in our own freedom and moral responsibility, the only way Kant saw himself able to do so was to foreclose the temporality of practical reason. One consequent was the unresolved bifurcation of Kant's Critical philosophy — an inability to reconcile and unify these two aspects of the human self, an inability to reconcile cognitive and moral experience. Of at least equal consequence is the fact that this detemporalization of a normative ethic, of freedom, of practical reason, rendered Kant's moral philosophy inoperable in his own terms.

Yet, on the face of it, this seems somewhat absurd. Any consideration of practical reason, of moral conscience, responsibility or decision, points to its temporal field. Morality arises, Kant had argued, from the distinction between inclination and obligation. Focused in the question "what should I do?", moral judgment looks from a present situation which poses a problem to the possibility of a resolution of that problem. Whether one proposes a teleological ethic of utilitarian means to chosen ends or a deontological ethic in which we seek supervening standards of right and wrong, it seems apparent that a temporally defined situation, a projection of moral possibilities into an as-yet undetermined future, and a foreseeable spread of time for requisite activity are all integral to any meaningful notion of moral responsibility and moral freedom.

Having perceived the centrality and fundamentality of time in cognitive experience, he seemingly felt it necessary to foreclose its applicability to areas of deeper concern. The concept of time with which he worked was intrinsically connected with the necessary determinism of the physical world and thereby would have subverted the whole of practical reason. To save the moral self, moral conscience, and moral freedom and responsibility (as well as the transcendental ego which underlies cognitive reason), he thus had to foreclose them from subjection to the only concept of time he really knew, the time of linear sequence.

He thus felt impelled to bifurcate human reason into the theoretical and the practical, one essentially to be described in time predicates and the other in their denial. Yet, something is seriously wrong, for Kant, himself, insisted on their ultimate unity: " ... if pure reason of itself can be and really is practical as the consciousness of the moral law shows it to be, it is only one and the same reason which judges apriori by principles, whether for theoretical or for practical purposes."[2] Indeed, Kant did not merely see them as unified; he insisted that in the distinction, practical reason had complete priority; it had "primacy ... [just] because every interest is ultimately practical, even that of speculative reason being only conditional and reaching perfection only in practical use."[3] Yet, Kant could not show *in his own terms* just *how* theoretical or speculative reason served practical interests, just because he could not show how a nontemporal reasoning could enter the sequential time of phenomenal events, in which moral reason must be applied.[4]

In a very real sense, Kant's discussion of time mixed two variant concepts of sequential time. When describing our observations of the phenomenal world which we seek to understand, Kant apparently accepted, without modification, that concept of time inherited from Aristotle, which seems to be used by all scientific observers of natural phenomena — time as the numbering of the motion of objects in

space in terms of before-and-after. But 'motion in space' refers to external phenomena and not necessarily to mental activity.[5] Consequently, when Kant introspectively observed the dynamic content of
consciousness, he seems to have used Locke's notion of primary time as the 'train of ideas' in consciousness. Indeed, this seems to be the working notion of time enunciated by Kant in the opening
pages of the *Critique of Pure Reason*. His essential argument for the fundamentality of the form of
time in cognitive consciousness, in fact, seems to be precisely this – that every idea, regardless of its
content or reference, is seen in consciousness to be in a sequence of ideas, before some, after others.

Aristotle's and Locke's notions of time – one referring to the external world and the other to internal
consciousness – have just this in common: they are descriptions of observed sequences from the point
of view of an observer supposedly external to them. Looking into a series of events, I see the present
scene emerge from that prior to it and in turn give way to a succeeding one. If, as a disinterested and
extraneous observer, I am asked to explain what is now before me, I do so in terms of an earlier scene.
My explanation, in terms of what I regard as an efficient causal sequence, finds the explanation of the
present scene in one that occurred earlier and is no longer. My explanatory scheme is essentially chronological and the present is explained in terms of the past. Insofar as the past is over and done with
and thereby unchangeable, we can see why this kind of explanation so easily suggests some kind of
determinism. For explanation is in terms of what is not now controllable and the farther back we
regress in the causal sequence, the more remote from controllability it appears to be.[6] If every event
is then observed in a sequence of before-and-after, no event contains its own explanation; each event
is seen as arising from some other event chronologically prior to it, and external from its present.
Once we apply some standard metric to this observational field, we are able to quantify and thereby
objectively describe the sequence of a train of events, as well as their durational spread, as they appear
in the observational field – in a numbered sequence of before-and-after.

For any such explanatory schema, chronology becomes crucial. The later is explained in terms of the
earlier; the present, as the product of the past which is beyond control, is thereby rendered determinate. All is then to be explained in the beginning, in a first cause or earlier state: "the future," as
Leibniz had once suggested then "is to be read in the past,"[7] and if there is an ultimate past beyond
which we cannot go, it must contain, in embryo, the entire future which is destined to emerge out of
it. Causal explanations in terms of prior states are expressed in statements of if-then relationships
which are statements of causal sequences structured in terms of before-and-after. Such causal explanations of the present in terms of earlier states external to what is being explained are necessary to any
kind of predictability. Such explanations arise out of that kind of disinterested observation which is
crucial to scientific inquiry; they record observed sequences without reference to the particularity of
the observer. They do not really offer any explanation of the reported sequence itself, but merely attribute efficacious necessity to selected chronology.

Kant was, of course, concerned to validate the new science of his time. Indeed, if one prime aim of the
Critique of Pure Reason was to establish the necessary universality of the principle that 'every event
has a[n efficient] cause', it is little wonder that Kant perceived the fundamental import of sequential
time to the explanatory scheme of a mechanistic science. For the notions of efficient causation and
predictability depend upon linear time, time plotted along the figure of a line in which the later is
explained by the earlier. Kant's validation of the new science was, then, a validation of the fundamental import of sequential or linear time in phenomena as they appear to us; it was, then, equally a validation of sequential determinism in the world of observable phenomena. Determination of an object,

in terms of the cognitive principles, means determination in a quantifiable causal-time sequence, in which what is 'after' is explained as the measured result of what came 'before'.

If universalized, this identification of explanation, causal sequence and time sequence, while facilitating predictability with regard to natural phenomena, effectively forecloses the possibility of morality. For morality — under which we include the notions of conscience, responsibility and decision — follows directly, as Kant claimed, from freedom which it immediately presupposes.[8]

Yet, Kant was so impressed with the import of causality for explanation that he repeatedly identified freedom as a kind of causality.[9] And causal explanation was so identified with chronological explanation of the present in terms of the past that we find him treating free acts of judgment and decision as nontemporal (because not determined), but yet somehow temporally prior conditions, along the analogy of an earlier mechanistic cause, for our deliberate entry into the time-world in order to act within it. The determined causality of the phenomenal order is describable in time predicates; somehow, noumenal freedom is able to provide a causal impetus into the time-world and is a kind of cause which must be prior[10] to the sensible effects which flow from its entrance into the world of sequential time. Yet, it is conceived as nontemporal and time predicates are not applicable to it. We must presuppose it in order to understand our experience of our own moral activity but we cannot understand it or explain it. To be free is to be "independent of determination by causes in the sensible world"[11] which appear in the form of linear time. Somehow, in a way that can be experienced but cannot be explained, freedom means an ability to be an original center of causation, an ability to initiate a causal sequence in the phenomenal sensible world of determined sequential order so as to manipulate or re-direct it as the judgment of practical reason demands. But practical moral reason, somehow able to enter and re-direct linear sequences, must somehow be outside of time and therefore must somehow be atemporal.

The ironic outcome is that Kant, setting out to reconcile the determinism he considered necessary to scientific explanation of natural phenomena with the freedom that is requisite to moral responsibility was so bound up by the notion of linear time that he had to detemporalize practical moral reason and morality — leaving the relation of the practical to the world of experience and action unexplained and inexplicable but somehow ultimately real. Somehow they are related — because we experience them as related — but we cannot understand how.

As with freedom and moral practical reason, so with the individual self that exercises freedom while examining conscience and making moral judgments, decisions and action commitments — we can have no explanation or understanding just because explanation and understanding are tied to efficient linear causality and thereby to linear time. But explanation in terms of such reduction of present to past cannot explain any act of freedom. Although Kant had identified freedom under the rubric of causality, he saw that it must be somehow exempt from the only concept of time he seems to have used, time as a linear series in which the present arises out of the past and in turn produces the future.

Perhaps one reason for the very strange detemporalized ontological setting in which he placed freedom, moral reason and the 'real' self, all of which seem to require a permeating temporality, is precisely this insight: linear time and linear causality go hand-in-hand to deny the autonomy of the self and the reality of that freedom which he was concerned to defend and establish.

Freedom meant, for Kant, self-determination, not determination by an earlier state of an external reality. Freedom *qua* freedom is not reducible to a past state or a mechanistic cause-and-effect relationship between two independent entities or events. Freedom *qua* freedom is not a predetermined reaction to an earlier or external stimulus, but the open option of choosing one's own responses to it. Freedom is characteristic — not of the things we apprehend as phenomenal appearances — but of persons who are centers of experience, who are experienc*ing* selves. This is to say that the linear time of what is observed is not necessarily the time of the observing, that the time-frame in terms of which we describe what is experienc*ed* is not necessarily the time-frame of the experienc*ing* itself.

All of our thoughts may, indeed, be observed to be in some kind of a linear sequence — after some and before others; our thoughts or ideas, as looked at externally as objects of observation, are, indeed, in that kind of 'train of ideas' which Kant seems to have borrowed from Locke. As such we may plot them along a line and number their order. But the import of a thought is not necessarily to be assessed by its sequential location; the import of a thought may more likely be found in its content, in what it is about, in that relationship in which it is found, in that to which it refers. The meaning and significance of a thought may be found, not by observing and numbering it as a point on a line, but by entering into it and looking out on the world, so to speak, from it. The meaning or significance of a thought may be found *in* the thinking itself, in the way it purports to relate itself to the world, in its intentional involvement. The meaning of a thought may best be assessed, not in terms of an external observation of it, but in terms of the monadological model wherein each act of thinking represents my way of reflecting my dynamic relationship with an aspect of my world from my peculiar point of view.

When we examine the content of our own thoughts we find that they are not primarily referent to a sequential location in a 'train of ideas' but to other ideas and to objects which are in a dynamic, shifting and overlapping perspective of present and past and future. The content of my present state of awareness is not merely a sensory observation of the actuality of the objects before me; it also includes my anticipations and recollections which are essentially conceptual in character. As Kant himself realized when he turned his explicit attention, however briefly, to the temporal cast of our experienc*ing*, as distinguished from the objects that are experienc*ed*, our most rudimentary awareness of a present apprehension is tied up in a synthesis that has melded into one cohesive judgment the presentness of sense-experience, the presence of the past that memory provides, the presence of the future that conceptual anticipation presents. Our experienc*ing* is not linear sequence but is a moving synthesis of past and future in a dynamic present that cannot be reduced to a point on a line.[12]

What Kant seems to have discerned but failed to develop is just this: an act of experiencing is not the product of the past. It is an undertaking in a present that brings pastness as selective memory into it, and brings futurity as conceptual anticipation into it as well. The shift here is decisively from the past to the synthesized present *as the focus of explanation* of whatever it is that is happening. The balance of this paper effectively welcomes this shift as proceeding in the right direction and, for all practical purposes, criticizes him for not having gone far enough. In any event, what Kant seems to have suggested here is that the perception of a train of events does not simply arise, or get reflected, in a parallel train of ideas, but is a structured interpretation arising out of a perception that is always a dynamic intellectual synthesis.

Indeed, a Lockean 'train of ideas' — just because it is the result of an analytic examination and not immediately experiential — is really pre-Critical in that it ignores Kant's own Critical (or Copernican) Revolution which insisted on the primacy of the structuring of our experiencing *and not* on the objects ostensibly observed by a passive receptor mind which did not contribute its own interpretive structures to the awareness of those appearing objects. The perception of a train of events in a numerable sequence of before-and-after is an activity of a constructed present which itself includes, internal to it, discriminated aspects of future and of past. The experiencing of time and of temporal passage is structured in dynamic temporal synthesizing which makes the frame of present-and-past-and-future (and not that of before-and-after) the fundamental time perspective of the experiencing self. The import of this point in Kant is but suggested by the fact that it appears in that part of the First Critique where Kant was concerned to establish the fundamental possibility of cognition itself in the integration of sense experience and conceptualization.[13] Unfortunately, Kant did not pursue or develop this insight. Quickly reverting to a traditional, pre-Critical, object-oriented focus on the static before-and-after sequence of observed things, he foreclosed the possibility of the unity of the Critical philosophy as he left it and thereby the validation of the possibility of morality he had hoped to establish.

PART TWO

Yet, if we examine an act of moral or practical reason — which Kant regarded as of greater philosophic import than scientific understanding — we can see that recognition of its temporal constitution saves it from the ironies that are generated by the essentially pre-Critical identification of temporal experience as linear, and thereby as efficaciously causal sequence. We should also see that a fully Critical examination of time — from the viewpoint of the experienc*er* — points to the mode of unification of practical and theoretical reason in accord with Kant's own intimations and hopes.

Although this paper argues its thesis in terms of moral reason — just because it presents the temporal argument most clearly and succinctly — the argument can be generalized: the time of mental activity, including cognitive activity, is discerned, not in the externality of observation, but in the internality of involvement.[14] Mental activity is not geared to the past but to the future; understanding and explanation depend, not on actual states in the past, but on the projection of future possibilities which are brought back in constituting the meaning of the present. If, indeed, it will be agreed that anticipatory projection and not causal determinism is the key to the understanding of the experiential present, then Kant's thesis concerning the priority of practical reason[15] starts to make sense, and finds confirmation precisely in that existential temporalization which he intimated but apparently never grasped.

If, then, we turn to the experience of moral reason, as the clearest example of mental activity for this purpose, we find that we can discern two theoretically separable 'moments' in its analysis. In any situation that can be dubbed a moral one, we can discern the 'moment' of judgment, and the 'moment' of decision to act. Ideally, each should be looked at in turn; but the essential points can be made by focusing particularly on the temporal frame incorporated into the second.

Take any specific moral judgment you have ever made. It was not made in a vacuum. It was made in a particular situation which posed a moral dilemma, a situation which posed a disagreeable prospect and was seen as urging interference to change its apparent course, or a situation which posed an attractive prospect and was seen as inviting action to assure its actualization or fulfillment. Facing alternative pressures, desires, values or demands, the dilemma was focused in the question 'what should I do?'. It could not be resolved in terms of desires alone because each one of several conflicting desires merely had set out a claim and each one of these conflicting claims had to be adjudicated. Rational judgment faces such a conflict of desires by asking 'which one should I honor?', 'which one should I ignore, sublimate or suppress?'. When I want, or feel called upon, to do three conflicting things and can only do one, which one should I do? Unless I am to be the slave of every passing inclination or external prod, I must judge and decide. *If I am to make a choice*, I have no choice except to invoke the requirement of the 'should': what should I do in this particular situation?

By raising the question of the 'should' or the 'ought', I have radically transformed that situation from a mere happening or flow of events in which there is an observable conflict of desires into a determinate situation which demands rational evaluation and deliberate interference. I have transformed it — from an observable event which I can dispassionately witness, behaviorly describe, or study — into a moral situation to which I call myself to *pre*-scribe a solution, a solution which looks beyond the present to a future resolution, a future resolution whose possibility gives meaning to the determinate situation to which I pre-scribe it. In seeking an 'ought', I am not only pre-scribing to the situation in which I find myself; I am prescribing to myself what I ought to do about it; I am anticipating the possibility of resolution and my participation in effecting that resolution. I am no longer a passive observer of a scene external to me; I have made myself part of the situation, defined the situation in terms of the 'should' I give to myself, and thereby I have made it also part of me. And I have presumed not only my temporal continuity but also my ability to synthesize the modes of time by joining that situation and myself together under the aegis of the future-referring 'should'.

How do I form my moral judgment? How can I determine what ought to be done in this situation? Even the most teleological ethic cannot avoid the question. For it must ask (a) which goal or value should be sought, and (b) which means to that goal or value-actualization should be pursued? No method of resolving the question of a moral dilemma can escape the mantle of some future-referring 'should'.

I need some criterion by which to determine what I ought to do in this, my situation. Moral philosophers have argued various and conflicting proposals about such criterion. For our present purposes we need merely note that advocates of a teleological ethic, such as Aristotle and Mill, have generally argued that present action should be guided by the anticipation of the realization of certain values and intelligent choice of means for attaining them. One of Kant's prime reasons for rejecting such an approach is that the future is not ours to see with certainty, that one needs to have a standard of judgment that does not depend either on a prophetic ability or the contingencies of the particularity of the occasion. He thus proposed what is generally termed a formal ethic focused on the principle of the categorical imperative.

To discuss Kant's reasons for a deontological ethic — an ethic of self-imposed obligation instead of one of prudential advantage — would take us far afield. For the present purpose we need but note two aspects of it: (a) as a completely apriori system of moral decision, it depends, not on sense-experience which he tied to linear time and causal sequence, but to pure conceptualization which he tied to anticipation; (b) although he argued that the criteria of right-and-wrong could not be found in a means-end discrimination, he did urge imaginative anticipation of certain kinds of possible consequence for the values grounding the decision as a test for the morality of a contemplated action. This is to say that, although Kant did not consider temporal modes or predicates in his moral philosophy, the only one that could be imported without distorting his prime moral theses is that mode of futurity which, I am urging, must be the prime temporal cast of practical reason.

One might conceivably resolve the question of the 'should' by the route of a teleologic ethic of means-to-end, or by Kant's route of the formal ethic of apriori moral reason. One might conceivably argue that moral judgment is essentially not temporal because it is an appeal to an ideal transcendent and timeless standard. However one chooses to resolve the question of moral judgment, it is clear that each route in some way leaves the final judgment — and the decision to act on the judgment — to what we call individual moral conscience.

Moral reason's call of conscience stands in marked contrast to all other forms of reason, and this uniqueness suggests why Kant had insisted on its preëminence. Alone, it turns my focus from the world to myself so that I may function with deliberateness in the world. It calls me from a preoccupation with outward things to the reality of my own being — so that I may utilize things rather than be utilized by them. In contrast to cognitive reasoning which focuses attention on the things it studies, moral reason, in the call of conscience, calls me back to me, to what Kant had called my noumenal self, my essential reality. In contrast to esthetic enjoyment which focuses on the things it contemplates, the call of conscience calls me back into my own reality. It asks me, not about the things in the world, but what *I* should do with them. It asks me, not about other people, but about *my* responsibilities for them. It forces me back into a radical self-awareness, for in facing the question 'what should *I* do?', I am facing my innermost being, my own values, my own commitments, my own essential temporal finitude, my own ability to synthesize the modes of finite time, my own ability to control my state of being in the world of other people and of things. It calls me back to myself, not in terms of the past or the immediate present, but in terms of the proximate future, of how my world and I should relate to each other in the next few minutes, hours, days, or years ahead. It calls me back to myself as a decider with a capacity for effecting relationships in the world. In the most literal sense, it calls me to re-form and re-orient that aspect of the world's development over which I am able to exercise influence and control. It arises out of a present situation which I see, interpret and understand in terms of moral possibilities. It calls me to decide which of these alternative possibilities, *I should* select to become actual. It calls me from my usual focus on things in the world to a focus on my own self so that I may act decisively in my world of other people and of things. It reveals to me the reality of my own being and continuity in time, the values to which I am truly loyal, the possibilities which are genuinely mine, the limitations which I must accept, the limitations which I freely choose. In facing the question 'what should I do about this situation?', I am facing no external scene but my own involvement; I am facing, in my own conscience, my own self in its existential reality as being time-bound and time-binding.

But the call to conscience is not only a call to judgment; it is an imperative for deliberative action. It is an imperative to moral action, to acting *because* my moral judgment demands it. What does this action involve? When I act in a deliberate way, I look ahead, anticipate certain possibilities for development and delineate the present situation in terms of them. Action means that I determine myself in a commitment to a course of activity. I foreclose other possibilities, values, and desires as distractions to be avoided. I block out, as it were, a portion of the future and commit it to the action to which I have pledged myself to hold. The resolution of the situation which I command myself to pursue involves, then, my own resolute action. I project myself into the situation in terms of the resolution that I see it as demanding. I seize the possibility on which I have determined. I assess what it is factually possible for me to do; I commit myself to a course of action by anticipating a sequence lying ahead of me and resolving my present understanding of my present situation and my course of action in terms of this anticipation.

That all deliberate action is to be described in terms of such anticipatory resoluteness is clearly apparent. Deliberate action can be imposed externally by, say, an employer or a military commander. The peculiar nature of the moral action lies in the call of conscience which initial anticipation initiates: within the realm of what it is factually possible for me to do, is there one course of action which I *should* do, not as a matter of prudence but of obligation to my own self? The call of conscience arises out of the question of the 'should' and joins the resoluteness of action in the self-imposition of the 'should'. Anticipation, as such, is then not an intellectual act of contemplation of an object but an existentially involved reading of my present situation in terms of the possibilities for me which I see suggested in it.

It is in terms of anticipatory resoluteness that my own essential temporality is most clearly revealed to me. In the commitment to resolute action, as emphasized in the imposition -submission of the 'should', I see that I am, in essence, a temporal being. My temporality is revealed in the necessity of making a moral choice: I cannot do everything that I want to do; I do not even have the time to do everything that I feel I ought to do. My temporality is most dramatically apparent in those cases in which I cannot 'take time' to decide — but must act at once. I cannot even repeat an action for any act is irrevocable; once done it cannot be undone. Any action or inaction contributes to the development of my situation, as of myself, and cannot be turned back. I may, indeed, conclude that my first decision was wrong, that I should try again. But I cannot try again. The original situation has been altered by my first, if mistaken, decision. I now face a new situation created out of my first attempt to act. Indeed, I may try to redeem it by a new, second decision. But this second decision must apply to the new situation which I have helped to create, not to the old one. Each judgment, each decision, each act is an unalterable historic fact of possibility-actualization which may, perhaps, be superseded but cannot be undone, repeated or replaced.

It is not merely that the situation, the decision, the act is each 'in time' as in some sort of neutral container. Time is also in them. Time gives them their reality as their transcience and their irreversibility. Time defines that they are and how they are — as time defines my own reality and my involvement with the situation with which I am concerned, whose requirements I judge, as demanding that I impose an obligation on my self. The urgency and pressing nature of my moral dilemma then points out its structuring in temporal terms, as my response to it is structured in temporal terms. My moral judgments and moral decisions, then, are judgments and decisions about how I should structure those aspects of temporal existence with which I, as a temporal being, become involved. My moral judgments and decisions are not merely judgments and decisions about approvable acts; moral reason's questions are one and all ontological — about how I should structure the time of my experience.

How can I do this? When I resolutely undertake a course of action in order to resolve a present situation, I reach into the future for an as-yet unrealized possibility and guide my present action in terms of it. I bring it back, so to speak, into the present and read the present situation in terms of its lack of this future possibility — which I resolve to correct. I invoke memory for precedents to guide me in the evaluation of this particular possibility and its attainment. But my evaluation of my present situation, which calls me to act, is in terms of what ought to be in it that could be in it which I can place into it. That aspect of my past experience which I call into the present, is precisely that memory selection which seems pertinent to the task lying ahead of me. I then see this situation, as my own involvement in terms of a future which is not-yet, but which I bring into the present, and a past which is no-longer, but which I bring into the present. The present situation is, then, no point on a line of before-and-after sequences. It is not the click of a full second, not the 'specious present' of some ten or twelve seconds. It is a *spread* of time, as I perceive and understand it, which takes my perspective of future and selected recall of the relevant past into constituting what I take to be the present situation.

When I discern a conflict of desires or values in the situation in which I am, I invoke a call to conscience and ask myself what I should do; I am invoking the mantle of the future-referring 'should' as a key to the meaning of the present situation and of my involvement in it. I may, indeed, compare it to similar experiences in the past, but the key to similarity is the call of selected moral possibility.

In each stage of decision, my orientation is toward what is not the case that can be the case if I seek to make it come into being. My orientation is not to the past, which I cannot re-make, but to that aspect of the future which lies within my grasp. In a very real sense, then, I am reading my present situation in terms of what is not-yet but truly might-be. I am reading it in terms of possibility for development which I take as genuine. I am reading my present situation as one whose being is not framed by the ticking seconds of a clock which gives notice of passing moments in a spatial idiom. I read the situation as a being who is able to transcend the limits of the immediate present: I can run ahead of myself into a future which is not-yet actual, and bring back a possibility which determines the meaning of the situation to me now, together with the nature of what I am to do about it. I am reading my present situation, not as an instance of atomically independent momentary actuality, but as a dynamic time-consuming synthesis of the specific possibilities of the future which I read into it, and the heritage of the past futures which brought it to its present state.

The situation as such is defined by me as standing out from its background of other happenings and events in terms of the future-pointing possibilities which I see it as suggesting. The problematic which the situation poses to me is the alternate course of possibility development in terms of which I define it as a situation. The resolution of the problem which defines the situation as *my* situation is undertaken by me in terms of the possibility which I grasp in my moment of vision into a not-yet future which I bring back into the living present.

Moral reason is then fundamental, as Kant saw it to be; but not because of a necessary presupposition that our moral selves are timeless. Moral reason is fundamental just because of our temporal nature. We are not each imprisoned in a series of atomic actual moments measurable by a clock. We are temporal beings just because of our need for temporal continuity, just because we are able to transcend an actual present and reconstitute it in terms of future possibilities that beckon us onward. We are able to structure present activity in a temporal field in terms of a selected future which we bring back to throw meaning on what is now present. We are able to take the immediate presentation, synthesize it with future possibility and selected recollection into a meaningful spread of temporal experience. We are able to read our interpretive canons, structured by the future possibilities brought back or retrieved for use in the present; such retrieved futures reveal certain lacks which we obligate ourselves to realize. We are able to structure our own selves in terms of the self-imposition and self-acceptance of the 'should', which charges us to redirect the flow of events by integrating new possibility into the structure of the actual. We are, in short, able to impose moral obligations just because we are able to bring future possibilities into the present as a spur to new activity leading us onward.

Practical reason does not only depend, as Kant seemed to think, upon the distinction between inclination and duty, between sensibility and reason. Practical reason does depend upon our ability to transcend momentary time, select a future, invoke a past, and synthesize them into the intelligible reality of the living present. Moral obligation depends upon our capacity to read the present in terms of what it yet can be and let the synthesis of the three modes of experiential time draw that constituted present into a future that will, in turn, continue to look beyond itself. Man is a being endowed with reason; this means that he can look beyond himself, because he is not enslaved by the actual, because he can actualize those possibilities he has made his own. Man has moral reason, which traditionally has been called practical reason, just because he has a temporal or time-forming perspective, not merely in terms of his cognition as Kant had argued, but in terms of his whole being, because he can deal with what is possible, because he can act today by determining himself in the light of those possibilities which he chooses for his tomorrow.

PART THREE

In this description of the actual functioning of moral reason, I have obviously followed or worked from Heidegger's analyses in a general way. If this kind of description has any validity, it suggests that a projective concept of time saves the coherence of Kant's ethics and the unity of the Critical system from the disintegration to which Kant's own notion of the primacy of linear time had condemned them. This projective description of experiencial temporality functions in terms of a priority of the future instead of the present or the past. It sees the focus of explanation of human practical reason to lie primarily in its view of its future rather than its inheritance from the past or the immediacy of the present. Whatever the situation in which we find ourselves, whatever the dilemma which we see posed for us, we read it in terms of what can be done about it, in terms of the possibilities which we see it as offering, in terms of the 'where do we go from here?'. But this is to see the meaning of the present primarily in terms of future-referring possibilities which we project ahead and retireve in order to discern the meaning of the present. This is to see the explanation of our own actions not as blindly determined but as necessarily posing options, requiring decisions, always with a reference, not to that which came before, but to that which may come after, always with reference to a future which gives meaning to the present act.

To take a projective notion of time as primary is to give a priority to our experience *as we experience it,* to the way in which we experience time, use time, incorporate time into our living. It is to be loyal to the authenticity of our own experience, which we experience by looking ahead and constituting the meaning of the present in the light of the envisaged future. It is to take the Kantian distinction between persons and things seriously by giving our personal experience of time priority over the time of the things that are the objects of experience. This means that our time is essentially cast in terms of moral reason's self-direction rather than in terms of the sequential time we cast over the things we observe – just because things are seen by us primarily in terms of the meanings and the needs we see them as offering in response to the demands we make on them.

We do, indeed, take the Locke-Kant notion of a 'train of ideas' in consciousness as an observable datum of our mental activity — provided, of course, that it is taken as a flow of interlocking ideas and not as truly separable, distinguishable and thereby numerable units. But their content, in terms of which we look out onto the world, is primarily referential to projecting possibilities of the future into what is given in the present as the source of its significance. The content of our ideas is primarily referential to possibilities for action, to alternatives for action, to the questions 'what can I ... ?', and 'what should I ... ?' do about the situation in which I find myself.

But this is to explain my actions as I perform them, not in terms of a past state which has yielded the present. It is to explain my actions in terms of my vision of the future seen as coming towards me, which I accept as genuine or desirable, and in the light of which I mold my present activity. This is to say that explanation in terms of futural possibility is explanation in full accord with the Leibnizian principle of sufficient reason, reason sufficient to explain my activity and to comprehend it, in its own integrity. This invocation of a principle of sufficient reason that is not reducible to efficient causality, resists the intellectual temptation to import an intellectual interpretive framework about inanimate things and impose it on my living experience as I experience it.

Projective temporality not only enables me to explain my present action in terms commensurate with my experience of it; it provides a temporal frame for the possibility of the freedom which marks my experiencing and is requisite for morality; it permits an account of the self in terms of that autonomy which Kant invoked but could not really explain.

Indeed, projective temporality permits that unification of theoretical and practical reason which Kant had sought but could not really achieve. For, if both theoretical and practical reason are, as Kant urged, but two aspects of the same reason, and if the practical side is the motive side, we can then see that the primacy of projective time, which grounds practical reason, permits or demands theoretical study of the phenomena of nature in terms of quantifiable and linear time constructs — for this theoretical activity always involves practical involvement and is undertaken to serve envisaged possibilities of future realization.

PART FOUR

This paper has been concerned to show that Heidegger's general notion of fundamental temporality as projective futural-possibility retrieving activity is the one temporal perspective that can ground morality together with the notions of freedom and responsibility inherent in it. It also saves Kant's Critical philosophy from the many crucial gaps he had struggled but failed to overcome.

Although this has its own importance, I suggest it as an instance of a more general proposal: that the concept of time is crucial to any systematic philosophy regardless of the role it is alleged to play in it. As a hypothesis, the paper urges that an examination of the particular concept of time involved can conceivably serve either to subvert or to reconstruct the system which it effectively grounds. This is to say that the concept of time is not an incidental ingredient of a philosophic system but its ground concept, no matter how disguised its fundamental role may be — as, for example, in Kant, where it was explicitly restricted to the functioning of theoretical reason in the understanding of the objects constituting natural phenomena without any explicit examination of the inherited notion of time that was invoked to see how it itself was experientially grounded or what it did or did not inherently involve.

NOTES

1. See Immanuel Kant: *Critique of Pure Reason*, tr. N.K. Smith (London: Macmillan & Co.; New York: St. Martin's Press) — hereafter referred to as *CPR* — A142 = B187, pp. 183-87 together with the chapter entitled "System of All Principles of Pure Understanding." Taken together, it becomes quite clear that the Kantian categories are derived from or rooted in what Kant described as the four possible modes of temporal experience and that the Principles of human knowledge are explicitly temporalized versions of those same categories.

2. *Critique of Practical Reason*, tr. Lewis White Beck (Chicago: University of Chicago Press 1949), hereafter referred to as *CPrR*, pp. 224-25.

3. *Ibid.*

4. The "Typic" (to which N. Lawrence has kindly redirected my attention) hardly serves as the practical counterpart of the Schematism in the First Critique; in fact, Kant explicitly distinguished the two. Yet what is needed is a 'moral schematism' which would function as the fount of a temporal "procedure of the imagination" (cf. *CPrR*, p. 177) for practical reason as the Schematism does for cognitive reason. For Kant's own explanations of his moral doctrines always involve a temporalizing imagination to bring and test possible imperatives *in concreto*. Yet apparently because the Schematism was tied, in the first Critique, to that notion of time which yields the determinism of the natural world, he eschewed its use and could not replace it in the realm of moral freedom.

5. Strangely enough, as Locke for one pointed out, Aristotle did not explicitly bring space into the discussion of time *qua* measure. But it was certainly presupposed. His essay on time in the *Physics* was quite clearly concerned with time in terms of the measuring of the motion of physical entities; this seems especially clear from section twelve of that essay. It even seems true in his brief (and generally overlooked) discussion of psychological time (cf. 448a-448b).

6. Cf. *CPR*, A532, 33 = B560, 61, p. 464.

7. "The Principles of Nature and of Grace, Based on Reason," *Leibniz Selections*, ed. P.P. Wiener (New York: Charles Scribner's Sons 1951), p. 530.

8. Cf. *Groundwork of the Metaphysic of Morals*, tr. H.J. Paton (New York: Harper & Row, Torchbooks), hereafter referred to as *GMM*, p. 115.

9. Cf. *GMM*, pp. 114, 126; *CPrR*, pp. 165, 175, 200, 236; also through the discussion of the Third Antinomy in *CPR*.

10. Kant obviously could only have taken 'prior' in a logical sense here; but, if the capacity for free decision, and motivating action, is not also, in some sense, temporally as well as logically prior, it is not clear just what could be meant in any concrete instance. But, if there must be some temporal meaning somehow implicit, then the strict alleged nontemporality of moral reason collapses.

11. *GMM*, p. 119.

12. Cf. *CPR*, A98-110.

     N.B. Kant never really pursued this although, in view of the principle of the Copernican Revolution, he really was obligated to do so in the Subjective Deduction, which was concerned with the dynamics of experiencing rather than with the so-called 'external world' as it is *already* structured in our interpretive knowledge of it. That he abandoned the Subjective Deduction in the Second Edition, after but sketching it out in the First Edition, serves as a prime example of a great thinker who did not see the import of his own insight.

13. Cf. *CPR*, A98-110.

14. For an indication of this temporal analysis of understanding generally, see my *Heidegger, Kant and Time* (Indianapolis & London: Indiana University Press 1971), esp. pp. 142-70.

15. See notes 2 and 3 above.

What Time Is Not

M. YAMAMOTO

The title of my paper may look odd. If one is questioned "What is a dog?", everyone could give a more or less valid answer. But to the question "What is not a dog?", there can be innumerable answers which may be correct, but mostly quite unhelpful. Provisionally, what I intend to do under this title is to consider what we should not hold time to be. Through the history of ideas we meet various opinions and theories concerning what time is. But it seems to me that most of them are not adequate enough, though each has its reason and often profound import as well. So I would like here to make a proposal for reconsidering how time appears in our life and thinking.

As a starting point I take up the problem concerning whether time is something objective or subjective. This issue should be regarded as an obsolete one, and its way of being asked may even be false. If so, however, we must make clear the reason why this is so.

1. ON THE OBJECTIVIST CONCEPTION OF TIME — EXISTENTIAL FEATURE OF TIME AND R -HYPOTHESIS

Ancient people identified time with the processes of nature, and their conceptions of time were often expressed in mythical guise. Since mankind began to distinguish "time itself" from the tangible and visible change of things, its existence has been put in question. And when Aristotle gave a theoretical expression to this problem, already "mind" or "soul" was called up as a witness for the existence of time. But after all scrutiny, because of the necessity to secure the objectivity and universality of time, the movement of celestial bodies constituted the existence of time (*Physica*, IV, x-xiv). This mode of thinking was, for a long time, widely held in the history of ideas about time.

I shall not discuss here the problem concerning whether or not Newton's "absolute time" was really necessary for his system of mechanics, but it is important to notice that "absolute time" played the role of substitute for the ancient "celestial movement." We can see a similar attitude in the philosophy of John Locke, the initiator of modern epistemology. In his empirical theory concerning the genesis of our ideas, Locke stressed the relativity of time measurement. But, on the other hand, he maintained the existence of "duration itself", which was "to be considered as going on in one constant, equal, uniform course." (*Essay*, II, xiv, 21; cf. xv, 11). And the notion of time which was symbolized by "t" has manifested its effectiveness in many domains of modern science. We have become accustomed to the way of thinking in which the flow of time itself, which is usually represented as a straight line, and on which all things and events should be arranged is first set up, that is, we have become used to the conception that time is something objective. I leave aside the problem whether time was conceived as an entity or as a pure relation, which I think is not relevant here.

However, when we conceive of time as something extended by itself, we are immediately faced with the difficulty which has been noticed since olden times, namely that the past is no longer, the future is not yet, and the present is nothing but a boundary between them, so that time does not exist at all. When we take the concept "now" a little loosely and conceive time as a series of successive "now"s, the situation is no better. That the past "now" existed can be denied without assuming any change in the present fact, and on the other hand, the present "now" will be driven away, strangely enough, by the future "now" which does not exist.

Such a conception of time as a series of "now"s which are each sandwiched between non-being is found typically in the philosophy of Descartes. He made use of it for the proof of existence of God, saying that because each instant is existentially independent of another, the conservation of the world means its being created continually at every instant (*Meditationes*, III). To insist on causal connections as the glue of successive instants is futile in this case.

These conditions concerning the existential feature of time are presented palpably in the argument which Bertrand Russell pointed out, once, in connection with the problem of memory. Even if the world had begun five minutes ago, everything would be the same as it is happening now, and nothing which happens now can disprove the hypothesis (*The Analysis of Mind* 1921, p. 159 f.). Russell himself put aside this hypothesis, only by saying that it is "logically tenable, but uninteresting." Mr. Wataru Kuroda, one of my colleagues, has referred to this hypothesis in connection with the problem of knowledge and claimed that it is empty because it is an assumption about which there could be no empirical procedure of verification (*Tetsugaku* II, Tokyo University Press, p. 143).

However, I think that the problem should be regarded from the reverse side. The above hypothesis (which I call hereafter the "R -hypothesis") is, so to say, a touchstone for a possible conception of time. The assumed beginning of the world in the R -hypothesis need not be located five minutes ago; it can be put equally one minute or one second ago. Therefore, if the R -hypothesis is simply put aside because it is neither provable nor disprovable, and if its falsity cannot be positively shown, then time cannot avoid being shattered into bits of "now"s and totally deprived of its reality. The usual conception of "objective time" is defenseless against this.

2.   ON THE SUBJECTIVIST CONCEPTION OF TIME — PERCEPTION AND ITS TIME-REFERRING CONDITIONS

Let us now turn our attention to the other side of the problem. To perceive something, for instance, to perceive a desk, and to say "Here is a desk," requires time-referring conditions. The desk is not only present to us, that is, actually given to our senses, but also we must know that it has kept the identity of being-this-desk through some extent of time, be it three days or only three seconds. And at the same time, we have expectations that the desk will keep in more or less the same state, if not subjected to violence from outside, that we can put things on it, that it will look a certain way if it is seen from the other side, and so on. If the desk appears one moment and disappears the next, or if it becomes a chair and then a heap of books, then we would never say "Here is a desk." That one thing exists means for us that we find a suitable interrelation and unity in its appearances through time, so that we can name it with some word. (This is what Kant called "synthesis of recognition in a concept" *K.d.r.V.*, A103).

One might ask, how about the case of an event or thing which is usually called "instantaneous": for instance, the click of a glass being struck by a pencil, or the life of a μ-meson which lasts about one-millionth of a second. But for these cases we can either repeat them, namely bring about events which are of the same pattern as they, or at least we can describe the circumstances under·which they occur. ("It was such a tone as this." "This line on the plate which was got by such and such devices shows the existence of a particle.") If no such procedure is possible, we may doubt our own ears, or do not necessarily trust the report, in either case we will not necessarily recognize the existence of the objects in question. And to repeat the same one, or to describe the circumstances, is to ascertain the identity of the object by putting it into the observable relations of before and after, namely into the temporal process of our perception.

On the other hand, one might ask, how about the case of ceaselessly changing phenomena, as in a kaleidoscope. There we can find no regularity nor unity, yet we certainly perceive them. But I dare to say that we grasp these phenomena in the mode of "this play-of-combinations-of-colors" that is some-how within one frame. If we are exposed totally to a sheer promiscuous change of sensory stimula-tions, we shall become dizzy rather than perceiving any change. Conversely, when we are dizzy, we experience such a sensory kaleidoscope. And then we lose our "sense" of time together with the other senses.

I avoid getting entangled here in a troublesome problem concerning the relation between saying that a thing exists and saying what it is, the necessary and sufficient conditions for the identification of a thing, or the structure of the conceptual framework of our perception and knowledge. I will confine our investigation to the temporal character of perception.

I may now perhaps take for granted that a certain breadth of time is presupposed in our perceiving anything, and that the inner structural unity of time constitutes the indispensable condition for our conscious experience. In this sense we can say that we do perceive the past. The degree of this breadth depends on the attitude or interest of the percipient. Let us consider the case of hearing a melody. It is not false to say that we hear each of the sounds of a melody one by one. But more usually and more properly we say that we hear the melody, that is to say, the melody itself as a whole. (It is the same in the case of hearing a man speak a sentence). Let us imagine that we are at the middle of the melody. The former sounds have passed by already. Nevertheless, we hold them vividly and immediately in our consciousness, that is, we have "retention" of them in the sense claimed by Edmund Husserl (in *Phänomenologie des inneren Zeitbewusstseins*), not that we represent these sounds by means of audi-tive images of past perceptions. The sounds which have already gone away are perceived in the same way as the present one is, and they have reality in the same way as the present one has.

The R -hypothesis discussed previously is rejected by this fact. Though the breadth of each percep-tional retention is more or less limited, yet one retention links with another successively. Therefore, We must accept the fact that an indefinite extent of the past is always necessarily presupposed as the stage on which our present perception comes to be realized.

Thus, by introducing the structure of consciousness we can prevent the difficulty of the pulverization of time into vanishing "now"s. We have, however, now the subjectivist conception, according to which the reality of time lies in the constitution of our consciousness. This conception has made up a main current of philosophical theories of time from Augustine to Heidegger and Sartre. I do not want to examine these doctrines in detail here. They are mostly very refined and often profound. Still, insofar as they tend to reduce time into a subjective source in one form or another, they cannot be entirely free from difficulties, because the human being to whom the very consciousness belongs was born and will die in the course of time.

3.   WHAT TIME IS LIKE – CLOCKS, ACTS, AND LANGUAGE

Thus far we have considered what time is not, that is, what we should not hold time to be. And now we may say at least that time is neither simply objective nor purely subjective. Can we proceed further and say positively what time is? I would like to offer here some suggestions concerning what time is like.

The hint is this: we know a problem which is similar in form to the case of time, that is the problem of meaning in language. Since ancient times philosophers have asked repeatedly what meaning is, and we meet both the objectivist and the subjectivist conceptions of it, but both lead us into difficulties.

On the one hand, some thinkers undertook to set up a separate sphere of meanings themselves in order to guarantee the objectivity and universality of meanings. That was conceived either connotatively as a kingdom of ideal entities, or denotatively as a community of classes of objects. But such enterprises necessitate changing laws and making revolutions endlessly to accommodate changes in our living language. At the same time, these thinkers had to extend extraterritorial zones unlimitedly for the sake of expressions such as "round square". On the other hand, other thinkers regarded consciousness as the bearer of linguistic meaning. When we understand the meaning of a word, we surely have a feeling which is different from that which we have when we do not understand it. Thus, even if a machine writes letters or pronounces words, we are inclined not to ascribe to the machine any understanding of meaning because the machine seems to lack such a state of consciousness. But if this is justified, then similarly we must say that we cannot feel the inner state of the mind of others, and thus we would have difficulty in ascribing the understanding of meaning of words to others. Therefore, if meaning were nothing but the subjective content of consciousness, then the publicity of language would not be accounted for, and men would have to lock themselves up within an incurable scepticism or stubborn solipsism.

These circumstances make me guess that there is some analogy and some parallelism between the problem of meaning and the problem of time. I will raise three following points.

(i)   When we ask what meaning is, we fall into difficulty if we first consider words as pronounced or written things on the one hand, and then seek something which should correspond to them and be called their meanings on the other, these meanings being conceived as objectively existing entities or as subjective contents of consciousness. I now borrow an idea of Ludwig Wittgenstein and say: The meaning of a word is the possible way of using the word in accordance with certain rules in some situation, and the meaning is actualized by the fact that we have used the word effectively in the given situation.

In the case of time, what corresponds to the pronounced or written word is the clock. We must not think that what we measure by clocks is time itself which exists by itself as something like an empty receptacle, nor should we think that the length of our memory or expectation precedes clock-reading which makes it public. We should say, not that a clock is made because time exists, but that time appears because we use a clock.

But I must hasten to add the following: What I call here a "clock" should never be understood only as the mechanical clock or watch which we use today, but should be understood in the widest sense. It includes things which existed prior to any devices which men have deliberately made to measure time such as the waterclock or the sundial. They include, for example, the alternation of day and night, the change of shape of the moon, the rotation of the four seasons, the birth and death of men, and the rise and fall of dynasties. When man began to regulate his life by using these phenomena as clocks, time appeared. Therefore, corresponding to the different ways of using clocks, time shows itself differently: linear or cyclical, homogeneous or value-loaded, continuous or discrete. What is important is that these types of time each have equal right, and consequently none of them has the absolute privilege to be time itself.

Let us take up, for instance, the idea of cyclical time. This is not a "false" and "curious" idea which comes from the "infantile" mentality of ancient or savage people. We think every day that "It is morning again." We do not think, for example, that "24 hours, or 1440 minutes have passed since yesterday morning." Only when we think about a program of one week (like a schedule of a conference), do we think that one day of the total course has passed away. Similarly, we think every year "The spring has come again," but not "Since last spring time has advanced by 365 days, which amounts to 8760 hours or to 525,600 minutes." Only when, for example, we think about the course of our own lives do we think that we have spent one more year of our given time. In comparison with these cases, the idea of an infinite homogeneous linear time is rather a remote and faint idea for our concrete life.

One and the same world is represented differently by different languages, and even in one language differently by different contexts of interest. We should not hold one among these different representations to be right, the others to be wrong, but that they together fill up our lives. Analogously, we are living in the woven combination of different types of time. Just as a word is located within the whole of language according to its differences and relations to other words, a denomination of time is located in the whole of temporal relations which is constituted by some way of using clocks. Just as the meaning of a word is checked by synonymity which is known through the agreement of ways of using words, time-checking is done by simultaneity which depends on the ways of using clocks.

ii)   What status has the concept of time which appears in physics? May we say that it makes the base of all representations of time? Speaking in the same analogy with the probelm of language, I think, it corresponds to the syntax of language or, rather, to the logical form of language which is expressed by formal logic. Just as there are various systems and various notations of logic, physical time also is variously formulated and manipulated mathematically, according to the variety of objects in question. The quantity of time can be differentiated, squared, in logarithm, reversible etc. The degree of convenience with which these formulations are used for description of phenomena is the criterion of their adequacy. However, they do not describe time as such. They make up a system of pure relations of before-and-after, but time in its reality is rather neglected from the beginning. Nowhere on a coordinate axis t can we find time. A movement of your eye or finger which traces the time-axis is surely a temporal phenomenon, but the straight line itself which lies there has nothing to do with time.

Syntax or logic shows the general constitutive rules of language, so that it sustains the publicity of it. Nevertheless, it does not constitute the reality of language. Similarly, time as it appears in the theory of physics does not constitute the reality of time, though it plays the role of being a universal frame of reference for all kinds of phenomena. What I call here "reality of time" is, if I may borrow an example from Henri Bergson (in *Essai sur les données immédiates de la conscience: · Time and Free Will* et al.), the duration which we experience in waiting for sugar to dissolve. This reality is lacking in the time of physicists because the theory of physics treats the syntax or logic of time, while, as it were, it neglects semantics.

(iii) What then does constitute the semantics of time? It is the interest in our lives, especially the intentions with which we act. We have seen before concerning perception that the concrete "now" has a breadth in proportion to our interest, and a similar truth is evident concerning our acts. In the case of our acts the stress is laid on the future. To act is to intend to attain or avoid something in the near or remote future. What we are doing now cannot be described except by referring to our concern for the future. A picture which shows a gesture of a man in an instant can represent his act, but it is because the beholder understands the scene by putting it within the context of before and after. On the other hand, to represent the movements of the body of a man on space-time-coordinates is not sufficient as a descirption of an act. Such a representation results in ambiguity. The series of these pieces of behaviour can be regarded as an act in a full and definite sense, only when they are related mutually and unified by the intention of the man in question. This corresponds to that fact that in the case of perception what unifies and identifies perception is that which enables us to say what the object is.

Let me illustrate what I mean. Imagine that I am walking on a road and that someone asked me "What are you doing?" I shall never answer "I am moving my legs," if not in jest. The reply "I am walking" also means nothing but my refusal to answer my interrogator. My answer becomes appropriate only if I tell the questioner the aim of my walking and say, for example, "I am going to the station to meet my friend" or "I am going to college to teach." But "I am only taking a walk" is also an appropriate answer, because an aimless walk has certainly its own aim.

Surely, there can be an answer which professes the very absence of any particular aim. Imagine that a man who is sitting in a sofa in a hotel lobby answers "I am doing nothing." As he replies, he smokes a cigarette. He does not want to count smoking as a doing, namely as an act. Yet, he intends perhaps to recover from fatigue, or he is waiting for the beginning of dinnertime.

Thus, what supplies reality and meaning to time is the direction to the future which is reflected by our interest in the present act. Therefore, the case in which the existence of time becomes an issue the most vividly is when we deliberate whether we can do something before a definite future time, or deliberate how to act till such a time, for instance, when a speaker asks the chairman "Is there yet time for me?" or when a man wants to go shopping just before the departure of his booked plane.

On the other hand, the reality of the past too comes over most urgently when we see it in the light of the interest of a present act. For instance, if I find I have lost some important documents, I examine scrupulously everything I have done since the time when I went out yesterday with them. In this case, I am less concerned about the particular events which I encountered yesterday, than about their succession and duration. Without such an interest in present acts, the past appears merely as a series of various scenes. But those scenes have also their background which is animated by my interest and acts in each case. They never stand regularly on one straight line, but each forms a mass apart, often overlapping and crossing each other.

Our concern about the remote past which we ourselves have not experienced is based on our historical interest. In this case, noticeable persons, events and epochs are each thought of in the form of a closed "world". The progress of time on a chronological table is drawn continuously, but historical time is originally discrete. By reason of it the problem of the demarcating of eras is not of mere convention, but of essential importance for our comprehension of history.

Physical time is the notion which we set up at the very limit when we abstract from our interests in life. It is surely objective, but to be objective does not mean immediately to be real. It means to have been set up as a framework of objects in a certain theory. Therefore, physical time has various characters according to the variety of theories in which it appears, for instance, time in classical mechanics, time in relativity theory, time in thermodynamics, time in cosmological speculations, time in the quantum theory of field and so on. Though they are useful instruments for our daily lives and practical acts, yet they in themselves are nothing but theoretical constructs. To say that they exist is the same as to say that the infinite set of natural numbers exists, or to say that absolute solidity exists.

Thus, we are living in miscellaneous contexts of time which are variously divided and superposed on one another. And we can give more or less exact definitions of time according to the ways of using a clock, very exact ones in physics, very inexact ones in children's talk, each kind has its own use and right in its universe of discourse. But if we had to give a definition of time itself, we could not say other than that "it is something which appears in various manners according to our interests and ways of using clocks." This is exactly the same type of claim as the definition of "substance" which annoyed John Locke, namely that a substance is something unknown which is supposed to support the concomitant appearances (*Essay,* II, xiii, 19-20; xxiii, 1-4). If we look for substance itself apart from all appearances, it is quite natural that we cannot find what substance is. It is like "looking for a knot in a rush," as Leibniz said in criticizing Locke's arguments (*Nouveaux essais*). In this sense, we should perhaps not question what time is.

I would like at the end to add the following point. In my discussion I have used the analogy between the problem of time and the problem of meaning in language. I think that this was not an accident. Both time and language come from the peculiar way in which human beings relate to the world. Our way of living does not consist of being buried within the stream of sensual impressions and reacting immediately against stimuli, whether by instinct or by conditioning. We set up our own framework to articulate experiences through which we represent and regulate actively the world and our lives. And what constitute such a framework are, among other things, language and clocks.

Since ancient times man has been characterized not only by having *logos,* but also by being mortal. The reason why mortality, which is common to all living things, becomes an issue only for man is that man is a unique being who knows his own death, namely the temporal limitedness of his own existence; hence, he cannot help being concerned about this fate. This situation makes us question the relation of ourselves with time again and again. But how should we put the question, if not in the form of what time is not? ――― that is the question.

VII. PHYSICS

A Non-Causal Approach to Physical Time

S. KAMEFUCHI

SUMMARY

It is contended that since causality is originally a metaphysical principle, it should be possible, in general, to formulate physical theories without recourse to this principle. The best known example for such an approach might be the theory of action at a distance, where the description of physical phenomena naturally takes an unfamiliar, non-causal form. The Machian aspect inherent in this approach is emphasized. The recent discovery of violation of time-reversal symmetry of the fundamental physical laws is discussed from the above viewpoint. It is hoped that this kind of non-causal approach will enable us to study some further aspects of time which otherwise are masked by, and intermingled with, causality.

1. TIME AND CAUSALITY

One of the main purposes of physics, or of physical sciences in general, is to find the regularities that may exist in the correlations between different physical phenomena. Such regularities, when quantitatively formulated in mathematical terms, are called physical laws. And it is customary, in most cases of dynamical theories, to express such physical laws in the form of differential equations, through which the physical quantities concerned, say F, are related to a continuous variable t that is called time.

It should be noted in this respect that this variable t used in physics has primarily not so much to do with any philosophical concepts, but is simply defined in reference to the dynamical states of some appropriate physical objects. Thus, the variable t represents, for example, the number of swings n of a pendulum, or the location l of a star in the sky. That is to say, t is defined as $t = t_1(n)$ or $t = t_2(l)$. The functional dependence of F on t, or $F = F(t)$, being determined by the above-mentioned differential equations, should therefore be interpreted as representing $F = F_1(n)$ or $F = F_2(l)$. In this sense, we may say that what is described by the physical laws for F is essentially the correlation between F and n or l. On the other hand, we know, of course, the correlation that exists between n and l, that is, $n = n(l)$ or $l = l(n)$, so that $F = F_1(n)$ can immediately be converted into $F = F_2(l)$ and vice-versa. Thus, to express the quantity F we may use any of the variables $n, l, \ldots$, whose representative is then denoted by t. That is to say, t is the variable abstracted from the concrete variables $n, l, \ldots$.

Causality is one of the basic principles which underlie all physical theories. Here I am using the word 'causality' in the usual, most naive sense, specifying that a cause will lead to an effect at *a later time*. This is a principle of a metaphysical nature, and in this respect, it should be clearly distinguished from other principles that are of a properly physical nature. I am particularly emphasizing this point since many authors in physics maintain an ambiguous attitude toward such a distinction. The causality principle provides us with the basic framework in which our thinking is formed and developed. And we humans, especially physicists, have become so accustomed to this causal way of thinking that it is nowadays quite rare to ask a question such as: "Which part of our knowledge in physical sciences is due to this particular way of thinking, and which part is really specific to the nature of the physical world?"

Such questions have been partly answered by Watanabe[1], who constantly emphasizes the importance of the particular role that causality plays in physical sciences. According to him, we have an inborn mental habit — being intimately bound up with causality — of formulating propositions in such a form as: "If the object is in state A, then it will be in state B at a later time", and physical theories provide the conditional probability for the process $A \to B$. Once such an attitude is taken, the symmetry with respect to past and future that originally existed in the physical laws will be lost: Prediction (of future) is always possible, but retrodiction (of past) is not. He goes on even to argue that the asymmetry or one-way-ness in the flow of time in the macroscopic world, which is exhibited, for example, by the increase of entropy, is also a natural consequence of this. I quite agree with him in this respect, except, however, that what he says might not perhaps be the only reason for the asymmetry in question.

The problem I would like to discuss mainly in this article is the following. In view of the above-mentioned nature of the causality principle, I believe that physical laws could alternatively be formulated without recourse to causality. Naturally, it should be expected from the outset that we would then have to change the pattern of our thinking in a drastic manner and to speak a language quite unfamiliar and unconventional. I believe, however, that in this way it would become possible to study some further aspects of time which otherwise are masked by, and intermingled with, causality. At this point I must hasten to add that when saying this I am not going to propose any new theory by myself, but merely to call attention to one of the old theories in physics which has remained rather unpopular in the orthodox school of physicists, that is, the theory of *action at a distance.*[2]

2.   CAUSAL DESCRIPTION : FIELD

In order to appreicate fully the unfamiliar features of a non-causal description of physical phenomena, let me begin with the conventional, causal description by specifically referring to the case of electromagnetism. The most characteristic features of the causal description of physical phenomena may be summarized as follows: The state of a given physical system at a certain instant t_0, to be denoted below by $S(t_0)$, can be uniquely determined by specifying the values of a set of quantities $q_1(t_0)$, $q_2(t_0)$, ... , which we call the general coordinates. It should be noticed here that to completely specify $S(t_0)$ we need not refer to those quantities at other times than t_0, such as $q_1(t)$, $q_2(t)$, ... with $t \lesssim t_0$. Thus, the prediction of the state of the system in the future, $S(t)$ with $t > t_0$, amounts to that of the quantities $q_1(t)$, $\ddot{q}_2(t)$, ... , and this is generally carried out by means of dynamical laws. As already mentioned in Section One such laws are usually given as a set of differential equations with respect to t, and all we have to do is to solve these equations under suitable initial conditions $q_1(t_0)$, $q_2(t_0)$, ... : When an initial condition $S(t_0)$ is given, the state in the immediate future $S(t_0 + \Delta t)$ is uniquely determined, and this in turn determines $S(t_0 + 2\Delta t)$ and so on, where Δt is an infinitesimally small time interval. Mathematically speaking, we are concerned here with the so-called *initial-value problem*.

I think that the concept of *field*, such as the electromagnetic field described by $\vec{E}$ (electric-field strength) and $\vec{H}$ (magnetic-field strength), is one whose theoretical necessity is justifiable only in connection with causality. Or in other words, if we gave up causality, then the field itself would become completely redundant. Let me explain the point by considering the following simple case.

Suppose that there are two charged particles: Particle 1 is at the origin O at $t = 0$, and particle 2 is at rest at another point P, which is at a distance r from O, and further that particle 1 is somehow accelerated at $t = 0$. The effect of this event propagates itself in the space around the point 0 as a wave of $\vec{E}$ and $\vec{H}$, the speed of propagation being equal to c, the velocity of light. By emitting this wave, the particle may lose (or gain) a certain amount of energy and momentum, which is carried away (or brought in) by the electromagnetic wave. When the wave reaches the point P, where particle 2 is situated at $t = r/c$, some of its energy and momentum are transferred to particle 2. This chain of events can be interpreted as particle 1 exerting a (retarded) force on particle 2.

In this kind of description, what has happened to particle 1 at $t = 0$ is being recorded or memorized, at any later time t ($< r/c$), by the electromagnetic field $\vec{E}(\vec{x}, t)$ and $\vec{H}(\vec{x}, t)$ of the same time t. Thus, our system now consists of particles 1 and 2 and the electromagnetic field, and the state of the system at t, $S(t)$, is specified by the coordinates and momenta of the particles, $q_1(t), q_2(t), p_1(t), p_2(t)$, and the field variables, $\vec{E}(\vec{x}, t), \vec{H}(\vec{x}, t)$. The energy and momentum possessed by the two particles at $t = 0$ are shared, during the time interval $0 < t < r/c$, by both particles and the field.

If, on the contrary, we did not introduce the concept of field at all, the entire picture of the process would be quite different. A certain fraction of the energy and momentum which were initially possessed by particle 1 would completely disappear from the scene during the interval $0 < t < r/c$, and reappear later at $t = r/c$. Thus, it would turn out that no causal connection exists between $S_p(0)$ (or $S_p(r/c)$) and $S_p(t_1)$ with $0 < t_1 < r/c$, where $S_p(t)$ denotes the state of the particles at t: Only between $S_p(0)$ and $S_p(r/c)$ does there exist a causal connection. As we have seen above, however, when the electromagnetic field is included as a constituent of the system, the causal connection is recovered between the states $S(t_1)$ and $S(t_2)$ with t_1 and t_2 being completely arbitrary: $0 \leqslant t_1, t_2 \leqslant r/c$. In this case we may particularly choose t_2 as $t_1 + \Delta t$ with Δt being infinitesimally small. The causality relation for such a case may then be called the *infinitesimal causality*. In this way, the causality, which in the above example holds true with respect to $t = 0$ and r/c, can be interpreted as the result of an infinite number of repetitions of the infinitesimal causalities with respect to the pairs of instants: $t = 0$ and Δt; $t = \Delta t$ and $2\Delta t$; .... ; $t = r/c - \Delta t$ and r/c.

Before concluding this section I would like to point out parenthetically that besides the concept of field which we have been discussing above, we can find in physics many similar cases in which causality is involved in an essential and inseparable manner. Whenever a new theoretical or experimental situation arises which at first appears to be inconsistent with causality, physicists try to save causality by generalizing the hitherto existing concepts or by introducing entirely new ones. For example, an apparently acausal motion of an electron from the present towards the past, which is theoretically admissible, can be reconciled with causality by introducing a new concept of anti-matter: The zig-zag motion of an electron with respect to the direction of time is reinterpreted as a causal motion of a system consisting of more than one electron and positron (anti-electron). See, for example, Fig. 3 in an article by Park.[3] A similar reinterpretation applies also to the problem of tachyons, or(probably fictitious) particles that can travel faster than light. Incidentally, this way of reinterpretation is sometimes referred to as the *reinterpretation principle.*[4]

3.   NON-CAUSAL DESCRIPTION : ACTION AT A DISTANCE

Let us now turn to an alternative theory of electromagnetism, the theory of action at a distance which does not employ the concept of field at all. In view of the above-mentioned close connection between causality and field, we may expect that a description by such a theory will naturally be of a non-causal character. In what follows we shall therefore refer to this as a non-causal description.

It has been known for a long time that electromagnetism can be formulated without recourse to the electromagnetic field: A theory of this type was considerably developed by Tetrode, Fokker, Schwarzschild and others in the early twenties, and revived 20 years ago by Wheeler and Feynman[2] and more recently by Hoyle and Narlikar.[5] As already mentioned, the theory is not popular among present-day physicists, simply because, I presume, they are so much used to the causal way of thinking.

The electromagnetic forces which give rise to electromagnetic phenomena in general are exerted by, and also received by, charged particles (or those with magnetic moment, etc.). Hence, it is possible to interpret such actions as direct interactions between charged particles which are not mediated by anything like fields. In this kind of description we may say that the electromagnetic field, or light in particular, does not exist as a primary physical entity. Of course, it is still possible to consider, as auxiliary variables, those quantities that mathematically correspond to the field variables $\vec{E}$ and $\vec{H}$, but this is not necessary.

Let us consider again the same example as in the preceding section, that is, the interaction between two charged particles 1 and 2. This time the whole process can be described in the following way. Particle 1, situated at the point O, tries at $t = 0$ to exert a force on particle 2. This action will come into effect, at $t = r/c$, at the point P where particle 2 is then situated. The force exerted by particle 1 acting in this way on particle 2 is called the retarded force $F_R^{(1)}(2)$. On the other hand, when viewed from the standpoint of particle 2, this action may be interpreted as a reaction which is exerted by particle 2 at $t = r/c$ and comes into effect at the point O at an earlier time $t = 0$. The force of particle 2 acting in this way on particle 1 is called the advanced force $F_A^{(2)}(1)$.

In the present description there appear only particles and no field. That is to say, there is no theoretical basis for distinguishing one particle from another; all particles must be treated on a completely equal footing, and consequently action and reaction must be treated in the same way. In other words, it is quite meaningless to inquire which particle exerts an action, and which particle exerts a reaction. Hence, we must conclude that both the retarded force F_R of a familiar nature and the advanced force F_A of a very unfamiliar nature equally play very fundamental roles. In fact, when formulating the theory by means of a Lorentz-invariant Lagrangian, we find that the elementary process that primarily takes place between any two charged particles is described by $(F_R + F_A)/2$.

The necessity of including F_A as well as F_R makes the description of electromagnetic phenomena quite different from the usual, causal description, since F_R and F_A come into play generally in a very intricate manner. The state of particles at present, $S_p(0)$, is related not only to the state in the past, $S_p(t)$ with $t < 0$, but also to the state in the future, $S_p(t)$ with $t > 0$, so that it is no longer obvious, from the outset, whether we can start by setting up an initial state $S_p(0)$ at $t = 0$ as we do in the causal description. In fact, physical quantities of the system of particles (such as the total energy, momentum and angular momentum) at $t = 0$ cannot be expressed as a function of the physical quantities of the individual particles (such as their coordinates and momenta) at the same time $t = 0$, but depend further on the latter quantities in the future $t > 0$ as well as in the past $t < 0$. This, I believe, is one of the most characteristic features of the non-causal description.

When various events influence each other in the 4-dimensional, space-time world, it is no longer possible to label each event simply as 'cause' or 'effect'. In other words, it is not possible to describe such events by referring to the one-dimensional, ever-increasing variable t. We recall in this connection that t is after all only a variable which is convenient for describing such simple phenomena as motions of a pendulum or of a star in the sky. We should not be surprised, therefore, if the variable t does not work in every case. The right attitude to take in such a situation is to observe all the events as a whole and to discuss the correlations between them. This kind of attitude, I believe, is the one that conforms to the original purpose of the physical sciences as mentioned in Section One. We may summarize the unusual features of the non-causal description by saying that the instant called 'present' loses here the privileged position it occupies in the causal description and is degraded to the same position as those of 'past' and 'future'.

Thus, determination of possible motions of a system of particles amounts to that of the over-all histories of individual particles, which are represented by the respective world-lines extending over the past, present and future. And this can be done by finding the conditions under which these world-lines are mutually balanced in a self-consistent manner. Mathematically speaking, we are concerned here not with the initial-value problem as in the usual theory, but with the so-called *self-consistent method*.

The reader may wonder why I emphasize so much the importance of the non-causal description, since the more familiar, causal one works very well in electromagnetism. I will discuss one of the main reasons in the next section, and make here only the following remark. It is well known that as far as ordinary electromagnetic phenomena are concerned, the two descriptions are completely equivalent. We know, however, that there are some conceivable cases, at least in theory, in which the causal description may completely fail. For example, there occurs, for tachyons[6] or for light paths inside a black hole[7], what we may call the acausal cycle; this is represented by a closed polygon $ABC \ldots XA$ in a space-time diagram, where for example, event B is the effect of event A, and is the cause of event C. If such a cycle is really found in nature, then we shall have to lean completely on the non-causal description.

Incidentally, I would like to mention that observations along a line similar to the above were made recently by Husimi[8] and Park[3]. The former noticed the usefulness of describing interactions between charged particles by 'eliminating' the field, whereas the latter discussed what he called the atemporal representation, which corresponds to one of the aspects of our non-causal description.

4.   THE MACHIAN ASPECT

An unusual feature of the present approach is that even when considering an electromagnetic phenomenon in a localized space-time region, we must take into account the effects from all other charged particles that may exist outside the region, or in the entire universe. For example, the action generated by particle 1 at $t = 0$ reaches particle 2, situated at a distance r, at $t = r/c$, thereby changing the state of this particle by exerting the retarded force F_R. The reaction by particle 2 then comes back to particle 1, and exerts in return the advanced force F_A on particle 1 at $t = 0$. In this connection we must take account of the following two important factors. First, the electromagnetic force is a so-called 'long-range force', meaning that the action of particle 1, for example, has influence on all far-off regions. Second, the time at which the above-mentioned advanced effect comes back to particle 1 is quite independent of the distance r between particles 2 and 1, so that the advanced effects arising from all the particles existing in the entire universe come back simultaneously to particle 1 at $t = 0$.

In this way we find that any local event is intrinsically intertwined with the rest of the universe. This makes a striking contrast with the situation in the causal description, where we can completely ignore whatever may be happening outside the space-time region concerned. We may summarize the above situation in the following way: When taking the causal approach we can confine ourselves to localized regions in space-time and thereby find the local physical laws, whereas when taking the non-causal approach we necessarily have to take account of the over-all structure of the entire universe and thereby find the relationships between the local and the global aspects of the physical laws. We may say therefore that the latter approach is along the line of what is usually referred to as the *Mach principle.*[9] In the pursuit of physical laws these two approaches are to play roles complementary to each other.

My personal preference for the Machian approach rests on the fact that there are certain things which become explainable (or obvious) only when viewed from such a standpoint. When saying this, I have in mind particularly the following two instances. The first again concerns the case of electromagnetism. As long as we remain within the framework of field theory, the Maxwell equation allows, as a solution, any linear combination of F_R and F_A, that is, $F = aF_R + (1 - a) F_A$, with a being an arbitrary constant such that $0 \leqslant a \leqslant 1$: The theory contains no principle by which to choose from among an infinite number of solutions F the particular solution F_R that corresponds to the case realized in nature. Or, in other words, the theory cannot explain the one-way-ness in the flow of time. It is only customary to choose the solution F_R by invoking causality which, however, as mentioned already, is not a physical principle.

In the theory of action at a distance, on the contrary, this kind of arbitrariness in the solutions does not exist at all. By suitably modifying the arguments originally developed by Wheeler and Feynman[2] Hoyle and Narlikar[5] have in fact shown that F_R is the only self-consistent solution for the effective interaction between two charged particles. The underlying assumptions by the latter authors are: i) the universe as a whole forms a complete absorber of electromagnetic effects, and ii) it is expanding in a way dictated by the model of the steady-state universe. They have further shown that the situation remains essentially the same when quantum-mechanical effects are taken into account. Their arguments, however, seem to need a re-examination in view of recent cosmological evidence which may not be favorable to the above-mentioned cosmological model. At any rate, it is interesting to note that in the non-causal description there exists a possibility of explaining the one-way-ness in the flow of time. (cf. Watanabe's explanation of this one-way-ness in the causal description, quoted in Section One).

The second instance concerns the theory of gravitation, the case originally considered by Mach himself.[9] By further extending the original idea of Mach, Sciama[10] has explicitly shown the possibility of relating the two kinds of masses, that is, the inertial mass m_i and gravitational mass m_g, which, incidentally, are regarded as completely independent quantities in the usual Newtonian mechanics. According to Sciama, the Machian theory provides the relation: $m_g = Gm_i$, with G, Newton's gravitational constant, being related to the cosmological parameters in such a way that $G\rho\tau^2 = 1$, where ρ is the mean (inertial-mass) density of matter in the universe, and τ the Hubble constant. (The matter at a distance r is receding from us with the velocity $v = r/\tau$). Furthermore, by making suitable assumptions about the gravitational force, he was able to derive, from Newton's first law of motion, the equation of motion (the second law) that contains various kinds of inertial forces. In view of the rather rough nature of the arguments, this theory also needs further refinement.

The most essential feature in the above two cases is that both the electromagnetic and gravitational forces are of long range; only this type of force can convey the response of the universe to our laboratory. According to quantum mechanics, such forces should be mediated by quanta or particles with vanishing mass. Thus, the photons and gravitons are those particles that are responsible for the electromagnetic and gravitational forces, respectively. We know, on the other hand, that there exist in nature some more species of particles with vanishing mass, that is, the μ -neutrino and e -neutrino, both being responsible for weak interactions of elementary particles. From the Machian point of view we may therefore expect that the two kinds of neutrinos should also be conveying to us some information about the structure of the universe. I believe that this is one of the most important problems to be studied in the future.

It is worth noticing in the above instances, especially in the case of gravitation that the local physical laws are determined, to a certain extent, by the over-all properties of matter in the universe. Conversely, of course, all material objects existing in the universe should obey the local physical laws. We may say therefore that a kind of *self-consistency* or *bootstrap relation*[11] exists between the local physical laws and the possible form of existence of matter in the universe. Lastly, it should be emphasized that the Machian approach enables us to realize the interrelations between different kinds of physical laws that govern the respective *strata* (or different structural levels)of the natural world.[12] Incidentally, the *Dirac principle*[13], through which certain fundamental constants of a microscopic nature are related to those of a macroscopic or cosmological nature, is akin in spirit to the present approach.

5.   SYMMETRY WITH RESPECT TO PAST AND FUTURE

Finally, I would like to call attention to one of the recent important discoveries in elementary particle physics, that is, violation of T (time-reversal) -symmetry ' or asymmetry of the fundamental laws with respect to past and future. A group at CERN in Geneva has recently obtained a direct evidence for the asymmetry in the decay of neutral K -mesons.[14] This is an especially important result in view of the fact that the conclusion is quite independent of whether the *CPT* -symmetry holds true in nature. Incidentally, it had been known for several years that a certain feature of the decay can be interpreted as indicating such an asymmetry if the *CPT* -symmetry is assumed.

Usually, in discussing basic problems on time, such as its statistical or cosmological aspect, we take it for granted that the fundamental physical laws are completely symmetric with respect to past and future. However, such an attitude should now be changed; we may expect for the reason mentioned above that such an asymmetry, although being a microscopic and rare local phenomenon, somehow reflects the overall structure of the macroscopic world. In this connection, it should be remembered that the increase of entropy, or *H* -theorem, can be proved, irrespective of whether the fundamental physical laws are symmetric with respect to past and future (cf. Section One).[1] Thus, in trying to find possible macroscopic effects of the microscopic asymmetry, we must study something else.

Certain attempts have already been made along this line by Ne'eman and his collaborators.[15] They considered a gas of K -mesons and their anti-particles, interacting with heat baths, in order to see how the *T* -asymmetric term in the decay affects its statistical properties. Their conclusion is that the Onsager relations between the coefficients of the heat currents from the gas to the bath are violated owing to the above-mentioned term. In other papers,[16] Ne'eman further discussed some cosmological implications of the *T* -asymmetric term. These works are, however, still at a very preliminary stage, and it is hoped that further investigations will be made in this direction.

I would like to take this opportunity to express my sincere gratitude to Professor S. Watanabe for valuable advice and helpful discussions.

NOTES

1. Watanabe, S.: In *The Voices of Time*, ed. J.T. Fraser, New York: Braziller 1966; *Progress of Theoretical Physics, Supplement*, Extra No. (1965), 135; In *The Study of Time*, vol. 1, eds., J.T. Fraser et al., Berlin/Heidelberg/New York: Springer 1972.

2. Kerner, E.H., ed.: *The Theory of Action-at-a-Distance in Relativistic Particle Dynamics*, New York: Gordon & Breach 1972.

3. Park, D.: In *The Study of Time*, eds., J.T. Fraser et al., Berlin, Heidelberg/New York: Springer 1972.

4. Bilaniuk, O.M.P., Deshpande, V.K. and Sudarshan, E.C.G.: *American Journal of Physics, 30* (1962), 718.

    Bilaniuk, O.M.P. and Sudarshan, E.C.G.: *Physics Today, 22* (1969), No. 5, 43.

5. Hoyle, F. and Narlikar, J.V.: *Proceedings of the Royal Society of London, A277* (1963), 1; *Annals of Physics* (N.Y.) *54* (1969), 207; *62* (1971), 44.

6. Newton, R.G.: *Science, 167* (1970), 1569.

7. Israel, W.: *Physical Review, 143* (1966), 1016.

8. Husimi, K.: *Shizen* (In Japanese), *28* (1973), No. 3, 30.

9. Mach, E.: *Die Mechanik in ihrer Entwicklung — historisch-kritisch dargestellt*, Wiesbaden, Brockhaus 1933.

10. Sciama, D.W.: *Monthly Notices of the Royal Astronomical Society, 113* (1953), 34; *The Unity of the Universe*, London: Faber & Faber 1959.

    Sciama, D.W., Wilson, P.C. and Gilman, R.C.: *Physical Review, 187* (1969), 1762.

11. Chew, G.F.: *Physics Today, 23* (1970), No. 10, 23.

12. Engels, F.: *Dialectics of Nature*, Moscow: Progress 1934.

    Sakata, S.: *Progress of Theoretical Physics, Supplement*, No. 50 (1971), 185.

13. Harrison, E.R.: *Physics Today, 25* (1972), No. 12, 30.

14. Winter, K.: *Proceedings of the 1971 Amsterdam International Conference on Elementary Particles*, eds. A.G. Tenner et al., Amsterdam/London: North Holland 1972, p. 333.

    Dass, G.V.: *Fortschritte der Physik, 20* (1972) 77.

15. Aharony, A. and Ne'eman, Y.: Paper presented at the 1972 Coral Gables Conference on Fundamental Interactions.

16. Ne'eman, Y.: *Proceedings of the Israel Academy of Sciences and Humanities,* Section of Sciences, No. 13 (1969).

Aharony, A. and Ne'eman, Y.: *International Journal of Theoretical Physics, 3* (1970), 437.

On the Origin of Indeterminancy

KEN-ICHI ONO

INTRODUCTION

For a long time physicists have treated time simply as an extension similar to space. However, though time and space are similar to each other in that they are both extensions, they are radically different in that time is an entity having an intrinsic direction.

When I say the direction of time, the question is not whether the fundamental equation changes or not under an inversion of time. Even if the fundamental equation lacks symmetry with respect to the inversion of time, it does not immediately mean that time has an intrinsic sense of direction.

These circumstances we can easily understand if we replace inversion of time by reflection of space. If the fundamental equation lacks symmetry with respect to a reflection of space, we can, by means of this fact, distinguish left-handed from right-handed systems. However, it does not mean that any difference in value arises between the two systems. The left-handed and the right-handed systems are still of the same dignity.

The situation with time is exactly the same. Even if the fundamental equation changes under an inversion of time, it does not mean any difference in value between the past and the future. It is no more than a problem of the symmetry of the fundamental equation.

In my opinion, the future is the future because it has an uncertainty in it. In the daily world, the future is of course uncertain. But it may not be so only for the daily world. If we successively simplify a process until it perfectly decomposes into elementary processes, still, I think the future will have uncertain parts as before. Indeterminacy is, so to speak, the inevitable destiny of the laws of nature.

However, as there is no form in classical mechanics to describe the indeterministic part of the law of nature, the people who constructed quantum mechanics had to work out a new form to describe spontaneous and indeterministic change of mechanical states. But in the forms used to describe the continuous and deterministic change of state they could follow in the steps of classical mechanics.

As Professor Schrödinger showed in the story of a cat,[1] the change of our knowledge as we learn something new occurs in the same way as the spontaneous and indeterministic change of a mechanical state. So the doctrine called the theory of observation was worked out and used to describe the indeterministic part of the laws of nature. By it we may understand observation in quantum mechanics as a concept referring to the cause in general which brings about a spontaneous and indeterministic change of state.

It is very unsatisfactory, however, that quantum mechanics requires two forms to describe the time-variation of a state: the wave equation and the theory of observation. The situation may not be independent of our instinctive feeling of resistance to admit indeterminacy from the standpoint of monistic realism. There are many who consider monistic realism directly bound to determinism. Professor Einstein shared this notion; he could not endure the dualism of quantum mechanics. And from his point of view there was no room for indeterminacy in elementary processes. Concerning the dualism of quantum mechanics, I am on the same side as Professor Einstein. However, I do not agree that monistic realism has no alternative but to lead us to determinism. In this paper, I would like to present a preliminary essay intended to unify the two forms used in quantum mechanics and also to reconcile monistic realism with indeterminism.

CRITICISM OF THE THEORY OF OBSERVATION IN QUANTUM MECHANICS

In quantum mechanics, observation acts to break the continuity of time-dependent change of mechanical states obeying the Schrödinger equation. It has a crucial effect on the results if, between the two given observations, there exists a third observation. Therefore, in quantum mechanics, in order to be able to construct a consistent theoretical system, it is necessary that all the observations are well ordered in time. If all the observations are made in one and the same Lorentz frame, this condition is automatically fulfilled. However, if the observations are made in different Lorentz frames, there exists no definite time order between them. In this sense, the structure of quantum mechanics is fundamentally non-relativistic.

Of course, there are cases where two observations in different Lorentz frames can be carried out without any contradiction. Let us denote by A and B the respective objects of two observations referring to different Lorentz frames. Then the two observations are compatible with each other when the observables A and B are not only commutable, but also their representation spaces are independent and, as a result, the wave function is the direct product of their representation vectors. In this case, at the moment t_1 when the observation of A is carried out in the first Lorentz frame, the representation vector of A performs a certain jump corresponding to the result of the observation. But this jump does not affect the observation of B. In the same way, at the moment t_2 when the observation of B is carried out in the second Lorentz frame, the representation vector of B performs a jump corresponding to the result of this observation (Fig. 1).

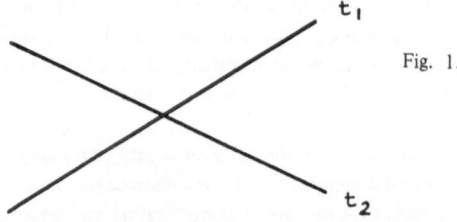

Fig. 1.

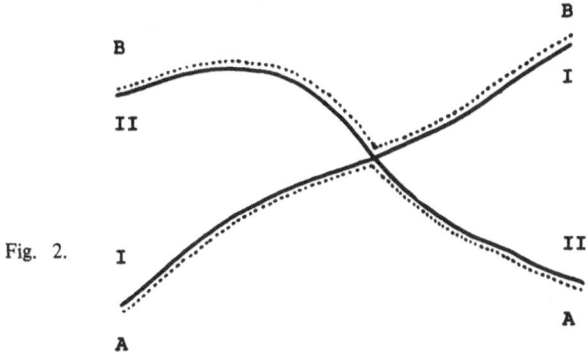

Fig. 2.

If one makes a complete set of commuting observables using only those observables which are in the relation mentioned above, and defines a state by the set of their characteristic values, thus constructing a certain theoretical system, then one succeeds in constructing relativistic quantum mechanics, where the word relativistic means that the theory does not refer to any definite Lorentz frame.

One example is the super-many-time theory.[2] In order to make a system of observables of above-mentioned character, the super-many-time theory restricts itself to the use of localized observables. Let the object A of the observation performed in a certain Lorentz frame be localized at the point (x_1, y_1, z_1) and let the moment when the observation is performed be t_1. In the same way, let the object B of the observation performed in the second Lorentz frame be localized at the point (x_2, y_2, z_2) and let the moment when this observation is performed be t_2. As long as the two points (x_1, y_1, z_1, t_1) and (x_2, y_2, z_2, t_2) are in space-like relation in time-space manifold, the two observables A and B are in the relation mentioned above. Then the two observations can be treated without any contradiction within the framework of ordinary quantum mechanics. Thus super-many-time theory appoints certain localized observables at every point on an arbitrary space-like surface, and defines a state with characteristic values of them. In super-many-time theory the state is not a function of time, but a function of the space-like surface.

However, even in super-many-time theory, it is also a premise of theoretical construction that the hypersurface corresponding to a cause does not intersect with the hypersurface corresponding to the effect. When the two hypersurfaces do intersect, we can always look at them as in Fig. 2. Therefore, it does not mean any practical restriction to demand that the two hypersurfaces do not intersect. However, considering that the reason why we can look in this different way is that every observable in use is localized in super-many-time theory, we must say that the super-many-time theory avoids answering the basic question, what is the situation when the objects of the two observations carried out on different Lorentz frames are not in a relation like that of A and B mentioned above.

When two observations refer to two different Lorentz frames, we may suppose that if the object observables are either non-commutable or their representation spaces are not independent, then those observations are impossible to carry out. If we postulate this impossibility, then quantum mechanics becomes relativistic for the time being.

Logically, this solution is available, but physically it is not satisfactory. For according to this postulate, it follows that as long as the relative velocity V of the two Lorentz frames in which observations are performed is not zero, those observations are impossible to carry out if the representation spaces of their object observables are not independent, even when the observables themselves are commutable, while as soon as V becomes zero, it becomes possible for those observations to be carried out. In other words, the situation at $V = 0$ is different from what we expect in the limit when V approaches zero. Logically, the fact is not inconvenient, but physically it makes us hesitate to accept the solution.

The origin of the non-relativistic character of quantum mechanics exists in the fact that in the theory of observation, the treatment is quite asymmetric with respect to time and space. In quantum mechanics, the moment when observation is carried out is localized at one point of the time axis and has no extension in the direction of time. On the other hand, the place where observation is performed is not localized in some finite volume, but extends throughout space. It is often considered paradoxical that in quantum mechanics the disturbance which observation exerts on an object spreads faster than light, or that the information of the result of observation does likewise. Such considerations tend to ignore the fact that in quantum mechanics the place where observation takes place is considered to extend throughout space.

In reality, it is likely that any observation requires a finite length of time, and that, on the contrary, it does not require all space as the place where it is performed. Yet, quantum mechanics is constructed on an approximation that considers the necessary time length as zero and the necessary space extension as infinity. And it is in this asymmetry in the treatment of time and space that the essentially non-relativistic character of quantum mechanics originates. On the other hand, the super-many-time theory tries to be relativistic by considering both the time length and the space extension of observation to be zero, that is, by considering observation as an act localized on one point and having no extension either in space or in time. It does this by restricting itself to localized observables. Even in super-many-time theory, observations are not localized in a finite volume, but as the extent of the object of observation is restricted within the localized observables, one can consider the place where the observable is defined as the place where the observation is carried out. In other words, in super-many-time theory, one succeeds in specifying the place where observation takes place by no longer treating the space coordinates as observables but degrading them to parameters; thus, one succeeds in constructing the relativistic theory.

Any observation has some object. Let us call it A. Then, accompanying the observation, the state of the system turns into one of the characteristic states of A. We obtain the characteristic value corresponding to that state as the result of the observation. All the characteristic states of A form one diagonal system.

Considering the observation as the cause in general of a spontaneous quantum transition, the fact that any observation has its object means that corresponding to any spontaneous transition there is one orthogonal system of states. What on earth is the way by which nature knows that orthogonal system? In other words, what conditions would we have to prepare in order to cheat nature of the object of our observation?

AN ASSUMPTION, OR PROPOSAL

As energy and time are complementary to each other, they are bound together by the uncertainty relation. Therefore, we cannot strictly identify the moment when an observation of energy is performed. In other words, in spite of the premise in quantum mechanics that considers the observation as an act done in a moment, an observation of energy cannot be finished in a moment but requires a finite length of time. This fact allows us to examine what happens during the observation, and, thus, to guess to a certain extent when and where the indeterministic time-variation of state takes place.

Let us denote the energy eigenvalue by E_s, the corresponding eigenstate by ϕ_s, and the initial state of the system by

$$\phi = \sum_s a_s \phi_s .$$

Then, if the time-variation of state can be described by the Schrödinger equation, the state will remain the same superposition of ϕ_s to the end. But the fact that the state jumps at the time of the observation means that in reality the state does not remain the same but transforms as time goes on to one of ϕ_s.

However, is it true that the jump of state which we see here is caused by the observation? If it is really caused by the observation, there must be certain material means for the observation to inflict such an action upon the object. But what can be the material means?

In order to measure the energy of a certain state of an atom, a molecule, or something like that, we measure, for instance, the wave length of the light they emit in a transition from the state ϕ in question to any other well known state ϕ_0. This is a destructive test, for when we know the energy of the state ϕ, the state is already transformed to ϕ_0. However, this transition is not what we are calling into question.

Observation tells us that the state ϕ is one of the energy eigenstates, and is not a superposition of them. For instance, if we apply a magnetic field to an atom, the degeneracy in magnetic quantum number will be removed, and we find the atom in one of the energy eigenstates which that atom has in the magnetic field, no matter what may have been the initial state at the moment when the magnetic field was applied to it.

The transition of state which takes place between the moment when the magnetic field is applied to the atom and the moment when we measure the energy of that atom is the very transition that we are calling into question. Can it be caused by the observation? My answer is no. I think it only shows the fact that the energy eigenstates are stable and the other states are unstable.

The situation is very much like that of a non-linear vibration. In non-linear vibration a superposition of solutions is not a solution. If we give a superposition of the characteristic vibrations as the initial state, there arises, as time goes on, competition between the partial vibrations, and in the end only one of them survives. Of course, it depends on the initial condition what partial vibration survives. For example, in the vibration of a violin string, what mode of vibration is excited depends on the velocity v with which the bow is driven, and on the force F with which the bow is pushed against the string. We can show on a graph with V and F as horizontal and vertical axes the experimentally obtained regions in which the fundamental vibration, the first higher harmonics, the second higher harmonics and so on are excited, or the intermediary unstable regions. For a bad player, it may look like a question of probability what mode of vibration is excited. However, since this competition between modes is fundamentally deterministic, for an expert player it is easy to excite the mode of vibration he wants.

It might be reasonable to assume that the time-variation of states in quantum mechanics looks like that of non-linear vibrations. There are many non-linear vibrations around us. The linear vibration is an approximation we obtain when we remove the impurities and leave only the essence of those vibrations. However, in this process of abstraction, we are liable to drop the characteristic aspects of actual vibrations. Strictly speaking, all actual mechanical vibrations are non-linear. The linear vibration is an extreme concept, like the ideal gas or the rigid body, which does not exist in the real world. In the same way, the Schrödinger equation would be an idealized description of the law of motion. Therefore, it seems to me reasonable to imagine that the variation of state deviates over a long period from what the Schrödinger equation would predict. It seems reasonable that the time-variation of an actual quantum state is linked in the meaning of correspondence to the time-variation of the actual non-linear vibration.

Straightening out the discussion, I would like to assume that

(i)   the time-variation of the superposition of energy eigenstates does not obey the Schrödinger equation. The superposition state transforms as time goes on to one of the component energy eigenstates.

(ii)   this transition is indeterministic. All we can do is to know the probability of any transition. The probability P for a state ϕ to transform to any eigenstate ϕ_s of energy is given by the following equation:

$$P = | \, (\phi_s, \phi) \, |^2 \ .$$

According to quantum mechanics, the result of an individual measurement is restricted to be one of the eigenvalues of the observable. Leaving the general story till later on, if we measure the energy of a dynamical system, we obtain as the result one of the energy eigenvalues, and we know that the system is then in the state corresponding to that eigenvalue. My assumption is that the state does not jump following the result of the observation, but that we obtain the result because the system has already performed the jump of state to one of the energy eigenstates, and is in that state at the moment of observation.

In reality, this assumption is not newly invented, but we have hitherto used it implicitly. Whenever we treat any quantum theoretical system, we are in the habit of considering it, without any reason, to be in an energy eigenstate. The above mentioned assumption is no more than the explicit representation of our implicit understanding.

The deviation of the time-variation of a state from that given by the Schrödinger equation may be a phenomenon which goes on slowly and continuously, like the competition between the partial waves in non-linear vibration. It is a kind of cumulative effect, like the motion of perihelion. The Schrödinger equation, though it does not correctly describe this process, is otherwise quite perfect. We must not overestimate its imperfection.

In order not to be misunderstood, I must add a little comment. I do not intend to insist that the fundamental law must be a non-linear equation. Someone may ask me to show the fundamental equation which takes the place of the Schrödinger equation. However, it seems to me doubtful that we can give it in the form of a differential equation. If the deviation from the Schrödinger equation is really a phenomenon fundamentally indeterministic, the correspondence between cause and effect is one to many, not one to one. It seems to me difficult to reconcile this fact with the uniqueness of the solution of any differential equation. Moreover, this deviation violates energy conservation. Of course, in the case when the initial state is an energy eigenstate, the energy is conserved, and in the case when the initial state is not an energy eigenstate, we cannot exactly define the energy itself. But even in the latter case there exists an expectation value of energy. And this expectation value is not equal before and after the spontaneous transition we are dealing with. On the other hand, at least in Hamiltonian formalism, if the Hamiltonian is invariant under displacement of time, the expectation value of energy is conserved. Therefore, it is certain that we cannot express the fundamental law in Hamiltonian form and very doubtful that we can do it in the form of a differential equation. It seems to me wise to use the Schrödinger equation as the fundamental equation for the time being, mutatis mutandis, under the recognition that there are cumulative effects of its small defect which may amount to such a magnitude as cannot be ignored.

RELATION TO THE THEORY OF OBSERVATION

If we repair the Schrödinger equation with the assumption mentioned above, then the law of motion includes the theory of observation, and we succeed in unifying the theoretical composition of quantum mechanics, on the premise that any observation is after all an observation of energy.

In order to observe a quantity A, one has only to devise an instrument in which the system has energy eigenvalues corresponding one to one to the eigenvalues of A. In other words, in that instrument, the eigenstate belonging to the eigenvalue A_s of the observable A is at the same time the eigenstate belonging to the eigenvalue E_s of the total energy. The relation between A_s and E_s is characteristic of the instrument, and is known from the first. Therefore, if one wants to observe A, one has only to observe the inergy in this instrument.

Thus, it is my opinion that all observation is performed as an observation of energy. For instance, one may call to mind the series of precision measurements of atomic and nuclear moments executed by Professor Rabi.

If this opinion is right, the convention in observation theory that the result of an individual measurement is restricted to being one of the eigenvalues of the observable finds its foundation in the fact that, as the energy eigenstates are stable and the other states are unstable, the system is always in an energy eigenstate when we perform the observation. And the probability of obtaining a certain eigenvalue is reduced to the probability of transition from the given initial state to the corresponding energy eigenstate.

Further, if all observation is finally observation of energy, to observe any quantity means to arrange the condition for the system to have its energy diagonalized at the same time with that quantity. Therefore, the problem whether or not it is possible to observe two given quantities in two different

Lorentz frames is reduced to the problem whether or not it is possible to arrange the condition for the system to have in those Lorentz frames the total energies corresponding to the given quantities. The latter problem we can answer by examining individual cases, and we do not have to assume possibility or impossibility independently of the other part of the theory. I said before that quantum mechanics can be made relativistic if we assume that observations are impossible to be carried out if they refer to different Lorentz frames, and if the object observables are either non-commutable or their representation spaces are not independent. However, from now on such an arbitrary assumption is neither necessary to invent, nor even allowable.

RELATION TO THE SECOND LAW

In short: for the time-variation of a state I am imagining that if the initial state is one of the energy eigenstates, that state will remain the same for ever, but if it is a superposition it will not remain the same but will transform to some one of the component energy eigenstates.

I think, though not with strong confidence, that when the vibration of a violin string turns into one of the component modes of vibration, not only the non-linearity but also the damping plays an important role. It is an irreversible process that the superposition of several modes of vibration gets straightened out as time goes on. It seems to me possible that the time-variation of quantum states is linked in the meaning of correspondence to an irreversible process.

If the time-variation of quantum states does not strictly obey the Schrödinger equation but behaves as mentioned above, the second law of thermodynamics becomes very elementary to prove. The second law is the law describing the tendency of nature to move from diversity towards uniformity. This tendency can be expected if the correspondence between cause and effect in elementary processes is not one to one, but one to many. For indeterminacy is a kind of diffusion process in phase space, and by this process distinctions are made even. If there corresponds only a single effect to a single cause, it is very idfficult to imagine the mechanism of the process from diversity to uniformity. Thus, the main difficulty in carrying out the proof of the second law is in the device of introducing something indeterministic into the deterministic development of things. In classical mechanics it is done by adopting as the initial state not a point but a finite volume in phase space, a procedure known as coarse-graining. In quantum mechanics, it is done by introducing the concept of disturbance which observation inflicts upon the object. But if the time-variation of a quantum state is like that mentioned above, then the difficulty does not exist any more because the fundamental law of motion itself has indeterministic parts in it.

RELATION TO THE THEORY OF RELATIVITY

Quantum mechanics is based on two fundamental pillars, the wave equation and the theroy of observation. They both describe the time-variation of states. Of the two, the wave equation is easy to form so as to satisfy the requirement of relativity. The difficulty is in the theory of observation because, as mentioned above, the theory is in principle of a non-relativistic nature.

However, if my assumption is right about the time-variation of states, then, as the theory of observation is included in the fundamental law of motion, the main difficulty in the relativistic formulation of quantum mechanics automatically vanishes, and the relativistic formulation becomes a mere technical problem.

In order that an energy eigenstate in some Lorentz frame be also an energy eigenstate in another Lorentz frame, it is necessary that not only the energy but also the momentum be conserved in the former frame. For instance, in the case of potential scattering, as momentum conservation does not hold, the energy eigenstate on the Lorentz frame at rest relative to the potential is not an energy eigenstate in other Lorentz frames. In reality, as the Hamiltonian on the first frame is not invariant under displacement in space, the Hamiltonian varies with time in any Lorentz frame moving relatively to the potential. Therefore energy conservation does not hold in such a frame, and we cannot think of an energy eigenstate on that frame. In such cases, we may think that what I said about the time-variation of states holds in that special Lorentz frame in which energy conservation holds.

CONCLUSION

It seems that the human being is an animal endowed with very dull sense of time. It may be reasonable to consider for a moment the adequate unit of length for our daily use as 1 meter, and the adequate unit of time as 1 second. In our sense, 1 meter and 1 second are sensed roughly of the same magnitude. However, in reality, 1 second is 300 million meters. Therefore, we may think that the accuracy of our sense of time is 300 million times lower than the accuracy of our sense of length.

This wonderful dullness of our sense of time leads us, it seems to me, to the mistake of considering the extension of observation in time direction to be zero, or of considering the time-variation of a state that slowly, almost unnoticeably deviates from the Schrödinger equation as a sudden abrupt jump.

NOTES

1.   *Naturwiss. 23*, 807 (1935).

2.   S. Tomonaga, *Progr. Theor. Phys. 1*, 27 (1946); J. Schwinger, *Phys. Rev. 74*, 1439 (1948).

Laws of Physics and Ideas of Time

D. PARK

In this paper I shall try to discuss time in nonmetaphorical language, and explore the contribution that physics can make to such a discussion. I take metaphorical language to be language that is not forced upon us. As Professor Lawrence shows in his paper in this volume, the metaphors of time serve us like a fly's many-faceted eye, bringing not the clarity of focussed sight but the largest possible area of awareness. Where there is no choice of language, where we know only one way to say something, then from some higher point of view it may still be a metaphor but we experience the situation differently, and I am inclined not to use the term.

I wish to formulate a nonmetaphorical definition of time that arises out of my experience as a physicist, an almost trivial definition out of an almost trivial subject, and then show that in following the path of inference we find that even such a beginner's trail leads into unexpected rocks and swamps. If I use only the simplest language, it is because I do not know how to use any other and convince even myself that I am making sense. The spirit in which I approach the problem of definitions is this: any definition that links words with experience is full of mysterious assumptions that flash and rattle and clank around us like a hostile army in the night. The most we can do is to remain awake and not take unnecessary risks. When all else fails we can comfort ourselves with the maxim of the Dutch physicist H.A. Kramers: "My own pet notion is that in the world of human thought generally, and in physical sciences particularly, the most important and fruitful concepts are those to which it is impossible to attach a well-defined meaning."

I. A DEFINITION AND ITS CONSEQUENCES

To formulate a definition of any concept that refers to our experience we must start by finding some assertions about it that we must all admit to be true. For example, the American poet T.S. Eliot has written that "April is the cruellest month." One may discuss this statement from several points of view, and it has been done, but one is not obliged to make such a statement. One *is* obliged to admit that April is the month that follows March. What must we admit about time? Useless to assert that time is "the moving image of eternity" or, in its American equivalent, "the river I go a-fishing in." These are poetic assertions full of truth but not to be taken literally, and in addition, all ideas of motion or flow require a prior concept of time if they are to be explained nonmetaphorically. The trouble here is that although we perceive events, we do not perceive time; it is an intellectual construction used to give us a language with which to deal with events, to interpolate between them and extrapolate from them. It is strongly influenced by linguistic and cultural conventions. In fact, I can think of no statement describing our experience of time whose truth must be admitted.

At this meeting of the Society we are intent on enlarging our comprehension of how man deals with problems of time. It is this concept that for this week of our lives organizes our thoughts on life, art, science, and the human personality. But the approach that my training forces me to follow forces me to adopt a definition of time based on publicly verifiable fact, even if it is a fact that lies rather far from our central areas of concern. I start with the fact that events in the universe of our direct physical experience require at least four numbers to identify them. Consider an event such as the arrival of a telegram. In this case, the four numbers serve to specify the place and time of its arrival. If one seeks to specify the event in some other way — to say it is the telegram that announced my brother's wedding, that can be done, as the telegraphic apparatus does it, by giving other numbers, but many more than four of them. Most events require more than four numbers to specify them, but it is a fact of experience that we cannot make do with fewer. There is no reason to think that there is anything arbitrary or man-made about this number four. It is datum of experience.

Three of the numbers refer to location in space and one to location in time.[1] It will be a proud day for physics when it can explain why there have to be four numbers rather than three or twenty. Of the four, we can identify the time dimension because of the permanence of objects, including our own bodies. Their extension in time is of quite a different kind from their extension in space. An object ordinarily occupies a definite position in space but not in time, unless it is like some short-lived atom that explodes the moment it comes into existence. Thus, the definition: *Time is the dimension in which duration occurs.* In writing this I wish to avoid a logical circle by assuming that the experience of duration, of the permanence of our bodies and of most of the objects around us, is an element of our wordless primitive existence. In her paper in this volume, Professor Green deals with experiments that show how children become aware of permanence and change before they have learned to speak.

If we pay attention to this fact which distinguishes time from the other dimensions we can save ourselves from those errors of analogical reasoning which run as follows: I am physically concentrated in a small region of space that I can call "here." By analogy, I must also be concentrated at some small interval of time which I call "now" (the technical term for this interval is the "specious present"). Since it is fairly obvious that solid bodies are extended in time in a way quite different from the way in which they are concentrated in space, many thinkers specify that this "now" manifests itself in or to our minds; how this is possible for the mind but not for the physical brain of which the mind is a function is considered to be a substantive but unanswered question.

If one gets used to imagining one's position in time as a point or small region analogous to one's position in space, one begins to ask what happens to this point. Here again the spatial analogy offers its treacherous aid: the "here" of our bodies moves from place to place in the course of time; therefore, by analogy our "now" must also change with time: the present advances towards the future. As time goes on, time passes, or vice-versa. The fact that as far as I can tell, such a formulation is devoid of logical consequences, is unsupported by what we know of physics, and does not seem to be universal in human cultures does not prevent its wide acceptance as a starting point for what purport to be factual discussions about the nature of time.

The notion that there is a "now" that in some intuitive though illogical way sweeps along from one "place" in time to another is reinforced by the feeling that we cannot "go back" in time to erase our mistakes. But what would such a going back amount to? What is imagined is not at all a return to an earlier point in time but rather a return to past circumstances with our present mind: it is like some-one living in New York who wishes to be taken perfectly literally when he says that his heart is in Paris. To say that *real* time travel does or does not take place is a statement without consequences. In fact, we are not anywhere in time and do not go anywhere in time; time is a dimension of experience like space, but the behavior of matter with respect to that dimension is so different from what it is with respect to the dimensions of space that almost no valid argument by analogy is possible.

II. TIME AND THE IDEAS OF PHYSICS

In the formulas of dynamics there is no "now." The letter t appears, meaning time, and "now" refers simply to any specified value of t. (How the usual convention of "now" can be introduced into this framework of ideas is shown in my contribution to *The Study of Time*, Volume I, entitled "The Myth of the Passage of Time.") A typical problem of dynamics is to determine the behavior of the solar system: nine planets and their satellites moving in orbits around the Sun. The Newtonian equations of motion do not determine the particular orbits followed; they determine all possible orbits, and which orbits are actually followed depends on how the system was set in motion.[2] A remarkable fea-ture of these equations is that they are reversible: to each solution of them corresponds another solution bearing the same relation to the first that a reversed motion picture does to its original.[3] This fact should warn us that the classical theory of dynamics, useful and conceptually simple though it may be, will not help us to clarify the human experience of time, for the activities in which we find ourselves never run equally well forwards and backwards.

> The Moving Finger writes, and having writ
> Moves on, nor all thy piety and wit
> Can lure it back to cancel half a line,
> Nor all thy tears wash out one word of it.

> (Omar Khayyam, tr. Fitzgerald)

The dynamical behavior of simple systems gives just as poor an analogy to help us understand our experience of time as does the geometry of space.

In using the ideas of dynamics to discuss physical systems like a human body and its surroundings we are confronted with a very great increase in complexity. To see that this is not, however, the main methodological point let us assume that it is not even a difficulty; that God has given us a computer of such stupendous power that it will solve all equations quickly. The equations are given; let us assume that they are Newton's equations of motion for each atom of ourselves and our environment. We set them into the computer and push the botton. The computer does not start. We have forgotten to put in the initial positions and motions of the atoms. But what can we do? How can we specify them all? On reflection, we realize that in order to understand a man's attachment to the scenes of his birth and childhood in Japan it is not necessary to know the positions and velocities of all the atoms in Japan. It is worse than unnecessary, it is ridiculously irrelevant. If laws of physics are to have anything to tell us about the human situation, they must be some other laws.

But we cannot just go on inventing laws and searching for relevance. In fact, to do so is to contradict the science's historical tendency, which is to decrease the number of its fundamental principles by reducing everything to the basic equations of dynamics. What is wrong is not the method of analysis, but the question that was posed in the first place. The understanding of human thoughts and actions, or of complex systems generally, does not rest on exact calculation. Nobody thinks that exact formulations or predictions in this domain are possible. The alternative to an exact prediction is the assignment of a future *probability*, and with the introduction of this single word an entirely new scene opens up.

The mathematical laws governing the calculation of probabilities in physics are derived from those of dynamics plus those of the theory of probability. The precedure is not very difficult — by this I mean that in the century that has elapsed since Boltzmann's pioneering work great progress has been made in understanding how questions should be posed and answered. We know, for example, that the paradox of reversible equations governing irreversible processes disappears: even though the elementary interactions of dynamics run equally well forwards and backwards, this is not so of probabilities; in complex systems their behavior is unidirectional, and this knowledge alone begins to give us a sense of the way in which, for us, the future differs from the past. The concept of entropy, which corresponds roughly to disorder in the universe, can be defined so that it belongs to the same family of ideas as does probability, and its perpetual increase towards a maximum then follows as a mathematical theorem.

Physics can never, I am sure, yield by itself any detailed insight into human psychology, but it can provide insight by the analysis of partial situations and by analogies that illuminate the whole. In general, analogies drawn from deterministic dynamics are bad analogies and tell us little; those drawn from statistical mechanics may be good and, within their limitations, tell us quite a lot. And since they contain the concept of time in a rigorous, quantitative, and thoroughly tested way, they may be illuminating on the subject of the human experience of time. If we are able to understand the behavior of some complex systems of interacting parts, their rhythms, their struggles for survival, their end in states of order or chaos, we may, by the use of imagination, achieve a little new insight into the nature of our human situations.

Everybody, I think, has the sense that random phenomena involving very large numbers are in a sense not very random: if I toss N coins at once, the fractional deviation from equality between the numbers of heads and tails will probably be about $1/\sqrt{N}$, so that if $N = 1$ million, the number of heads will probably differ from the number of tails by about one part in a thousand. For the huge numbers involved in counting atoms, the obvious signs of randomness are lacking, and it is almost as if we were dealing with laws of strict causality — in fact, in the last century many people thought that that was what they were. But it is necessary, in today's science at any rate, to make the jump from a strictly deterministic to a probabilistic formulation of theory before we can even hope to establish a scientific basis for understanding very complex systems. Our normal experience of the world is with systems containing huge numbers of particles; here the uncertainties are small and we ordinarily deal with situations in which the probabilities are very close to 0 or 1. The other situations have a special character for us. Waiting for a taxi can be entertaining or frustrating according to circumstances, and we have a special category of amusements, card-playing and other games of chance, in which procedures are carefully regulated so that for a while we have the experience of living in a world in which the probabilities do not lie close to 0 or 1.

I mentioned Newton's laws a moment ago. Of course, these are not the laws one uses nowadays to understand the behavior of atoms. The laws of quantum mechanics fit into this discussion perfectly, however. They introduce a new kind of randomness, the quantum-mechanical indeterminacy first discussed by Heisenberg; their only contribution to this discussion is to remove even the last vestige of a conceptual possibility of a strictly causal analysis of complex situations. There are serious questions as to the correct physical interpretation of the laws of quantum mechanics. We shall have to think about these later in this paper, but I do not think that they are important at this point.

III. ON A RATIONAL FORMULATION

Let me sum up what I think to be the situation. It is that if we try to define time as a form of the mind or a special kind of human experience we are in danger of ending up talking only about our own languages and cultures and not about time. Time is a form of the mind or a kind of experience to the same degree that water is, and a discussion of water on this basis could end with a thoughtful and by no means empty mixture of fact, fancy, and folklore, but it would hardly lead us to understand anything we did not understand before. I think it is possible that 100 years from now everything that we today regard as objective scientific fact will be regarded as forms of mental activity, but we are not yet ready to think that way; what works for us at present is to think of water as a collection of molecules, and the molecules in turn as collections of smaller structures, governed by laws of statistical mechanics. The difference between water and time (I am embarrassed to say this, but it must be said) is that water comes in the category of objects. An object exists somewhere in space (we say nowadays), it has certain properties, it does things. Time is a dimension of our experience, especially our experience of the external world of things. The fact that we can formulate theories of objects in terms of the dimensions of space and time encourages us, though it does not compel us, to conclude that concepts of space and time have an objective character analogous to the objective character of objects, but of a different kind. Space and time do not do anything. They do not move, or change, or endure, or pass away; these are what objects do, and those very words can only be defined with reference to the dimensions of space and time.

Our license for insisting that this is at present the right way to think about these matters is the success of the physical science based on them. As regards human consciousness, physical science has hardly begun. We understand some of the biochemistry, some of the physical chemistry, some of the electrical motions in nerves. And there is no sign of methodological difficulty here; no sign that the foundations will have to be strengthened or extended or changed in order to support the superstructure, provided that we can learn how to pose the questions and interpret the answers.

IV. DIFFICULTIES IN THE INTERPRETATION OF PROBABILITY

To people not trained in physical science it may seem strange that I am so confident that we know exactly how to use concepts of space and time in the analysis of natural phenomena, yet at the same time am hopeful but a little doubtful that we can learn how to pose the questions and interpret the answers.

The equations of theoretical physics consist of mathematical symbols written on paper. These symbols have perfectly precise mathematical meanings. They are also supposed to have something to do with our experience of nature; if this were not the case, the equations would remain as pure mathematics. The mathematical and physical definitions of the symbols have nothing to do with each other; they are logically entirely independent. Mathematical definitions are conventional in nature; physical definitions assume the truth of physical hypotheses. I have made this point above in talking about time. To give but two other well-known examples, the definition of the term "solar system" supposes that the sun and planets form a system of objects and not a collection of little lights in the sky; the scientific definition of temperature assumes the truth of the second law of thermodynamics. The equations of the special theory of relativity were more or less known before Einstein to Fitzgerald, Lorentz, and Poincaré; Schrödinger wrote down the profound and essentially correct equation of quantum mechanics that bears his name with almost no idea of its physical significance. There are people, of whom I am one, who solve the equations of general relativity in their work but sometimes do not have a perfectly clear idea what the solutions mean.

The most notorious difficulties of interpretation occur in quantum mechanics, but even in the remarks on statistical mechanics that I have made earlier there is a question of meaning to which I must draw your attention. I have pointed out that the theory of complex systems, especially biological systems, demands a statistical approach. This theory yields its answers in the form of a statement that there is a certain probability that a certain event will occur. I have mentioned that the probabilities evaluated in this way are commonly very close to 0 or 1 — they very nearly amount to certainties, but a question of principle remains. This question is already inherent in the simplest possible probabilistic situations, so let us look at one of them. Suppose I throw a well-balanced coin. The probability that the head will appear is known to be very close to $1/2$. What does this tell us about what will happen when the coin is thrown? Nothing. It will come up one way or the other; that is all. Suppose I throw two coins, or one coin twice. Again, probabilities tell us nothing about what will happen. All they do is to tell us what will probably happen, and there is no escape from this logical circle. The theory of probabilities says nothing about the world of facts and things. Then how is it that a gambler who understands the subject can beat one who does not? One answer to this question is that he will not necessarily beat him, only probably. (The only gambler who is sure to win is the one who leaves nothing to chance.) But this is an incomplete answer, for it does not lead out of the logical circle. We know that there is good reason to evaluate probabilities correctly whether in gambling or in living our daily lives, but I do not think anybody has shown that there is a logical reason why this is so. Perhaps I can best express the situation in the form of a Kantian imperative: always *act* as if the laws of probability controlled the situation in which you find yourself. This is, in fact, what sensible people do. But what a strange monster our science of statistical physics has become! It is interpreted as making statements not about the world, but only about how we should act in it. The most elementary postulate of science as ordinarily understood, that it refers to a world that exists independently of ourselves, is violated in the science of physics that claims to be the most fundamental one. Unless I have simply made a mistake in what I have been telling you, the claim that the results of physics verify the correctness of an objective description of events taking place is without foundation, and so, therefore, is the claim that physicists' space and time are validated by this description. The most I can claim, and I do claim it, is that we ought to lead certain parts of our lives as if this description were correct — and this, as I have said, is what physicists and most of the rest of us do.

V. DIFFICULTIES WITH THE CONCEPT OF AN EVENT

I will give one more example meant to serve as a caution against a naive belief in an external world understood in terms of common sense and objectivity supported by physics. Let us consider the idea of an event, for this idea is of fundamental importance when we think about time. Events are the milestones of time. But exactly what are they? One may say that the disintegration of an atomic nucleus millions of years ago, producing an alpha particle whose track we can observe today, was an event, or that a disintegration registered as a click in a Geiger counter and perceived by a particular human observer now is an event. Both are reasonable examples, but they present somewhat different conceptual problems, for the time of the one event is reconstructed from evidence, whereas the other, one is inclined to say, is observed as it occurs.

What determines the instant at which a nucleus will disintegrate? One might think that if one knew enough about the inner state of the nucleus at any given moment one could solve some equations and predict the moment of disintegration. But according to quantum mechanics, the best theory we have, this procedure is impossible, since the theory does not contain, even in principle, the quantity and kind of information required in order to make the prediction. The disintegration is therefore a random event, impossible to predict, and the theory can specify only the probability that it will have occurred by the end of ten minutes or ten years or ten centuries. This probability is a smoothly increasing function of time, very nearly but not quite exponential in form. There are other events that the theory can predict more definitely than this, but none perfectly definitely.

What are events as we experience them? The essential character of simple events like the registration of a nuclear disintegration is suddenness, often there is an element of surprise. The counter flashes or clicks and that is that. The theory contains statements about probabilities; it does not contain accomplished facts. We can only put them in by an artifice — a very natural one but an artifice all the same. Suppose I toss a coin. It falls to the table and lies there. I do not look at it. The probability of a head is 1/2. Now I look at it, and see that it fell with the head up. At this moment the probability jumps from 1/2 to 1. If we were describing the process in terms of probabilities, we see that nothing changed at the moment the coin came to rest on the table, but only at the moment I looked at it. The event, as far as a theory of probability is concerned, occurs at the moment I take cognizance of what has happened. Now we all understand very well that the critical moment was the moment the coin fell onto the table. From then on, concepts of probability are really irrelevant, since the position of the coin may then be publicly verified by anybody at any time. The fall of the coin is an event; my looking at the coin is also an event, but it is an event of another kind, and one feels that it is less important. Perhaps one could not argue very strongly for this lesser importance, since there is no point in playing the game of heads and tails if nobody looks at the coin, but I am sure I do not have to explain to anybody what I mean.

The strange thing about quantum mechanics, which is, essentially, theoretical physics as at present understood, is that it does not contain the concept of an event. Since it deals with probabilities, the only event in this theory occurs at the moment we look at what has happened. In our earlier example of the disintegrating atomic nucleus, the theory predicts only the probability that we will have seen the Geiger counter register on or before a certain moment, but when we do see it, the probability changes abruptly to 1. This abrupt change is not part of the dynamical description of the nuclear process, but is merely analogous to the change that occurs at the moment I look at a tossed coin.

What are we to say about a theory that does not contain the concept of an event? Einstein thought that it is incomplete, and that one should try to complete it. I think it is fair to say that most contemporary physicists feel the incompleteness just as strongly as Einstein but are convinced by various theorems proved during the last few years that within very general assumptions the completion is impossible.

Whatever position one takes, it is clear that the central element of experience around which we organize our understanding of time is the event, and one may well ask why a physicist, brandishing a theory that does not contain such an element, should presume to offer advice as to how time should be discussed.

Of course, physicists experience events and use the word "now" like other people, even though the present structure of physical theory seems to be logically complete without them. It would take a radical shift of the subject matter of theoretical physics in the direction of extreme subjectivity before these ideas could be accommodated, but this century has brought great changes in what is meant by an objective description of nature, and I do not think that the possibility of such a shift can be excluded. At present, physicists tend to explain their science by talking about what will be perceived by an "observer," a being equipped with senses, memory, instruments, and some elements of a human personality. The "observer" does not indulge in the subjectivity of "now", and what is real to him is not facts and events but his perceptions of them. It is not necessary to interpret physics in these terms but it is convenient for most purposes. To those for whom physics is a strange affair I can promise that nobody wanted it to come out this way. Physicists are people who look for absolutes; perhaps we are almost the last ones who do. It is not by our choice that the perceiving and willing subject is in the theory along with the perceived object; perhaps it will not always be so. But the part of this theory on which I base my defense of the relevance of physics is its dynamical laws. In the great majority of situations in which they are actually used, we do not have to get involved in any of the fine questions of interpretation I have been discussing. These laws have enabled us to understand for the first time the atomic level of reality, and without this understanding there is hardly any natural phenomenon that can be said to be understood. When we consider the huge body of evidence in many sciences that, whatever they may ultimately turn out to mean, the laws correspond with our experienced reality at very many points, we can at least understand why a physicist should claim that in this broad context time is easily dealt with and perfectly understood.

The main point I have tried to make in this paper is that there remain difficulties, of principle more than of practice, which have denied us a clear and generally accepted understanding of what the theory really means. And it seems that these difficulties have to do with the temporal concepts that are embedded in the concept of probability and that of an event — and doubtless other concepts as well. It is as if the confusions and ambiguities that cloud our understanding of our subjective experience of time had contaminated that part of physics that deals with experience. If this is so, then physicists have something to learn from the study of time. I maintain also that in any case, the great success and clarity of modern physical theory show that physics has something to contribute to the study of time.

I think of a Mayan temple platform that rises out of a trackless jungle. It invites us to climb and look around. There are perhaps many mysteries about it — as to how it was made and what purpose it served, but it supports our footsteps all the same, and from its broad and solid surface we may be able to discern signs of order in the surrounding country to which we can attach meanings that would have escaped us had we remained below.

NOTES

1. I remind the reader interested in relativity theory that even in the fusion of space and time that occurs there this separation can still be made in infinitely many ways.

2. It could also be determined by a specification of how the planets are going to end up at some moment far in the future, but one has the impression that this mathematical fact is not part of physics.

3. One sometimes hears this reversibility of motion referred to as the reversibility of time. I shall not comment further on this abuse of terminology.

Causality and Time

M.S. WATANABE

1. WILL, CAUSATION, TIME, AND BEING

The time was one day in the latter part of the Second Century before Christ. The place was some-where in the Northwest part of India. The interlocutors were Buddhist Priest Nagasena and Greek King Menandros or Milinda.[*1]

Nagasena:	O, Great King! Are the flame at early night and the flame at midnight the same?
Milinda:	O, Sacred Teacher! They are not.
Nagasena:	Are the flame at midnight and the flame at late night the same?
Milinda:	O, Sacred Teacher! They are not.
Nagasena:	O, Great King! Are the flame at early night, the flame at midnight and the flame at late night all different?
Milinda:	O, Sacred Teacher! They are not. The flame continues to burn all night through constituting the same lamp light.
Nagasena:	O, Great King! The individual continues in the same fashion. That which is created and that which perishes are different, yet they neither precede nor succeed and they continue as one.

The essence of this passage is that that which maintains the identity of a person, like that of a flame, is not a substance but a causal chain. The same is true, not only of a life, but also of a chain of re-births. This is not a continuation of the same soul, but a chain of causation carried by "karma." Nagasena denounces the concept of Atman, the Soul, which being an individuation of Brahman, constitutes the Substance in Brahmanism. Buddhism does not recognize the everlasting, permanent substance underlying the inconstant existence whose identity is only sustained by causality. Causality takes the place of being. Buddhist literatures abound with different theories about time, but one thing is common: Everything is inconstant; there is nothing that is selfsame; if we see an identity of being in events in the course of time, it is due to the causal chain that connects them. Thus, in Buddhism, causality, time and being form an inseparable trinity.

---

[*]I translated from a Japanese version. I give in the bibliography the corresponding place in the English version made by T.W. Rhys Davids.

Buddhism, being a passive religion, does not discuss how man uses causality to achieve his own ends, but it does discuss the way to discontinue the causal chain and stop the circular time of rebirth. Nagasena, at variance with later Mahayanistic Buddhist philosophers, states: "For those who have entered nirvana, there is no time."[2] What then is nirvana? Nagasena's answer is: "O, great King, nirvana is cessation [of desires]"[3] because "those who have desires (kilese) will result in new lives, but those who have no desires will result in no new lives."[4] In other words, presence and absence of desires imply respectively continuation and discontinuation of the temporal chain of causation. We may therefore add a fourth element: desire or volition, to the trinity of time, causality, being. So much for Buddhism. I would like to thank my wife, Dorothea, not only for calling my attention to, but also for discussing with me, all these interesting passages of Buddhist literature.[5,6]

This ancient dialogue brings us right up to the basic propositions that I should like to use as the metaphysical backdrop for this paper: Life is will; will engenders desire; desire has an end to achieve or an obstacle to avoid; the end-seeking desire can be satisfied by utilizing causal laws, end becoming effect and means becoming cause; the separation of end = effect and means = cause creates time. In animals, the first term end = effect and the second term means = cause are not consciously separated, and the concept of time does not seem to be awakened. The conscious separation of the two terms creates at least the subjective aspect of time. In this sense also, which is slightly different from the above-mentioned Buddhist view, time is bound up with desire and causality. The bi-unity of teleology-causality and the creation of time by will were the two foundations on which I based my paper entitled "Creation of Time" that was included in my book called *Time* published in 1949.[7]

The Western philosopher who most emphatically advanced the theory of the trinity of will-time-causality was probably Marie-Jean Guyau who wrote, among many other interesting books, one entitled *La gènese de l'idee de temps,* published in Paris in 1902.[8] He says in this book:

> Un être qui ne désirerait rien, qui n'aspirerait à rien, verrait se fermer devant
> lui le temps ...... il faut désirer, il faut vouloir, il faut étendre la main et
> marcher, pour créer l'avenir. L'avenir n'est pas ce qui vient vers nous, mais ce
> vers quoi nous allons ...... L'intention, avec l'effort qui l'accompagne, est le
> premier germe des idées vulgaires de cause efficiente et de cause finale.

Although it is not indicated how we can verify these statements, everyone would be inclined to agree with Guyau, in particular, when we recognize that the genesis in Guyau's sense need not necessarily coincide with the development of the concept of time in children in Piaget's sense.[9] As was implied by Nagasena and more explicitly maintained by later philosophers,* being and time were often identified in Buddhism, forming the four-unity desire-causality-time-being. This can be roughly compared with Heidegger,[10] Bergson,[11] and some of the Western philosophers of life of the Nineteenth Century. But, in the case of the substance-denying Buddhism, the role of causality is particularly important as a link between time and being.

A much limited application of the four-unity, namely, the reciprocity of teleology and causality in the sense of duality: end = effect, means = cause, was referred to in the analysis of causation by such contemporary philosophers as R.G. Collingwood[12] and D. Gasking.[13] These philosophers failed, however, to see the essential role played by time in linking teleology and causality.

---

*See particularly quotations from Dogen in Reference 6.

In the present paper, I should like to examine some of the modern theories of causation and try to formulate what seems to be an adequate explication of the whole complex of ideas surrounding causality, placing a renewed emphasis on the relationship between time and causation — an aspect of causation many modern philosophers tend to ignore or minimize.

2. CAUSALITY AND TELEOLOGY

Let us conform to the good old tradition of starting a discussion of causality with the four main causes of Aristotle. In paraphrasing in our own words, we may say that the effective cause is the intentional action taken by an agent initiating a certain process, the final cause is the object or state of affairs intended by him in this process. The material cause is the physical system in which the process takes place. The formal cause is the universal category to which the end product of the process belongs. It is often stated that the idea of material cause is archaic and no longer worth discussing in the present state of philosophy and science. I do not agree with this statement because there is nothing wrong about attributing the cause, at least partly, to the physical system that brings about the end product. A statue is standing there because of the material, bronze, that is keeping the shape of the statue.

It is interesting to note in passing that the Buddhist view of phenomena implies that the material cause has no substance and is reducible to a non-material causal chain. The traditional macroscopic physical science, of course, acknowledges the existence of matter, but in its fundamental formulation modern physics takes the view that matter is a state of vacuum or nothingness.* Matter is reduced to the laws that determine the behavior of the vacuum. In curious coincidence, Buddhism and modern physics have come to take a very similar view about the material cause.

Mention of the formal cause as distinct from the final cause is, apart from its reference to the "form" of the resultant, particularly important because the intended object or state of affairs and the resulting object or state of affairs are not always the same.

As I have stated elsewhere,[7,14,15] science or prescience is a bridge between the final cause and the efficient cause. Science in its primitive form gives to a man an advice of the form: If you want to have A, you should do B, because B will probably lead to A. The man follows this advice and performs B. Then A is the final cause and B is the efficient cause. The teleological view considers B as produced by A and the causal view considers A as produced by B. The first production process reflects the temporal order in human decision-making. The second production reflects the temporal order in the physical development. There is no conflict between teleology and causality in this double process. A teleological explanation can start to interfere with the causal explanation only if the desired end A is assumed not only as being effective in determining B in the decision-making, but also as having continued effect in guiding or changing the course of development of the process from B to A that is determined by causality.

---

*The ether has been declared by Einstein to be non-material. The elementary particles have gradually been deprived of more and more of their material attributes and now are considered to be special "slates" of an primordial particle which does not have material attributes at all.

There are two important modifications or elaborations to make on the concept of final cause and effective cause. First, the intended final state of affairs A and the actual state of affairs A that is obtained are not necessarily the same, because scientific advice is seldom deterministic or infallible. Furthermore, men often do not follow the scientific advice faithfully. In addition to this, the scientific advice usually assumes the isolation of the physical system from the exterior during the process, but this isolation is often violated. Second, a science in its pure and advanced form does not say: if you do this (B), you will get that (A), but does say if the initial condition or state is B , then the final condition or state will be A. Therefore, the decision of action B and the initial state B must be separated.

This conceptual separation brings out one of the most important facts about human existence. Man has, within inevitable limitations, the freedom to choose the initial state B . If this freedom is not available, the whole causal thinking has no significance for man. The existence of this freedom is an undeniable fact of life that we have to accept. We called it elsewhere: the Postulate of Freedom.[16]

Now we have introduced four independent moments that constitute the teleological-causal behavioral ring: (a) Intended end, (b) decision of action, (c) initial state, and (d) final state. The process of (a) producing (b) is the teleological decision-making in the presence of a goal and scientific knowledge. The process of (b) producing (c) is made possible by selectability guaranteed by the Postulate of Freedom. The process of (c) producing (d) is governed by scientific causal laws in conjunction with the given initial conditions. The resultant (d) can be either identical or not identical with the goal (a). In the former case the teleological-causal ring is successful and in the latter case it fails. Sometimes, similarity instead of identity between (d) and (a) is considered as acceptably successful. See Fig. 1.

The final state is determined by the initial condition and the scientific laws. The former is a contingent element and the latter is a nomological element. These two are both necessary. The nomological element cannot be changed by human intervention, but it plays an important part in determining the final state. As a matter of fact, I have shown elsewhere that in determining the final state, the amount of information furnished by the laws is infinitely larger than the amount of information furnished by the initial condition, at least in the case of differential equations.[16] In spite of this, we usually do not mention scientific law as part of the cause. This is mainly due to the habit of only counting those factors as part of the cause that are not uniquely predetermined and are selectable or changeable. This habit obviously originates from the fact that the concept of cause is intimately bound up with the Postulate of Freedom.

This portion of the process, namely from (c) to (d) is so important that it deserves a special name, say, scientific causation. The starting point of scientific causation, i.e., (c), can be considered, in a sense, as taking the place of Aristotle's material cause, with the understanding that scientific laws can be included in (c), if we consider any possibility of the scientific law being different from the actual one. The decision of action (b) when considered as producing the final state (d) may be roughly equated with Aristotle's efficient cause. The term (a), as I see it, is clearly equivalent to Aristotle's final cause. The notion of formal cause has a connotation of designating the "universal" to which the final cause belongs, but if we ignore that aspect of the concept, the formal cause may be made to roughly correspond to the actual final state (d) which may or may not be identical with the intended end or final cause (a). When the term causality is used in ordinary language, it usually means either scientific causality (c) → (d) or efficient causality (b) → (d) or some confusion of the two.

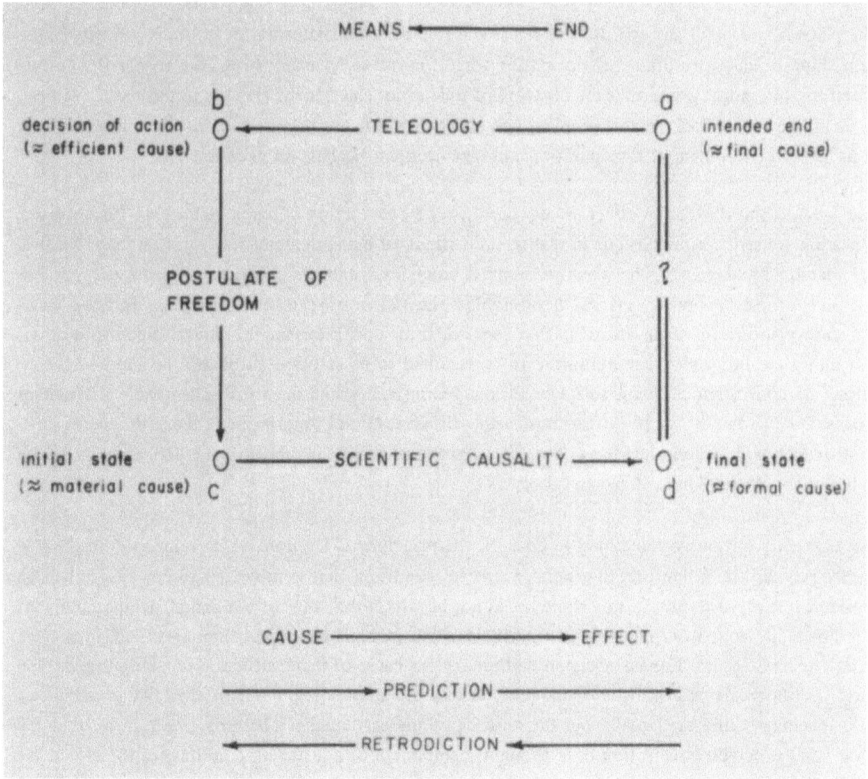

Fig. 1

The inferential determination of the final state on the basis of the initial state is prediction, while the inferential determination of the initial state on the basis of the final state is what I called retrodiction.[17] The asymmetry of prediction and retrodiction will be discussed later more in detail. If the intended end (a) and the actual product (d) are identical, and if (b) and (c) are merged as one, retrodiction goes parallel with teleology. The only difference will then be that the former is considered as relation to be found in the outside world whereas the latter is to be found in mind.

3. CAUSE AS SUFFICIENT CONDITION

From now on, let us limit our discussion to scientific causality, occasionally making reference to its relation to the efficient cause. There are two states of affairs: (c) and (d). According to their role in life, these two must be related in such a way that if (c) then (d). Therefore, the primary definition of cause (c) and effect (d) is that the cause is a sufficient condition for the effect, or what amounts to the same, the effect is a necessary condition for the cause. In the analysis of the last section, (c) and (d) were designated as the initial and final conditions, implying that a certain duration of time must take place between (c) and (d). But, as far as the role of scientific causality in life is concerned, the main thing is that we obtain (d) no matter when, or rather, in many cases, the sooner the better. Therefore, one might argue that the duration of time is not an essential element in the causal relation. We shall later come back to this problem and show that it is not that simple. But for the time being, let us ignore the problem of time duration and take up the logical aspect of causation.

A sufficient condition for an effect usually consists of a conjunction of many conditions. For instance, to get fire we strike a match, but in reality the striking of a match alone does not guarantee the fire. The match should be dry; the wood piece must be combustible; the match box should be provided with an appropriate abrasive surface; there should be sufficient oxygen in the air, etc... . But, we usually do not mention them all, assuming that some of these conditions are usually satisfied if not otherwise indicated. But which conditions are to be assumed to be satisfied automatically and which conditions are considered as changeable depends on circumstances, and one of the changeable alternatives is considered as the cause. If we are considering different kinds of atmosphere, the existence of sufficient oxygen will be considered as a cause. If we are considering the question of whether we strike the match or not, the striking will be considered as a cause.

This usage has obviously something to do with the principle of freedom which is involved in the role of causality in life. Therefore, in general, whatever condition that is subjected to our choice will be considered as a cause. But,   this choice need not be within the range of our actual manipulability. If the choice or election of change can be considered as a possible alternative in our thought, it can also be counted as a cause. For instance, an earthquake is a cause of the destruction of a building, not because we can create or avoid the earthquake, but we are considering in our thought two possibilities: the happening of the earthquake and the absence of the earthquake. Therefore, we can say that the cause is a "selectable" (in action or in thought) component of a sufficient condition of the effect. We usually do not mention this, because the other components of the sufficient condition are understood to be fixed and satisfied.

It is now important to specify the nature of the implication involved in the statement that the cause implies the effect. It is from the beginning clear that this is not logical implication. The well-known argument of Hume[18] with regard to "conjoined occurrence" is justifiable if interpreted as maintaining that the causal relation is an empirical one, inferred by an inductive process. Consequently, the cause implies the effect when relevant empirical laws are taken as part of the postulates in deduction. The empirical laws that are supposed to play roles in such a deduction are causal laws, whereas causality or universality of causality means the existence of such causal laws for every phenomenon. When we talk about a causal implication, we assume such a causal law, but we may not necessarily be able to spell it out.

Now, it is often overlooked that there are two kinds of empirical laws: those that are fundamental and those that result from the fundamental laws combined with contingent facts. Newton's laws are fundamental laws, but the law that the sun rises approximately every 24 hours is a semi-contingent law dependent on the contingent fact about the earth's rotation. The fundamental laws are usually considered as fixed and unchangeable, therefore they are not mentioned and constitute part of the silent postulates for deduction. Since, in many considerations, the contingent facts are regarded as changeable or selectable, they are often not included as part of the silent postulates or the fixed condition.

After having defined the cause as a selectable portion of a sufficient condition for effect, I have to hasten to add an important additional requirement. Man is an economical animal in the sense that he prefers to spend as little effort as possible to achieve the desired effect. Hence, if c_1 and c_2 are both sufficient conditions for d, i.e., if $c_1 \to d$ and $c_2 \to d$ and if c_1 implies c_2 and not vice versa, i.e. if $c_1 \to c_2$ and not $c_2 \to c_1$, then we prefer c_2 to c_1 because in the case of c_1 we would have to do additional effort x such that $c_1 = c_2 \cap x$, i.e., c_1 is equivalent to c_2 and x combined. For instance, we prefer to eliminate a superfluous superstitious ceremony which does not affect the process of producing_ a certain effect. Reflecting these human conditions, the cause should be as little "restrictive" a sufficient condition as possible, when we agree to say that β is less restrictive than a when $a \to \beta$ and not $\beta \to a$. Of course, logically speaking, the least restrictive sufficient condition for the effect d is d itself. But, we usually cannot realize d directly, and we stop at a c which is feasible or realizable as a preparation for d.

In view of the fact that we do not count a sufficient condition as a cause if it contains superfluous elements (such as x in the foregoing), some thinkers jumped to the conclusion that the cause must be a necessary and sufficient condition of the effect. Even Hume defined a cause as "an object, followed by another and where all objects similar to the first are followed by objects similar to the second, ... , in other words, where if the first object had not been, the second never had existed."[19] Whatever the real intention of Hume might have been, Richard Taylor is right to point out[20] that whereas the first half of this passage seems to indicate that the cause is a sufficient condition or at least a constant in sequence, the second half considers the necessary condition as a cause. Stegmuller makes an interesting suggestion that when we want to avoid something undesirable β, we take its necessary condition a as a cause.[21] This may be due to a kind of usual logical confusion which is also seen in Hume. In reality, what is important for life is that not-a is a sufficient condition, i.e., a cause for not-β, and by a loose use of the word a is said to be the cause of β.

But a simple reflection makes it clear that the requirement of a cause being a necessary condition is certainly unfounded. For instance, in order to increase the humidity of a room we can do several things: turn on a humidifying machine in the room, or put an evaporator of water on the radiator, or let the shower run in the adjacent bathroom. None of these conditions is a necessary condition for the increase of humidity. The disjunction of these three is not a necessary condition either. Yet, anybody will agree that any one of the three is a legitimate cause of higher humidity. But, on the other hand, the turning-on of an electric bulb in the bathroom at the same time as the turning open of the shower valve constitutes, in conjunction with the latter, a sufficient condition, but it can be eliminated, therefore does not count as a part of the cause. Of course, the increase of humidity itself is the least restrictive sufficient condition, but it cannot be done directly, therefore it does not count as a cause either.

Having defined, so far, the cause as a selectable component of a sufficient condition, that is free from superfluous elements yet feasible or realizable, let us examine some of the examples that are discussed in connection with causality.

(A)   Is "its being day" a cause of "its being night" 12 hours later?
        People are usually inclined to answer no. The reason people give the negative answer, however, can be varied.

(A1)   There is no selectability in a proper sense of the word. Day and night are given and, the day cannot be made night and the night cannot be made day.

(A2)   Day is not a cause of night because it is conceivable that day continues and night never comes. Day is not a sufficient condition for night 12 hours later if we include only the fundamental laws as the postulates of the inference. The 24 -hour cycle is in fact a consequence of a semi-contingent law which could be violated from a more fundamental point of view.

(A3)   Day is not a cause of night because day has no "power" of bringing about night.

       It is not clear what "power" precisely means, but we feel that there is something to this statement, particularly, when we compare the present example with such examples as "the sun causes plants to grow" or "the poison was the cause of his death." Nobody, so far as is known to the author, has succeeded in explicating the idea of "power", but I shall try in a later section my own way of interpreting the idea of power.

Another example:

(B)   Is an animal's being a biped the cause of its being a talking animal?

      The above mentioned three comments are valide here, too.

(B1)   An animal is either biped or not, and we can hardly imagine that we can change this property without changing all other conditions of the animal.

(B2)   There is nothing against the acceptable fundamental laws to imagine a non-talking biped.

(B3)   The fact that an animal is a biped does not seem to have the power to make the animal talk a language. In this example, we can add another reason to answer the question (B) negatively.

(B4)   Being biped is not a cause of a talking entity because the two conditions are simultaneous. The cause must precede the effect.

From these discussions, we can see that our definition covers the reasons (A1), (A2), (B1) and (B2). But to incorporate the thoughts expressed in (A3), (B3) and (B4) we shall have to add further restrictions to the concept of causality. Namely, we have to formulate in a proper way the two conditions:

(i)   The cause must have a power in some sense to produce the effect.

(ii)   The cause must precede the effect.

4. LAPSE OF TIME AND FREE ENERGY

It is usually taken for granted that the cause temporally precedes the effect. But there are two kinds of serious argument against this assumption. One consists of theoretical arguments that the cause and effect must take place simultaneously. The other consists of actual examples in which the cause and effect seem to take place simultaneously showing that a lapse of time between cause and effect is not a necessary factor for causality.

The most famous case of the first category is B. Russell's paper called "On the Notion of Cause."[22] He advances two arguments to establish the simultaneity of cause and effect, one claiming the perfect symmetry of cause and effect, and the other arguing that any lapse of time contradicts the definition of cause and effect. The first argument is based on the following observation:

> The law makes no difference between past and future: the future 'determines' the
> past in exactly the same sense in which the past determines the future.[23]

This statement of Russell's, however, is valid only in the classical physics applied to microscopic phenomena assuming that all the detailed states of atoms are maximally measured. This, of course, in
practice, can never be done. The statement is simply false for quantum mechanics and also for classical macroscopic physics. This is essentially what I called irretrodictability of these two theoretical
schemes. The details were explained in some of my papers, including the one in *The Study of Time*.[15]
The main reason for this asymmetry lies in the basically probabilistic nature of the latter two kinds
of theory. Although Russell drew a wrong conclusion, this consideration indicates clearly that causality has something to do with the asymmetry of time. For this reason, Russell's analysis should be
appreciated.

The second of Russell's arguments runs as follows:

> Philosophers, no doubt, think of cause and effect as contiguous in time, but this,
> for reasons already given, is impossible. Hence, since there are no infinitesimal
> time-intervals, there must be some finite lapse of time τ between cause and effect.
> This however at once raises insuperable difficulties. However short we make the
> interval τ, something may happen during the interval which prevents the expected
> result. I put my penny in the slot, but before I can draw out my ticket there is an
> earthquake which upsets the machine and my calculation.

This objection of Russell's does not occur if we define causality, as we did, as valid only for an isolated
system, i.e., as valid only on condition that no disturbance originating from outside the considered
system takes place. In fact, any scientific prediction about an isolated system is valid only under such
a proviso. If we consider this as invalidating prediction in general, it amounts to denouncing science in
general. This last point is closely related to my argument that no prediction is possible except by
"gambling" on external interference.[15] Other than this, as far as the discussion of causality is concerned, we are not bound to take the "predictive" view of science. We can take an "explanatory" view
of science and look at the process after both cause and effect have taken place. Then we need not be
troubled by the unpredicted disturbance. If the earthquake jammed the machine, the earthquake is to
be considered as a cause of the failure, and this does not interfere with the validity of causality.

Let us now consider a few examples that apparently do not require any lapse of time between cause
and effect. Take, for example, the case of magnetic field generated by a direct current running along a
straight (resistanceless) wire. We know intuitively that the current is the cause and the magnetic field is
the effect. But, according to Ampère's law, the magnetic field at a point is $B = 2I/(cr)$ where I is the
current and r is the distance of the point from the center of the wire. This equation does not involve
the time variable, and everything seems to happen simultaneously. But this picture is an oversimplifi-
cation and the equation is valid only after the system has reached an equilibrium. If we start to let the
current flow, the magnetic field starts to build up around the wire and gradually spreads out to farther
points. During the period of the field build-up (which theoretically continues infinitely if the space is
infinite), the energy of the magnetic field is supplied by the electric current and it flows out of the
wire with finite velocity. It takes some time until the current reaches approximately its final maximum
value. This may be explained in terms of the self-inductance too.

On the other hand, when we switch off the battery supplying the current, the energy stored up in the magnetic field will continue to flow after the battery is cut. Thus, obviously during the build-up period, the cause of the magnetic field is the electric current, in two senses: (1) the (free) energy comes from the current, (2) the current starts to flow prior to the build-up of the magnetic field. If we apply the same criterion to the period after the battery is switched off, we can say, in agreement with the common sense view, that the magnetic field is the cause of the electric current in the sense that (1) the (free) energy now comes from the magnetic field and (2) this energy-flow takes finite time.

The above explanation made an unrealistic assumption that the wire is resistanceless, for in reality such a wire does not exist. Actually, there is always a certain amount of resistance which causes the free energy of the electric current to become heat energy. This means that, in reality, the time lapse between cause and effect is accompanied by degeneration of free energy or entropy increase. In this sense, the cause as having "power" can be interpreted as having more free energy. The example of magnetic field around the electric current has often been discussed in the past, and, for instance, Korch[24] explained adequately that a time lapse is necessary between cause and effect, but nobody has related the idea of "power" of a cause with the free energy or negentropy.

We have seen that, when carefully examined with adequate knowledge of physics, there is no such thing as simultaneous cause and effect. But it may be pointed out that a layman to whom the effect seems to take place instantaneously, will nonetheless correctly decide which one is the cause and which is the effect. This should not be a cause for surprise since his intuition is probably helped by the fact that the battery is the energy source in this experiment, supplying free energy.

Another example is a locomotive pulling a train. When the locomotive starts to move, it takes a certain duration of time before the entire train starts also to move, since there is no such thing as an absolutely firm and rigid connection. This can be interpreted as showing that the locomotive is the cause of the motion of the train, again for two reasons, (a) the locomotive starts to move first, (b) the free energy comes from the locomotive. Of course, when the train has reached a constant speed, it is difficult to see the time lapse, but the energy flow will show what the cause of the motion is. The free energy obviously comes from the locomotive and is spent partly in accelerating the train and partly in overcoming the friction on all the wheels of the train. The locomotive has the "power" to pull the train, and the entropy is generated (the free energy is spent) in the direction of cause to effect. When the locomotive engineer stops the steam supply to the piston, the kinetic energy of the entire train will continue to move the train but will gradually be spent in overcoming the friction and the train will stop eventually. In this phase, the locomotive is no longer the cause.

Another example is the application of the Boyle-Charles law: $pV = RT$. We can take one of the three variables as a function of the two remaining variables, for instance, $V = V(T,p)$. If we increase p keeping temperature T, the volume V will decrease. As far as this equation is concerned, there is no explicit mention of time and the equation seems to be the relationship among simultaneous values of the three variables. But this is because the equation is valid only after an equilibrium has been readied. When the system is in equilibrium, only two variables are independent and the third one is dependent on them. If the system is not in equilibrium, temperature, pressure and density are not uniform in the system, hence it requires an infinite number of variables to describe the system adequately, and there is no simple relation among the variables either. But, if sufficient time is given, the system will reach an equilibrium state under the given constraints and boundary conditions. During this process of reaching equilibrium, the entropy of the entire system will increase.

If we increase p while keeping T constant, we must put a heavier weight on the piston, under which the gas is confined and kept in contact with a heat reservoir (sometimes called thermostat) of temperature T. But when we first change the weight corresponding to the desired value of p, the pressure, heat and density within the gas will become nonuniform and the volume will oscillate at the beginning. It will take some time before we can talk about the values of p, V, T that will obey the Boyle-Charles law. The initial state is the beginning state where the new weight corresponding to p is placed on the piston, and the final state is readied when the variables p, V, T are settled and take the values prescribed by the Boyle-Charles law. In between, the entropy will have increased in the entire system. It is an elliptical but not entirely false way to say that the pressure change is the cause of the new volume. But the important thing is the fact that time lapse was necessary and the irreversible process has taken place. This classical example was first discussed by Philipp Frank[25]; he emphasized that time lapse was needed but he did not mention the entropy increase of the entire system between cause and effect.

In the first two examples that we discussed, we could assign cause and effect to different subsystems. For instance, the locomotive is the cause of the moving train, etc. But this is an anomalous way of describing a causal relation. According to the schema we gave earlier of causality, we should take the combined system of subsystems that are in mutual interaction. For instance, in the case of current and magnetic field, the correct way to describe would be to take the combined system of electric circuit and the magnetic field and take the states at two different instants as initial and final states. For instance, as the initial state we can take the instant when the battery has just been connected,   and as the final state we can take the instant when the magnetic field is fully built up.

This remark is particularly important in cases where the distinction is unclear between two subsystems that are to be called cause and effect. Take two blocks of metal, one, say A, at temperature T_1 and B at temperature T_2 ($< T_1$). If we put them in contact, the temperature difference will disappear after some time. But the physical process may be considered as aimed at cooling the hot block A by bringing it in contact with a cool block B or at warming the cool block B by bringing it in contact with a hot block A. In the former case, the coolant body B may be better characterized as the cause of cooling of A. In the latter case, the heater A may be better characterized as a cause of warming of B. The energy flows in this case from A to B, but this does not determine uniquely which of the two subsystems should be called the cause of the process. The better description is to call the combined system A + B with temperature difference as the initial state and the same combined system A + B without temperature difference as the final state. There is, of course, a finite (in theory, infinite) lapse of time between the initial and final states and there takes place an obvious entropy increase.

There is another kind of law which does not show asymmetry of time. It is well-known that Maxwell's equations allow two apparently contradicting solutions, one representing advanced potential and the other retarded potential. But if the emission and absorption of the electromagnetic field by electric charges are both finished, the description in terms of advanced potential and the description in terms of retarded potential are entirely equivalent. The only thing is that when we solve the equation by the initial condition of charged bodies, we have to use the retarded potential. The emission and absorption of electromagnetic waves by themselves involves microscopically no irreversibility, but given usual initial conditions, the process, macroscopically interpreted, will represent an entropy increase. That is the same as in the case of dynamics. The law of dynamics is reversible, but given a certain initial condition it can describe a thermodynamically irreversible process such as diffusion. The entropy decreasing process can also be represented but such an initial condition is very seldom available.

5. IRREVERSIBLE TIME IN NONPHYSICAL CASES

From the above consideration, we may conclude that:

(i)   The cause must temporarily precede the effect.

(ii)   Between cause and effect, there occurs an irreversible process characterized by entropy increase or free energy loss.

(iii)   The "power" of a cause can be interpreted as the expendable free energy or negentropy that is found in the initial state.

In the examples discussed in the last section, there is obviously an entropy increase (or loss of free energy) between the initial state and final state. A question remains: If there is no irreversibility characterized by an entropy increase between initial and final state, do we or do we not call them cause and effect? The answer is that we could call them so, but it does not appear right. Suppose we have a pendulum without friction, swinging with a constant amplitude; would you call one position of the pendulum a cause for another position of the pendulum? We would not do so, probably because the second position can be called cause for the first position too. We seem to expect an irreversible process between cause and effect. Probably the same kind of psychology is working when we are reluctant to call day a cause of night, because night can also be called cause of day. Thus, irreversibility is expected of causality.

So far we have only considered physical phenomena. As an example of a nonphysical case, let us consider the interesting case invented by Chisholm and Taylor.[26] They consider three events C_1, E and C_2 that have happened in their order. But C_1 and C_2 are sufficient conditions for E. An example is that C_1 stands for Mr. X taking poison, E for Mr. X dying, and C_2 for Mr. Y taking over the position held by Mr. X before his untimely death. Their purpose is to show that the common types of argument for the temporal priority of the cause to the effect are inconclusive, basing themselves on the assumption that "A causes B" is equivalent to "A is sufficient for B." Since they base themselves on this symmetrical assumption, they cannot find anything against calling a later event the cause of an earlier event if the later event is a sufficient condition for the earlier event. These authors, however, have to admit that there is an asymmetry in the sense that future events are sometimes "up to us" in the sense that we can influence their happening or non-happening, preventing us from identifying cause and effect merely as sufficient and necessary conditions.

This example involves two important factors characterizing causality. First, when we say that a cause must be selectable we expect that depending on the selection the effect changes. If B is not affected by any choice of A, A cannot be considered as a cause of B. In this example, an alternative to C_2 can be considered, but this possibility does not change the "effect" E. Therefore, C_2 is not selectable in our sense. To make this point still clearer, we could characterize the cause not only as "selectable" but also "influential". This brings us to the second point. The reason why E cannot be changed by C_2 or anything we do instead of C_2 is essentially irreversibility of time. In a nonphysical case like this we cannot define the entropy or free energy in a way relevant to the events involved, yet in human or societal systems irreversibility of time is even stronger than in physical systems in the sense that we cannot imagine a development in which the cause event and the effect event are interchanged. As a result, the fact that the cause can give rise to the effect but not vice versa gives the idea that the cause has potency or power. The idea of power is thus closely related to irreversibility of time and the cause has to be temporarily prior to the effect.

6. FEASIBILITY OF CAUSAL DESCRIPTION

Philosophers and scientists alike would consider predictive or causal description as a legitimate enterprise but not retrodictive or counter-causal description. But they never question why. I venture to maintain that there is no *a priori* reason for this preference and that the former is only useful and feasible in experience whereas the latter is not. The utility of causality in goal-seeking human behavior was already explained in Section 2 of the present paper. The feasibility of causal description, as we shall see in this section, derives from the entropy-increasing tendency of our environment. This may be the profound reason why living matter was born and lives in such a way that its past-future direction (manifested in its goal-seeking behavior) coincides with the entropy-increasing direction of its environs.

Why the predictive or causal description is feasible and the retrodictive or counter-causal description is in general not feasible in the entropy-increasing system can be seen easily by an example. Suppose we have a U-shaped rail on which a ball can freely move. See Fig. 2. Suppose we start from a position anywhere on the rail; the position of the ball after a sufficiently long time is unique and predictable, namely at the bottom of the rail. Now on the other hand, suppose we find the ball at the bottom and we try to infer the earlier position of the ball. Unless we use extra information obtained from other sources, this retrodiction is impossible. Any answer could be correct and no answer is sure to be true. The source of this asymmetry is that due to the inevitable friction (which causes entropy-increase), many initial states end up in a single final state. This makes the prediction or causal description not only possible but unique and retrodiction or counter-causal description impossible.

I introduced the term retrodiction[17] in connection with this impossibility. Some philosophers borrowed the term and attached an almost opposite meaning to it stating that the past is recorded and fixed.[27]

To discuss the matter connected with prediction and retrodiction it is more incisive to use the probabilistic language, and the reader is referred to my previous paper for details.[15] We limit ourselves here to mentioning briefly some of the main points. Let C and E be the cause and the effect.

1.   If C is earlier than E, the causal probability $p(E|C)$ exists. This may be considered as corresponding to the principle of universality of causality.

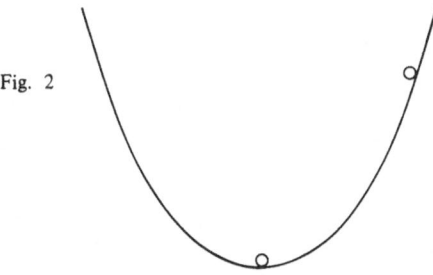

Fig. 2

2.   In the description of C and E, we can eliminate many superfluous or accidental conditions without changing the value of p(E|C). We can thus obtain a law-like status for p(E|C). This is the basis of the principle of uniformity of causality.

3.   The probability p(E|C) in many cases is not 0 or 1 but not very far from 0 or 1, so that black-and-white type of causality is approximately adequate. This is the basis of the effectiveness of causality.

4.   In physics, the process C → E involves an entropy increase or loss of free energy. In nonphysical cases, C → E is possible but E → C is impossible.

5.   The probability p(E|C) loses some or all of the above properties if E is prior to C.

It is important that causal description is possible and effective because our environs are entropy-increasing systems. This leads to a striking conclusion, that to describe an open system where entropy is decreasing, the counter-causal description may be possible and effective.

7. CONCLUSION

Born in a corner of the universe where the entropy is increasing in one direction of time, life is to survive and expand. But in which direction of time is it to live, grow and age? The only possible direction is one in which the future is foreseeable and controlable, so that by adapting and programming itself and acting suitably it can satisfy its needs and desires. That foreseeable and controlable direction is one in which the entropy is increasing.

To say that the future is predictable and controlable is to say that causality works with its cause-effect direction coinciding with the present-future direction. The most important fact about the world is its predictability and controlability. What the world is made out of is of minor importance. The idea of matter – the attenuated remnant of the idea of "substance" – was once the central concept in science. But, with the relativity theory and quantum theory, the idea of matter has now been replaced by vacuum or nothingness. Thus, substance and function that once made up the physical world are now reduced to pure function. The principle of causality is the law of this function, the law that governs functional processes. In a manner of speech the world has become causality. But causality is time, since neither time without causality nor causality without time is conceivable. Thus, we may conclude: Life needs causality. The world provides causality. The world is causality. And causality is time. Thus, we have come back to the four-unity from which we started in Section 1. These four items indeed cover such a vast territory that the reader will not be surprised to encounter in this volume on time any special topic within this four-unity.

NOTES

1. T.W. Rhys Davids: *The Questions of King Milinda,* Vol. 1, being Vol. 35 of F. Max Müller (ed.): *The Sacred Books of the East,* Oxford: Clarendon Press 1890, p. 64.

2. *Ibid.,* Vol. 1, p. 78.

3. See *ibid.,* p. 106.

4. See *ibid.,* p. 50.

5. Dorothea W. Dauer: *Buddhist Influence on German Literature to the end of Nineteenth Century,* Dissertation, University of Texas, 1953.

6. Dorothea W. Dauer: "Nietzsche and Concept of Time," in this volume.

7. S. Watanabe: *Time* (in Japanese), Tokyo: Hakujitsusyoin 1948.

8. Jean Marie Guyau: *La genèse de l'idée de temps,* Paris: Félix Alcan 1902, p. 32.

9. Jean Piaget: *The Child's Conception of Time,* New York: Basic Books 1969.

10. Martin Heidegger: *Sein und Zeit,* Tubingen: M. Niemeyer 1953.

11. Henri Bergson: *L'évolution créatrice,* Paris: Félix Alcan 1912.

12. R.G. Collingwood: *An Essay on Metaphysics,* Oxford: Clarendon Press 1940.

13. Douglas Gasking: "Causation and Recipes," in *Mind,* Vol. 64, No. 256, Edinburgh: Thomas Nelson & Sons, Ltd. 1955, pp. 479-89.

14. S. Watanabe: "Time and the Probabilistic View of the World," in *The Voices of Time,* J.T. Fraser (ed.), New York: Braziller 1966.

15. —————. "Creative Time," in *The Study of Time,* Fraser, Haber, Muller (eds.), Heidelberg: Springer-Verlag 1972.

16. —————. *Knowing and Guessing,* New York: John Wiley & Sons 1969, p. 116.

17. —————. "Symmetry of Physical Laws," Part I, II, III in *Reviews of Modern Physics,* Vol. 27, No. 1, January 1955, pp. 26-39; 40-76 and Vol. 27, No. 2, April 1955, pp. 179-86.

18. David Hume: *An Enquiry Concerning Human Understanding,* Section VII, Part ii, and *A Treatise of Human Nature,* La Salle, Illinois: The Open Court Publishing Company 1963.

19.   ————. *An Enquiry Concerning Human Understanding*, L.A. Selby-Bigge (ed.), Oxford: Clarendon Press 1894, Section VII, Part ii, p. 76.

20.   Richard Taylor: "Causation," in *Monist*, Vol. 47, La Salle, Illinois:The Open Court Publishing   Company 1963.

21.   Wolfgang Stegmüller: *Wissenschaftliche Erklärung und Begründung*, Berlin: Springer-Verlag 1969, p. 435.

22.   Bertrand Russell: "On the Notion of Cause," in *Mysticism and Logic*, London: Longmans, Green and Company 1925.

23.   *Ibid.*, p. 189.

24.   Helmut Korch: *Das Problem der Kausalitat*, Berlin: Deutscher Verlag der Wissenschafter 1965.

25.   Philipp Frank: *Das Kausalgesetz und seine Grenzen*, (Schriften zur wissenschaftlichen Weltauffassung, Vol. 6), Wien: J. Springer 1932.

26.   Roderick Chisholm and Richard Taylor: "Making Things to Have Happened," in *Analysis*, Vol. 20, No. 4, Oxford: Basil Blackwell, March 1960, p. 73.

27.   Adolf Grünbaum: *Philosophical Problems of Space and Time*, New York: Knopf 1963.

VIII. POLITICAL PHILOSOPHY

The History of Political Philosophy and the Myth of the Tradition

J.G. GUNNELL

I. POLITICAL PHILOSOPHY AND THE HISTORY OF IDEAS

In a discussion of methodological problems in the study of the history of ideas in which the focus is primarily on literature dealing with the history of political theory or philosophy, Quentin Skinner has addressed himself to what he contends are a number of conceptual difficulties related to various mythologies, mental sets, paradigms, and other types of preconceptions which have informed interpretations of writings from the past and explanations of the development of ideas.[1] He argues that these difficulties, which are the product of the often opposed but equally mistaken beliefs that it is possible to rely exclusively either on an analysis of a text itself or on an investigation of its social context to determine its meaning, suggest a need for establishing adequate procedures of inquiry and criteria for determining the meaning of such works. Skinner's thesis is provocative and has drawn considerable critical comment.[2] But despite the importance of the issues he raises about the recovery of historical meaning and knowledge of the past as well as the possible merits of his positive arguments about textual interpretation, he tends to treat the history of ideas as if it were a concrete activity distinguished by a common goal, i.e., a search for meaning and understanding, and to assume that the history of political philosophy may be differentiated principally in terms of the particular subject matter to which it addresses itself, i.e., political ideas. Not only does he assume that it is possible, and valid, to discuss methodological problems in the history of ideas in general in terms of examples elicited from the research of a variety of individuals concerned with the history of political philosophy but that it is feasible, and legitimate, to analyze and categorize the work of historians of political philosophy in terms of methodological commitments to either textual or contextual modes of interpretation. To treat the history of ideas and the history of political philosophy as a discipline and subfield respectively and to employ such a textual/contextual typology violates one of Skinner's own principal methodological prescriptions, i.e., the need to attend to the actual intentions of an author and the particular character of his argument in order to understand the meaning of his work.

It is doubtful that the "history of ideas" can be assumed to refer to a distinct activity or that it is anything more than an analytical category for subsuming certain general concerns, subjects, and problems characteristic of a diverse number of fields such as the history of religion, the history of philosophy, and the history of science. In addition, it is a mistake to assume that the various specialized scholarly communities that may be in some way involved with the history of ideas have the same degree of coherence and compactness. For example, while the history of science, despite sharp controversies over approach, may constitute a relatively well-defined field of study, the history of political philosophy, despite its recognition as a sub-field of political theory in political science, is not a comparable discipline. The extent to which the literature that is often considered as comprising this field constitutes a distinctive genre is far from immediately apparent. It is misleading to suggest that the work of historians of political philosophy in general can be intelligibly viewed as part of an activity termed the history of ideas, which can be defined in turn by the aim of an antiseptic search for meaning and understanding in historical texts as presupposed by Skinner's analysis. It may be relevant to argue that this is what they should be doing, but it is first necessary to determine what in fact they are doing. It is even more misleading to suggest that work in this area can be generally assumed to embrace exclusively either a textual or contextual approach to interpretation or that these terms designate actual methods of procedure. Although many of the problems to which Skinner refers regarding the criteria of adequate interpretation are necessarily raised in some form by any undertaking which makes claims to knowledge about the meaning of texts and the history of ideas, and may be present in various forms in studies of political ideas, these problems must be approached in terms of an examination of the enterprise in which the particular "historian" is involved, the kinds of arguments he is making, and the methodological assumptions he actually employs.

There is no doubt that research in the history of political philosophy has been characterized by a lack of any clear criteria for evaluating the adequacy of both interpretations of the meaning of texts and explanations of the development of ideas. This is in part a consequence of the fact that the history of political philosophy (or theory) is only a vaguely circumscribed activity and does not constitute a discipline with a distinct theoretical orientation, common conception of its subject matter, and generally accepted procedures of analysis. However, the absence of standards for undertaking and judging work in this area may also be explained in part by the fact that many of the studies which have been most influential in defining the content of this field and establishing approaches to inquiry, although nominally directed toward an objective understanding of historical texts and the development of political ideas, have been primarily informed by a concern with developing a critique of the philosophical foundations of contemporary politics through an examination of its intellectual antecedents. This practical concern with the political thought and practice of the modern age has contributed to the emergence of, and maybe essentially given rise to, a number of related premises which together may be termed *the myth of the tradition.*

This myth is reflected in one way or another in much of the scholarship in the history of political philosophy and in the characteristic textbook assumption that there is a series of major political philosophers whose works constitute an actual historical tradition which in some important respects is the source, for better or worse, of modern political ideas and contemporary politics. As J.G.A. Pocock has noted, "a canon of major works had been isolated by academic tradition, running from Plato to Aristotle to Augustine to Aquinas to Marsilius to Machiavelli to Hobbes to Locke to Hume to Rousseau to Burke to Hegel to Marx."[3] Gradually, this canon has come to be conceived, despite great disparities between the works chosen for inclusion but because of what has been understood at a certain level of abstraction as their concern with a set of common problems and ideas, as the product of a concrete activity with a recognizable temporal career. Accordingly, "the intellectual similarities between these systems were supposed to constitute the continuities, the dissimilarities between them, the processes of change, of a historical order,"[4] and it was assumed that this order could be isolated, studied, and understood as a unity culminating in the present. Notwithstanding very great differences in substantive argument about such matters as the precise composition, meaning, direction and impact of the tradition, the assumptions of its existence and its relevance for understanding the present have produced a syndrome in the study of the history of political philosophy which is characteristic of what is usually understood to be some of the most important work in the field.

Specific arguments about meaning and change in the history of political ideas tend to be developed in terms of presuppositions about politics, the activity of the political philosopher, and the tradition of political philosophy. These retrospectively applied analytical constructions govern representations of the intentions of the theorist and the choice of figures who constitute this tradition. The history of political philosophy emerges as a virtual history consisting of a conventional repertoire of epic works arranged chronologically. The development of political ideas becomes the transition from the thought of one epic figure to the next, and the meaning of a work is viewed as a function of its role in, or contribution to, this reconstructed tradition. The tradition is understood as authoritative in the sense that it is considered as a source of knowledge about the present and as containing within it a moment or moments of revelation about politics which may be extracted for evaluating prior and subsequent developments. But in terms of this dramaturgical image of the tradition as a movement of generation and degeneration, accumulated increments of thought over time also constitute a veil which must be penetrated if the revelations are to be re-appropriated. The study of the tradition is conceived as a venture in the discovery of error and the recovery of truth. The history of political philosophy becomes an enterprise which is primarily a critique of the present couched in the form of a mythopoeic *aitiologia*.

Although this syndrome is common to many commentaries on the history of political philosophy, the difficulties in treating this body of scholarship as if it were a well-integrated field of study create a problem for attempting to bridge between the arguments of particular individuals. It is possible to point to certain individuals who are representative in many respects of research in the area and who have had considerable impact on the scholarship in the field, yet to move too freely between their arguments is to risk the danger of an unwarranted attribution of the views of one to another. Although there are numerous general family resemblances with regard to patterns of approach and argument which are characteristic of a large portion of the field, there are also many obvious differences, and even severe disagreements, on many issues which make it awkward and misleading to attempt to discuss these individuals as a group.

The strategy in the following pages will be to examine the character and function of the myth of the tradition in the work of one influential scholar in the field of the history of political philosophy. The arguments of Leo Strauss will be discussed, since he has done considerable research in the history of political philosophy and has addressed himself in a deliberate and self-conscious manner to what might be termed methodological issues in this field and to the problems of defining the subject matter and character of the enterprise. To choose Strauss as a paradigmatic case is not to suggest that he speaks for the field, yet his scholarship has had a significant impact and has influenced, both directly and indirectly, many students who have come as close to constituting a school as any group of scholars in this area and who have written many pages fleshing out or working within the confines of Strauss's basic themes concerning the character of the tradition, the figures who comprise it, and the manner of studying it.

II. THE STRAUSSIAN INTERPRETATION

For Strauss, the impetus to engage in research in the history of political philosophy does not derive from any strictly scholarly concern with historical explanation and puzzle-solving but from a confrontation with what he alleges to be the "crisis of our time, the crisis of the West".[5] Politically, this crisis involves the external threat of "Communism" and the internal threat of liberal "permissive egalitarianism" which together endanger democracy in an unprecedented manner.[6] Modern liberalism, Strauss maintains, has relinquished all absolutes and become "entirely relativistic",[7] and, consequently, it can no longer defend its own principles. However, the political crisis, in both its dimensions, is the result of a more fundamental intellectual crisis which is rooted in the decline of political philosophy. The task of the history of political philosophy is to examine the process which led to this decline and to attempt to regain what has been lost in the hope of finding a solution to current problems.

Strauss defines political philosophy as that activity which attempts "to replace opinion about the nature of political things by knowledge of the nature of political things", which most essentially is knowledge about the right or good society and the best political order.[8] This activity, he argues, "was originated by Socrates" and has continued, at least in an attentuated form, until the modern age, but "today, political philosophy is in a state of decay and perhaps of putrefaction, if it has not vanished altogether."[9] The decline of political philosophy is attributed to the growth of the complementary doctrines of historicism and positivism which have undercut the tradition of political philosophy by denying the possibility of knowledge of values and asserting the cultural and historical relativity of all values. Positivism and historicism have created a climate of "unqualified relativism" in Western thought and tend to lead to "nihilism" or "the inability to take a stand for civilization against cannibalism".[10] Strauss suggests that these doctrines find their most characteristic expression in contemporary social science with its emphasis on the heterogeneity of facts and values and its view of values as merely preferences which are outside the realm of demonstrable knowledge. Social science is not only the product of these characteristic ideas of modernity but tends to reinforce these ideas. It thus has a "corroding" effect on society,[11] and by failing to provide either a defense of democracy or a critique of communism, it contributes to the political crisis.

Strauss insists that political science, in particular, as the heir to the tradition of political philosophy, is the existential manifestation of the fate of political philosophy in the modern age, and parallels, as well as contributes to, the crisis of the West in all its dimensions. Strauss argues that political science in its contemporary positivistic form is not only an academic discipline, but an educational institution which "wields very great authority" in society.[12] His indictment of political science includes the charges of "atheism", the "denial of a common good", a rejection of the idea that "man has natural ends", a lack of recognition of a "difference between men and brutes", and the fostering of "relativism".[13] He maintains that the discipline "reflects the most dangerous proclivities of democracy" by accepting the equality of all desires and values.[14] It tends to depreciate the differences between democracy and communism by treating them as simply two incommensurable ideologies and thus exacerbates the external threat to the West by weakening the internal constitution of its political order.

Since Strauss explains the modern crisis as a function of the decline of political philosophy and the denial of its very possibility, any solution to the crisis presupposes an understanding of the process of this decline as well as a knowledge of the original character of political philosophy, and "we are therefore in need of historical studies in order to familiarize ourselves with the whole complexity of the issue."[15] Such historical studies are not conceived as an end in themselves but as a therapeutic endeavor necessary for a diagnosis of modern ills and the rediscovery of political philosophy. Strauss argues that "political philosophy is not a historical discipline," since it aims at trans-historical knowledge, and that it "is fundamentally different from the history of political philosophy"[16] which, as often practiced, tends to treat all past philosophies as so many ideological reflections of their historical period rather than consider the possibility that some such philosophy may have discovered a truth relevant for all time.

Strauss maintains that the modern crisis not only demands research in the history of political philosophy but that the break-down of the tradition has "the accidental advantage of enabling us to understand in an untraditional or fresh manner what was hitherto only understood in a traditional or derivative manner."[17] The character of the tradition becomes accessible at the point of its demise when it is possible to detach oneself from this tradition sufficiently to gain the objectivity required for "solid knowledge" of the past and "to study the political philosophies as they were understood by their originators" and achieve an interpretation that "understands the thought of a philosopher exactly as he understands it himself."[18] Once the ideas of the past have been understood on their own terms rather than through the screen of the tradition and modern ideas, they can be submitted to a "philosophic critique concerned exclusively with their truth or falsehood" and the recovery of true knowledge about politics.[19] The goal of the history of political philosophy is "the restoration of political philosophy" through a critical or destructive effort aimed at stripping away from the true understanding of politics the distortions of "inherited knowledge" which separate the modern age from this understanding.[20]

According to Strauss, the "great tradition" of political philosophy consists of two fundamental periods: ancient and modern. He contends that politics and classical political philosophy emerged together in the context of the Greek polis and that this original philosophy apprehended political phenomena in an unmediated manner. Thus, "classical political philosophy is the true science of political things," and the "essential character of all political situations was grasped by the old political science."[21] Modern political philosophy began with the deliberate transformation of the classical teaching of Socrates, Plato, and Aristotle by modern philosophers such as Machiavelli and Hobbes, and it can be understood as at once a derivation from and rejection of those ancient principles. The spirit of the modern age is grounded in the suppositions of modern political philosophy, but the latter was built upon classical political philosophy, despite its divergence from it. Consequently, classical philosophy becomes the starting point for understanding the entire tradition as well as the basic truths about politics.

The purpose of the history of political philosophy is to penetrate the tradition, peel away the accretions, and reach that point where the meaning of classical political philosophy can be clearly perceived, just as it viewed political phenomena directly, rather than "through the lenses of modern political philosophy and its various successors."[22] Since the apostasy of modern political philosophy marked the beginning of the decline of political philosophy, this "old quarrel between the ancients and the moderns"[23] must be re-staged in order to undercut the foundations of liberalism, communism and allied doctrines which are responsible for the crisis of the West. The tradition of political philosophy has been marked by a progressive deterioration, and just as Socrates was the founder of the classical teaching, "the founder of modern political philosophy is Machiavelli" on whose work the entire corpus of modern political philosophy is based.[24]

In Strauss's view, Machiavelli deliberately transfigured political philosophy by excluding from its scope the questions which had previously guided inquiry such as those relating to how men ought to live and the character of the best political order. Instead, he taught "that one should take one's bearings, not by how men ought to live but by how they actually live," and this led to a "lowering of the standards of political life" which continues to infect modern attitudes.[25] Strauss characterizes Machiavelli as a "teacher of evil" who by inverting the tradition and transmitting his teaching to posterity produced the "first wave of modernity".[26] This wave crested in the work of Hobbes who softened the "revolting character" of Machiavelli's teaching in order to make it more palatable but also more capable of realization, and it surged into modern political thought and action through Locke who further mitigated the teaching in order to make it more acceptable.[27] Locke, usually acknowledged as the founder of liberalism, becomes, in Strauss's interpretation, the propagator of Hobbesian principles under the guise of traditional natural law teachings.

This thread of the tradition was complemented by another attack on classical doctrines stemming from Rousseau who set in motion the "second wave of modernity: the wave which bore both German idealistic philosophy and the romanticism of all ranks in all countries."[28] Strauss argues that Rousseau, by his acceptance of the general will as the standard of political and moral judgment, fostered a relativism which has become characteristic of the modern age. The philosophy of history typified by Hegel and Marx attempted to locate the standard of value in the historical process itself, but the failure of this vision left nothing but a view of history as a succession of ideologies. This prepared the way for the emergence of "the third wave of modernity — of the wave that bears us today. This last epoch was inaugurated by Nietzsche," who Strauss designates as "*the* philosopher of relativism" who laid the foundation for the "radical historicism" of modern thought which has destroyed political philosophy and produced the crisis of our time.[29]

III. HISTORY AND PRAGMATICS

Strauss maintains that he is engaged in historical studies with regard to both his reconstruction of the tradition as a whole and his interpretation of the works of various figures in that tradition, but, even though the criteria for specifying historical studies may be notoriously vague, it is necessary critically to examine this contention. It might seem that little can be gained by attempting to decide whether or not Strauss is really "doing" history, but although he is concerned with the past, his claim to be recounting the history of political philosophy requires scrutiny. Without putting too much emphasis on arriving at a definitive concept of history and a decision about what properly constitutes historical method or argument, it might be useful to view Strauss's work in terms of Michael Oakeshott's distinction between a "historical" and a "practical" attitude toward the past.[30]

Oakeshott argues that while the historian's concern is with the past as an independent object of investigation worthy of study in its own right, a practical approach to the past is characterized by a concern with how it relates to, or explains, the present. Although Oakeshott stresses the "disinterested" posture of the historian, the distinction does not turn simply on the question of objectivity or the presence of bias but on more general attributes. While the historical attitude involves a careful examination of concrete events and the connections between them, an attempt to eschew evaluation and produce a detailed narrative that illuminates change by producing a full account of change, and a lack of concern with the past as a source of authority for contemporary political and religious beliefs, the practical attitude is marked by the tendencies to approach the past in terms of abstract categories, to read events backward and make sense of the past as it is perceived in relation to the present, to select what is useful for describing and solving current problems, and to employ a language of praise or blame and engage in justification or condemnation. Oakeshott's idealizations are certainly not free from difficulties as criteria for characterizing and judging arguments about the past, but if taken as a basis for sorting out activities rather than as a principle for distinguishing good and bad historical work, they do point to important differences in approach and distribution of emphasis.

Although Strauss employs what might be considered the language of historical description and explanation, his attitude toward the past is basically practical or instrumental. It is difficult, and maybe even misleading, to attempt to understand and evaluate his arguments on the assumption that his enterprise is a historical one in the sense of Oakeshott's definition, despite his claim to understand past political philosophers as they understood themselves. This is most apparent in Strauss's discussion of the tradition. He claims to be answering questions about the existence and constitution of the tradition and how it developed, how concepts have changed, the relationship between political philosophy and politics, and the criteria for understanding and explaining the works of past thinkers, but the status of these claims is problematical.

If Strauss's arguments about the tradition are taken as strictly "historical", one could only conclude that they are inadequate even if viewed in terms of the most general standards for assessing statements about meaning and change in the history of ideas. The existence of a tradition is not something that Strauss sets about demonstrating but something that he assumes or asks his audience to assume. He neither explicates precisely what sort of a thing a tradition is as such nor makes clear the basis on which he argues that the diverse works that he deals with under the category of political philosophy constitute a tradition. A determination of the character of this putative tradition of political philosophy would seem to be crucial to any evaluation of Strauss's thesis about the modern crisis, since he is unequivocal in his assertion that the crisis is a product, in a direct causal sense, of this tradition. Yet, he offers little in the way of a concrete account of either the actual historical connections between the figures who supposedly comprise the tradition or the relationship between the tradition and contemporary politics.

Strauss begins with an analytical reconstruction of the activity of political philosophy, drawn largely from his interpretation of Greek political philosophy, which serves as a paradigm for both selecting and judging works belonging to the tradition. The activity of political philosophy as described by Strauss is a model or ideal type which is reified and discussed as if it were a self-consciously defined enterprise existing for over two thousand years despite important transformations. His arguments seem to be predicated on an acceptance of the common assumption that the chronologically arranged works which constitute the subject matter of most textbooks on the history of political philosophy may be legitimately treated as a tradition reflecting the development of a historically delimited activity. Strauss takes this canon as a datum and proceeds to infuse meaning and structure and render it as a continuing dialogue over time about certain perennial issues, concerns and problems. But despite Strauss's particular thesis about the overall direction of the tradition and his emphasis on the crucial role of certain figures in that tradition, his account is not far removed from that of received opinion, i.e., the assumptions that it began with the Greeks, that Machiavelli was the founder of modernity and the precursor of modern political science, and that there is a significant division between classical and modern segments of the tradition. Such notions are the basis on which Strauss develops his argument, but they, along with many of his emendations, are either in principle systematically undemonstrable or without significant support in his analysis as it stands.

To speak of political philosophy as a tradition begun by Socrates, transformed by Machiavelli, and atrophying in the modern age or to talk about waves of modernity and the forces of historicism and positivism as causes of current political problems is to enter into a realm of discourse which, despite its informational and narrative form, seems to be outside any context of rational discussion and criticism and expressly designed to avoid refutation. To state, for example, that Machiavelli was the founder of modern political philosophy is not, notwithstanding the grammatical similarity, like stating that Plato was the founder of the Academy or, even, that Marx was the founder of communism. In addition to his generalizations about the tradition, even Strauss's characterization of the present crisis, which is the focus of his concern, lacks the kind of concreteness that would give it relevance and contact with actual events. It is presented in terms of such abstract categories as his catalogue of "isms" — positivism, historicism, liberalism, relativism, communism, egalitarianism, nihilism — which would seem to sharply conflict with his own insistence on understanding political ideas on their own terms.

Despite Strauss's warnings about historicism and its tendency to conflate philosophical and historical questions and his suggestion that research in the history of political philosophy should be viewed as basically only an *ad hoc* endeavor required by the peculiar conditions of our time, it seems that he employs a mode of analysis in which a philosophical understanding and critique of the present become inseparable from an argument regarding its historical development. The authority for his condemnation of modernity involves an invocation of its origins and the postulation of a truth about politics that lies hidden beneath the dross of the tradition. The question is whether there is any other way to describe Strauss's argument than as a philosophy of history in the very sense that the concept is usually applied to much of the historical work of the 19th century with its notion of an immanent meaning in the unfolding epochs of history, except that now, in Strauss's case, the governing principle is not progress but decline. His suggestion that historical analysis is merely a preliminary sublimating operation required in order to recover the true character of political philosophy is difficult to accept on its face, since it constitutes virtually all of his scholarship.

Most of the issues which Strauss purports to solve by his historical account are not, in any apparent manner, adequately solved at all or even directly confronted. Most significantly, changes in ideas are not explained except in the most cursory way such as the attempt to account for major changes in the Western tradition by asserting that Machiavelli and the moderns rejected the claims of the classics. For the most part, neither a general analysis of conceptual change nor an examination of concrete instances of such change is presented. The ideas of various thinkers are merely chronologically juxtaposed and treated in such a way as to suggest that a conscious dialogue within the conventions of a common activity was taking place between individuals who were actually engaged in very different enterprises and separated by great temporal and cultural distances. A consideration of the whole range of issues surrounding the question of the relationship between political thought and action is not only neglected in principle but constantly begged in his accounts of how modern political philosophy transformed modern politics and triumphed over the classical tradition, how historicism and positivism have caused the weakening of liberal democracy, and how political science contributes to the ills of contemporary society. There is a persistent confounding of political ideas and political events or institutions so that often it seems that Strauss is talking about the development of political ideas and political forms at once. Political philosophers are sometimes treated as if they were legislators for an age and for the tradition and sometimes as if they were representatives of an age or stages in the tradition. Precisely how some of these individuals, such as Machiavelli, have had the great impact on society which Strauss invariably attributes to them is never clear. In general, Strauss's arguments about the tradition are structured in such a way that it is impossible to discover anything that would constitute disconfirmatory evidence. He constantly moves back and forth between the ideas of political philosophers, intellectual movements and assessments of politics, making spurious allusions between them which are informed by his general preconception of the historical pattern of the tradition. This is evident in his discussion of the relationship between modern political philosophers, the contemporary political crisis, the academic discipline of political science, and doctrines such as positivism and historicism.

Similar difficulties are apparent in Strauss's approach to the interpretation of particular works in political philosophy despite his apparent concern with historiographic and hermeneutic principles. Although he repeatedly stresses the need for understanding an author in the way in which he understood himself, he never sets forth the specific criteria of such understanding, and he brings numerous presuppositions to his analyses which would seem to preclude any interpretation on this level. He approaches an author's work by ascribing to him the paradigmatic intention of engaging in the activity of political philosophy as Strauss defines this activity, and the specific substantive intention attributed to an author is often largely a function of the place and role Strauss assigns to him in the general scheme of the tradition rather than something elicited from a study of the text and the historical context in which the individual wrote. Strauss insists on an objective interpretation of texts and an understanding of an author on his own terms, yet these resolves, although something which ultimately must be examined in light of Strauss's particular interpretative work, would seem to be very difficult to sustain in view of his general approach. Although he rejects interpretations of an author which merely "relate his thought to his 'historical situation'"[31] and emphasizes the requirement of relying on an author's own statements to discover his intentions, he never works through the manifold problems of deriving such intentions and the complicated issues revolving around the relationship of texts to contexts in interpretation. Strauss continually links intentions to the place an author occupies in the virtual context of the reconstructed tradition, and so it becomes, for example, Machiavelli's intention to lower the standards of political action or Hobbes's purpose to ameliorate the teaching of Machiavelli which would seem to indicate fallacies of the most transparent sort. He enjoins interpreters to avoid, and claims that he takes care to avoid, "extraneous information", modern hypotheses", and "conventions" of scholarship and, instead, moves entirely within the "'circle of ideas'" of the author himself,[32] but it is difficult to see in what way Strauss carries out, or even in principle could carry out, this program in view of the many preconceptions and instrumental aims he brings to his work.

Strauss condemns contemporary politics and the intellectual climate of the modern age which he argues has emerged from the decline of the tradition, and it appears that he deliberately sets out to impeach those individuals often accepted as the intellectual fathers of modern political philosophy and political institutions. It is difficult, for example, to read Strauss's argument about the underlying similarities between the thought of Locke and Hobbes (no matter how well taken this might be in some respects) as merely the product of an objective analysis of Locke's intentions when it seems so obviously informed by a prior design to undermine certain aspects of the ideology of liberal democracy by suggesting that it rests on premises which, if clearly understood, one might be less inclined to accept. Strauss's reasons for interpreting an author can in no way automatically serve as a basis for impugning his interpretation, but any general examination of his work makes it difficult to escape the conclusion that although he chides the historicist for confusing interpretation and criticism, the figures he selects to construct the tradition are already meaningful within the vision of the tradition he wishes to impart before he sets out to uncover the meaning of their work. He insists on the need for a rigorous understanding of an author's intentions, and certainly one of the most significant problems of interpretation is to develop procedures for understanding an author's intentions in his own terms and to develop the theoretical basis on which such a possibility might be predicated as well as the criteria of successful execution. But in Strauss's case all this remains at the level of a general maxim for historical investigation. Despite his emphasis on the autonomy of the text for understanding what an author is saying, his theory of interpretation is empty as far as providing a substantive conception of the phenomena of literary artifacts and the criteria of their intelligibility.

Whatever assessment might be made of Strauss's efforts at the exegsis of particular historical texts, it would seem to be a mistake to evaluate his enterprise, either his general reconstruction of the tradition or his description of the role of particular thinkers, as a straightforward "historical" endeavor in the sense, for example, that Oakeshott employs the adjective or even in the sense that it might be usually understood in ordinary language. If one wishes to determine what Strauss is doing, it may not be possible to find a conventional category that provides an adequate designation or explanation, and it may be possible to convey the character of this work only by pointing to modes of discourse with which it shares certain kinds of family resemblances. It is necessary to realize that at least his general arguments about the tradition are so completely incapable of supporting any sustained scrutiny that it would strain the imagination to assume that they are precisely what they may seem to be in terms of their ostensible form. Strauss himself at one point may be providing an indication of the genre of this apparently anomalous kind of argument when he suggests that "only because public speech demands a mixture of seriousness and playfulness, can a true Platonist present the serious teaching, the philosophical teaching, in a historical and hence playful, garb."[33]

What Strauss's "serious teaching" may be, whether that of Aristotle or something which even the interpretation of Aristotle becomes a "playful" way of alluding to, is not of particular concern here. What is of concern is to note that one would have to attribute extreme credulity to Strauss to assume that he takes a great deal of his historical argument seriously and not realize that the purpose of his history of political philosophy is instrumental not only in the sense of being a means to the end of understanding and critically examining the modern crisis but in the sense that the very form of the argument is instrumental and serves certain rhetorical functions. One hypothesis might be that Strauss employs a historical form of argument because he believes himself compelled to do so in an age in which only historical arguments about politics carry authority.

Strauss's reconstruction of the tradition is more a dramatic tale, a myth, calculated to evoke certain attitudes toward contemporary issues than a historical exericse, but what is the case with regard to his work may also be characteristic of much of the literature usually considered as belonging to research in the history of political philosophy. Sometimes, it is assumed that this field of study has been nearly exhausted, but, in fact, particular interpretations of the works of political philosophers as well as general accounts of the historical development of political thought have been governed for so long by various forms of the myth of the tradition that a re-examination of this subject matter may well be overdue.

NOTES

1. Quentin Skinner: "Meaning and Understanding in the History of Ideas," *History and Theory,* 1 (1969), pp. 3-53.

2. For critical discussions of Skinner's argument, see Margaret Leslie: "On Defense of Anachronism," *Political Studies,* 4 (1970); Charles Tarlton: "Historicity, Meaning and Revisionism in the Study of Political Thought," *History and Theory,* 3 (1973); Bhiku Parekh and R.N. Berki: "The History of Political Ideas," *Journal of History of Ideas,* 2 (1973).

3. J.G.A. Pocock: *Politics, Language, and Time* (New York: Atheneum 1971) pp. 4-5.

4. *Ibid.,* p. 9.

5. Leo Strauss: *The City and Man* (Chicago: Rand McNally 1964) p. 1.

6. *Ibid.,* p. 3; Strauss: "Political Philosophy and the Crisis of Our Time," in *The Post-Behaviorial Era,* eds. George J. Graham, Jr. and George W. Carey (New York: David McKay 1972) p. 222.

7. Leo Strauss: "Relativism," in *Relativism and the Study of Man,* eds. Helmut Schoeck and James W. Wiggin (Princeton: D. Van Nostrand 1961) p. 140.

8. Leo Strauss: *What is Political Philosophy?* (Glencoe: Free Press 1959) pp. 11-12.

9. Leo Strauss: "Introduction," in *History of Political Philosophy* (Chicago: Rand McNally 1972) p. 2; *What is Political Philosophy?* p. 17.

10. Leo Strauss: *Natural Right and History* (Chicago: University of Chicago Press 1953) pp. 2-5; "Social Science and Humanism," in *The State of the Social Sciences,* ed. Leonard D. White (Chicago: University of Chicago Press 1956) p. 422.

11. Strauss: *What is Political Philosophy?* p. 19.

12. Leo Strauss: "Epilogue," in *Essays on the Scientific Study of Politics,* ed. Herbert J. Storing (New York: Holt, Rinehart and Winston 1962) p. 307.

13. *Ibid.,* pp. 322-326.

14. *Ibid.,* p. 26.

15. Strauss: *Natural Right and History,* p. 7.

16. Strauss: *What is Political Philosophy?* p. 56-7.

17. Strauss: *The City and Man,* p. 9.

18. Strauss: *What is Political Philosophy?* p. 66-7.

19. *Ibid.*, p. 66.

20. *Ibid.*, p. 76; "Political Philosophy and the Crisis of Our Time," p. 218.

21. Strauss: *The City and Man*, p. 10; "Epilogue", p. 313.

22. Strauss: *The City and Man*, p. 9.

23. Strauss: "Political Philosophy and the Crisis of Our Time," p. 217; *What is Political Philosophy?* p. 172.

24. *Ibid.*, p. 40.

25. Leo Strauss: *On Tyranny* (Ithaca: Cornell University Press 1963) pp. 110-11.

26. Leo Strauss: *Thoughts on Machiavelli* (Glencoe: Free Press 1958) p. 13.

27. Strauss: *What is Political Philosophy?* p. 47-9.

28. *Ibid.*, p. 50.

29. *Ibid.*, p. 54-5.

30. Michael Oakeshott: *Rationalism in Politics*, p. 137-67.

31. Strauss: *On Tyranny*, p. 24.

32. *Ibid.*, p. 25.

33. Leo Strauss: "Farabi's Plato," in *Louis Ginzberg Jubilee Volume* (New York: The American Academy for Jewish Research 1945) pp. 376-77. I owe this citation to Professor Eugene F. Miller of the University of Georgia.

IX. PSYCHOLOGY

Events are Perceivable But Time Is Not

J.J. GIBSON

For centuries psychologists have been trying to explain how a man or an animal could perceive space. They have thought of space as having three dimensions and the difficulty was how an observer could see the third dimension. For depth, as Bishop Berkeley asserted at the outset of the *New Theory of Vision* (1709), "is a line endwise to the eye which projects only one point in the fund of the eye." Space was its dimensions. It was empty save for a collection of objects or bodies. For an observer, the objects were in different directions at various distances and the question was how these distances could be detected. For two hundred and fifty years we have tried to answer this question and failed. The explanations have been controversial, contradictory, and confused.

I have been arguing since 1950 that it is a false question and therefore unanswerable. There is no such thing as depth perception, or the perception of distance, or the third dimension, or in fact of the perception of space. There is only the perception of textured surfaces and what I call the "layout" of these surfaces (Gibson, 1950). A surface is an interface between a substance and the medium. The fundamental surface is the ground. The earth and the sky are the main components of the world, but the earth is cluttered by its furniture. The true question is how we perceive all these surfaces with their inclination to one another, and their curvatures and their edges. Above all, since they are generally opaque, how do we perceive the hidden surfaces that lie behind the ones that face us? This problem is the crucial one, not the problem of the third dimension. How do we perceive that portion of the layout of the world that is temporarily *out of sight*? This question, please note, has reference to time as well as space. Actually, what it refers to is the motions of objects, the locomotion of an observer, and the fact that whatever goes out of sight during one displacement will come into sight during the opposite displacement.

I now want to argue that the perception of time is a puzzle of the same sort that the perception of space has been — an insoluble one. There is no such thing as the perception of time, but only the perception of events and locomotions. These events and locomotions, moreover, do not occur in space but in the medium of an environment that is rigid and permanent. Abstract space is a sort of ghost of the surfaces of the world, and abstract time is a ghost of the events of the world.

ECOLOGICAL EVENTS

This line of argument rests on a neglected level of optics and physics. I call it ecological optics and environmental physics (Gibson, 1966b). Ecological optics begins with the ambient optic array at a moving point of observation. Environmental physics begins with substances, surfaces, and the medium. Environmental events consist mainly of changes in the layout of these surfaces.

A change in the layout of surfaces may be either, first, a re-shaping of those that are more or less plastic or, second, a re-positioning of those topologically "closed" surfaces that we call objects. Both re-shaping and re-positioning require mechanical forces in accordance with the laws of Newtonian mechanics, but events are not reducible to the Newtonian motions of bodies in space. The terrestrial environment does not consist of bodies in space.

A very radical change in layout, a third type of event, occurs when a surface goes out of existence. Going out of existence is not to be confused with going out of sight, or becoming hidden. When a surface goes out of existence it ceases to reflect light to any point of observation whereas when a surface goes out of sight it is still reflecting light to some other point of observation. The failure to distinguish these cases is a source of confusion. The surface of a cloud can dissipate; the surface of a puddle can evaporate; the surface of a solid can sublimate, or melt, or dissolve, or disintegrate; the surface of an organism can decay. When a substance goes into the gaseous state, its substance becomes part of the medium and its surface ceases to exist (Gibson, Kaplan, Reynolds and Wheeler, 1969).

The destruction of a surface is a highly significant ecological event but not a significant physical event or, rather, its importance is not recognized in physics. The processes of dissipation, evaporation, dissolution, disintegration, and decay are not reversible in the way that the elementary processes of physics are reversible. Although surfaces are formed, the processes are not the opposite of those by which surfaces are destroyed. This fact can be illustrated by a motion picture film in which shots of commonly observed events are compared with the film run forward and then backward. It becomes evident that the breaking of a whole object is not simply the reverse of the mending of a broken object, and that when the breaking is artificially reversed on film, the event is perceived as magical, that is, as violating the laws of ecological physics (Gibson and Kaushall, 1973).

It is sometimes said that the professional magician, the conjuror, causes us to perceive events that violate the laws of physics. But I suggest that the laws violated are not to be found in the textbooks. They are not laws of theoretical physics but of carpenters' physics. They do not involve explicitly scientific knowledge but *tacit* knowledge, and this is why children are fascinated by conjuring. The rules are learned by perceptual observation, not by mastery of an intellectual discipline. But they are nonetheless interesting and worthy of study. The fact that magic is mysterious is proof that these rules are implicitly known.

Thus, a very great difference between ecological physics and physical physics is that things can be destroyed and created at the ecological level of analysis but not at the physical level. Physics proper does not admit of destruction, for matter is conserved. The most it admits is the rearrangement of particles. But this concern for order and disorder at the level of atoms, together with the theory of entropy, is not relevant when we are concerned with the ecological level of analysis. Surfaces cease to exist or begin to exist with a change in the "state" of matter, and these are things on the scale of animals; this is the world at *our* level.

Summary: Three Kinds of Events

The re-positioning of objects, the re-shaping of surfaces, and the annihilation or creation of surfaces — these are, I think, the main types of environmental events. (I leave out of account changes in the chemical composition of substances and the resulting changes in their surfaces, their colors for example, because it would take the discussion too far afield). These include the natural events of the terrestrial world, where pebbles roll, the leaves fall, the streams flow, and the animals scurry about. Things wax and wane, organisms grow and die, animals behave by deforming their surfaces, human faces smile and frown, and the sun moves across the neavens, always rising on one side of the habitat and setting on the other.

How shall we describe these events if we cannot adequately do so in terms of space and time, of bodies and their motions? I suggest that we can do so in terms of the reciprocal concepts of permanence and change, more exactly, *persistence* and change. The substances and surfaces of the environment persist in some respects and change in others. The *basis* of the environment is the flat earth, the surface of support. This surface is permanent in shape. The streams and lakes and swamps are changeable in shape. The plants and animals are even more changeable. The earth is a solid substance with a surface that is rigid and resists deformation. The surfaces of plastic and liquid substances have less resistance to deformation. The surfaces of plants and animals are visco-elastic, and they even undergo *spontaneous* deformations, by growing and behaving. Nevertheless, in all these cases there is some permanence underlying the change, and this is true of the stream, the swamp, the growing tree, and the behaving animal.

The three kinds of ecological events listed above all involve some persistence and some change. First, the re-positioning of a rigid object such as a falling apple or a rolling stone leaves the shape and size of the object invariant, and the background surfaces invariant, but alters the pattern of the environment. When a housewife rearranges the furniture in her living room, for example, the room is still the same but the objects have been displaced and rotated. Second, the re-shaping of surfaces involves a somewhat more radical change, with less persistence. When a lump of clay is transformed into an image of Venus, for example, the two surfaces are topologically equivalent but the shapes are very different. Third, the annihilation or creation of a surface involves the maximum change of environmental layout. Nevertheless something persists, the remaining surfaces of the environment, to preserve a basic invariance in the world.

THE PERCEIVING OF EVENTS

The argument I have given above is not consistent with the usual theory of sense perception, that is to say, sensation-based perception. It is only consistent with what I call a theory of information-based perception (Gibson, 1966b). I believe that sensations or sense data play no part in the explanation of perception, and that the channels of sensation are incidental and irrelevant. Perceptual systems take the place of the traditional senses, and they actively seek stimulation of the receptors proper in order to extract information from the ambient energy that bathes the individual. Perception is a direct pickup of this information, without the necessity of mediating sense impressions that have to be interpreted. Ambient light, sound, odor, and mechanical contact all provide *stimulation* but not a set of punctate or momentary *stimuli*. Stimulation comes in an array that is also in flux; the retina, the skin, and the cochlea are all stimulated by a pattern which changes from moment to moment.

According to this theory, there are in fact environmental events and they are in fact perceived, although to different degrees by different animals. They are visually perceived by means of the information in the ambient array of light, and only by such means, although different observers pay attention to different aspects of this information. Events can be well or ill perceived, and there is no assumption that there must exist a sequence of phenomenal events corresponding to the physical events and running parallel to them. Events are nested within longer events, which is to say that there are subordinate and superordinate events, and it would be impossible for any observer to perceive the whole of them in complete detail. Perception is selective.

Events of all types, as I said above, have a component of change and a component of persistence. The environment is neither frozen at one extreme nor a complete chaos at the other. It is never wholly the same after an event, as it was before, and it is never wholly different afterwards from what it was before. Similarly, the optical information to specify an event must also have two components, one of varying optical structure and one of invariant optical structure. The optic array itself is neither frozen nor chaotic; only so could an observer detect what is changing and what is persisting in the world at the same time. The coexistence of variance with invariance in a changing pattern is perfectly easy to define mathematically.

Finally, the information-based theory of perception asserts that when an event has been perceived there are two kinds of concurrent *awareness*, one of variation and one of non-variation. That is to say the observer perceives *both what is altered and what remains unaltered in the environment*. This sounds strange to anyone who accepts the traditional theory of sense-perception for it seems to say that the same "stimulus event" can yield two different percepts at the same time. A percept is supposed to *correspond* to its stimulus; how can one retinal motion yield two percepts?

Nevertheless this is what happens, as I discovered many years ago when I was experimentally investigating optical transformations as stimuli for visual perception (Gibson, 1957). In one experiment a shadow was cast on a translucent screen and made to undergo cycles of perspective transformation, that is, foreshortening and the reverse. The form projected could be either rectangular or potato-shaped, either regular or irregular. What the observers perceived in this experiment was a rigid form being turned, slanted back and forth. That is they always saw both *something rotating and something rectangular*. In the other case it was *something rotating* and something *uniquely potato-shaped* (Gibson and Gibson, 1957). The duality of the perception was unmistakable; the observer could accurately judge both the degree of rotation and the shape of the object.

Was it the case that one optical "motion" had aroused two percepts? No, for the optical foreshortening should not be called a *motion* and should not be thought of as a *stimulus*. It was information for the perception of an event in the environment, and sensations of motion had nothing to do with it nor, of course, did sensations of form. The transformation of foreshortening included both variants and invariants that specified both the turning of the object and the shape of the object. The "motion" in the light to the eye was nothing like the motion of the object, and the "forms" in the light coming to the eye did not match the form of the object, but nevertheless the transformation carried information about a turning object. The slant of the surface altered while the shape of the surface did not, and both facts were clearly perceived.

TIME AND SPACE IN AN INFORMATION-BASED THEORY OF PERCEPTION

It should now be clear how, in an information-based theory of perception, the detection of change and non-change substitutes for the perception of time and space. In the traditional theories of perception the ideas of time and space are necessary in one form or another to make possible the interpreting of sensory data. In Kant's formulation time and space were supposed to be intuitions prior to experience, or innate ideas. In other formulations they were supposed to be learned ideas but in any case they were required for perception. But if perception is not based on meaningless raw sensations, time and space are not required.

Time and space are concepts, abstracted from the percepts of events and surfaces. They are not perceived, and they are not prerequisite to perceiving. They do not give meaning to percepts and they are not imposed by the mind on the deliverances of sense. Time and space are intellectual achievements, not perceptual categories. They are useful in the study of physics but not in the study of psychology.

Isaac Newton's famous assertion that "absolute, true, and mathematical time, of itself and from its own nature, flows equally without relation to anything external" was the postulate of a physicist trying to simplify his problems. It did simplify physics but that does not mean it will simplify ecology and psychology. It leads to the idea of empty time which, like the idea of empty space, brings with it insoluble problems for ecology and psychology. This implies that events are what "fill" time, as if time were a container into which events can be put. But this metaphor is surely wrong for the psychology of event perception. Time is not a receptacle for events, just as space is not a receptacle for objects. A better metaphor would be to suggest, as already mentioned, that time is the ghost of events and that space is the ghost of surfaces.

William James in the *Principles of Psychology* (James, 1890) wrote a chapter headed "The Perception of Time" but he understood very well, despite the title, that the mere passage of time, empty time, is not perceived. The fact is that our experience is never empty. A sequence of external stimuli or, at the very least, the rhythms of the observer's body, provide a flow of *change*, and it is this we perceive rather than a flow of time as such. James made the same point in an even more famous chapter on "The Stream of Thought." It seems to me that he here foreshadows the theory of the concurrent perception of persistence and change.

PAST, PRESENT, AND FUTURE IN THE NEW THEORY OF EVENT PERCEPTION

The seemingly innocent hypothesis that events are perceived has radical implications that are upsetting to orthodox psychology. Assuming that shorter events are nested within longer events, that nothing is instantaneous, and that sequences are apprehended, the usual distinction between percpetion and memory comes into question. For where is the borderline between perceiving and remembering? Does perceiving go backward in time? For seconds? For minutes? For hours? When do percepts stop and begin to be memories or, in another way of putting it, go into storage? The facts of memory are supposed to be well understood but these questions cannot be answered. Equally embarrassing questions can be asked about expectation.

Event perception implies a rejection of the division of the stream of awareness into a past, a present, and a future, with borderlines between (Gibson, 1966a). The puzzle of where to draw the lines does not arise. The feelings of past, present, and future are merged or, more exactly, the activity of perception is acknowledged to be retrospective and prospective. It is necessarily so, since perception is not confined to an instant. As James put it, consciousness is a *stream*, and consciousness of the present is not a traveling razor's edge. Even a thunderclap, says James, involves a "feeling of silence just gone." There is always a "consciousness of whence and whither." All this comes from the chapter on the stream of thought. James implies, as I have here asserted, that the perception of what *has* occurred, what *is* occurring, and what *will* occur are all of a piece.

I reject the assumption that present sensations are supplemented by memories of the past and by images of the future. I reject the assumption that perceiving is brought about by a sequence of discrete stimuli to which the brain makes a contribution of some sort — a "subjective" contribution. I assume instead that the brain responds to variants and invariants in the flow of stimulus information. The difficulties that arise with a theory of traces, where they are laid down, how they are stored, and how they are retrieved or recalled when needed later are avoided.

The puzzle of past, present, and future is not relevant to the problems of event perception. The feeling of *now* is nevertheless often strongly experienced, and we often speak of the *present moment*. Whence comes this compelling experience? I suggest it comes from proprioception, that is, from the perception of the body of the observer himself as distinguished from his environment. It comes particularly from locomotion, and very largely from the visual perception of the locomotion of the observer through the environment. One who makes a journey sees himself moving relative to a stable and rigid world. The flowing perspective in the ambient array of light is ordinarily not noticed. The visual sensations of motion are paid no attention and the underlying invariants that specify the layout of surfaces are what gets perceived. But the traveler also sees his body in the environment and its momentary position. He perceives *here* and, in fact, the very perception of the environment entails the perceiving of *here*. The traveler perceives the path to be traveled if he looks ahead, the path that has been traveled if he looks behind, and the position in between is called *here*.

The traveler is tempted to think of the linear path as the dimension of time and to see the path traveled as the past, that to be traveled as the future, and the division point as the present. The point *here* and the moment *now* coincide. One has "foresight" of the future and "hindsight" of the past. It is from this analogy, I suggest, that we get our conception of time as divided into past and future by the instantaneous present. But this does not mean that the dimension of time, pure time, is experienced during locomotion. What is experienced is a moving self in a stationary environment.

BIBLIOGRAPHY

Gibson, J.J.: *The Perception of the Visual World.* Boston: Houghton Mifflin 1950.

Gibson, J.J.: "Optical Motions and Transformations as Stimuli for Visual Perception." *Psychological Review* 64 (1957): 288-95.

Gibson, J.J.: "The Problem of Temporal Order in Stimulation and Perception." *Journal of Psychology* 62 (1966): 141-49. (a)

Gibson, J.J.: *The Senses Considered as Perceptual Systems.* Boston: Houghton Mifflin 1966. (b)

Gibson, J.J. and Gibson, E.J.: "Continuous Perspective Transformations and the Perception of Rigid Motion." *Journal of Experimental Psychology* 54 (1957): 129-38.

Gibson, J.J., Kaplan, G., Reynolds, H., and Wheeler, K.: "The Change from Visible to Invisible: A Study of Optical Transitions." *Perceptions and Psychophysics* 5 (1969): 113 - 116.

Gibson, J.J. and Kaushall, P.: *Reversible and Non-Reversible Events.* (Motion Picture Film). Psychological Cinema Register, State College, Pennsylvania, 1973.

James, William: *Principles of Psychology.* New York: Henry Holt 1890.

Time Experience and Memory Processes

J.A. MICHON

1. INTRODUCTION

The experience of time, and more particularly of duration, has been studied rather separately from its functional fundament: the memory process. Yet, in the past few years some rather intriguing patterns of connection have emerged. Especially the effect of the usual distinction between immediate memory (IM), short term memory (STM) and long term memory (LTM) (Shiffrin and Atkinson 1969; Norman 1970) seems to provide some conceptual cement to link the two fields: time and memory.

At the first Conference of this Society in Oberwolfach I elaborated on a number of aspects of the relation between time and memory (Michon 1972). On that occasion I formulated a triad of basic assumptions which may be considered as the unavoidable underpinnings of any reasonable cognitive theory of human time experience.

a) Equivalence of Temporal and Non-Temporal Information

We assume that the duration of an event, i.e., a time interval, may present itself as a stimulus in its own right: human beings as well as animals can react in distinctly different ways to stimuli that differ in duration only. Furthermore, duration is scalable, as we shall see, and interactions between duration and other characteristics of stimuli may be observed. Time is most certainly not a "neutral" variable. It claims, on the contrary, equal rights as a genuine source of information; equal to other stimulus dimensions such as brightness, loudness and pitch.

b) The *Re*constructive Aspect of the Experience of Duration

Construction of percepts on the basis of prior analysis of distinctive features of the stimuli, and a *re*construction of memory contents on the basis of a plan or program "how to remember", have proved to be highly fruitful process descriptions for perception and retention. As such they appear to provide a tremendous potentiality for the development of coherent psychological theory (cf. Lindsay and Norman 1972).

In the psychology of time too this framework has become manifest: the cognitive theory of time experience, as formulated for example by Ornstein (1969), Michon (1972) and Vroon (1972) rests on the assumption that the structure imposed on the events within an interval will determine the estimate of duration.

c) The Relation with Memory Processes

The relation between time experience and memory is relatively self-evident. If a subject is to produce a judgment about some time interval $t = T_1 - T_0$, he must at least have a mental representation of the whole interval t at the instant T_1 and shortly thereafter. If the task is to compare two intervals $t \approx t_2$, the time during which the first must be kept in memory will be of the order of t, since the two intervals, each of approximately length t, are to be presented in succession but to be compared simultaneously with one another. Several authors (a.o. Creelman 1962; Treisman 1963; Michon 1967; Wing 1973) have studied the effect of this temporary memory storage on the precision of reproducing or comparing intervals. Surprisingly there is an almost universal agreement about this precision: the standard deviation of the estimates and judgments will increase with the square root of the average interval length: $\sqrt{t}$. Admittedly, there is more to it than this extremely simple relation. Most of the authors require one or more additional parameters to account for their results, but the fact remains that they all ascribe the square root relation to the memory factor as such. This pleasing unison is also supported by recent findings from research on memory, as we shall see later (Wickelgren 1972).

Taking the plausible assumption about an intrinsic relation between memory and duration evaluation seriously, we are led to ask what is the role of each of the partial memory processes that can be distinguished in the operation of human memory: primary or immediate memory (IM), short term memory (STM) and long term memory (LTM). IM is considered a buffer process that enables us to retain incoming information for about 500 msec so that preprocessing, such as extracting   distinctive features and coding, can take place. STM serves as a working memory which will hold a limited amount of information for approximately 30 sec while it is processed. LTM, finally, is a storage process for coded information of practically unlimited capacity and permanence (Shiffrin and Atkinson 1969; Norman 1971; Wickelgren and Berian 1971).

2. THE INTERNAL REPRESENTATION OF THE TIME SCALE

2.1. Estimation of Invervals

Several authors have determined psychophysical scales for time intervals (see Fraisse 1967; Michon 1967b). In most cases the results of these scaling experiments will be described by Stevens' psychophysical law, relating the physical magnitude, ϕ, of a stimulus with its perceived magnitude, ψ:

$$\psi = a\phi^b$$

in which a is a scaling constant, and the characteristic exponent b differs from one dimension to the other. For duration b is commonly found to exceed the value 1 by a small fraction, its average value being $b = 1.1$, which implies a slight overestimation of the longer intervals. However, Fraisse (1967) argues that the deviation from $b = 1$ is so small that we do better considering the mapping of the physical into the subjective time scale to be veridical.

The usual range for which this value $b \approx 1$ has been established reaches from a few tenths of a second to a few minutes. Longer intervals have been neglected because they tend to be unpleasant both for subject and experimenter; shorter intervals, however, have received a stepmotherly attention in research programs without apparent reasons. Strictly speaking, the relation $\psi = a\phi^{1.1}$ has been established only for the range between 1 sec and 2 min , and consequently it appears to be worthwhile to pay somewhat more attention to the short as well as to the long end of the scale.

2.2. Short Intervals

In an experiment in which the range between 50 and 2000 msec was probed in a very detailed way, the present author showed that the relation $\psi = a\phi^{1.1}$ does not quite hold for intervals of less than 500 msec (Michon 1967b). For durations between 50 and 500 msec the relation $\psi = a\phi^{0.6}$ appears to give a more adequate description of the results (Figure 1). This finding appears to support the idea that different processes are operating when short intervals (< 500 msec) than when longer intervals (> 500 msec) are to be judged (Woodrow 1951; Fraisse 1956, 1967).

In the context of the relation between time experience and memory processes we can even sharpen this assumption a little more:

> *Hypothesis 1: The transition of the parameter b in Stevens' law from value*
> *b = 0.5 to b = 1.0 at 500 msec is to be ascribed to the transition from immediate*
> *memory to short term memory.*

This 'sharpening' refers to three specific points: (1) we assume that interval estimates below 500 msec increase with the square root of the physical duration ($b = 0.5$); (2) we assume that for intervals longer than 500 msec the estimates are linearly proportional with physical duration, and (3) the judgments are explicitly related to two memory processes which can be distinguished experimentally.

Fig. 1. Relation between magnitude estimation (t_{est}) and physical duration (t_f) for short intervals. (From Michon 1967b).

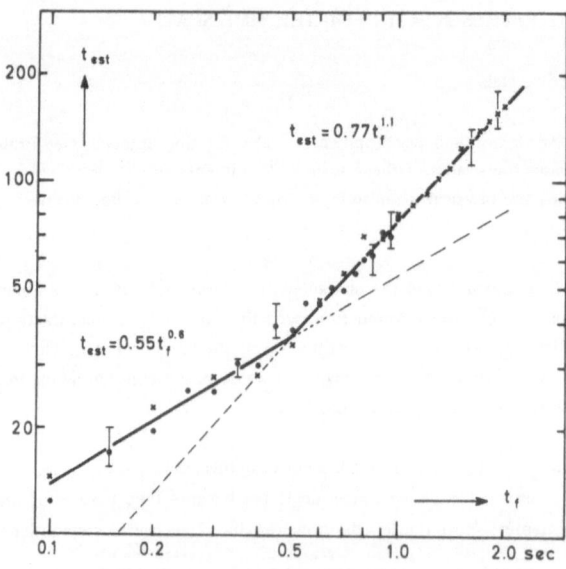

2.3. Long Intervals

It is not an established fact that the linear relation between physical and subjective duration will also hold beyond the 1 or 2 minutes for which it has been established. There is no direct supporting evidence available. However, Crombag, Roskam and Meuwese (1973) present some results which make such a linear relation quite plausible. On the basis of pairwise comparisons by students of estimated periods spent on certain curricular events they were able to demonstrate that even for intervals of more than 80 hours the linear relation between subjective time and its physical reference interval is maintained. (Figure 2).

It appears on the face of the available evidence as if a rather elegant assumption we would like to make cannot be made after all. We ought to have found a noticeable change in the relation between subjective and physical duration between 20 and 30 sec , as a result of the transition from short to long term memory (Shiffrin and Atkinson 1969; Wickelgren 1972: Wickelgren and Berian 1971).

Such a change is evidently not taking place, and this can again by summarized in a "sharpened" hypothesis.

> *Hypothesis 2A: The transition between* STM *and* LTM *at* 20 *to* 30 *sec is not observable in the time estimations of normal subjects.*

The crucial part of the statement here is the word *normal*.

In 1957 Scoville and Milner described a serious case of memory loss. This, now famous, patient H.M. is still participating in experiments and apparently is unable to retain any new experience for more than a few minutes; in the 15 years that have passed since the bilateral removal of his medio-temporal cortex almost no new information has been added to his memory (Scoville and Milner 1957; Wickelgren 1968; Deelman 1972). Evidently H.M.'s LTM is not functionally operating anymore.

Fig. 2. Relation between estimated duration and reference time for very long intervals. (After Crombag, Roskam and Meuwese 1973).

Fig. 3. Relation between reproduced and physical duration in a patient with deeply disturbed permanent memory function. (After Richards 1973).

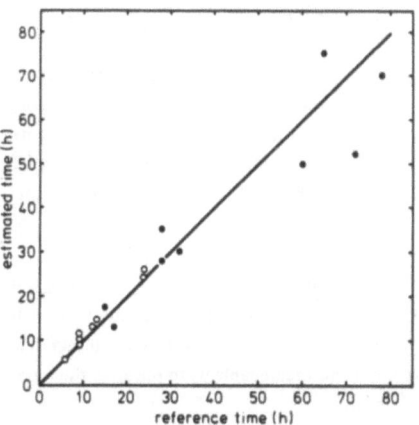

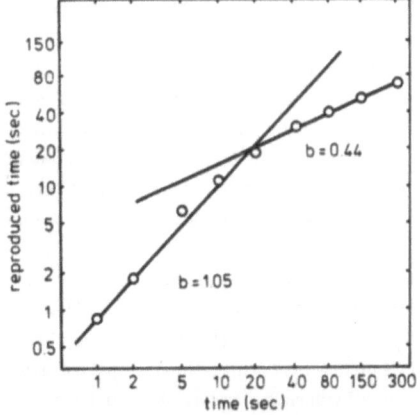

Recently Richards (1973) published the results of an experiment in which H.M. was asked to reproduce time intervals lasting up to 300 sec. The results shown in Fig. 3 once more show an almost rectilinear relation between stimulus and response intervals. However, at 20 sec we notice a sharp transition. The slope of the line $\psi = a\phi^b$ becomes less steep, and the value of b for longer intervals is very close to $b = 0.5$. (See Figure 3).

We may now supplement our hypothesis 2A with

> *Hypothesis 2B: If* LTM *does not function, the transition between* 20 *and* 30 *sec does occur. In that case the experienced duration of intervals whose length requires the availability of* LTM *will approach proportionality with the square root of physical time.*

Provisionally we may conclude that the relation between experienced and physical duration, as we observe it in estimation and reproduction experiments, does not run counter to the assumption that experienced duration is determined by IM, STM and LTM respectively when we are dealing with intervals shorter than 500 msec , 0.5 to 30 sec , and longer than 30 sec.

The question now is if we can lend more credibility to this assumption.

3. PERCEPTUAL AND COGNITIVE JUDGMENT OF INTERVALS

In fact, the connection between time experience and memory can be elaborated further along the following line.

Our IM acts as a pre-cognitive mechanism. Apparently, it plays the role of a very temporary store in which incoming information may be kept ready for further selection, analysis and encoding, for at most one second.

Given this perceptual nature of IM we expect that time intervals falling within this range will, in a way, be "perceived" even though we lack a time *sense*. Judgments about their duration will be highly pre-analytic, emotional and "immediate", in short, they will qualify as *sensations*. This quality may indeed be attributed to intervals up to about 500 to 700 msec (Woodrow 1951; Fraisse 1956, 1967).

The implication is that the assumption of cognitive theory that time evaluation is based on the reconstruction of patterns of events (Ornstein 1969; Block 1972; Michon 1972; Vroon 1972) requires some adaptation at short intervals. It does seem tenable for longer intervals, however, and one could envisage that the events occurring during the intervals are ordered and valued by reference to our experience with clock time in such a way that the estimate will approach the physical duration of the interval very closely. This would explain the rectilinear relation between experienced and "real" duration.

Also, the absence of a transition between the range of STM and LTM may be placed in this framework. It is generally assumed that LTM is a "passive" mechanism and that LTM contents must be brought back into STM before they may be actively processed (compared, transformed, integrated). Our lifelong experience with clock and calendar (social time) will then enable us to linearize the reconstructed sequence of events which took place during the interval.

When LTM is no longer accessible, as is the case with the patient H.M., the process of reconstruction and linearization will be impossible for time intervals exceeding the range of the STM. H.M. is unable to base his evaluation of such long intervals on the reconstruction of LTM contents and consequently has to rely on vague emotional impressions, and perhaps on a LTM for proprioceptive information (which appears to have suffered much less). The finding that time experience is proportional to the square root of physical time supports the suggestion that H.M.'s evaluation of long intervals indeed has a quasi-perceptual, emotional and immediate character.

4. A UNIQUE INTERNAL REFERENCE SCALE

There is another important question to be answered. In the preceding sections we proposed that the experience of intervals less than 500 msec has a distinctly perceptual character. This was ascribed to the pre-cognitive characteristics of IM. Consequently we may foster the expectation that psychophysical relations that can be observed in "genuine" sensory dimensions will also become manifest in time perception.

This may be tested by way of Teghtsoonian's recent analysis of the psychophysical scaling process (Teghtsoonian 1971). Here the fundamental proposition is that subjects will map stimulus dimensions into a unique internal reference scale of constant length. The length of this internal scale is between 1.5 and 2 log units, or between 50 and 100 discriminably different steps (subjectively just noticeable differences). Physical dimensions which vary over a large natural range of several log units, such as light (day vs night), or sound energy, must be compressed quite considerably in order to fit the internal scale. Other stimulus variables, such as electric current, in which the difference between threshold intensity and a noxious level of intensity is small, have to be stretched to fit the reference scale. This Procrustes-phenomenon is reflected in the differences between the values of the characteristic parameter b in Stevens' law $\psi = a\phi^b$. In general it is found that log (range) x $b \approx 1.53$ (Poulton 1973).

Teghtsoonian went a step further: if there is one single internal scale ψ, then one is forced to accept that the internal difference threshold $\frac{\Delta\psi}{\psi}$ (Ekman's fraction) is identical for all physical scales ϕ that are mapped into ψ. Since, according to Stevens' law

$$\frac{\psi + \Delta\psi}{\psi} = (\frac{\phi + \Delta\phi}{\phi})^b ,$$

we expect that

$$\frac{\Delta\psi}{\psi} = (1 + \frac{\Delta\phi}{\phi})^b - 1 = \text{constant}$$

if Teghtsoonian's assumptions are valid.

And indeed, Teghtsoonian (1971) does show that $\frac{\Delta\psi}{\psi}$ is constant and has a value of approximately 0.030, at least for dimensions for which estimates of b and $\frac{\Delta\phi}{\phi}$ are available.

If we can show that $\frac{\Delta\psi}{\psi} = 0.030$ also holds for time intervals, we would have further support for the supposition that duration has the properties of a sensory dimension such as brightness, loudness or electric shock.

In the literature we find values for $\frac{\Delta\phi}{\phi}$ ranging between 0.03 and 0.08, with a distinct clustering around $\frac{\Delta\phi}{\phi}$ = 0.06, for intervals up to about ca.1 sec (Fraisse 1967; Michon 1956, 1967; Treisman 1963). For the short intervals (< 500 msec) where we observed a value b = 0.5 or 0.6 (Fig. 1), we obtain an estimate for $\frac{\Delta\psi}{\psi}$ between 0.030 and 0.035, which is clearly in agreement with Teghtsoonian's hypotheses.

For intervals of more than one second it is much less clear which value should be adopted for $\frac{\Delta\phi}{\phi}$. Some authors maintain that Weber's law $\frac{\Delta\phi}{\phi}$ = c still holds (Treisman 1963), while others provide evidence showing that the Weber fraction $\frac{\Delta\phi}{\phi}$ will increase as a function of interval length.

Whatever the precise relation, none of the studies suggests a decreasing Weber fraction with increasing interval length. At the same time $b \approx 1.0$ for intervals longer than 500 msec. Hence we may draw the conclusion that for such intervals

$$\frac{\Delta\psi}{\psi} = 1 + (\frac{\Delta\phi}{\phi} \mid \frac{\Delta\phi}{\phi} > 0.06)^{1.0} - 1 > 0.06 .$$

This inequality contradicts the assumption of a unique internal reference scale of fixed length, but lends support to the multi-process hypothesis.

On the basis of the assumption that severe disturbance of LTM will lead to a quasi-perceptual, impressionistic evaluation of long intervals, we would expect that in the patient H.M. the Weber fraction $\frac{\Delta\phi}{\phi}$ for such intervals would be of the order of 0.06. No such data are available at present.

5. THE DECAY OF THE MEMORY TRACE

The square root of t has played an important role in the previous sections: it looks as if experienced duration under certain conditions does not increase with physical duration t, but with $\sqrt{t}$. Also it has been established experimentally that the dispersion in judgment or *reproduction* will grow with $\sqrt{t}$. We need some elucidation on these two effects, and this requires a closer look at some properties of memory, in particular the decay of memory traces.

First, we have to decide what is meant by memory trace when we are dealing with temporal information. Ornstein (1969) has provided a framework for the cognitive theory of memory for time intervals, by defining the memory trace in terms of number and complexity of events. Actually this thought reaches quite far back in time, but Ornstein showed that the experimental manipulation of size and complexity of the memory contents relating to a particular series of events will produce predictable changes in *remembered* duration. In the present context the question therefore is, whether the spontaneous decay of such cognitive memory-elements will proceed with $\sqrt{t}$. In more direct terms: if a

Fig. 4. Experimental paradigm for recency judgments. Whenever a letter appears a second time in a sequence of "randomly"
chosen letters, the subject estimates the number of letters
presented in between.

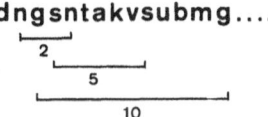

time t elapses between the presentation of an element X and the moment at which the retention of this element is tested, will the probability that X is being recalled be proportional with $\sqrt{t}$?

The answer to this question seems to be contained in a number of studies about the so called judg-ment of recency (e.g. see Hinrichs 1971). These investigations deal with the ability to judge how many elements of a series *(n)* are inserted between the first and second presentation of a particular element (Figure 4).

Hinrichs (1971) has developed a model to account for the results of these experiments, in which the strength of a decaying memory trace is tested against interval time criteria, and in which this trace strength is expressed in terms of the dispersion in the judgment of the number of interspersed items n. The author undertook to fit to his data two common decay functions, one in which memory strength $S(n) = Me^{-kn}$ and one in which $S(n) = a \log n + b$ (it will be clear that n in this case is equivalent with time t). Both functions, by and large, provide an almost equivalent fit to the observed data (Figure 5).

Fig. 5. Relation between memory strength and number of interspersed letters in a recency judgment experiment. Results (o) from
Hinrichs (1971) and predicted decay according to three models.

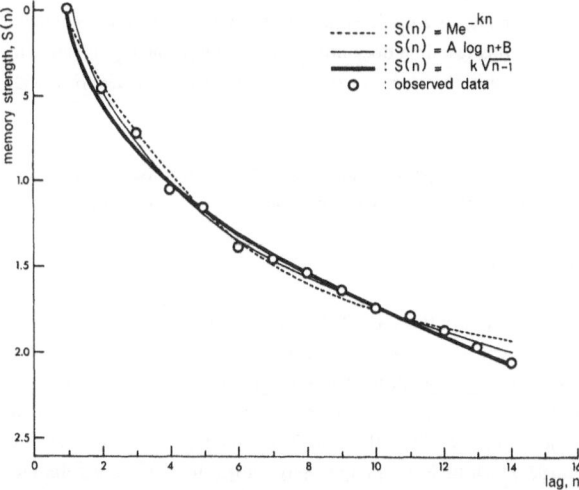

However, there is a different model which will fit the data almost as well. As is shown in Fig. 5 $S(n) = k \sqrt{(n-1)}$, with $k = 0.57$ will describe Hinrichs' results quite well.

In the light of Hinrichs' own assumptions the latter model is, in fact, quite plausible. Recency judgments are evidently not always correct, and Hinrichs' model takes this variability explicitly into account. His assumptions require that the variation in judgment will increase with time (i.e. the number of intervening items n), and inversely proportional with the decay function adopted.

If we assume that this decay function is $S(t) = k \sqrt{t}$ (or $S(n) = (k\sqrt{n} - 1)$) then, by implication, the variability of judgments about intervals of length t (or n) will increase by $\sqrt{t}$. And that implication is supported quite extensively. Partly the evidence has been mentioned before: Creelman (1962), Michon (1967), and Wing (1973), among others, use a memory factor $\sqrt{t}$ to account for the variability of estimation, production or reproduction of intervals. More direct and definitive evidence derives from an extensive study by Wickelgren, who demonstrated that memory decay for learned (symbolic) information will proceed with $\sqrt{t}$ for periods in the range of 40 sec to several years (Wickelgren 1972). In his study he covered a vast array of tasks from the learning of nonsense syllables to the retention of vocabulary in a foreign language.

6. SUMMARY AND DISCUSSION

The preceding arguments can be summarized to present the following picture.

The evaluation of the duration of time intervals by subjects is determined by properties of human memory. Intervals of less than 500 msec are under the influence of immediate memory, a pre-cognitive buffer store for incoming information. Consequently, evaluation of intervals in this range bears a distinctly sensory character. For longer intervals the determining factor is the storage and interpretation of the events which took place during the interval. The distinction between short and long term memory will only become apparent when the normal functions of memory are disturbed. In normal subjects rescaling of longer intervals (> 500 msec) will take place on the basis of the experience with the social-physical time frame. As a result estimates of duration will approximately be proportional to physical duration. When this rescaling does not occur a proportionality of estimates with $\sqrt{t}$ will be observed. This relation is also observable in the decay of memory traces for a wide range of learned materials.

The connections between time experience and memory we have discussed are admittedly speculative to some extent. In particular, three of our assumptions may raise some doubts.

6.1. Effects of Stimulus Range

In several recent publications Poulton (1969, 1973) has pointed out that repeated measurements on the same subjects in one experiment will introduce a judgment reference scale whose length is dependent on the range of the set of stimuli used in the experiment. As a result differences in the value of the parameter b in Stevens' law will be observed, depending on whether stimuli are close or far apart on the physical scale. If this is indeed the case, Teghtsoonian's analysis is undermined: the unique internal reference scale which he is ascribing to physiologically determined limitations of the human perceptual system is degraded into an experimental artifact. The discussion regarding this point is not yet closed, and Teghtsoonian (1971) presents a number of counter-arguments that appear to soften the acute criticism of Poulton.

It should be added that the assumptions we made about rescaling of long intervals in normal subjects touch upon the same problem.

6.2. The Memory Trace Decay Process

The status of memory trace theory is somewhat unclear. Many recent models consider memory decay in terms of an all-or-none process, in which any memory event will suddenly become inaccessible with a given probability, instead of looking at it as a gradual weakening of a brain process. Trace theory as treated by Hinrichs (1971) or Wickelgren (1970, 1972) however, admits both conceptions. The remaining question is: what status as a model has the relation $S(t) = k \sqrt{t}$? Models postulating an exponential decay $S(t) = Me^{-kt}$ predict a constant proportional reduction of the memory contents present at any particular instant. Such models do not predict a square root relation for estimated duration or to the variability of the estimates. Wickelgren's model for long term memory, however, does incorporate square root decay in a rather plausible way (Wickelgren 1972). In his model the strength of the memory trace is determined by two parameters: the *decay* due to "detrimental" factors such as interference, and a continuously increasing resistance of the trace against further decay. In fact, this model may predict the observed values in recency judgments quite well, as indicated by Wickelgren.

6.3. Effects of the Experimental Paradigm

In our discussion we have neglected to make a distinction between the various ways in which a judgment about an interval can be achieved. Recently Vroon (1973) showed what the consequences are that follow from particular experimental paradigms. In a further and more detailed analysis of the relation between time and memory processes it will be necessary to take such consequences into account.

REFERENCES

Block, R.A.: *Memory and the Experience of Duration in Retrospect.* Dissertation. University of Oregon 1973.

Creelman, C.D.: *Human Discrimination of Auditory Duration.* Journal of the Acoustical Society of America. 34 (1962) 582-93.

Crombag, H.F.M., E.E.Ch.I. Roskam, and W.A.T. Meuwese: *Het meten van studie-belasting.* Leiden: Bureau Onderzoek van Onderwijs 1973. Memorandum Nr. 212-73.

Deelman, B.G.: *Etudes in de Neuropsychologie.* Dissertation. Rijksuniversiteit Groningen 1972.

Fraisse, P: *Les structures rythmiques.* Louvain: Studia Psychologica 1956.

————. *Psychologie du temps.* Paris: Presses Universitaires de France 1967. 2nd Ed.

Hinrichs, J. V.: *A Two-Process Memory-Strength Theory for Judgment of Recency.* Psychological Review. 70 (1970) 223-33.

Lindsay, P.H. and D.A. Norman: *Human Information Processing.* New York: Academic Press 1972.

Michon, J.A.: *Timing in Temporal Tracking.* Assen: Van Gorcum 1967.

————. *Magnitude scaling of short durations with closely spaced stimuli.* Psychonomic Science. 9 (1967b) 359-60.

————. *Processing of temporal information and the cognitive theory of time experience.* In: J.T. Fraser, F.C. Haber and G.H. Muller (eds) *The Study of Time.* Heidelberg: Springer 1972.

Neisser, V.: *Cognitive Psychology.* New York: Appleton-Century-Crofts 1967.

Norman, D.A. (ed.): *Models of Human Memory.* New York: Academic Press 1970.

Ornstein, R.E.: *On the Experience of Time.* Harmondsworth: Penguin Books 1969.

Poulton, E.C.: *The new psychophysics: six models for magnitude estimation.* Psychological Bulletin. 69 (1968) 1-19.

————. *Unwanted range effects from using within-subject experimental designs.* Psychological Bulletin 80(1973) 113-21.

Richards, W.: *Time reproductions by H.M.* Acta Psychologica. 37 (1973).

Scoville, W.B. and B. Milner: *Loss of recent memory after bilateral hippocampal lesions.* Journal of Neurology, Neurosurgery and Psychiatry. 20 (1957) 11-21.

Shiffrin, R.M. and R.C. Atkinson: *Storage and retrieval processes in long-term memory.* Psychological Review 69 (1969), 179-204.

Teghtsoonian, R.: *On the exponents in Stevens' law and the constant in Ekman's law.* Psychological Review. 71(1971) 71-80.

Treisman, M.: *Temporal discrimination and the indifference interval: implications for a model of the "internal clock".* Psychological Monographs. 77 (1963) Nr. 13 (whole Nr. 576).

Vroon, P.A.: *Enkele psychofysische en cognitieve aspecten van de tijdzin.* Dissertation. University of Utrecht 1972.

———. *Some process variables in sequential time estimations.* Psychologisch Laboratorium Utrecht (1973) rapport nr. 7.

Wickelgren, W.A.: *Sparing of short-term memory in an amnesic patient: implications for strength theory of memory.* Neuropsychologia. 6 (1968) 235-44.

———. *Trace resistance and the decay of long-term memory.* Journal of Mathematical Psychology 9 (1972) 418-55.

———. *Multitrace strength theory.* In: D.A. Norman (ed.) *Models of Human Memory.* New York: Academic Press 1970.

Wickelgren, W.A. and K.M. Berian: *Dual trace theory and the consolidation of long-term memory.* Journal of Mathematical Psychology. 8 (1971) 404-17.

Wing, A.M.: *The Timing of Interresponse Intervals by Human Subjects.* Dissertation. McMaster University, Hamilton, Ontario 1973.

Woodrow, H.: *Time perception.* In: S.S. Stevens (ed.) Handbook of Experimental Psychology. New York: Wiley 1951.

Time and the Structure of Human Cognition

M. TODA

Occasionally, it hits me how marvelous a creature Man is. Among all his numerous talents, the one that fascinates me most is his ability to create abstract notions out of something which he does not really understand. The emergence of the abstract notion of time is obviously the case in point. No one, apparently, can claim to know what time is. Nevertheless, there is this brave breed of people called physicists, who used this elusive notion as one of the basic building blocks of their theory, and miraculously, the theory worked. When one of the leading figures of the clan, by the name of Albert Einstein, quietly mumbled his secret incantation which sounded like "Combine time with space in such a way that nothing can travel faster than the speed of light, then mass is equal to energy," lo and behold, atoms exploded ever so noisily.

Is it just Nature's favoritism? Has she ordered Time to play possum when it is under the hands of physicists? Because our intuitive feeling certainly tells us that time is something we live with, something far livelier than the abstract mathematical entity that physicists refer to in their theory by a single letter label "t". Still, no one who lives in the present technological world would be able to deny the tremendous power that this simple t possesses. If it is not just due to a divine grace bestowed upon physicists, then what or who is responsible for the miracle? I would not give a damn to the claim some physicists occasionally make: *the victory of logical reasoning!* How can one be logical about something which he does not really understand? Or, ...

Maybe I am wrong. Maybe one can be logical only with something he does not fully comprehend. Just because he does not know what it is, he can take a detached attitude toward it, and may try cool-headedly to confine it within a sterile, break-safe capsule, which is alternatively called the *abstract notion*. If he is lucky and succeeds in trapping something really important within the capsule, then all the power of logic is at his service; he may stretch the capsule from infinity to infinity, map it onto the number continuum and build an atom-shattering monstrous theory upon it. There is obviously no doubt that physicists succeeded in trapping some really important ingredient of time within their capsule labeled t , but equally certainly it is not all of the time that is captured within their capsule. Our intuition is crying out to tell us that time is something that flows unlike the physical time which is frozen still. Not only just being frozen still, the physicists' "t" would be quite at a loss which way to flow if it is thawed and ordered to flow.

On the other hand, try and see if you can make your child into an abstract notion. Or your wife, husband or anyone you know very well, for that matter. We cannot do it just because we know them too well. And, in natural consequence, we can hardly deal with them with pure, cold logic alone.[2]Even though to a lesser degree, knowing the object of investigation too well on an intuitive level has been an unfailing source of difficulty in psychology.

The purpose of my writing this paper is neither to solve the mystery of man's ability of capsule-making or to clarify what is left out by the physicists' t-capsule, which are both undertakings far beyond my capability. What I intend to do in this paper is to tackle a problem concerning the primary functioning of the human cognitive system, and its temporal characteristics in particular. Such a problem contains a mystery of no lesser degree than those in the above-mentioned two; however, we have more clues here. And, through this attempt, I hope possibly we may obtain some useful hints relevant to the others.

The notion of *cognitive system* is now being used in many different contexts, not necessarily with the same definition. There seems to be, however, a general understanding that it means an internal representation of man's external environment, including himself at the center. There is no doubt that every individual has one such *system,* a coherent (or nearly coherent) body of structural internal schemata, without which behavior can hardly be organized in a coherent way, and that is an essential feature of consistent adaptation. Note, however, that nothing like a set of still "images" (stereoscopic or not) of the surrounding world or a more abstract Weltanschauung type of construct should be identified as the cognitive system. These things belong, if they do belong to one's cognitive system at all, to a functional level high in the organizational strata of the cognitive system, located at the exhibition gallery of what a cognitive system can ultimately create, where what you can see are mostly capsules and very little activities. As you go down the levels, you will see gradually more and more activities, and, naturally, the busiest part will be found in the basement section, the place which will be called the *primary cognitive subsystem* or the *primary system* for short.

The primary system is, of course, a conventional categorization, as it is entirely impossible to draw a sharp borderline between subsystems of any system as tightly organized as the cognitive system. (Hence I have here another unmistakable example of an abstract notion created out of something beyond true comprehension.) Anyway, by the primary system I mean roughly the part of cognitive system which is in charge of moment-by-moment *actions* of a person, covert as well as overt actions. Perception is one such action, and as we do not know *how* we are perceiving, the actual functioning of the primary system is beyond the reach of direct conscious probes.[3]

As a rule, with any smoothly running system interminably performing basic functions, we are seldom aware even of its existence. The primary system seems operating even in the dreaming stage though in a certain lax way. But there is an observation made by the clinical psychologist Shor, which seems to describe how it is when all the noises made by the primary system are somehow turned off:[4]

> "I had been asleep for a number of hours. My level of body tonus was fairly
> high and my mind clear of dream-images so that I believe I was not asleep but
> rather in some kind of trance-like state. At that time I was neither conscious
> of my personal identity, nor of prior experiences, nor of the external world.
> It was just that out of nowhere I was aware of my own thought processes. I
> did not know, however, that they were thought processes or who I was, even
> that I was an I. There was sheer awareness in isolation from any experiential
> context. It was neither pleasant nor unpleasant; it was not goal directed, just

sheer existing. After a time, 'wondering' started to fill my awareness; that there was more than this, a gap, an emptiness. As soon as this 'wondering' was set into motion there was immediately a change in my awareness. In an instant, as if in a flash, full awareness of myself and reality expanded around me. To say that 'I woke up' or that 'I remembered', while perhaps correct, would miss the point of the experience entirely. The significant thing was that my mind changed fundamentally in that brief instant. In re-discovering myself and the world, something vital had happened; suddenly all specifications of reality had become apparent to me. At one moment my awareness was devoid of all structure and in the next moment I was myself in a multivaried universe of time, space, motion and desire [Shor, 1959]."

Such is a rare experience indeed, but we need not directly experience it ourselves to learn from this observation that we are constantly living within a cognitive *context* which supplies the identification and continuation of oneself in the given environment, the context updated at every moment. Even though the updating may occasionally fail, the context-providing function *of the primary system* is only temporarily suspended, and will usually be resumed soon enough with the help of higher cognitive subsystems to avoid any serious discontinuity. It appears that the strong intuitive conviction of the continuity of oneself is maintained only by virtue of this extremely efficient operation of the primary system.

Although it is easy to say that the primary system constantly provides a quite veridical context, I wish the reader would appreciate first the immensity of this task, taking into account, in particular, the high degree of complexity of the human environment today with which we still seem to be able to cope as individuals. The apparent success of the primary system in carrying out this task seems to suggest that it possesses, at least, the following three major facilities: 1) an enormously good environmental simulation system, 2) a fairly large processing capacity for the oncoming information, and 3) a precise real-time clock. Obviously, 1) and 2) hold a complementary relation; with a greater information processing capacity, one can tolerate a less precise simulation. But there is a limit. Information cannot be processed without a context, or an ensemble for the possible messages, which must, in the first place, be provided by the cognitive simulation. And there are good reasons to believe that, in fact, the information processing capacity of the primary system is not tremendously large.[5] That leads us to a conclusion that the precision of the simulation schemata of the primary system is unbelievably good, a feat that can be achieved only through quite exhaustive processing and utilization of the redundancies observed in our environment − or, to be more precise, the redundancies contained in our sensory inputs coupled with our own bodily motions.

So far, what I have said are not much more than a list of statements. In order to substantiate my arguments, let me reflect for awhile on how the working of the primary system expresses itself in our visual information processing.

While we are awake, we usually keep our eyes open, and they are notorious in not staying still, known to shift their object of fixation about three times a second on a rough average. According to Jacobson, a single eye can handle a potential maximum information of about 4.3 million bits per second [Jacobson, 1951]. So a flood of visual information must be coming all the time through this pair of live windows, an amount enough to drown any honest − by which I mean unprepared − information processing device. How much of this sheer bulk of potential information do we actually utilize? Search your memory. How much of today's visual experience do you retain in your memory? In my case, just a few fleeting fragmentary images like the thinning grey hair of a man I was introduced to, a vague shape of a pretty girl sitting in front of me in the subway, the creamy whiteness of mashed potato at the dinner table, and so on. Perhaps, I could recall more if I really try hard, but not too much more. Apparently, these images are not all that I had to use my eyes for today, since I have survived the day. I went to my office, attended a meeting, then went to a concert and came back home safely without being hit by a car, without falling down the staircase, without bumping into more than one person. Therefore, I must have used the visual information I received, just barely as much as the circumstantial needs dictated, since otherwise something more substantial must remain in memory other than these whimsical snapshot images.

Yesterday, at my graduate seminar, someone mentioned an old piece of experimental apparatus which had been sitting for a long time in the corridor near my office. Surprisingly, no one remembered if it was still there, even though everyone passes the place at least once a day. After a lengthy argument (it was a seminar on cognition, and I assure you that we were not wasting time), we reached the conclusion that it must still be there, as no one noticed anything suggesting otherwise. So we all went to check our hypothesis, and it was confirmed. Needless to say, we must have been *seeing* it everyday, and utilizing the visual information, as no one apparently has bumped into this bulky apparatus lately.

Obviously, this is the type of visual information processing we are engaged in almost every moment without appreciable effort. Automatic. And an information processing can be carried out automatically only if a fixed number of alternative outcomes are spelled out in advance, and for each of which a fixed maneuver is predetermined. In other words, each time we see, we are *expecting* what to see, and what is really seen is just checked against the expectation. To be precise, expectation is a word a bit too strong to describe the operation. It does not reduce uncertainty to zero. You must positively see if a car is not coming before you cross a street. If you see a car coming, or a bicycle, or a pedestrian, or none at all, it will trigger an appropriate action, and you will usually forget the visual experience immediately. If it is some exotic vehicle, however, you may perhaps start a more detailed information processing, and remember the occasion. If it is a dinosaur – It is a pity that we live in a dull, redundant world in which we seldom encounter anything we don't expect at all. But just by virtue of this tremendous redundancy, we can let the primary system handle chores of most of our visual information processing, and it carries them out with utmost ease. Only at occasional moments of encounter with a novelty, meaning low expectation as well as no expectation, the primary system transfers the task up to a higher cognitive subsystem for a real processing.

The "expectation" considered here as an alternative label for the context-providing activity of the primary system should never be mixed up with the more conscious, or higher level, activity of guessing or predicting. As evidenced by the obvious unreliability of the outcomes of guessing or predicting, in sharp contrast to the high reliability of the primary system's expectations, the former is an activity categorically different from the latter, usually dealing with a much longer temporal period extending into the future. Of course, if you come down to the basis of these secondary activities, you will find out that they are, as well as all other cognitive activities, deeply rooted in the functions of the primary system, as I shall discuss later. Note also that we do guess or predict only occasionally, whereas the expectation of the primary system is a constant activity. The difference is somewhat like the one between smoking and breathing. As the activity of breathing is rarely brought into awareness even though it is our body's vital activity of constantly supplying oxygen to the blood stream, we are often even more ignorant of the context-providing activity of the primary system, without which we would immediately suffocate mentally, since no oncoming information can be processed without a context![6]

Of course, it is extremely difficult to prove this point directly, since as long as we are awake our primary sustem will never stop providing a context except in a very rare case, such as reported by Shor. It can be shown, indirectly however, by considering cases when the contexts provided continue being irremediably wrong for a certain period. This is a situation also difficult to create, as the cognitive system as a whole is resourceful, with diverse means to explain away apparently incongruent information. But let me tell you a piece of my own experience.

When I was a graduate student, I went out with my friends to an amusement park one day. We found a little hut there named Surprise House and decided to try it. The inside of the hut was a small room the shape of a cylinder laid horizontally. The walls were painted so as to give the place the appearance of an ordinary room; we all sat on a pair of benches placed there face to face and waited. Then, suddenly, it started. "Surprise" was too modest a word; that instant, I panicked as everyone else apparently did. Perhaps, I was the first among the group to regain my mind. I looked around and saw a girl who wore a beautiful kimono grovelling on the floor quietly mourning, with her fingers feebly scratching the floor panel. I felt a pain on one of my knees and found that a male student sitting next to me held it with all his strength, his face devoid of expression, eyes wide open with pupils dilated. Perhaps I was in panic for only a few seconds, although I felt it an indefinitely long period during which my cognitive system must have been in a frantic turmoil to make out the utterly incomprehensible sense date. I began to "see" things normally again only when it finally sorted out an answer: The cylinder turned constantly with a moderate speed. The benches were a swing system which moved synchronously with the cylinder for awhile, and then swung back in the opposite direction. But the impression you received inside this simple apparatus, unprepared, was just beyond description. It was either that Hell-broke-loose or that the universe-collapsed-around-you. Interestingly enough, my memory of the experience is closely associated with the sense of darkness (the illumination was constant) and an ear-shattering noise (the apparatus made no noise, even though we must have shouted at the top of our voices).

Now let me turn to the problem of time and consider first the temporal mechanism with which the primary system is driven. If we pay attention only to the fact that the primary system constantly provides quite veridical expectations, we might conclude that it must be a work of a very effective real-time simulation device controlled by an extremely precise internal clock. It is true that the primary system must operate on a real-time basis, but note that the exact *real-timeness* may be produced by a constant feedback from the real world as well; one may let the real world keep the timing for the system. Apparently, this is economical, and we may well assume that it is the primary system's way. The updating through feedback would not be sufficient for the primary system entirely to dispense with an internal clock, however. Various known facts concerning perception lead us to believe that the operation of the primary system is in discrete steps, consisting of a few distinct phases, each of which requires a finite time for its completion. For instance, first, the sensory input requires an integration over time; secondly, identification of a sensory message against a given context may be easy or difficult, and the information processing time may vary accordingly; and thirdly, the providing of the next context according to the identified feedback message would consume some finite time. In all, the updating through feedback can be done only at discrete moments with varied intervals, where the variability is further enhanced by movements of sensory organs and shifts of attention. As precise timing is crucial in some types of response control and also in the perception of movement, too much variability in the cognitive processing cycle must be counterbalanced by some kind of smoothly operating timing device. If one calls this an internal clock, then man and animals do have a very good internal clock, though good only for timing of brief moments. Apparently, there seems little theoretical difficulty in assuming some neurophysiological processes which can perform this function.

It is interesting to observe that even at this basic level the primary system's redundancy processing appears to exert a strong influence upon the perception of spatio-temporal events. Many experimental results concerning apparent movement, [e.g., Helson, 1930; Abe, 1935; Cohen, 1967; Aiba, 1973], may lead us to suspect that some laws of conservation must play a vital part in the primary system's operation, which may be one of the system's inborn functions comparable to Hebb's primitive unities [Hebb, 1949]. In contrast, the *secondary system,* by which I mean collectively all the strata of cognitive system higher than primary, comes to acquire conservation laws during each individual's development as demonstrated by Piaget.

When we come to much longer time intervals than those handled by the primary system, there is no evidence for dependable internal clocks. It is well known that many animals possess certain, fairly precise *timers* which set off predetermined activity patterns at the prescribed moments in the recurring cycles of time, and these timers are even adjustable to external cycles of events. But we should not mix up timers with clocks, even though the same word "internal clock" is often loosely applied to both types of internal timing devices. A clock is something that allows the user a context-free evaluation of the passage of time of virtually *any* length within a limit, while a timer is taken to be a context-bound, fixed-interval signaling system.

By saying that we have no dependable internal clocks for long intervals, I do not imply that we have no internal clocks at all for these time intervals. On the contrary, as much of the literature on "psychological time" suggests, we indeed have explicit feelings about the passage of time and can make intuitive temporal judgments. Even rats can make temporal discriminations and control their behavior accordingly as the results in the DRL (Differential Reinforcement of Low rate) experiments clearly show [Farmer & Schoenfeld 1964]. The fact that the findings obtained with man are messy and mutually contradictory indicates only that we have a set of rather ill-built internal clocks whose outcomes are unreliable. This very unreliability, among other things, strongly suggests that these clocks are products of the secondary system, and the secondary system makes them upon demand out of bits and pieces — mostly out of lousy physiological timers. Note that there must be plenty of physiological processes, like breathing, which repeat themselves, and they are lousy timers because they are not originally built as timers and they are context-bound. Still, when necessity arises, the secondary system can make a makeshift clock out of them and count off their cycles. Of course, this is pure speculation, but for that matter, everything is speculation concerning the secondary system, which, in the case of man, is such a tremendous system and hardly any trustwrothy theory has been established concerning its internal structure. However, it might occasionally be worthwhile to let our speculation fly freely, as I allow myself to do in the rest of this paper.

As we move into the realm of the secondary system, I want to make the following points explicit: First, note that a system can have an internal clock without possessing a notion of time. So these two entities, the internal clock and the notion of time, must be kept separate. However, there is a certain relationship between them, and I conjecture that our secondary system would never have acquired a notion of time if the primary system had no internal clock or if the secondary system had a perfect internal clock. To show the plausibility of the first part of this conjecture, let me consider space instead of time. There is no doubt that the primary system is endowed with some kind of "spatial alignment" operator (three-dimensional, though not necessarily Euclidean) to apply to sensory messages and to produce the "spatial" perception. To me, it seems very unlikely, if not totally impossible, for the secondary system to come to bear a fullfledged notion of space without being originally guided by the primary system's spatial alignment operation. Though comparatively less obvious, I think the same must be true with the notion of time.

The second part of the conjecture, that a perfect internal clock must have inhibited the birth of a notion of time, appears more straightforward. Taking for granted that the secondary system develops "notions" and builds conceptual schemata with them, let me enquire into the purpose underlying these activities. I am a holder of the view that the information processing of the cognitive system is not an end itself, but, at least originally, has been meant to serve the purpose of control. This relationship is quite obvious with the primary system, and there is no reason to doubt that the same holds with the secondary system, too. So, if the secondary system happens to come by a perfect internal clock, what needs can there be for a notion of time?[7]

Whether or not these brief arguments satisfy the reader, I must confront an inescapable question: What exactly is it, upon requiring a control by the secondary system, that led the system to create a notion of time? Apparently, everything that comes after external clocks and external calendars is beside the point, as external clocks and calendars cannot be made without a notion of time. Another possibility of the need to regulate interpersonal activity patterns seems also off the mark, since such a need must be preceded by the need to regulate one's own activity pattern over time, i.e., *planning* his behavior beyond the immediate future. To me, this seems the most plausible candidate for the need responsible for the birth of the notion of time, if only one could somehow explain how such an "over time" need could have existed in the first place without a notion of time.

In order to attempt this difficult explanation, I need to discuss briefly the role of (long-term) memory in the cognitive system. As I have mentioned earlier, the primary system needs a powerful simulator to provide veridical contexts nearly all the time. A context it provides offers some, not too large a number of alternative messages about the state of the environment against which the given sensory input is to be matched.[8] If the resultant matching is good enough with one of the alternatives, the primary system goes on to the next cycle. The identified sensory state and the current state of bodily movements become the joint input both to the response control system and to the simulator. Then the simulator automatically goes about providing the next context.

Obviously, the simulator must have an access to the data for simulation, and the data must consist primarily of one's past experiences, processed or unprocessed, which are assumed to locate in a place traditionally called the memory. Note, however, that for the smooth, rapid functioning of the primary system no time can be wasted in the elaborate memory search for information retrieval. How does the primary system's simulator overcome this difficulty? The answer seems simple to me: The memory *is* the simulator. For there is no reason to assume that the human memory is something structureless, a mere storage house. To me, it is all the more plausible to consider memory as a dynamic system itself rather than as a silent heap of information, and we often get a wrong impression about the true nature of memory only because we tend to view it from the angle of the secondary, conscious use of it. Note that the primary system must be much older than the secondary system in evolution, and the memory system must have evolved originally with the evolution of the primary system. Nature would certainly have disfavored the inefficiency of developing two systems separately when it could have achieved the same goal with one. At least conceptually, it seems not too difficult to work out a neural model for dynamic memory, essentially a tree-like structure, which we may hereafter call the *context-tree*.

Now let me return to the role of the secondary system. When the input matching by the primary system fails to produce an exact matching, the secondary system is called in to take over the information processing. There appear to be two cases for the primary system's failure.[9] The first kind is obtained when the matching is not exact, but there exists a general, overall agreement between the given and the expected. In this case, the primary system is versatile enough to keep providing contexts while the secondary system works in parallel at the "novel or ambiguous perception." The second case is obtained when the given sensory message is totally out of context, in which case the primary system's context-providing function is suspended and the perception disrupted as in my experience inside the Surprise House, until the secondary process succeeds in finding a proper flow of context congruent with the given sensory process.

In either case, the secondary system's task is to *reset* at the right place the state of the primary system's context-providing function. How does the secondary system go about this task? One thing is clear. The secondary system must be able to trace external processes faster than real-time. Note that sometimes the failure of the primary system may have been caused by making a wrong branching at some node of the context-tree. In this case, the secondary system must retrace the context-tree, find the wrong branching and hurry back to the present moment on the corrected path. Obviously, there is no logical difficulty in this faster than real-time tracing, as the secondary system need not pause on the way to wait for each intermediary context being confirmed. Of course, the secondary system would not always be able to enjoy a job as easy as this one, and it should often have to engage in a wild hunt for clues. In my speculation, the roles of notions, concrete or abstract, and conceptual schemata built with them are primarily to make this wild, rapid hunt systematic and efficient. If I am allowed to use a clumsy analogy, which I am afraid might be misleading rather than enlightening, the fundamental structure of memory, or of the context-tree, is like streets and roads upon which the primary system, an eternal pedestrian, always treads on foot, guided by a floating spirit called sensory feedback. On the other hand, the secondary system, a more modern contraption evolutionarily speaking, is a much faster traveller, can flit around the world either by driving or flying, and is called up by the primary system every once and awhile to hunt for the lost guiding spirit. What I want to imply by this analogy is that, even though the hyperstructure of the conceptual schemata covers the exact same area as the primary structure of memory does, these two structures belong to two separate distinct systems. I want to emphasize, particularly, that we are likely to be greatly mistaken when we deplore the awkwardness of our long-term memory, since it is tantamount to complaining about the inefficiency of the secondary system's wild goose chase utilizing its highways, airline systems, and also, I must add, the almost instantaneous signaling system called association. Not only is this secondary use of memory by the secondary system worth an applause, but also we should always be aware of the almost unfailing efficiency of the primary system's primary use of memory.

So much for the reinstatement of memory. Note that the secondary system's activity is not always tied up by the requests from the primary system. In times of leisure, it will attend to the business of its own, building notions, building schemata, and sometimes making a free run over the context-tree. It is free in the sense that its course is not bound by the sensory feedback. To assume the secondary system to own this last faculty is surely natural, as a more constrained version of this activity is one of the secondary system's routine works for the primary system. The point I want to draw the reader's attention to here is that this free ride is a rudimentary form of prediction.

At this point it is vitally important to recognize that, even though the context-tree is a fairly veridical representation of the spatio-temporal configuration of the external environment, it needs to be veridical only in a topological sense. For any system that works under a constant external feedback, the correct branching is all that is required for taking the right path. Imagine, for example, a map you scribble over a pad of paper to hand to your friend visiting your house. As a rule, such a map is a mess metrically. But as long as the routes are correctly represented topologically, your friend will reach your house all right. Only when your friend asks you how long he has to be on Route so-and-so, do you need to add to your memo an approximate distance.

Indeed, there are occasions when it is desirable to add to your topological cognitive map a supplementary note on spatio-temporal distances. The need to turn the above-mentioned rudimentary prediction into a real one will very well serve as a motivation to create a systematic format under which these supplementary notes acquire consistency. This motivation, *then,* might have created a notion of time somewhere in the long history of evolution. (Note that this "then" stands for a long and elaborate argument that has not yet been made.)

Instead of concluding this paper (no one can conclude time), let me float to a new topic on time. Before writing this paper, I was chatting with my friend, Imai, over a glass of brandy. As usual, I was teasing him with seemingly unsolvable questions, and I asked, "Why do we use metaphors like flowing river and shooting arrow for time, the things which can hardly be used as external clocks?" He mused for awhile, as usual, and a cryptic answer returned to me: "That's it!"

"What's it?"

"External clocks utilize cyclic events, you see. In order to make cyclic sequence of events cyclic, you must have something non-cyclic, uniformly flowing, beneath it. A figure needs a ground. Time is the ground for the clock."

After some information processing, my secondary system decided that it liked this idea. So I bought it, and poured another brandy into his glass.

NOTES

1. The author wishes to acknowledge his gratitude to Dr. Imai, from whom he obtained many important suggestions, to Dr. Aiba and Miss Nakamura for their kind help in preparing this manuscript, and to many of the participants in the Conference for their valuable comments on the oral version of this paper.

2. One should not, however, take too lightly the power of abstract notions in human affairs. People have invented quite a few abstract notions about the other groups of people whom they do not really know personally. Unfortunately, these abstract notions are very often loaded with emotions, and, inasmuch as they are effective as emotion manipulators, they are much less useful for-scientific purposes.

3. I am still in some doubt whether we should characterize consciousness as a subsystem of the cognitive system. It is plain, however, that consciousness is closely related to the executive, high order activities of the cognitive system, and these activities are perhaps too slow to directly interfere with the rapidly, constantly-running basic machinery of the primary system.

4. I found this citation in Neisser's book [Neisser, 1967], and to the book I owe much in writing this paper.

5. It is estimated that it takes about 10 milliseconds to visually identify an English letter [Sperling, 1963]. Though efficiency cannot be judged by such figures alone, this figure is not too impressive.

6. This is one of the most important axioms of information theory, though often overlooked in its application.

7. A similar argument also applies to the notion of space. Note that the object identification is apparently more important to the primary system than the distance estimation so that we have perceptual constancy which facilitates the former and impairs the latter. Therefore, the secondary system is in a similar situation with both space and time; the primary system supplies to it no ideal materials with which to build a perfect measuring device for either space or time.

8. I am speculating here only about dominant features of the primary process, neglecting details. Especially, note that by "matching" no specific cognitive theory like "template matching" is implied.

9. There are other types of failures as well, though not discussed here.

REFERENCES

Abe, S.: "Experimental Study on the Correlation between Time and Space." *Tohoku Psych.* Folia. 3 (1935): 53-68.

Aiba, T.S.: *Apparent Radial Motion in Stroboscopic Illumination.* Paper to be read at U.S. – Japan Seminar on Space and Motion Perception, Honolulu, 1973.

Cohen, J.: *Psychological Time in Health and Disease.* Illinois: Charles C. Thomas 1967.

Farmer, J. and W.N. Schoenfeld: "Inter-reinforcement Times for the Bar Pressing Response of White Rats on Two DRL Schedules." *J. Exp. Animal Behavior* 7 (1964): 119-22.

Hebb, D.O.: *The Organization of Behavior.* New York: Wiley 1949.

Helson, H.: "The Tau-effect – an Example of Psychological Relativity." *Science* 71 (1930): 536-37.

Jacobson, H.: "The Information Capacity of the Human Eye." *Science* 113 (1951): 292-93.

Neisser, U.: *Cognitive Psychology.* New York: Appleton-Century-Crofts 1967.

Shor, R.E.: "Hypnosis and the Concept of the Generalized Reality-orientation." *Amer. J. Psychother.* 13 (1959): 582-602.

Sperling, G.: "A Model for Visual Memory Tasks." *Human Factors* 5 (1963): 19-31.

X. SOCIETY

Time Structuring and Time Measurement: On the Interrelation Between Timekeepers and Social Time

H. NOWOTNY

I. INTRODUCTION

At first sight the interrelation between the two main themes of this paper, time structuring and time measurement, seems to be simple enough. Time is something that we measure and that we measure *with*. But what is it that we measure and how is it constructed that we come to think of it as being measurable? As Leach has pointed out, in any society the prevailing ideas about the nature of time and space are closely linked up with the kinds of measuring scales which are thought to be appropriate. If we alter the scales and dimensions with which we measure, we seem to alter the nature of that which is being measured, as well.

Our scientific culture makes us believe that we stand on firm, objective ground insofar as we know what it is that we measure. The prevailing conception of time is largely one of a unitary time, which can be broken down into sub-units of equal, measurable duration. If we follow Leach and wish to draw conclusions from the nature of the scales that we use to measure time with, to our prevailing conceptions about time, then it appears to be significant that we operate today with more precise scales, but with far fewer dimensions than ever before. The history of timekeepers, as told in the prior papers, suggests that this has not always been the case. In the absence of a common denominator or medium in which all values can be expressed in terms of interchangeable numbers, men have operated with many time dimensions and with different scales of measurement.

The history of time-keeping, however suggests something more: in their time-keeping operations, men have always used a model, a hypothetical, external reference scale upon which they could map their conception of what time is. Examples are the movements of heavenly bodies, the natural cycles of agriculture, and the presumed activities of the gods. Today, as befits an age dominated by scientific thinking, precision and unambiguity,   we have agreed to the definition of a standard second in terms of a spectroscopic frequency. The impact of this uncritical quest for physical precision is evident in many facets of our daily lives.

The preference for the physical-scientific methods of time-keeping notwithstanding, there remains current among users of clocks an intuitive feeling about the measurement of time based on personal and

---

*There is a certain addictive quality in the preoccupation with time. Like any true addict, I encountered others whose visions I was fortunate to share and whose intellectual generosity I wish to acknowledge here: Edmund Leach, Otthein Rammstedt, and especially Manfred Schmutzer.

social experience. From a sociological point of view, the physical, "external" time, to which industrialized society tends to refer, becomes a social construct whose nature is somewhat arbitrary and which can be shown to vary in accordance with certain other, dominant features of the general socio-economic conditions of a society. We shall call this inter-subjective, common, time experience which is shared by individuals living under similar social circumstances, social time. Apart from the subjective psychological experiences, grounded perhaps on some information processing mechanisms common to all human organisms (Michon, Toda, Ornstein), there exists however an inter-subjective, social experience of time. We can even go one step further: social time comes into existence through the processes of social interaction on the behavioural as well as on the symbolic level (Sorokin & Merton, W. Moore, Munn). In its most obvious manifestation, the social component becomes evident through the need to coordinate social behaviour and interaction. Indeed, many of our actions are not meaningful in terms of time except when other people are involved. But human interaction is not carried out on the behavioural level alone. What is additionally involved and interwoven into the behavioural fabric are processes of symbolic interaction. In the history of social time, these processes of symbolic interaction can be exemplified by the different conceptions of inter-generational time cycles, and by norms regulating interpersonal temporal behaviour, such as 'moving ahead of each other' (Munn). Not only are relations among individuals, but also relations between individuals and their gods and other idols are involved. Here the link to timekeepers is quite obvious. Very early clocks were probably not used primarily to keep time to facilitate and coordinate social activities, but served as devices made to reflect the orderly sequence of cosmic events, partially reflected by happenings on earth (de Solla Price).

The functional components of social time, born out of necessity and developed in accordance with the needs of modern society, appear to be obvious. But the symbolic processes of time structuring, which are more intriguing, more subtle, and far more consequential in their social manifestations are unobvious. We wish to explore, therefore, some of the mechanisms through which social time is structured. The term "social time" refers to the experience of inter-subjective time created through social interaction, both on the behavioural and symbolic plane. In accordance with its communal nature, different societies and groups within a society develop specific, variable forms of social time. The underlying social processes through which the forms of social time are generated are referred to by the literature as 'time structuring'. It is of interest to note that certain distinctions, identified by psychologists as operating on the level of individual consciousness, appear to have social correlates, moulded by social factors. Thus, for instance, past, present and future may be found on the individual as well as on the social level but with differing significance. In their social forms, the drawing of boundaries between these segments of time, and the meaning attributed to them, as well as the relative weight they are accorded in general, can be shown to be the outcome of social conditions and processes. Societies, or groups within societies, can be found that are primarily past- or future-oriented, as will be shown below. Short-term and long-term time perspectives or time horizons are another example of time categories which appear, but differ in their individual and social significance. There seems to exist a physiological basis for a distinction between short-term and long-term memory (Michon, Deffenbacher & Brown). But whether a society, or a group within it, possesses primarily a short-term or long-term time perspective is a socially and not a physiologically determined fact.

In the context of the present paper we shall content ourselves with an examination of some of the underlying social mechanisms illustrated by these two examples: how do societies, and groups within them, structure social time with regard to (a) the distinction between short-term and long-term time perspective, and (b) the distinction and relative weight accorded to past, present and future. After a brief critical review of the literature, we shall turn to a discussion of some fundamental aspects of time structuring as exemplified by these distinctions and will return later to the interrelation between time structuring and time measurement.

II. THE POOR HAVE LESS TIME: ON TIME HORIZONS AND TIME ORIENTATIONS

An impressive number of research findings on the study of different time horizons and time orientations, especially among different groups within a society, could be summarized in these words: the poor have less time; they tend to live in the present. Other findings suggest two other simple conclusions: industrialized societies tend to be future-oriented; non-industrialized ones are past-oriented.[1]

In one of the earliest studies on time and society, LeShan has given the following description of differences in time horizons among different social classes in the United States:

> In the lower-lower class, the orientation is one of quick sequences of tension and relief. One does not frustrate oneself for long periods or plan action with goals far in the future. The future generally is an indefinite, vague, diffuse region and its rewards and punishments are too uncertain to have such motivating value ...

> In the upper-lower, middle and lower-upper classes, the orientation is one of much longer tension-relief sequences. As the individual grows older, he plans further and further into the future and acts on these plans ...

> In the upper-upper class, the individual sees himself as part of a sequence of several or more generations, and the orientation is backward to the past ... (LeShan 1952: 589)

These findings, by and large, have been confirmed by others, as may be seen in the thorough survey by Moenks. The differences which have been found seem to run straight through various areas of cognitive performance. Fraisse quotes LeShan's report that middle class children who have been asked to invent stories tell stories that cover a longer period than those invented by working class children. Bernstein has demonstrated that the time span of anticipation of the working class child differs from that of a middle class child. These differences are already embedded in the sophisticated ways in which linguistic codes are employed by each social class. Barndt and Johnson have shown that delinquent boys have shorter future time perspectives than a control group of non-delinquents. Nuttin, in a general attempt to link differences in future perspectives to differences in sex and age, has proposed as a working hypothesis "that the depth of the future time perspective in human motivation is not primarily related to age and to differences in age as such, but rather to the nature of the behavioral plans and tasks and to the social structure in which these plans and tasks are embedded." (1973: 72)

It must be pointed out with Rammstedt, however, that many studies on time differentials among different social groups, and on deferred gratification, can be criticized from a sociological point of view, not only on the basis of their often inadequate and superficial methodology (a criticism which cannot be pursued further here), but also on the ground that their authors often became the victims of their own position in the social structure. This finds expression in their frequent assumption that a long-term perspective, until recently found mainly in industrialized countries in a middle-class milieu, constitutes the 'normal' socio-psychological standard against which 'deviations' can be measured. In the literature on deferred gratification (e.g. Schneider & Lysgaard, Singer, Wilensky & McGraven), it is too often assumed that there exists a unitary future, one which is equally valid for all social classes and totally unrelated to the specific socio-economic and political situation in which the individual finds himself. It has been shown by more sensitive social scientists (e.g. Lewis; Miller, Riesman & Seagull ; Ortiz) that a short-term time perspective, such as is held by certain cultural minorities, the working class and some segments of the peasantry is far from exhibiting the 'irrationality' often imputed to them by middle class social scientists. On the contrary, it often constitutes the only rational strategy to survive in an environment which is to a high degree uncertain, loaded with risks that are beyond the individual's control and influence, and about which only a minimum of information is available. Where the future prospects are dim and the very idea of future induces nothing but a state of anxiety, as was the case for a large proportion of the working class in the 19th century (Rèzsohazy 1957) – what need or reward is there to lead one to an expansion of the temporal horizon?

Yet, a reference to the objective conditions alone does not constitute adequate explanation for the phenomenon of different time structures. Socio-economic conditions, like any other 'objective' feature of the environment, are mediated through cognitive structures and representations as well as through symbolic processes of interaction and may lead to the creation of what appear to be exceptions. Sufficient empirical evidence exists to show that neither working class individuals nor members of cultural minorities lack the ability to transcend the immediate present with its needs and frustrations. Among the many counter-indicators, we find, for instance, that elaborate provisions for specific events are sometimes undertaken by members of these groups. Such events may be the individual's own funeral or a daughter's marriage, which are often connected with great expenditures; such preparations can bring forth attitudes toward savings that rival those extolled by the Protestant Ethic (Holmberg). Another example would be that of revolutionary movements which can take on the character of millenarian movements, especially for the lower strata of a society (Talmon, Worsley). This again suggests that members of the lower social strata are capable of developing long range time perspectives.

Nor can it be maintained that traditional, agricultural societies show no concern with their future, although this concern is likely to take on a form which differs from the one familiar to an industrialized society. Bourdieu (1963), in a study of North African peasants, had asked: how is it possible that the inhibition against scrutinizing the future sanctioned by custom and religion alike is reconciled in practice with such economic necessities as making provisions for a bad harvest or making decisions which inevitably lead to activities located in the future. Bourdieu's research led him to the conclusion that economic decisions and planning activities among the peasants of Kabylia are based on traditional models of behaviour which have been transmitted from the past and which serve to organize the future as well. It cannot be said, therefore, that future orientation does not exist in this type of society, although it is likely to take on a different form and is closely modelled after experiences that have been made in the past.

In a similar vein, evidence accumulated by anthropologists suggests that the time spans covered by belief systems in traditional societies are often rather extended ones, even if predominantly they are oriented towards the past. But the function of this orientation is also often tied to the present. Genealogies, for instance, dating back several generations, are minutely organized and cared for, but they have been shown to serve not primarily as a store for oral history, or as a backbone for societal identity, but mainly as a basis for legitimizing the claims to power of those who live now (Goody & Watt).

This brief excursion into the literature warns us against jumping too rapidly to unwarranted conclusions. The evidence does not permit any simple causal explanation, such as the reduction of differences in time perspectives and orientation directly to socio-economic con: tions. It confirms, however, that such differences exist, both with regard to time orientation of different types of societies (notably between industrialized and non-industrialized ones) and time horizons of different groups within a society. These, however, are not given once and for all, but tend to vary. In the following we shall attempt to provide a more adequate explanation for these variations by raising first some fundamental issues regarding the mechanisms through which social time is structured.

III. THE VALUE OF TIME: ABUNDANCE AND SCARCITY

The most fundamental dividing line between modern, industrialized societies, and traditional, non-industrialized ones, appears to lie in the *value* accorded to time as such. Under certain conditions, time may be judged as valuable or as having no intrinsic value, as being relatively scarce or abundant. Comparative studies on the uses of time, especially as carried out with the help of time-budget studies (Szalai), have brought forth evidence that shows that in economically less developed countries time is not being accorded the same value as in highly developed countries. Research carried out on the Ivory Coast by Vercauteren illustrates this very well. In Korhogo, an urban center with predominantly commercial and administrative functions, the merchants are well aware of the value of money, but not of time. They display a considerable amount of commercial spirit and employ astute means to attract clients. Thus, they do understand the causal linkage between persuasion and earnings. However, no evident link may be found between the way they organize their activities in terms of time, and the amount of money they can earn. This is in sharp contrast with comparable attitudes found in industrialized countries. This work appears to constitute for them still a multi-functional activity not (yet) geared exclusively towards productivity.

This and similar evidence (Szalai) suggest a link between the value accorded to time and the experience of its relative scarcity or abundance. In a traditional society, time does not possess much value as such. The members of a traditional society can be said to "possess" plenty of time, especially as compared with the harassed individuals of an industrialized society. This observation is supported by studies which have focused on certain "losses" or "gains" in value of time. In a classical study on the effects of unemployment (Lazarsfeld et al.) it was shown how all of a sudden time lost its value for the unemployed, although they had now plenty of it – or maybe because of this. In a study on the effects of prolonged hospital treatment, Roth showed how time gained suddenly in value and how 'time out' acquired new qualities. It became a sort of prized commodity, for which patients would start to bargain with their doctors.

We are thus led to the tentative conclusion that time will be accorded value whenever it comes to be considered rare from the point of view of the individual. But this is likely to remain a purely subjective experience, as long as the meaning attributed to time is not shared by other members of the society or group. Temporal experiences, as Bourdieu (1972) has pointed out, have to be brought together developing correlations between temporal symbols and other mediating devices before they can be used in subtle and highly skilled ways as strategies to build up social relationships. To the extent that these are temporal experiences taken as given, and are regarded as external constraints, they may acquire the character of something sacred as Durkheim has shown. The sacred, however, cannot be meddled with,

or be acquired by humans. As long as the society is the "keeper of time" through its macro-level institutions and its "high priests", the question of the individual member according value to time does not
even arise. All that human activities can do then is to conform to these sacred, societal patterns and to
fit themselves into the "temporal niches" that have been provided for them. In most societies strong
social norms can be found which regulate the use of time when and how it is to be filled and with what
kind of activities. Our present day calendars are relics of what used to be elaborate societal time-
tables, telling the individual what the right moments were for the performance of certain acts, to be
embedded in the constant flow of sacred, macro-societal time. Many elaborate precautions and efforts
were undertaken, in order to find out precisely what these right moments were — be it in order to
wage war, to marry or to beget children.

Needham, Ling and Price show the extensive efforts made by the Chinese ruling dynasty in the 11th
century, designed to identify the right moment for proper succession. The Chinese emperor not only
possessed the most sophisticated clocks available at the time, but he lived according to a schedule putting forth the order in which he was to spend the 365 nights of the year with his 121 wives. On more
mundane levels, the same concern reappeared in many different forms, all directed to find out how
human activities could best be woven into the pre-determined, temporal patterns or schedules of societal time. As long as society is the keeper of time, and time was bound up with the whole life of a
society, there can be no question of time's being valuable or valueless for an individual, for the determination of the value of time is beyond his control.

We have noted above that the value accorded to time separates industrialized from non-industrialized
societies. We have seen now that in societies in which a conception of sacred, societal time dominates
the question of an individual according value or no value to time, does not pose itself. Rather, the
relation of individual to societal time is one of following the right order, of finding the temporal niche
into which one could place the activities over which individual control could be exerted. But how,
then, did the change come about, leading to the phenomenon of regarding time as a prized commodity,
one which is on the verge of replacing money?

IV. TIME AND ECONOMICS: THE EXTENSION OF THE SOCIAL PRESENT

There is much convincing evidence to support the hypothesis that time acquired its value, in the sense
in which this is understood today, through that change in the history of economic development, when
time was discovered as a factor in productivity. It was the moment when time came to be used as a
medium in which more of something could be produced (Rezsohazy 1970). The secular value accorded to time was first, and primarily an economic value, or a value that could easily be translated into
economic terms. It was not only the fact that the cognitive representation of the act of production
presupposes an ordering of activities in a sequence. This alone would not have been sufficient to enhance the value of time. When the "multiplier effect" of the machine was added, time came to be seen
as a productivity factor and hence something that was — at least in principle — bound to become
scarce. Time became the medium in which human activities, especially economic activities, could be
stepped up to a previously unimagined rate of growth. The accounts of social historians in the early
days of the industrial revolution are very telling in this respect and vividly demonstrate the rapid emergence of a whole new set of attitudes vis-a-vis time. Time became a symbol for the production of
economic wealth and was treated like a valuable object in itself, as Benjamin Franklin so bluntly put it

in his advice given to a young tradesman. Time could be acquired by an individual like any other object; it could be saved or unwisely spent, and it was to be invested properly, so that it could not get lost. The notion of progress that cropped up in this connection pointed to the possibility of unlimited wealth and betterment of human life in the future. Time was no longer considered to be something sacred and given, at best reproducable through the "myths of eternal return", as Eliade has described them, but became equated with an economic object, whose production it symbolized; it became possible to "make more out of one's time". Timekeepers, according to Rezsohazy (1970), became at that time the regulators and controllers of action. Work was paid according to the time spent at it, and timekeepers were the quantifiers which transformed an activity into its monetary value.

Today, in most industrialized societies, time is highly valued and considered to be more scarce than ever. Scarcity of time is on the verge of replacing the scarcity of money, not only for those individuals who have enough money already, but for the economic system of whole societies. Economic activity is increasingly measured by the number of hours it takes to produce certain goods, and this serves as an apparently sound basis for comparing the economic standards of different countries. Every sense of crisis, so often connected with the fear of scarcity, also induces in us a feeling that "time is running out", as though it were limited, just as money used to be before inflation became rampant. If one inquires into the reasons why time has become so scarce and highly valued, the answer is this: time has become the medium in which the results of production may accumulate. Yet, a strange, complementary relationship seems to exist: the more one wants to produce, the more time becomes scarce. Those who produce most have the least of it. It is also no longer necessary to limit time as a medium of production to the production of economic goods alone. It has become a medium of production of all sorts of human, especially social, activities. We want to get to know more and more people; we want to do more and more things. Until recently, we were led to believe that we live in an age of abundance economically-speaking as well as with regard to social activities. So the question arises: is it possible to increase the amount of time available in order to produce more? Is it possible to "produce" time?

Measured against any kind of external reference scale it is quite obvious that the amount of time available to an individual or a society can neither be increased nor decreased. It is possible to produce *in* time, as a medium in which activities take place, but not time itself. And yet, the desire to produce more and to accommodate more and more activities in the time span available has led to a curious phenomenon: we appear somehow to "borrow" time by extending our time horizon into the future. As there is a surplus of wishes and desires, plans and activities that are to be realized and that cannot all be accommodated in the present, the temporal horizon is widened by extending it into the future. We "borrow" time by living already partly in the future. It thus becomes clear under what conditions time is experienced as something which is scarce. Faced with a great number of possibilities, activities and plans that are to be realized, the temporal medium of present time appears as insufficient, hence scarce. It is therefore the discrepancy between the relative abundance of possibilities and the relative scarcity of present time which determines this particular experience of scarcity.

In traditional, non-industrialized societies these problems do not arise. Time appears as given on the societal level, where it is generated through certain macro-temporal, often cyclical activities. Since such activities cannot be altered by the individual, available time will neither increase nor decrease. However, there is a parallel to the wish to expand the present: time, as the medium in which activities are to be realized, is not extended by "borrowing" from the future, but is being "reproduced" by bringing back elements from the past that have been encountered before. It may well be that the myths of eternal return offer an opportunity similar to the one which industrialized societies have created by borrowing from the future, namely, to transcend the immediate present. It has been maintained that to speak of a present makes sense only because there is the individual experiencing it subjectively with his consciousness (Fraisse). To this we might add, that to speak of societal past and future makes sense only because a society or group cannot accommodate all its members' social consciousness in the immediate present. If our assumption is correct, that an extension into past or future occurs because the present cannot accommodate more than a certain share of possibilities, we must then ask why it is that some societies extend their present into the future, and why others extend it into the past?

We have seen that under those conditions in which there exists the prospect of a great number of possibilities to be realized, present time will be experienced as being relatively scarce, and time will be considered as a medium in which at least some of the perceived possibilities can be actualized. In societies where activities and economic goods can be produced more or less at will, where time is bound to become an extremely scarce "commodity" that can be increased only by opening up towards the future. This is achieved by partly living in the future already, by planning it, and through attributing linearity to it. This extension towards the future has its roots in the need for more time, which is in turn a necessary pre-condition for the production of more goods and activities. By contrast, in a non-industrialized society, the possibilities of producing goods and activities do not exist in abundance, nor is production perceived as itself "producable." Hence, time will not be experienced as being scarce, for in relation to what should it be scarce? However, in non-industrialized societies, time is also regarded as a medium, but not one which serves mainly to produce economic goods. It is a medium in which social relations are to be created, maintained and structured. These relations exist among humans as well as between individuals and divine beings. They are structured in a temporal sense insofar as great importance is attached to the performance of acts at the right moment, of knowing not only one's place in the social structure, but also "one's time" in the temporal structure of the society. It may be as important in such a type of society not to be too fast, not to be too much ahead of each other, or not to interfere with the inter-generational cycle, as Munn has shown for the Trobriand Islanders, as it is for us to work harder, in order to have more money with which we can do more things.[2]

V. TEMPORAL STRATEGIES: ACCOMMODATION AND DISPLACEMENT OF SOCIAL NEEDS AND CLAIMS

In our inquiry about the mechanisms of extending the social present into past and future, it is useful to introduce a distinction here which is well established in sociology. I will describe this distinction in an over simplified way. If we ask what it is precisely that human beings want to accommodate in the social present, the answer boils down to three issues: (i) the satisfaction of economic needs, which is largely accomplished through the production, distribution and consumption of economic goods; (ii) the satisfaction of needs for social recognition and prestige that enables an individual to acquire or hold certain positions in a social structure; and finally, (iii) social claims with regard to power, i.e., control over one's fellow men and their activities. These are, in a nutshell, the issues usually summed up under the ideas of class, status and power. It is a truism that a social structure usually does not secure equality for its members in these three domains, for only a fraction of the members will be able to satisfy their needs in a given social present. The others can either resign to their fate of losing out, or they will tend to strive for a higher degree of satisfaction. Another possibility would be that they are granted more satisfaction partially, i.e., in certain periods of their lives, like in their economically productive years. It is impossible to discuss here in detail how the temporal accommodation of needs and claims can be accomplished, and what results different structures of inequality in this respect might have. Instead, we have to concentrate on a few consequences of different ways of accommodating some of these needs and claims in the temporal structure of a society.

Let us begin by inquiring about the conditions under which the social claims to power are displaced from the social present. It is a peculiarity of such claims that they need a legitimizing basis, often located in interpretations of the past, or systematically connected with it, through laws that are to guarantee their continuity. Only power which needs no further legitimation because it is based on coercion exerted in the present, is exclusively rooted in the present. It can be observed that the past of any society is usually arranged in such a way that it can serve to legitimize and support the power claims of those groups or individuals that dominate in the present. If these claims are challenged, if a new group or an individual acquires power, or if other major shifts in the political configuration occur, the past is likely to be re-arranged accordingly. If, on the other hand, there is no room in the present social structure to accommodate the power claims of a new group by assigning them the positions they wish to attain, this group is likely to find itself also prevented from establishing a legitimizing basis in the past. One strategy for such a group may be to abandon the placing of their claims in the future, for they have to skip the present. This may be accomplished, for instance, by referring to certain features of the society's past, claiming that they are neglected in the present and demanding or promising that they will be re-installed in the future. This is a strategy commonly found among revitalistic and restorative movements. Another possibility is to adopt a new course which negates and devalues the past and is thus explicitly directed to the future. This is a strategy found among utopian and radical movements of all sorts, as they want to cut themselves loose from the past and attempt to create a New Man in a New World.

One of the first historical records for the displacement of power claims by a group that lost out in the present may be found in certain writings of the Near East from 300 B.C. onwards (Holl). In these quasi-official writings, the idea of an after-life in which distributive justice takes place appears for the first time. It can be shown that these ideas originated in a class of scribes who lost political influence at the time and who apparently took revenge by painting a future modelled after their ideas. Similar displacement strategies into the future have been observed in millenarian and revolutionary movements, as a rich literature on the subject testifies. The common pattern shows that whenever social claims to power cannot be enacted in the present, a strong tendency exists to transfer them under various forms into the future. Under such circumstances the future also becomes the basis for legitimizing present political action.

An interesting case for displacement into the future is that of the charismatic leader who builds up a political following. Such a leader who may be an empire builder or chief of a tribe may lack the resources to put his plans and visions into practice. He may be observed to embark on a course of action in which he trades the resources possessed by his followers for future promises of rewards (Bailey). This is an intricate game of strategizing in which an immediate reward for support offered by the followers is not possible, so that the charismatic leader has to offer future rewards. It therefore becomes necessary for him to construct a vision of a future, in which the promises that he makes occupy a credible place. Thus, the future becomes a surrogate for a present.

Let us now turn to the satisfaction of economic needs and claims of social recognition. As in the case of the charismatic leader, we shall encounter again an interesting mixture of scarcity of resources – symbolic and material – and their interconnections, as the basis upon which displacements occur. In any traditional society the means of economic production are, in general, very limited; this means that the satisfaction of economic needs has to be largely accommodated through immediate consumption in the present. At all times, it has been possible, however, for a small group to extract an economic surplus which was invested by those in a position to do so, in ways that would strengthen their own claims to power and/or their claims to social recognition (B. Moore). The relics of the resulting conspicuous consumption can still be admired in many parts of the world today, where palaces, temples and cathedrals were not only built to serve the glory of the Almighty, but also to perpetuate the social prestige of the persons of highest status. This brings us to the question of status preservation. The scarce factor in this case is that of symbolic resources, which have to be kept scarce if they are not to lose their symbolic value. Even in traditional societies where status is said to be largely ascribed and not achieved, strategies must be employed in order to preserve positions of high status, as these are constantly threatened by others who also want to attain high status. This need not necessarily occur in an outright, status-competing fashion. In the rigid Indian caste system, where status is normatively prescribed and strongly enforced through religious sanctions, threat of status decline in the highest caste appears in the form of increase of numbers of their own members (Douglas). The Indian Brahmins are led to practice a form of population control largely for reasons connected with status preservation and the intricate ways of manipulating status symbols. Among other groups having high status, one can also often find a pronounced concern with safe-guarding a highly traditional and status-maintaining way of life. What this means in terms of time structuring will be discussed below.

VI. TIME AND SYMBOLIC RESOURCES: LONG-TERM AND SHORT-TERM PERSPECTIVES[3]

It was argued above that those who enjoy positions of high status tend to work to preserve their privileges or power for the future. This can only be accomplished by developing strategies with regard to the preservation of status or power. Such strategies in their turn necessarily depend on and generate long term time perspectives. On the other hand, we also observe that many consumption needs do not require a long-term perspective, as the need vanishes upon being satisfied (though it might return later). Before we can explain how these strategies intermix, we must briefly discuss how status is gained in the first place and how economic goods and status are interrelated.

We start from the assumption that every society consists of a system of unequal positions, that is, some positions are more highly valued than others, and are endowed with different privileges, rewards, and recognition. Status systems appear to be universal phenomena; generally, the status that an individual holds is only partly dependent on the positions where he finds himself, and partly determined through the objects with which he affiliates himself. Objects are to be interpreted here in a broad sense – they can be persons, goods, and even ideas. When an individual associates himself with an object that is highly valued in a particular society or group, symbolic resources are put at his disposal, and he will gain in status. In case he affiliates himself with an object which is negatively valued, he depletes himself in symbolic resources and his status will decline. Nor does the valuation of objects remain static. It is subject to the dynamics of social valuation which itself depends partly on affiliation with other objects or persons, and partly on the frequency with which they appear. Rare objects have great symbolic value, as their information content is high (Schmutzer *v.* Foerster). The probability that they will be co-opted by a person of high status is, therefore, also high.

Thus, we may observe that an abundance of symbolic resources and of objects, carriers of symbolic resources, leads to their inflation. They offer little attraction for those who want to preserve their high status positions, and will therefore either be passed on rapidly or excluded from affiliation with high status persons. This is the result of a very general process of symbolic devaluation that occurs whenever objects with symbolic value, even if this value is initially high, tend to lose it by becoming more frequent. Only rare objects have high symbolic value that can be preserved at the price of keeping them rare.

Based on these considerations, we can now return to the question of how long-term and short-term perspectives are associated with different positions in the social structure. The upper strata, generally composed of persons occupying high status positions, must employ strategies directed to two goals. First, they must find rare and potentially valuable objects with which they can affiliate themselves (search strategies); secondly, they will tend to employ strategies that permit them to keep those objects rare that have high symbolic, status-conferring value. Such strategies lead to various devices through which the symbolic resources can be manipulated. But the creation of strategies necessitates the development of long-term time perspectives. In a situation of economic surplus production, consequences will arise for symbolic resources also. As more new goods appear on the market, the rarity of objects monopolized by the upper strata will constantly be threatened, as these objects will filter down gradually to the lower strata and become widely distributed. The upper strata thus will develop an interest in finding or producing novel objects with high symbolic value with which they can affiliate themselves first.

But, as the process of economic surplus production is not to be halted, the lower strata will be induced to develop a constant desire to obtain new objects which also will confer status on them. This can be achieved by the development of long-term perspectives with regard to the means with which to obtain such goods. The typical case is described by sociologists as that of rising expectations, i.e., the belief in betterment not only in terms of the material existence, but also with regard to a betterment of their status position. This attitude manifests itself in new aspirations for higher education, and also in other ways in which industrialized societies have normatively prescribed means of status achievement. But the lower strata will always lose out in such a process, as long as the upper strata are in a position to manipulate symbolic resources in their own favour. Such manipulation is achieved by keeping status-conferring objects which, at least for some time after their initial appearance, will tend to remain rare.

We can specify now under what conditions non-induced, originally short-term perspectives are likely to be developed in different strata of a society. This is generally the case in all those situations in which there is either no feasibility or no necessity to develop strategies for attaining or maintaining status-conferring symbolic resources. The latter case of no necessity arises among the upper strata, whenever their power or status positions (which we always assume to include the possibility of extracting economic surplus from the lower strata and of converting it into either power or status) are held to be so secure and unchallenged that no need is felt to be concerned with its preservation. To be sure, even in highly hierarchical societies this will happen only for short periods and never wholly so.

The lower strata are left then with short-term time perspectives, as they do not perceive, nor do they objectively have a chance to develop reasonable strategies in order to obtain convertible symbolic or material resources. Only in those circumstances where the means of status gain are prescribed within their own group (as is the case where status could be gained, e.g., by having a beautiful funeral or a lavish marriage ceremony) can we expect to find long term time perspectives. With regard to a status gain relative to, and often at the expense of, the upper strata, however, conditions generally are such that strategies are to no avail, hence not likely to be developed. Thus, it appears that the lower strata have only two possibilities open to them, if they want to develop longer time perspectives. Either they can adopt the long-term perspectives induced in them by the upper strata, thereby voluntarily or involuntarily taking part in the process of symbolic devaluation which works to their own disadvantage; or, they can develop long-term revolutionary strategies, whereby they place their claims to increased status into the future. In content, such revolutionary strategies amount to attempts to abolish the rarity of status-conferring goods, or simply, to occupy the positions held previously by the members of the upper strata. It is, however, difficult to visualize a society in which the rarity of status-conferring goods is completely abolished. If new goods are introduced, they would have to be introduced in great numbers and, preferably, also at the same time. These are conditions which are difficult to meet, as it is doubtful whether productive and distributive mechanisms in any society can be synchronized to a degree that would eliminate time-lags in their distribution. Such truly revolutionary strategies would amount to the abolishment of value-creating mechanism, at least as they are now controlled by those who have high status. However, while probably no society can exist without some mechanism which produces symbolic resources, the invention of more equal production mechanisms does appear possible.

VII. ABUNDANCE OF ECONOMIC RESOURCES: PROLONGED PRESENT AND SHORTENED FUTURE

After this excursion into the realm of the production and management of symbolic resources and their impact on short-term and long-term perspectives, we want to return to the condition which characterizes most our present-day industrialized societies: the relative abundance of economic resources. It was explained earlier that in a situation of a surplus of resources the future, and the extension of the present into the future, acuqire new meaning, for it holds out the promise that more goods may be produced and more activities can be performed than the present can hold. We have seen that this is likely to lead to an extension of the present into the future, into which plans and expectations can be placed. Future possibilities are investigated in more and more detail, bringing the future closer to the present. Compared with the vast domain of possibilities that the future contains, the present appears to shrink to a short interval which passes with increasing rapidity. The future is no longer seen as a mere extension from past and present, as was the case in a situation where resources were still scarce, but turns instead into an open-ended, linear progression which holds out the promise of more, better and even "denser" times. In short, we are describing progress. A common way of experiencing this conception is expressed in the readiness to "believe" in the future and the willingness to exchange the present for it (Bell & Mau).

Yet, at almost the same time, one can observe the appearance of countervailing tendencies which result in shortening again the long-term perspective that has just been developed. The future comes to be seen no longer as an extension of the present, but rather as an already overloaded present, multiplied as it were by some growth factor. The reasons for this change in the qualitative nature of the future seem to be the following. The preoccupation with the future as an extension of an overloaded present leads to a closer interconnection between the two, at least with regard to the near future. The extension of the present into the future also leads to an extension of this future backwards into the present. The more the future is planned, the more it is filled up with activities, wishes, plans and desires, the more it comes to resemble the overloaded present which it was intended to expand. In such a situation new modes of perception are likely to be developed which no longer permit a clear-cut distinction between present and future. As with a busy man's schedule that has been already filled up for months in advance, a society intensely preoccupied with its future is likely to find itself "booked up" in advance and thus has to learn to re-code its present. Thus, we have a two-fold process at work. First, the present becomes extended in its boundaries by "borrowing" time in which more can be produced. This leads necessarily to long-term societal perspectives. However, as the future is filled up by planning, new boundaries appear that mark off a more extended present from a future that fuses with it. This fusion results in a new kind of extended, yet short-term societal perspective.

In this connection it is instructive to cite a finding of a survey conducted by Galtung (1970) in several countries on the images of the year 2000. It was found that in highly developed societies the year 2000 is seen as something far away, while the prognosis of what the general state of affairs will be then was a highly sceptical one. By contrast, in developing countries the year 2000 was seen as being relatively near; it was believed that a rapid rise in economic development will be reached by then. This may not only be interpreted as a typical case of rising expectations, but fits also well into the pattern described above, illustrating the two-fold process at work. The developing countries find themselves at that state of the process today where a long-term perspective has just been acquired, made possible by a newly experienced situation of relative abundance. The developed countries find themselves at a stage where they have already acquired a conception of the future as a newly extended present, one which necessitates new cut-off points experienced as (new) short-term perspectives. It is very probable that the re-coding of the present is facilitated by the fact that the extended present appears much more problematic than was formerly believed. It would thus require a higher degree of attention and more intensive information processing on the societal level. These requirements are also likely to lead to a shortening of the time span which can be attended to. Some of the problems, formerly believed not to occur before the year 2000, are with us now.

VIII. ON TIME-KEEPERS AND TIME-MEASUREMENT

The process of symbolic valuation described before can be taken as paradigmatic for the changing symbolic value accorded to timekeepers over the ages. As the possession and use of clocks filtered down from the aristocracy to the bourgeoisie, to be later almost forced upon the working class, their history came to illustrate the processes in which persons or groups affiliated themselves with timekeepers as status symbols. Likewise, the introduction of clocks to non-European civilizations, such as those of China and Japan, where clocks remained for centuries the playthings of the wealthy and the mighty (Bedini), illustrates the same process. In the modern world where economic production necessitates time-keeping and facilitates the wide-spread possession of timekeepers, clocks have been replaced by watches. The formerly communal symbol, prized for its rarity and affiliation with high status persons, including those inhabitating the heavens, has yielded its place to utilitarian, functional objects that are individually owned and are in practically everyone's possession. According to our hypothesis, such a mass-diffusion process is invariably accompanied by symbolic devaluation, at least of the object that has been chosen to represent the concept of timekeeper. But if timekeepers can no longer be represented by mass-produced watches and be a highly valued object at the same time, where is the new status symbol of a timekeeper to be found? Apart from fancy variations of the product, it might be that the close scheduling itself, and perhaps the human operators who serve it (such as secretaries or service institutions) will replace the clock as status symbol. The modern high status person is one whose time is extremely valuable and who, correspondingly, possesses very little of it. He has to spend it wisely, distribute it economically among his many activities. There is also a concept of time-economizing behind this pattern; time which can be saved can also be spent elsewhere, presumably at more rewarding activities or with more rewarding persons — those that may serve to increase one's social status even more.

To have too much time at one's disposal has, paradoxically, become an indicator in our society of having nothing of importance to do and, consequently, of being of no importance. The concept of time behind such an attitude is no longer based on the movement of stars, planets or gods, and their meaning for human beings, but reveals rather a time concept based on economic activities and how they can be transformed into various social amenities and privileges. We have moved towards a state where man keeps, as it were, his own time. But, specifically, whose time? Should the high status person's time or the busy man's time be more valuable than that of anybody else? Is there no common time left in which everybody can share and participate — a communally experienced and valued time?

At this point we want to return to the questions raised at the very outset of this paper. To what extent do the measurement scales that we use to measure time and the underlying conceptions of dimensions on which they are based, influence our notions about time? If one compares today's time-keeping operations, especially in measurement aspects, with those of former times, one difference emerges very clearly. In former times, different scales were used for keeping or measuring the times of different activities. In a non-scientific world all scales, including those that were used to measure time, are adjustable to circumstances. It appears to be a peculiarity of a scientifically-minded society, however, to prefer scales which are unambiguous, exact and universally valid. As Leach has pointed out, we seem to operate today with scales of great precision but of fewer dimensions than in earlier epochs. The reasons for this are to be sought in our preference for all values to be expressed in interchangeable numbers. Just as in economics the value of different products, of food, labour, or land, can be expressed in terms of a single numerical dimension (quantities of money), so we can now convert formerly different time scales and the activities that were measured by them into one standardized, uniform and uni-dimensional time scale. This trend may be interpreted as the result of an increase in complexity on a large and broad scale of social organization. Yet, the question remains — and it is an open question which I want to leave open — whether this does not imply a loss of richness in structure, and a loss of the creative power that lies in differentiation. To set up multiple scales destined for different occasions and purposes poses no difficulty whatsoever from a mathematical or technical point of view — multifunctional and multi-dimensional timekeepers could easily be constructed. If this would happen, human activities would have to adjust again to different circumstances and take on a newer and richer quality, as a consequence. It would be another example for the interrelation between time structuring and time measurement that we have observed throughout this paper. Only this time, time measurement would re-influence the structuring of time. As Bourdieu (1972) put it when he referred to the cycle of social relations generated through gift-giving in North Africa: to abolish the interval is to abolish strategy. I would like to paraphrase him: to abolish a uni-dimensional time concept, is to restore the richness of social life.

NOTES

1.   The literature on the subject is vast. See especially Moenks (1967) and Luescher (1970).

2.   This point may be illustrated by the following quotation: "I'm working so hard that I'm killing myself and wrecking my family, but I'm making so much money that I can afford it." T.C. Schelling, Foreword, Symposium: Time in Economic Life, *Quarterly Journal of Economics, 4* (1973), 627.

3.   The following section is heavily indebted to the work of Manfred Schmutzer.

340

REFERENCES

Barndt, R. and Johnson, D.: "Time Orientations in Delinquents." *Journal of Abnormal and Social Psychology,* 51 (1955), 343-45.

Bailey, F.G.: *Strategems and Spoils.* Oxford: B. Blackwell 1970.

Bedini, S.: this volume.

Bell, W. and Mau, J., eds.: *The Sociology of the Future.* New York: Russell Sage Foundation 1971.

Bernstein, B.: *Class, Codes and Control.* Vol. I. London: University of London Press 1971.

Bourdieu, P.: "La société traditionelle, attitude à l'égard du temps et conduite économique," *Sociologie du Travail,* 1 (1963), 24-44.

Bourdieu, P.: *Esquisse d'une théorie de la pratique.* Genève: Librairie Droz 1972.

Deffenbacher, K. and Brown, E.: "Memory and Cognition: an Information Processing Model of Man." *Theory and Decision,* 4 (1973), 141-78.

Douglas, M.: "Population Control in Human Groups." *British Journal of Sociology* (1966), 263.

Eliade, M.: *Myths, Dreams and Mysteries.* London: Harvill Press 1960.

Findlay, J.N.: "Time: a Treatment of some Puzzles." In *Problems of Space and Time.* Ed., J.J.C. Smart. New York: Macmillan 1964. (1941).

Fraisse, P.: *The Psychology of Time.* New York: Harper & Row 1963.

Galtung, J.: "Images of the World in the Year 2000." Vienna 1970. Private communication.

Galtung, J.: *Members of Two Worlds.* Oslo: Universitets-forlaget 1971.

Goody, J. and Watt, I.: "The Consequences of Literacy." *Comparative Studies in Society and History,* 5 (1962-63), 304-26; 332-45.

Holl, A.: Personal communication.

Holmberg, A.: "Age in the Andes." In *Aging and Leisure.* Ed., R. Kleemeier. New York: Oxford University Press 1961.

Lazarsfeld, P.; Jahoda, M. and Zeisel, H.: *Die Arbeitslosen von Marienthal.* Leipzig: S. Hirzel 1933.

Leach, E.: "Some Anthropological Observations on Number, Time and Common Sense." Paper presented at the Second International Congress on Mathematical Education. Exeter 1972. Mimeo.

LeShan, L.L.: "Time Orientation and Social Class." *Journal of Abnormal and Social Psychology*, 47 (1952), 589-92.

Lewis, D.: *Five Families: Mexican Case Studies in the Culture of Poverty*. New York: Basic Books 1959.

Luescher, K.: "The Social Psychology of Time." Paper presented at the Colloquium in Social Psychology at Cornell University. Cornell 1970. Mimeo.

Michon, J.: "Processing of Temporal Information and the Cognitive Theory of Time Experience." In *The Study of Time*. Vol. I. Eds, J.T. Fraser et al. New York: Springer-Verlag, 1972.

Miller, S.M.; Riessman, F.; Seagull, A.A.: "Poverty and Self-indulgence: a Critique of the Non-Deferred Gratification Pattern." In *Poverty in America*. Eds., L.A. Forman, J.L. Kornbluth and A. Haber. Ann Arbor 1965.

Moenks, F.: "Zeitperspektive als psychologische Variable." *Archiv für die gesamte Psychologie* 119 (1967), 131-61.

Moore, B.: *Social Origins of Dictatorship and Democracy*. Boston: Beacon Press 1966.

Moore, W.: *Man, Time and Society*. New York: Wiley 1963.

Munn, N.D.: "Symbolic Time in the Trobriands of Malinowski's Era: an Essay on the Anthropology of Time." In N.D. Munn: *Essays in Social Symbolism* (To be published.)

Needham, J.; Ling, W. and de Sola Price, D.J.: *Heavenly Clockwork: the Great Astronomical Clocks of Medieval China*. Cambridge: Cambridge University Press 1961.

Nuttin, J.: "The Future Time Perspective in Human Motivation and Learning." in *Proceedings of the 17th International Congress of Psychology*. Amsterdam: North-Holland Publishing Company 1963. Pp. 60-82.

Ornstein, R.: *On the Experience of Time*. Middlesex, England: Penguin Books 1969.

Ortiz, S.: "Reflections on the Concept of Peasant Culture and Peasant Cognitive System." In *Peasants and Peasant Society*. Ed., T. Shanin. Middlesex, England: Penguin Books 1971. Pp. 322-36.

Pocock, D.: "The Anthropology of Time-reckoning." *Contributions to Indian Sociology*, 7 (1964), 18-29.

Price, D. de Solla: "Clockwork before the Clock and Timekeepers before Time-keeping." This volume.

Rammstedt, O.: *Revolution und das Bewusstsein von Zukunft*. Universität Bielefeld 1972. Manuscript, private communication.

Rezsöházy, R.: *Histoire du mouvement mutualiste chrétien en Belgique*. Paris/ Bruxelles: Erasme 1957.

Rezsöházy, R.: *Temps social et developpement*. Bruxelles: La renaissance du livre 1970.

Roth, H.: *Time Tables: Structuring the Passage of Time in Hospital Treatment and Other Careers*. Indianapolis: Bobbs-Merrill 1963.

Schmutzer, M.: "Social Crystallization: Variations on a Structural Theme." Essex University. Internal Memo, 1974.

Schneider, L. and Lysgaard, S.: "The Deferred Gratification Pattern." *American Sociological Review*, 18 (1953), 142-49.

Singer, J.L.; Wilensky, H. and McGraven, V.G.: "Delaying Capacity, Fantasy and Planning Ability." *Journal of Consulting Psychology*, 20 (1956), 375-83.

Sorokin, P. and Merton, R.K.: "Social Time: a Methodological and Functional Analysis." *American Journal of Sociology*, 42 (1937), 615-29.

Szalai, A., ed.: *The Uses of Time*. The Hague: Mouton 1972.

Talmon, Y.: "Millenarian Movements." *Archives Européens de Sociologie*, 7 (1966), 159-200.

Toda, M.: "Time and Space in the Structure of Human Cognition." This volume.

Vercauteren. P.: *Cahiers d'observation de paysans ivoriens*. Louvain: dactylographié 1965.

Von Foerster, H.: "On Self-Organizing Systems and their Environment." In *Self-Organizing Systems*. Eds., M. Yavits and S. Cameron. Oxford, England: Pergamon Press 1960.

Worsley, P. *The Trumpet Shall Sound*. London 1968.

An Analysis of Future Orientation and Some of its Social Determinants

G. TROMMSDORFF AND H. LAMM

INTRODUCTION

Time and the Experience of the Future

Time will be understood here from an internal point of view, that is, as a person's subjective experience. The psychological experience of time contains the experience of the past, the present and the future. This experience can be cognitively represented in the mind of a person; thus, cognitive abilities and intellectual training would be relevant for the way a person conceives of time. Furthermore, the motivational structure of a person determines the experience of time to some extent; past, present and future contain events which are more or less desirable for a person. Hence, the motivational tendency to approach or to avoid specific events is relevant for a person's psychological experience of the future.

A person's experience of the future will be analysed in terms of his image of the future, that is, what he expects the future to be like, what events he would like to occur in the future and what events he rather fears.

Future orientation will be analysed here as a set of subjective expectations and beliefs held by a person about his future. The future includes positive and negative goals which a person believes he may attain or rather avoid, and the expectation of positive or negative actions of others. Thus, experience of the future implies a goal-related structuring of time, judgments and evaluations of future problems, and planning for possible future behavior.

Since we view human beings as "social animals", and since experience of the future is understood here as a psychological process which has been learned in specific social settings, we assume different ways of future orientation for persons from different social groups.

Future Time Perspective

People's experience of the future has partly been studied in the literature under the label of "future time perspective". In these studies mainly cognitive variables were taken into account while the analy-

*This study was supported by the Deutsche Forschungsgemeinschaft and was conducted in the Sonderforschungsbereich 24, Universität Mannheim. We wish to thank Siegfried Streufert for his valuable comments on the manuscript. We also thank Rolf Werner Schmidt and Henning Eckel for the ideas they contributed to this paper.

sis of affective and evaluative components was largely neglected. Future time perspective also has primarily been treated as a unidimensional variable which could either be measured by its structure or by its extension into the future (Barabasz 1970; LeShan 1952; Teahan 1958; Wallace 1956). Some authors have defined *structure* in terms of the density or frequency of concerns (the number of future events mentioned by the subject) (e.g., Kastenbaum 1961). Wallace (1956) measured structure as the coherence of time ordering of events. He thus stressed the aspect of consistently ordering several events into a time sequence. However, neither the content of the events and their order in terms of a sequence of preconditions and consequences, nor the duration and the relative importance of the future events for the person were taken into account by this method. It was assumed that a coherent ordering of future events — in terms of *when* they occurred — is an indicator for a person's logical structuring of the future. Coherence, as measured by Wallace (1956), is the correlation coefficient between subjects' indication of how old they would be when certain future events occurred (one variable) and the ordering of the same future events into a sequence of time (second variable). This method seems to replace one variable for another rather than account for the coherence of future time perspective in terms of its potentially multidimensional basis.

Beside the frequency of concerns and the coherence of future time aspects, the *extension* of future time perspective was measured in several studies. Wallace (1956) defined extension as "length of future time span which is conceptualized" (Wallace 1956, p. 240) and as " ... the range of years included between subject's actual age and the most distant event given by him" (1956, p. 241).

There have been numerous studies on future time perspective using the "classical methods" of measuring the variety, coherence, or extension of the future but neglecting other important variables which theoretically account for further relevant cognitive aspects of future time perspective such as logical and time structuring, judgment of probabilities of occurrence, and the evaluative assessment of future events.

In the following, an attempt will be made to discuss people's experience of the future including various relevant cognitive and evaluative variables which so far have rarely been taken into account in the analysis of future time perspective and have not yet been integrated as dimensions of the psychological experience of the future.

ANALYSIS OF FUTURE ORIENTATION

The term "future orientation" (FO) rather than future time perspective will be used here systematically to investigate what people experience when focussing their attention on the future.

Future orientation thus will broadly be understood here in terms of a person's expectations and concerns — his hopes and fears — which will be analysed through cognitive, evaluative and affective concepts.

Cognitive Aspects

Logic. Persons may anticipate the future as a complex of events[1] — distinguishable entities — occurring in a process of change. The future may be structured as a set of more or less logically interrelated future events. A logical ordering of the future may focus on *preconditions* as well as on *consequences* which are precisely defined and coherently ordered into a consistent system of interrelated events. Such a logically structured conceptualization of the future probably takes place in relation to those future events for which a rather general agreement exists as to how they can be attained or be avoided (preconditions) and what they are likely to entail (consequences) (e.g., educational requirements for attaining a specific job which in turn offers a certain amount of income and prestige). Planning for the future probably depends to some degree on a logically structured FO.

It would be interesting to study the amount of coherent (logical) future structuring relative to a complex but rather incoherent and non-integrated view of the future. Studies on cognitive complexity (see Streufert 1970) would be helpful for a further analysis of this problem.

Occurrence, duration. The structure of FO may also be analysed in terms of *when* the expected events will occur and *how long* they may last. The expected *occurrence* and *duration* of future events is closely related to the logical sequence of preconditions and consequences (see above). An education which is broken off before the required time (duration, occurrence) will not be a sufficient basis (precondition) for attaining a good job (hope) and thus makes the person renounce the consumption of some desirable goods (consequences).

Extension (time span). Another cognitive aspect is the time span a person takes into account with regard to his future. Beside considering when the future events will occur and how long they will last, a person has to think ahead to foresee the range of future events following the expected ones. The longer the time perspective is extended into the future, the more consequences of future events could theoretically be taken into account for adequately structuring one's future. A long extension of the future, however, does not necessarily mean that more future events and consequences are taken into consideration than in a shorter extension of the future orientation. The future events may be of different duration and thus determine the total time span of a person. It may be more realistic to have a short FO — depending on the kind of future events taken into account. In some respects the FO of older people necessarily is shorter than that of younger people.

Variety. Structuring and — possibly because of learning experiences — re-structuring one's future in accord with one's values and beliefs implies the cognitive ability to process information concerning a variety of concerns. Thus, a sufficient variety of future events (see footnote 1)   should be taken into account in order to structure one's future adequately. It is therefore important to measure the number of different *kinds* of future events taken into consideration by a person. Some persons will envisage future events (hopes and fears) which belong to the same category of events (e.g., several job-related concerns); others will envisage the same number of events but each constituting a different kind of future-related concern. (Incidentally, it is problematic to assume that a specific category of event — as defined by the investigator — has the same meaning for different persons.)

As stated above, extension and variety of FO are different aspects and should be treated as variables which may correlate only under certain circumstances.

Probability judgments. A basic requirement for structuring one's future is to assess the expected probabilities of occurrence of the future events. Persons may estimate the probability in terms of their certainty as to whether the events will occur. Time (near or far future) and extension (short/long) may also be the focus of probability estimations.

Affective and Evaluative Aspects

A person's attitude toward the future may contain positive and negative feelings. The uncertainty implied by the future and the expectation of dissatisfying future events may support negative feelings toward the future for some persons, while other persons may view the uncertainty rather as a challenge or expect rewarding events to occur (*affective aspect*).

An analysis of *evaluative aspects* focuses on the relative rewards and costs of future events. Certain future events relate to preconditions which are rather costly for the person while the consequences are rather rewarding. Hence, the relative weight of preconditions and consequences in relation to future events should be taken into consideration as another factor influencing the cognitive work of structuring the future. Assuming too costly preconditions for attaining a desirable future event, the person may rather re-structure his future and envisage another desirable future event with less costly preconditions.

Content of future-related concerns. Hopes and fears are related to different needs and motivations of people. Hence, one cannot analyse FO independently of the *content* of the future-related concerns (although future time perspective has mostly been viewed under the structural aspect and as a content-irrelevant variable). The investigator of FO should take into account the meaning and relative importance of people's future related concerns and study the values underlying people's different hopes and fears.

If the *personal* future has a different meaning to people than the future of *public life*, one can expect that people have different FO's relating to different kinds of future events.[2] Hence, logical structuring, expected occurrence and duration of events, and extension into the future will be different depending on the kind of people's future-related concerns.

Maximization of rewards in the future. Persons evaluate future events in terms of their relative rewards and costs. Persons structuring their FO "rationally" would follow the rule of maximizing expected rewards and minimizing expected costs in the future. Insko & Schopler (1972) distinguish between three types of psychological hedonism: hedonism of the past, hedonism of the present and hedonism of the future. "The distinction among these three types of hedonism is necessitated by the fact that responses which maximized reward in the present may not be responses which maximized reward in the past or will maximize reward in the future" (p. 32). Hedonism of the future would be an attempt of the individual to maximize the expected reward of events and minimize the expected occurrence of unpleasant events in the future – a requirement of rational planning and future-orientated decision making.[3] This attempt may imply a great deal of cognitive work on the part of the individual. The person has to look ahead, collect information on possible decision alternatives and their potential consequences with respect to his present situation, and relate the overall rewards and costs of his different behavioral alternatives to the desired future state.

"Irrational" processes influencing FO. Future-orientated decision making and the structure of FO which is guided by hedonism of the future will be referred to as "future-oriented rationality".

The "rational" structuring of future events could be disrupted by primarily considering rewarding preconditions and consequences of desirable future events instead of taking into account the negative or costly preconditions and consequences. For example, a desirable future event, such as marriage, may be followed by costly behaviors, such as giving up some freedom and investing time and energy for maintaining the partnership. A "rational" structuring of the future would take into account all influential causes and consequences of an event, not distorted by their evaluative character.

The *occurrence* and *duration* of future events may also be judged non-rationally. Events which are expected to occur in the far future may have a different motivational impact on structuring the future than events which will occur very soon; furthermore, the motivational impact will depend on the relative negative or positive value of these future events.[4]

Expecting to retire and to have less money in the far future may induce people to save some money and attain other forms of economic security in the meantime – a case of rational planning. Equally familiar is an example of an irrational case: A heavy smoker may expect to get cancer some time from now if continuing to smoke. However, since the negative future event is rather distant in time, the smoker may tell himself to continue smoking for a little while and to stop sometime in the future. Both events, getting cancer and stopping smoking, are not fixed in a time schedule of FO and thus may be used for psychological defenses. The dissonance experienced by the future threat and the present reward is reduced by the vagueness of the time schedule lying between now and the far future.

Irrational structuring of the future may also result from the expectancy of a threatening event in the very near future. This closer threat may prevent the person from rationally structuring the time following this event. This case often arises in crisis situations when people panic.

Similarly, the motivational impact on a person's *time span* extended into the future may have irrational effects. Persons expecting negative events like sickness and economic crisis in the far future may prevent themselves from thinking this far ahead and rather extend their time perspective into the near, more rewarding future. On the other hand, a dissatisfying present and the expectancy of a costly near future may stimulate persons to hope for rewarding events in the far future and thus extend their time perspective far ahead   too far to rationally decide on how probable those desirable future events (e.g., a revolution or the rebirth of the savior) are.

The variety of possibilities seen for one's future may be restricted by avoiding to envisage potentially costly and/or threatening future events. It can be assumed that a greater variety of future events is seen when these events are desirable, if it is supposed that people prefer to think about desirable, as compared to undesirable, events.

Finally, the probability estimate may be distorted because of motivational factors. A highly undesirable future event may be viewed as less likely, while a highly desirable event may be viewed as more likely (tendency for "wishful thinking", see McGuire 1968).[5] Persons with a negative attitude toward the future would tend to expect primarily negative events to occur in the future.

Cognitive, Affective and Evaluative Aspects of FO: Optimism/Pessimism

Optimism/pessimism is a largely neglected variable in the study of FO. Vaughn and Knapp (1965) considered optimism/pessimism as a personality variable.

Evaluative probability estimates. Optimism/pessimism may be defined in terms of the judgment people make about the probability of occurrence concerning the timing and duration of future events. An optimist — in contrast to a pessimist — would expect a desirable event to occur in the near future and would expect it to have a long duration. Likewise, an optimist would judge rewarding events to occur with high probability, and costly events with low probability. Though a pessimist's probability judgments may be symmetric to those of an optimist, the motivational background of these judgments may be different. Wishful thinking tendencies would stimulate some people to make unrealistically optimistic judgments. Unrealistically pessimistic judgments may result from uncertainty and insecurity. Some people underestimate in order to prevent too high expectations of other people. An insecure student would rather tell his parents he will fail; in case he really fails, the parents may not punish him severely since they have already lowered their expectation level; if he does not fail, his parents will be delighted and reward him possibly more than if he had not made such a pessimistic judgment.

Hedonistic learning experience. Cantril (1965) and Galtung (1970) conceived of optimism/pessimism as an important variable of people's concerns about the future. They measured optimism in terms of where a person places himself on a self-anchoring scale indicating his best and his worst possible life at the present time, five years ago, and 5 years from now. The distance from the point where a person places himself in the present to the point where he expects to stand in the future, served as an indicator for optimism/pessimism.

If the expectation of future events is seen on a dimension of maximizing one's satisfaction (hedonism of the future), it makes sense to define optimism as being present when the expected outcomes of future events are judged to be greater than the outcomes of past and present experiences; a pessimistic judgment would be given if future outcomes are judged to lie below this comparison level of past and present outcomes. Optimism/pessimism thus can be defined as the discrepancy between the expected outcomes and the actual future outcomes. (A person's subjective experience of the present and expectation of the future are the basis of this measurement of optimism/pessimism.)

One may assume that other structural aspects of FO such as time extension can influence optimism/pessimism. Conversely, an optimistic view of the future can stimulate a person to extend his time perspective more into the future since rewarding outcomes are expected.

Optimism was defined so far (a) as evaluative probability estimation concerning the occurrence, time and duration of future events, and (b) as the perceived discrepancy between past and present outcomes and expected future outcomes (distance from comparison level of present standing). The first definition depends on the expected reward and cost of the future event; the second definition depends on the relation between how the present is perceived and what is expected from the future in terms of maximizing one's satisfaction (hedonism of the future). Both definitions implicitly take into account people's beliefs about who controls future outcomes, as will be discussed later.

Behavioral Aspects of FO

Mainly from studies on attitudes (see Abelson et al. 1968) one can expect the attitudinal component of FO to have some behavioral implications. FO also can determine decision making and behavior due to a person's motivation to achieve some future goals and try to avoid certain other fear-inducing future events (see Heckhausen 1967). Decision making and behavior partly depend on a person's belief in his own abilities to master the future or, alternatively, his attribution of outcomes to other forces (see Weiner 1971), and on his adequate cognitive structuring of present and future. Present decisions are usually made on the basis of future outcomes. On the other hand, the present situation and past experience determine the setting of goals and the evaluation of possible future events. The kind of future events which people expect to occur, people's judgment of these events, and the people's choice can often be extrapolated from the present situation.

An important behavioral aspect of FO is a person's choice between immediately rewarding vs. later rewarding events. A person's ability to postpone immediate smaller rewards in exchange for later larger rewards may be predicted from the cognitive and evaluative character of his FO (see Mischel 1974).

Analytical Schemata of FO: Summary

Our intention is to point out that FO is a multidimensional complex. The cognitive dimension can be analysed in terms of structural variables such as logical ordering of future events, probability judgments concerning the expectation of occurrence, time and duration of future events, frequency, and variety of future events. The affective and evaluative dimension may be analysed in terms of the reward value of future events. An investigation of FO should focus on the different aspects of FO and how they relate to each other (e.g., estimation of probabilities as a cognitive variable and desirability of future events as an evaluative variable).[6] It is not adequate to use one aspect of FO (such as future time extension) as an indicator for another aspect (such as variety) or for the whole complex of FO.[7] The dimensions of FO and the relationship between any of the variables of FO may not be the same (a) with respect to any kind of future-oriented concerns, and (b) for all persons.

In the following, the general analysis of FO as described so far will be treated with respect to several factors which are assumed to account for individual differences among people on the various dimensions of FO.

SOCIAL DETERMINANTS OF FUTURE ORIENTATION

Theoretical Background and Derivation of Hypothesis

One may assume that FO is determined by social learning experience which probably is not the same for everyone but similar for persons with the same life style, background, traditions, values and child-rearing experiences — for persons belonging to the same social groups.

Social variables, such as status and role, are rather complex, but may be used for grouping people into some broad categories which take into account differences in cognitive development, learning and motivation. Social status partly accounts for a specific learning experience affecting the cognitive, evaluative and affective style of a person. Social roles determine people's expectations and behavior to some degree so that this variable may be used also to predict differences on various aspects of FO. In the following, we will only analyse the influence of low and high social status on FO and select age and sex as bio-social roles to study their relative impact on people's FO, that is, on the cognitive, affective and evaluative aspects of FO.

Derivation of Hypotheses

Future orientation can partly be seen as goal-related structuring of the future and planning for possible future behavior. Achievement motivation, as has been shown, is related to a person's assessment of future events, that is, to hopes and fears he wants to attain or to avoid (see Doob, 1971 ; Heckhausen 1967; Knapp & Garbutt 1965; Weiner 1971). Thus, some of the theoretical considerations from studies on the impact of social class and socialization experience on achievement motivation may be invoked for generating hypotheses about social determinants of FO.

On the other hand, FO can be seen mainly as a cognitive variable. Research dealing with the impact of social variables on cognitive abilities and cognitive development (see Kagan & Kogan 1970; Frohlich 1972) could be used here to derive some hypotheses concerning the social determinants of FO.

Cognitive aspects. Cognitive abilities — such as complexity — are assumed to be partly dependent on the development and learning stage of the individuum. These in turn are presumably influenced by social status (e.g., education level) and by bio-social factors such as age (the time and length of learning experience) and sex (the socialized role).

Impact of social status. It seems rather obvious to assume a better cognitive learning experience for persons of high than for persons of low social status. Studies on cognitive abilities of persons from different social status have pointed out that the verbal code, the ability for abstraction, differentiation and long-term planning is better developed in middle than in low status people (see Oeverman 1969).

A relation of present social standing and future time perspective[8] has been demonstrated by Teahan's (1958) finding: Children who had more success in school thought more about the future and had wider temporal perspectives than children at the bottom of the class. However, the question remains whether this effect is due to the influence of social status since the data are correlational.

Schneider & Lysgaard (1958) found that middle class children are more willing than working class children to postpone a reward; they preferred to save a larger part of a fictitious winning. This greater capacity of middle class children to delay rewards may be attributed to their ability to look further into the future than lower class children, and may be related to high achievement motivation and the ability to tolerate frustration (see Mischel, 1974).

LeShan (1952) found that children from the middle class invented stories covering a larger period of time than working class children.[9] The author assumed that middle class children are more oriented toward distant projects than working class children, who are taught not to look for distant events which they probably cannot attain. He thus explained the above finding by the class-related socialization experience of these children.

The line of reasoning for hypothesizing a relation between the cognitive dimension of FO and social status could be as follows. The socialization patterns of high status persons presumably induce a high need for achievement. High need for achievement implies high ability to postpone immediate rewards (Mischel 1966). High capacity to delay gratifications requires an extended FO.[10] This, a more extended FO can be expected for high status people rather than for low status people. Evidence for the assumed impact of variables like socialization experience on FO does not yet exist. Data indicating a possible relation between these variables are only correlational.

One could readily assume a better cognitive development for persons of high education than for persons of low educational level. To the extent that better cognitive development involves greater cognitive complexity, a greater variety of FO may be expected to accompany higher educational levels. A better cognitive education should strengthen the ability for abstraction and logical thinking. Thus, possible sequences of logically interrelated future events will presumably be viewed with greater precision and frequency by high than by lower class persons.

The presumably better cognitive development of higher class persons may also lead one to expect that their FO is more extended. The cognitive ability to envision a high variety of future events may induce a person to extend his time perspective and place these future events in a long-run sequence. Thus, leaving aside assumptions from research on achievement motivation, we may expect a more logical, precise, consistent and coherent FO, greater variety, and a more extended FO[11] for persons from high than for persons from low class — on the assumption of better cognitive education for higher than for lower class persons.

Assuming that people's FO consists of various aspects relating to a variety of future events, one should analyse the impact of social variables on the structure of FO in relation to the variety of interests, needs and values people have. High status persons are known to participate more in public affairs than low status persons (see Popitz 1968). Assuming that present concerns affect future-related concerns, one may expect a better structured FO for high status persons than for low status persons with respect to public concerns.

Impact of age. The learning experience of an adult will be greater than that of an adolescent. His ability to differentiate and his access to information will be more developed than for adolescents. It would be in line with this assumption to expect a more complex FO for adults as compared with adolescents.

The extension of time perspective into the future presumably is affected by the expectation of dramatic changes during life — the next following period of change in the near future will probably be of greatest concern, eventually followed by changes further away. Young people generally envisage more problematic events occurring in the near future than adults of 30-40 years old; a greater frequency of changes occurs in the life of adolescents, like finishing an education, getting a job, getting married, and becoming independent. These dramatic changes will probably occur in the near future of adolescents, while comparable drastic changes (such as illness, retirement etc.) for adults presumably occur only in the more extended future, like one decade or two from now. In this line of reasoning, one may assume that adolescents have a less extended time perspective than adults since more dramatic events may happen to them in the near future.

Impact of sex. A different cognitive orientation toward specific future events can by hypothesized for persons of different sex. Research on sex-role learning (Maccoby 1966; Myrdal & Klein 1956) has shown that sex-typed behavior and interests are socialized since early childhood. Females are more oriented toward social-emotional (expressive) behaviors while males are more oriented toward instrumental roles. The traditional socialization experience directs the concerns of females more toward family-related problems and the concerns of males more toward public problems (Trommsdorff 1968). Again, assuming an effect of present concerns on future-related concerns, females presumably envisage more private concerns than males.

Optimism/Pessimism

Belief in internal/external control. A generally optimistic view of the future may derive from trust in one's own abilities and from the belief in being able to master difficulties oneself ("internal control") as opposed to low trust in one's own abilities and the belief in external forces (other people; external circumstances, fate) ("external control").[12] Would a person believing more in internal control be also more pessimistic about the future? Gore & Rotter (1963) indicate a positive correlation between belief in external control and pessimism. A person believing more in external than in internal control of the future presumably has less concrete information as to how these external forces (like fate) work and what they will bring about. The less information a person has about how outcomes are affected, the more likely he probably is to make non-realistic[13] judgments about the future, judgments which may be extremely optimistic or pessimistic.

Belief in internal/external control[14] will be understood here as an intervening variable presumably accounting for differences in the extent of optimism/pessimism.

Why should some people be more optimistic or pessimistic[15] than other people? Again, persons will be seen as members of social and biological groups, characterized by certain socio-psychological features. Social status and role again are analysed as determinants of FO.

Impact of social status. A person's setting of goals can be assumed to be related to his beliefs concerning the attainability of these goals. These beliefs probably are dependent on the given present situation and on the available information concerning possible future outcomes and how they are to be achieved. A relatively high standing in the present (and past) could affect a person's belief that his future standing will be at least as satisfying as now. One may thus hypothesize that high status people – who presumably enjoy a better social situation (and who also have more access to information on how to reach certain goals) – are more optimistic about their future than low status people are.

On the other hand, one may as well assume that low status people expect their relatively unsatisfying present standing to change into a more positive direction. Low status people have to lose less than high status people and may expect rather to win than to lose. They may thus be more optimistic than high status people expecting changes from their currently dissatisfying position into a desirable direction. Which of the two hypotheses about optimism and social status is correct must be tested by empirical data.

A relation between pessimistic expectancies, fatalism and low social status has been suggested by sociologists (Popitz 1968). One may assume that people of low status have less information – because of their lesser cognitive development – about how goals may be attained and how fears may be countered. The lack of information as to how one may work on the realization or avoidance of future events and the lack of past success in this respect may make these people more prone to believe in external forces. They may become resigned to their fate and rather leave things happening. However, the lack of information and the belief in external forces may induce unrealistic but not necessarily more pessimistic expectations — e.g., when wishful thinking tendencies work in the direction of optimism.

Social class probably also determines the *kind* of future-related concerns (hopes and fears) according to the values and goals people have. From the literature on social participation we know that people of higher status participate more in public affairs than people of lower status (Popitz 1968). Thus, a stronger belief in internal control for high class than for low class people can be hypothesized for public concerns.

Impact of age. Assuming that young people have more possibilities ahead and have less experience and information about possible future outcomes than adults, a wishful thinking tendency resulting in high optimism could be expected for young people more than for adults. Adolescents, furthermore, are farther away than adults from negative aspects of the future, such as old age, retirement, sickness and death; all this may help to discourage pessimism. Depending on their time span, their belief in external control, and on the future concerns conceived of, adolescents may — on the other hand — be more pessimistic about their future than adults. The amount of change to be expected in the near future is greater, more crucial, and possibly less certain than for adults. The uncertainty in regard to highly important future events may be threatening to some people — presumably to those with low belief in their own abilities (and low self-esteem) — and may induce pessimistic judgments.

Impact of sex. Probably more negative future events are waiting for females than for males, such as the menopause and the loosening of family ties which are more important to females than to males (see Lehr 1968). One may thus expect females of a certain age to be more pessimistic about their near future than males.

Empirical Evidence

In the following, some findings of a recent study, conducted together with R.W. Schmidt and Henning Eckel (unpublished data,[16] Sonderforschungsbereich 24, Universität Mannheim, 1973) will be reported in which we tested some of the hypotheses discussed. A multidimensional concept of FO as described above was the basis of this study analysing the impact of social status and role on some of the relevant dimensions of FO.

We asked persons[17] of low and high social status[18] to list their hopes and fears (open-ended question), to indicate their age when these future events will occur, and to indicate whether these events will depend on own abilities or on external forces (see footnote 1).

Cognitive structure of FO.

Impact of social status. Persons of high status named significantly[19] more concerns (hopes and fears) and had a greater number of different categories[20] of concerns than persons of low status. They also had a more extended[21] FO than low status persons. Thus, the FO of better educated persons is characterized by a larger number and a greater variety of concerns, and a longer time extension into the future than is the case for less educated persons.

Impact of age. No difference in the variety of future-related concerns was found between adolescents and adults. However, persons from these two age groups differed significantly in the distribution of their concerns to specific categories of future events. Adolescents named more private and fewer public hopes and fears than adults: More than 80% of adolescents' future-related concerns related to private events.

Impact of sex. Females were more concerned about family-related than about job-related problems — as we had predicted. Adolescent males were more concerned about their job than about their family; adult males had about the same amount of concerns for family as for job-related problems.

Females were not only more concerned, they also had a more extended future time perspective, with regard to family-related than to job-related concerns. For males (with the exception of low status males who had a more extended FO with regard to job-related concerns), the family and job-related concerns were viewed with about the same future time extensions.

It is interesting to note that only adult *females* were equally interested as adolescents of both sexes in personal problems, such as private life, personal satisfaction and development of personality traits.

Conclusions. These results demonstrate the effect of status, age and sex (presumably determined by learning experience) on specific future-related concerns. If we had not differentiated FO as a set of various categories of concerns, we would not have found adolescents to differ from adults. Viewing FO as a multidimensional phenomenon, we have shown that *adolescents*, when asked, name more concerns relating to their private life (family, job and personal satisfaction) than concerns relating to their future public life (economy, political development, environment). *Adults*, on the other hand, entertain the same number of future-related concerns for private and for public life. Adolescents may be more concerned about their future private life as long as they have not achieved stability (a completed education, a job, and marriage).

We also found a more individualistic FO for adult females and adolescents in contrast to a FO attached to public problems in adult males. Probably adult males believe they have solved their individual problems; having achieved a fair degree of private security, they now have room for shifting their interest to more public concerns.

A more structured FO[22] for private than for public concerns in low status adolescents, and in female adults of both status, indicates a tendency to retreat from the complicated problems of public life and to get settled in a well structured private life.

It is necessary to view FO as comprising various variables. We should not extrapolate from the knowledge of one dimension to the total FO. The general prediction that better educated people have a more structured and longer time perspective than less educated people was only partly supported. People in both status groups have about the same number and the same variety of concerns, with the exception of adolescents. People of higher education, however, have a more extended FO than people of lower education in *all* concerns.

Optimism/pessimism.[23] Generally, people of high status were more optimistic than people of low status, but this was only true for adults.

While low status females of both age groups were very optimistic, high status girls were rather pessimistic and had more fears than boys. Low status females and high status girls believed in external control of future outcomes. High status women (all of whom had a job) believed more than high status girls in internal than external control and were more optimistic.

Possibly, females (only high status ones in the case of girls) have been frequently discouraged in initiating and completing things on their own, and thus were encouraged to believe that future outcomes are rather controlled by external forces. While low status females adopt a wishful thinking tendency, high status girls are extremely pessimistic about their future. The high status women are more optimistic than high status girls, possibly because of satisfying past attainments. The high status girls presumably have not had similarly encouraging experiences and thus they adopt a more passive and pessimistic view of the future.

The (more or less implicit) assumption of sociologists that low class people believe in external control and have a fatalistic outlook into the future (see Popitz 1968) could not be supported in our study for males. Other variables like sex should be taken into account when making assumptions about social status, belief in external control and pessimism.

Concluding remarks.

Differences in people's FO may be predicted partly from their social status and role. It seems probable that socialization experience for certain roles and cognitive development account for some of the variance in FO among people.

Socialization experiences in higher social class probably are related to the attainment of influence and to social responsibility; thus, more concerns for public events — compared with lower class persons — are acquired. Socialization for public influence and responsibility, *and* the cognitive development which increases the ability to structure complex problems, may account for a more extended and better structured FO with respect to public events in high status adults.

Socialization experiences — in the way of passive, family-oriented and individualistic role patterns on the one hand and active role patterns oriented toward public concerns on the other hand — presumably account for the different future concerns of females and males.

If the expectation of a better future situation is based on a presently satisfying social situation which also gives enough reason to believe that hopes could be realized and feared states of affairs could be avoided without external help, then moderate optimism is the result. An unsatisfying present social situation, not offering adequate means for change by employing one's own abilities, gives a basis for belief in external control and extreme optimism (wishful thinking) or pessimism.

The focus of this paper was to investigate FO as a multidimensional variable, comprising cognitive, affective and evaluative components. The expectation and evaluation of the future has to be studied in relation to the concerns (hopes and fears) people have, that is, the various future events that may happen to them. Social variables may be used as indicators for specific social experiences determining FO. Mediating variables — such as belief in internal versus external control — constitute additional determinants of people's optimism/pessimism.

The knowledge of a person's FO may help to predict his patterns of future behavior. Too little is known yet about the impact of FO on behavioral styles. One may assume that a clearly structured FO, based on rational judgment concerning future events, may help a person to make plans and to arrive at decisions which improve the quality of his future life.

NOTES

1. In the following, *future events* or future related concerns are understood as distinct entities of the future which a person refers to as his hopes or fears or as an expected state of being in the future.

2. The meaning of private and public concerns for different people may not coincide with the investigator's categorization of concerns. Thus, for some people, family and public-life-related concerns lie at the extremes of the same dimension, while for other people these concerns imply different dimensions.

3. The classical models in decision theory are based on the assumption that probabilities and values are *independent* of each other (see Edwards 1962). For a discussion of conditions under which the independence assumption does or does not hold, see Feather (1959) and Kogan & Wallach (1966).

4. For example, an individual engages in wishful thinking when he expects the future occurrence of a desirable event that has an objectively smaller than 50% chance of occurrence, and when he expects the respective state of attainment to last longer than is objectively likely.

5. Crandall, Soloman & Kellaway (1955) and Irwin (1953) showed the same tendency for estimates in a chance context.

6. See optimism/pessimism as indicator for cognitive *and* evaluative aspects of FO.

7. Several studies on future time perspective use one variable, e.g., future time extension, as an indicator for future time perspective.

8. There are quite a few studies on time perspective and social class; however, they should be critically evaluated in the context of measurements used. For an overview see Doob (1971).

9. While Lessing (1968) confirmed these findings of LeShan (1952), Ellis et al. (1955) and Judson & Tuttle (1966) could not find social class differences in future time extension of children.

10. Doob (1971) assumes the following relation between temporal motives and behavioral tendencies: "Gratification is likely to be deferred when the temporal orientation is toward the future and vice-versa," (p. 93). However, when the attainment of future rewards is doubted (pessimistic judgment), immediate gratifications are not likely to be postponed.

11. This description of FO only points to the cognitive style of the person – no predictions are made whether such a person *behaves* according to this style by rationally planning the future, responding in a flexible way to new information, or coping with "his future" in a realistic, non-rigid way. The actual implementation of a person's plans may involve departures from his original FO and this, in turn, may induce him to change his behavior accordingly (see "concomitant change" in Doob 1971, p. 403 ff).

12. We thus use the term internal/external control somewhat differently from Rotter (1966).

13. Non-realistic judgments may be based on the mutual interdependence between probabilities and values (see footnotes 3 and 5 on the previous page).

14. Belief in internal/external control could be measured by asking subjects to indicate on a scale how far future outcomes are believed to be controlled by oneself (own abilities) or by external forces (other people; external events; fate). This measure was used in our study reported later. The formulation for the measure used in the study to be reported later was worked out by H. Eckel and R.-W. Schmidt (diplom thesis, Universität Mannheim 1973).

15. The term optimism and pessimism is not used to indicate non-realistic judgment, unless stated otherwise. An optimistic judgment would be realistic depending on where the person stands in the present and what his actual outcomes will be in the future.

16. The diplom thesis of H. Eckel and R. W. Schmidt was based on this investigation.

17. Randomly selected persons: 100 adolescents of each sex between 14 and 16 years and 100 adults of both sexes between 35 and 45 years. In each age group half of the subjects were low and half were high status persons. Thus, for each independent variable (status, sex, age) there were two levels.

18. Social status was measured by educational level. Several studies have shown that level of education is a sufficiently valid indicator for social status in West Germany. Education correlates significantly with income and social status (Daheim 1970).

19. All differences reported here are at $p < .05$ or below and have been obtained by analysis of variance (social status, age, sex as factors with two levels each).

20. The concerns were categorized by two independent raters into several private and public categories. The total number of different categories was computed for each subject. (Private life: family, job, personal satisfaction; public life: economy, political and social development).

21. Extension was measured as the distance between subjects' actual age and the age at which the subjects thought the future event would occur. Only the most distant event was used for this measure.

22. Here: A greater variety of future-related concerns (number of different categories of concerns).

23. The person indicates on an eleven-point scale, where he stands on the ladder today (present), where he has stood in the past, and where he thinks he will stand in the future. The top of the "ladder of life" is the best possible life as defined by the person, and the bottom is the worst possible life, again, as defined by him (see Cantril 1965, p. 22).

Judgments were made for the total of the hopes and fears named by the subject; a list of ten areas of private concern (e.g., health, interpersonal relations) and ten areas of public concern (e.g., peace in the world, crime in the Federal Republic of Germany) were presented to the subject. Optimism is given if the difference between the future and the present is positive.

BIBLIOGRAPHY

Abelson, R.P.; Aronson, E.; McGuire, W.J.; Newcomb, Th.M.; Rosenberg, M.J. and Tannen-baum, P.H. (Eds.): *Theories of Cognitive Consistency: A Sourcebook.* Chicago: Rand Mc-Nally 1968.

Barabasz, A.F.: "Temporal Orientation and Academic Achievement in College." *Journal of Social Psychology, 80* (1970), 231-2.

Cantril, H.: *The Pattern of Human Concerns.* New Brunswick, N.J.: Rutgers University Press 1965.

Crandall, V.J.; Solomon, D. and Kellaway, R.: "Expectancy Statements and Decision Times as Functions of Objective Probabilities and Reinforcement Values." *Journal of Personality, 24* (1955), 192-203.

Daheim, H.: "Soziale Herkunft, Schule und Rekrutierung der Berufe." *Kölner Zeitschrift für Soziologie und Sozialpsychologie,* Sonderheft 5 (1970), 200-17.

Doob, L.W.: *Patterning of Time.* New Haven: Yale University Press 1971.

Edwards, W.: "Utility, Subjective Probability; their Interaction and Variance Preferences." *Journal of Conflict Resolution, 6* (1962), 42-51.

Ellis, L.M.; Ellis, R.; Mandel, E.D.; Schaeffer, M.S.; Sommer, G. and Sommer, G.: "Time Orientation and Social Class. An Experimental Supplement." *Journal of Abnormal and Social Psychology, 51* (1955), 146-7.

Feather, N.T.: "Subjective Probability and Decision under Uncertainty." *Psychological Review, 66* (1959), 150-64.

Fröhlich, W.D.: "Sozialisation und kognitive Stile. Einige Denkmöglichkeiten und Befunde." In C.F. Graumann (Ed.), *Handbuch der Psychologie, Vol. 7,* p. 2. Göttingen: Hogrefe 1972.

Galtung, J.: "Images of the World in the Year 2000. A Synthesis of the Marginals of a Ten Nations-Study." *European Coordination Centre for Research and Documentation in Social Science.* Wien 1970.

Gore, P.M. and Rotter, J.B.: "A Personality Correlate of Social Action." *Journal of Personality, 31* (1963), 58-64.

Heckhausen, H.: *The Anatomy of Achievement Motivation.* New York: Academic Press 1967.

Insko, Ch.A. and Schopler, J.: *Experimental Social Psychology.* New York: Academic Press 1972.

Irwin, F.W.: "Stated Expectations as Functions of Probability and Desirability of Outcomes." *Journal of Personality, 21* (1953), 329-35.

Judson, A.J. and Tuttle, C.E.: "Time Perspective and Social Class." *Perceptual and Motor Skills, 23* (1966), 1074.

Kagan, J. and Kogan, N.: "Individual Variation in Cognitive Processes." in P.H. Mussen (Ed.), *Carmichael's Manual of Child Psychology. Vol. 1.* New York: John Wiley & Sons 1970.

Kastenbaum, R.: "The Dimensions of Future Time Perspective. An Experimental Analysis." *Journal of General Psychology, 65* (1961), 203-18.

Knapp, R.H. and Garbutt, J.T.: "Variations in Time Description and Need Achievement." *Journal of Social Psychology, 67* (1965), 269-72.

Kogan, N. and Wallach, M.A.: "Risk taking as a Function of the Situation, the Person and the Group." In G. Mandler, P. Mussen, N. Kogan and M.A. Wallach (Eds.), *New Directions in Psychology III.* New York: Holt, Rinehart & Winston 1967.

Lehr, U.: "Zur Problematik des Menschen im reiferen Erwachsenenalter – eine sozialpsychologische Interpretation der 'Wechseljahre'." In H. Thomae and U. Lehr (Eds.), *Altern, Probleme und Tatsachen.* Frankfurt/M: Akademische Verlags-gesellschaft 1968.

LeShan, L.L.: "Time Orientation and Social Class." *Journal of Abnormal and Social Psychology, 47* (1952), 589-92.

Lessing, E.E.: "Demographic, Developmental and Personality Correlates of Length of Future Time Perspective." *Journal of Personality, 36* (1968), 183-201.

Maccoby, E.E. (Ed.): *The Development of Sex Differences.* Stanford: Stanford University Press 1966.

McGuire, W.J.: "Theory of the Structure of Human Thought." In R.P. Abelson et al (Eds.) *Theories of Cognitive Consistency: A Sourcebook.* Chicago: Rand McNally 1968.

Mischel, W.: "Theory and Research on the Antecedents of Self-imposed Delay of Reward." In B.A. Maher (Ed.), *Progress in Experimental Personality Research. Vol. 3.* New York: Academic Press 1966.

——————. "Processes in Delay of Gratification." In L. Berkowitz (Ed.), *Advances in Experimental Social Psychology. Vol. 7.* New York: Academic Press 1974.

Myrdal, A. and Klein, V.: *Women's Two Roles: Home and Work.* London: Routlege & Kegan 1956.

Oevermann, U.: *Sprache und soziale Herkunft.* Frankfurt: Suhrkamp 1969.

Popitz, H.: *Prozesse der Machtbildung.* Tübingen: Paul Mohr 1968.

Rotter, J.B.: "Generalized Expectancies for Internal versus External Control of Reinforcement." *Psychological Monographs, 80* (1966), (1, Whole No. 609).

Schneider, L. and Lysgaard. S.: "The Deferred Gratification Pattern: a Preliminary Study." *American Sociological Review, 18* (1953), 143-49.

Streufert, S.: "Complexity and Complex Decision Making." *Journal of Experimental Social Psychology, 6* (1970), 494-509.

Teahan, J.E.: "Future Time Perspective, Optimism and Academic Achievement." *Journal of Abnormal and Social Psychology, 57* (1958), 379-80.

Trommsdorff, G.: "Kommunikationsstrategie sechs westdeutscher Frauenzeitschriften: Einkommenshöhe der Leserin als beschränkender Einfluss auf ihre sozialen Orientierungsmöglichkeiten." *Kölner Zeitschrift für Soziologie und Sozialpsychologie, 1* (1969), 60-92.

Vaughn, J.A. and Knapp, R.H.: "A Study in Pessimism." *Journal of Social Psychology, 59* (1963), 77-92.

Wallace, M: "Future Time Perspective in Schizophrenia." *Journal of Abnormal and Social Psychology, 52* (1956), 240-45.

Weiner, B.: "New Conceptions in the Study of Achievement Motivation." In B.A. Maher (Ed.), *Progress in Experimental Personality Research. Vol. 5.* New York: Academic Press 1970.

XI. SPECIAL SESSION ON TIMEKEEPERS AND TIME

Clockmaking — The Most General Trade   J.T. FRASER

These papers were read and discussed at a special symposium on *Timekeepers and Time,* sponsored by the Bulova Watch Company, Inc., and held in connection with the Second Conference of the International Society for the Study of Time. To appreciate fully the significance of these invited papers, I would like to place their themes in the perspective of the larger enterprise of which they are part, namely, that of the study of time.

The study of time concerns itself with various issues that stem from man's awareness of temporal passage: men, animals, and plants live and die, generations follow upon generations, the seasons change, the earth ages, and even the universe evolves. This passing of time is beheld, as it were, by two of our alter egos: Homo sapiens, or "man the knower" and Homo faber, "man the maker."

Homo sapiens reflects upon the facts of time and, inspired and awed, searches for value, truth, wisdom, beauty, and abstract knowledge. However, to paraphrase a hackneyed though still beautiful Biblical utterance, it is not by value, truth, wisdom or abstract knowledge alone that man does live. While Homo sapiens searches for meanings, man the maker organizes his daily activities – in fact, his whole life – with the assistance of natural processes which he regards as regular and predictable. Such processes constitute the means of timekeeping and are the functional bases of calendars and clocks. Both of these devices – calendars and clocks – are tools which assist man in labor directed toward the domination and control of the environment, of other men and the control, if possible, even of himself.

Historically, the determination of which specific processes are suitable for timekeeping has been a matter of changing judgments made in the light of changing world-views. The papers of this symposium deal with a few of the very many ideational determinants of clocks, as well as with clock technologies, but only to the extent that techniques of time measurement relate to intellectual, religious, political and social attitudes and preferences. Since the papers are concerned only with a period beginning in the centuries just preceding the birth of Christ, let me speculate about the much older origins of man's concern with time.

There are good reasons to believe that man's awareness of time emerged simultaneously with – and is related to – such other capacities as those of extended memory and expectation, the identification of the self as an individual, the realization of the inexorability of death, and even the ability to separate the intelligible from the emotive in the structure of language. It is pure speculation, but reasonable to assume, that man's rapid ascent and dominion over other living organisms was substantially assisted, or perhaps made possible, by his capacity to prepare for long range future contingencies by using information which he could retain by means of his long term memory of past experiences.

This particular gift, the knowledge of time, manifests itself through such skills as the purposeful imitation of natural cycles, and the derivation of certain advantages from the use of such models. Watching the changing shadow of sticks and stones is one example; regulating the rate of flow of water or sand is another, as is regulating the rate of burning. Calendars are also models: they image the daily and yearly rotation of the skies. We still use timing devices that employ the flow of liquid through a nozzle (such as the carburetors of our cars); the flow of sand (in hour glasses); or rate of heat generation (in automatic sprinklers). The photographic zenith tube of the United States Naval Observatory in Washington D.C. is a distant progeny of the sun stick. The dripping faucet that reminds the sleepless sleeper of the passage of time is a latter-day clepsydra. More familiar, however, are the controlled oscillations of large bodies (pendulums or torsional balances) and the controlled vibrations of very small bodies (quartz crystals or tuning forks). In any case, the historical parade of timekeepers is an impressive one.

Reverting now to our two alter egos, I note that for almost two millenia the pride of Homo sapiens, man searching for meanings and values, has been philosophy. And philosophy is classically defined as that department of knowledge which deals with the most general causes and principles of things. In a way, we may claim that timekeepers are the most general machines: they connect each clockwatcher with the cosmic processes of the universe at large, and with the earth, locally; and they connect man with man in making organized social action possible, for better or for worse. Every watch dial is a miniature planetarium, with the sun as its single star. But if this evaluation of timekeepers is correct, then clockmaking is the most general trade — the pride and joy of Homo faber, man the maker. Certainly, the influence of clocks as models of reliability of production has been profound in the industrialized West, and through it, in the whole modern world. The reader might wish to convince himself of the appropriateness or inappropriateness of these claims by enjoying and studying the fine papers in this volume, and following up some of the references appended to the papers.

Now, after these remarks about the more general framework into which the substance of the present volume may be placed, let the Authors do the talking.

J.T. Fraser

Clockwork Before the Clock and Timekeepers Before Timekeeping

D. DE SOLLA PRICE

The object of this paper is to bring new evidence to bear upon the origin of timekeepers, and to show from this that a particular technology, the making of astronomical models, has been crucial. The techniques of these models have brought about a change in Man's dealing with time at the deepest philosophical and personal levels. Our thesis, in brief, is that timekeeping was invented from and because of a prior available technology during the first century B.C.

Some explanation is necessary at the outset about what is meant by timekeeping in this context. A great deal of damage and impediment to clear scholarship has been caused by the widespread and facile idea that man has always "kept time" by primitive astronomical means. There seems no firm evidence to support such a contention. There certainly exist in the evidence several calendar-like series of omens in an interesting sequence such as those of heliacal risings and settings of the bright stars, new and full moons, equinoxes and solstices, and there is an obvious reliance on the cycles of such occurrences and events that correspond in the sequences of rural agriculture and religious observance. What there is *not* is the generalized measurement and statement of time of day, nor even that of time of month or of year except for the particular discrete omen points. Thus, in a sense we have the quite strong development of astronomy, long before time is used as a concept, out of this one context of special omen events which occur in the heavens and are reflected by happenings on earth.

A consequence of the belief that Man had always "kept time" is the erroneous supposition that prior to the mechanical timekeepers, which dominated so much of the medieval and later practice, there must have been various alternative non-mechanical timekeepers such as sundials, water- and sand-clocks and similar devices. This is a typical example of the error of historicism in which all that happens in the past is seen by the historian as leading with a single arrow to the present. In fact, there are often lines of development that lead to traditions that have died and are no longer with us or of such dominating importance as they were in the past. Because the measurement of time is of such significance today, and because we know what the traditions of astronomical practice in observation have been during the past few centuries, we have a strong tendency to project such considerations back into a past where they are not necessarily valid. In particular, it would appear from the history of astronomy and from the known development of sundials and other such devices that virtually all of the evidence for shadow- and water-clocks being used as timekeepers before about the fourth century B.C. is fictional. What clearly exists is a set of megalithic monuments, hopelessly glamourised today, like Stonehenge and the other ancient stone circles which have something to do with directions of sunrise and sunset at least, more doubtfully moonrise and moonset also, and with extremely dubious claims to be also involved with eclipse cycles. There are also water-clock pots from Old Kingdom Egypt, and several star ceilings. I take all this evidence not as that of time-determining practices, but rather as devices for reflecting the orderly sequences of the omen events – heliacal risings and settings, new moons, equinoxes and solstices. Similar histories were rehearsed in all the ancient river-valley civilizations, certainly in Egypt, China, Mesopotamia, the Mayan area, and perhaps also in India. Much later, shortly before the time of Christ there is a new movement, almost synchronously in China and in Greece, which establishes a new trend leading to the modern practice. What did not happen was that man wanted to measure time and so devised new ways of doing it. What did happen is that in the course of following an old trend, not quite yet extinct, he developed quite sophisticated techniques, important for their technological brilliance, that gave him for the first time the possibility of doing something he had not wanted before it was readily available. This product, timekeeping, caught on, and it is due to this ancient fashion that time became a matter of the deep philosophical and scientific importance it has today.

The ancient trend with which this historical sequence began may be identified as the making of idols, that is images or models which duplicate all natural things of creation. Probably behind it all is the idea of sympathetic magic working through a liaison between the real thing and the model, and perhaps too the power of being a sort of "do-it-yourself" creator of part of the universe. At all events, from the beginning there appear two different lines of such models of creation, corresponding on the one hand to the animate world where the models are of living beings and animals, and on the other hand to the inanimate world of the celestial bodies – Sun, Moon, planets and stars, and the bowl of the sky which they inhabit. These lines correspond precisely to the two most anciently successful sciences, medical physiology and astronomy. There is good reason why they should, for the essential element in scientific achievement is the making of a successful model, be it an idol or a modern piece of computer simulation or analytical mathematics, or the behaviour of twisted strands of macromolecules. It is worth noting that the ancient sciences might well have developed more from the stimulus of the ease with which they could be modelled than by any practical utility of applying the knowledge. A case might well be made for the practicability of medicine, but ancient physiology was much more explanatory than of use in treating the sick, and as for the astronomical theory, superbly developed and sophisticated as it was, there is an air of utter artificiality in all the historians' explanations that try to make it seem as if the chief purpose had been the needs of the calendar or even of divination of the future; the sorts of theories that were successfully formed were just not those that are the most practically useful.

In the earliest times all the models of both lines were evidently static like the dolls, the painted astronomical ceilings, and the megaliths that sat and brooded on their ceremonial site and waited for the Sun to ascend to his place on the appointed throne. What seems to happen round about the fourth century before Christ is the beginning of attempts to rise above the static character of the old models and introduce elements of movement which seemed to have been the very essence of the creations.

It is the dynamic modelling of the inanimate world with which we must be concerned in the story of timekeeping, but it is worth noting that the other line of creation was also most important in its evolution. What happened is that the dolls became automata (figures that move by themselves) and this produced over the ages the chain that leads to robots, to the physiological theories and modellings of pneuma and element theory, and to some extent to the modern cybernetic machines and automatic tools that are near the heart of the newest scientific technologies. It may seem a long way from the toy singing-birds of the Automaton Theater of Heron of Alexandria to such advanced automation. In the history of technology nonetheless, the line of evolution is clear, and it involves such interesting items as the punched card Jacquard weaving loom, the automata of Vaucanson who invented rubber tubing in the eighteenth century while simulating the digestive gut of a model duck, the cuckoo on the Black Forest cuckoo-clock, and the cock on the Great Clock of Strassburg where the two halves of simulated creation are brought together in a technological altar piece erected to the glory of God. The clock, as we shall see, is a sort of fallen angel from the world of astronomy, but the jackwork commonly set on earlier clocks was an essential part of the design that this was a model of creation, a tangible means of expression human comprehension and of extolling the skill and ingenuity of the Creator, rather than a mere time-telling device.[1]

The first piece of evidence for this interpretation of the other line of creation models, those concerned with astronomical phenomena, comes from the recent work of Dr. Sharon Gibbs[2] who has, for the first time, given a catalogue raisonné of the complete corpus of Greek and Roman stone sundials. Now that one can see the complete series of some 300 known specimens rather than the sparse individual publications previously available, some lines of evolution and interpretation become clear. The dials range in date from the late fourth century B.C. to the end of Roman affluence in about the fourth century A.D., and thence after some gap and a few poor dials in the scientifically derelict Byzantine civilization the thread takes up again in Islam and is thence transmitted much improved to the Christian middle ages and the prolific diallers of the Renaissance and Reformation.

The most significant characteristic of the early dials before the time of Christ is that they have no hour numerals, and the rather good dial from Chios dating from the 2nd century B.C. has not even the hour lines marked. From the very beginning, however, the dials carry the zodiacal marking of the annual cycle of the solar calendar, and frequently there are inscriptions of names of tropics, solstices, and zodiacal signs referring to this cycle. In later dials there may be some tendency for such marking to disappear just as the numbering of the hour lines begins to be common. Another very telling feature of these early dials throughout the series is that they display great geometrical ingenuity in the way in which they use the shadow of the gnomon pointer to mirror the heavenly motion of the Sun onto the surface of a cone or a sphere. The very common use of the cone rather than the sphere which one might expect to find may indeed be the motivation that led Greek geometers towards the treatment of conic sections, a most important area for the evolution of early mathematical analysis.

It seems evident, then, both from the evolution of the markings and from the geometry, that in the beginning the sundial was devised as a means of modelling the path of the sun in its annual rather than its diurnal rotation. Probably the ingenuity lavished on the better dials by geometricians did much to improve the general art of dialling, and the gradual familiarity of these "ritual objects" led shortly after the time of Christ to their actual use as indicators of the time of day. There exist, it is true, several literary allusions to sundials that antedate the series of extant sundials but it seems, on detailed examination of each, that one cannot support the contention that this technology was used for non-astronomical timekeeping. In particular the Skiotheron of Anaximander ca. 580 B.C. is almost certainly a device for the determination of equinoxes and perhaps solstices (though this is so difficult it was seldom achieved with any reasonable accuracy in antiquity). The mention of the polos and gnomon by Herodotus ca. 430 B.C. and by the Babylonians is very probably restricted to technical astronomy, and even the use of seasonal hours in Egypt ca. 1800-1200 B.C. is primarily due to their division of the year calendar into decans of 10 days. There may well have been a folk practice whereby the hungry man told when it was dinner time by the pacing of his own shadow — a similar method is still testified by modern anthropologists[3] — but there does not seem much evidence of this being a widespread practice in the earliest times. The position is then from all this evidence that whatever exotic practices might have existed at various times, there is a transition in the common practice of sundial use. Before the time of Christ there is a strong indication we are dealing with the astronomical models illustrating the annual cycle of the Sun, after we have a timing device which is actually used for telling both the time of day and the time of year.

The line that leads to the modern mechanical clocks is quite distinct from that of the ancestry of sundials, the two being related only by the allegedly similar "time-telling" function. A reasonable starting point is a pair of inventions that seem to arise about the time of Archimedes and are inevitably associated with the name of this mechanical and mathematical genius. The one invention seems to be that of the toothed gear which as we will see may have arisen naturally from a modelling of the calendar cycles. The other is that of using a clepsydra — a leaky water pot with a float in it — to drive a rotating star map, globe, or planetary model at a slow rate in time (approximately) with the actual rotation in the heavens above.

Until now we have known very little of the evolution of the toothed gear wheel and the practice has been to follow the absence of literary allusions and artifacts by supposing that the very minimum existed in Greek antiquity. This was an idea which appealed to vogue, for many have supposed that the Greeks were uninterested in things technological and mechanical perhaps through general inaptitude but more probably as some corollary of a slave society in which technology was reputed to be menial work unbefitting the gentleman and scholar.

After the Archimedian use of a simple gearwheel meshed with a worm to give a very large gear ratio, one knows nothing of the gear until the only slightly more sophisticated use by Heron and Vitruvius of pairs of gears meshing together in simple ratios for measuring the turns of a carriage wheel or raising a heavy load. After that, in ninth century Islam the gear is attested in more complicated ratios, in geared astrolabe mechanism recognizable as clockwork ancestors of all modern machinery. In these Islamic cases the gearing is used as a simulacrum for the calendar cycles. Then in thirteenth century Europe this technology is applied to the rotating water clock dials to produce the first great astronomical clocks, a fashion which swept through the European cathedrals and filled people with wonder from the thirteenth century onwards. From those first wondrous clocks built by Richard of Wallingford and by Giovanni de Dondi there is a clear line through to the Great Clock of Strassburg and those of Prague and Lund, to all the simpler blacksmith clocks of the Middle Ages, the ringing of chimes for the monastic rule, and the general devolution of this Zeiss Planetarium of the ancient world into a mere commonplace time-telling wrist-watch.

Some other side-shoots of the story deserve at least brief mention. There is a marvellous example of possible cross-cultural diffusion in comparing the evolution of this line of model construction in the West with that which existed in China. Beginning with the apparently quite independent invention of a water-powered model celestial sphere by Chang Hêng at just about the time of Heron, the Chinese developed a complicated water-wheel escapement mechanism. It was devised by I- Hsing ca. 725 A.D. and reached high perfection in a fine pagoda tower built by Su-Sung in the eleventh century A.D. This structure contains both the astronomical models and the jackwork just as in the West. The important feature of the water-wheel type of escapement might well be the stimulus by means of travellers'

Fig. 1. Astronomical clocktower built by Su Sung, A.D. 1090. Artist's reconstruction by John Christiansen, courtesy of Cambridge University Press.

Fig. 2. Interior mechanism of the Tower of Winds in Athens, as reconstructed by Noble and Price.

tales for the construction in the medieval West of the crucial device of the mechanical escapement which measures time by chopping it up into uniform and countable intervals. Alternatively, it is possible that the Western escapement is quite independent and an adaptation of the bell-ringing machinery which accompanied the early rotating water-clocks.

Returning to the main story it would seem that there exists a crucial period of development around the time of Christ. By this time one had taken the astronomical models to the point where the sundial was ready to become a timekeeper, the technology of gearing was available for modelling and ready to become "clockwork" and the rotating water-clock was ready to be complicated by melding with the other techniques to become a showpiece and then a practical time-teller. It is just at this point of time that there have recently come to light two remarkable new pieces of evidence that clear up many points in the story and extend it to the point already made in outline.

The first of these new contributions concerns a reconstruction of the important structure called the Tower of Winds in the Roman agora of Athens. It is one of the few buildings of classical antiquity which have never been buried or destroyed and stands complete with its original roof. The interior has, however, been gutted, but in 1965 Joseph Noble and I made a careful analysis of the remains of channels and other fittings on the inside floors and walls and from this and the literary evidence we were able to reconstruct the interior mechanism with reasonable confidence in the main outlines.[4] The tower had been built by a Macedonian geometer, Andronikos Kyrrhestes, who is also known for one of the most complex and geometrically ingenious sundials, preserved on the island of Tenos. It was erected ca. 75 B.C. and has been famous ever since for its many sundials, wind-vane, water-clock, and other scientific marvels. We were able to show that the water-clock in question had been of a type that incorporates a rotating disc with star map and model Sun rotated by clepsydra power behind a wire grid representing the horizon and lines of equal azimuth and altitude. Because of the existence of this clock and the set of nine elaborate sundials, it seemed that we were dealing with a structure that had functioned as a public clock at the entrance to the market place.

Later research, starting from curiosity about the role of the wind-vane and the very careful geometric design of the octagonal base of the structure made it clear that more was involved than practical time-telling.[5] Further evidence from a somewhat garbled eye-witness account of the tower by a Turkish traveller in 1668 supplied the clues which lead to the present conclusion that the prime purpose of the Tower of Winds was to manifest in a model the fundamental four-element theory on which ancient science was based, together with a panoply of illustrations of the consequences of this theory in various directions. The octagonal plan of the structure is motivated by the fact that the four qualities, hot and cold, wet and dry, taken in pairs define the four elements, air, earth, fire and water. The diagrammatic representation of this is had by setting a square diagonally within another square of the same size, and on the basis of this theory there *must* be eight distinct winds, an alignment with the zodiacal signs of the heavens, and thence to the cycle of the ages of man, the bodily humors, the properties of stones and herbs, and indeed a complete concordance of scientific theory. Thus, the water clock showing a rotating heavenly model set behind a representation of the earth, decorated by fountains of water and lit by fiery torches, was not al all a mere timepiece. The tradition is indeed that of the model of creation, and this is the noblest extant example from antiquity.

The Tower of Winds, though a model universe, is clearly close to the point where such a structure could be used for time-telling. Highly visible in the center of town, it may have been the very cause of an enormously increased public awareness of the passage of time and the easy availability that made it possible to keep appointments and schedule events. Gradually, over the few generations, when the Tower and its devices were maintained and in good repair, its effect would cumulate and its fame spread so that the existence of a clock, in the *modern* sense, would change the usage of all sundials and water-clocks thereafter. I do not think, however, that this period can have lasted very long, probably not more than a century or maybe much less, for there is little in later writings to indicate that any author knew much more than the external fabric of the device rather than its impressive working interior. An interesting side-point worth comment in the context of a book about time is the fact that in this Tower, as indeed in all anaphoric water-clocks, it was necessary to "wind" the device, probably daily, by letting a new supply of water into or out of the measuring tank. Every day a custodian would have to enter and turn the appropriate tap, and at this point, to an outside observer, it would look as if the whole model created universe was in the course of rapid motion in the opposite direction to that of the normal steady progress. It may well be the familiarity with such a water-clock phenomenon that lies behind Plato's Myth of Cosmic Reversal and other such philosophical notions in cosmology.

The second piece of new evidence, which has come to hand even more recently, concerns the enigmatic Antikythera Mechanism. This Hellenistic object was recovered by sponge-fishers off the island of Antikythera between the Peloponnesus and Crete in the first great underwater discovery just after 1900, and it is preserved with many bronze and marble statues from the same sunken treasure ship in the National Archaeological Museum in Athens. It is not just the earliest, but the *only* existing scientific artifact of any mechanical complexity to have been preserved from antiquity, its four fragments of bronze plate being full of the remains of many gear wheels, inscribed scales and pieces of astronomical inscription. As so frequently in the history of scientific technology (i.e. instruments and clocks) it illustrates, perhaps better than most items, the "fallen angel" principle that the earliest object in the development of the series happens to be by far the most complicated. In this case, particularly it has led those ignorant of the rest of the history to suppose that it is one of the artifacts that might be relics from some alien visitors to the planet. Fortunately, we no longer need such a wild hypothesis now that it has been possible to solve the longstanding puzzle presented by the visible gearing being too fragmentary to determine the precise nature and working of the device and thereby its probable origin and provenience.[6]

A couple of years ago, thanks to the effective cooperation of Dr. Karakalos of the Greek Atomic Energy Commission and the Museum we were able to take fine X-ray photographs of the fragments. This not only showed that very many more gears were preserved inside the fragments, but it also enabled us to count the gear teeth over long stretches instead of the few places where they were barely visible. To my great joy I found that the fitting together of the four extant fragments – the fourth small piece long lost in the museum store coming to light dramatically only at the end of the investigation – showed that the gear-trains were very much more complete than had been supposed. This enabled me to find the ratios involved with the gearing and thereby to find the purpose of the dials and of the complete mechanism. This being done, it was possible to find its place in the evolution of similar devices and make strong conjectures about its origin. We knew already from scale markings that the device had been constructed in 87 B.C. and that it had been used for at least two years during which time it was broken in several places and repaired. Then some time shortly after a very similar device was seen by Cicero when he was studying philosophy in the island of Rhodes in 79-77 B.C. it was lost at sea with a cargo of other art objects that were in process of being shipped

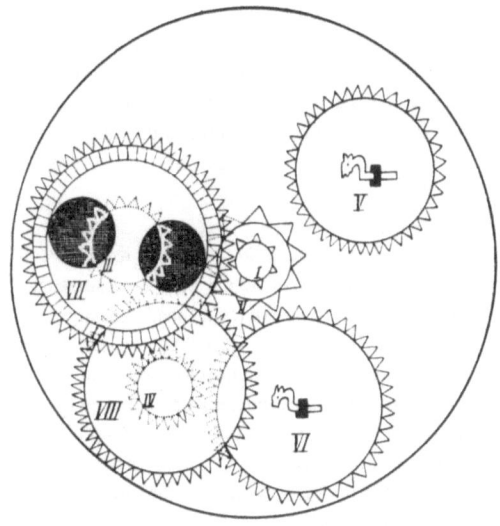

Fig. 3. Design for a geared calendrical computing device by al-Biruni, ca. 1000 A.D. This and the following three illustrations are from "Gears from the Greeks, the Antikythera Mechanism – a Calendar Computer from ca. 80 B.C.", *Transactions of the American Philosophical Society*, Philadelphia, Pa., December 1974. For detailed explanation not given in this text, the reader is referred to that article.

Fig. 4. Casing of the Antikythera Mechanism.

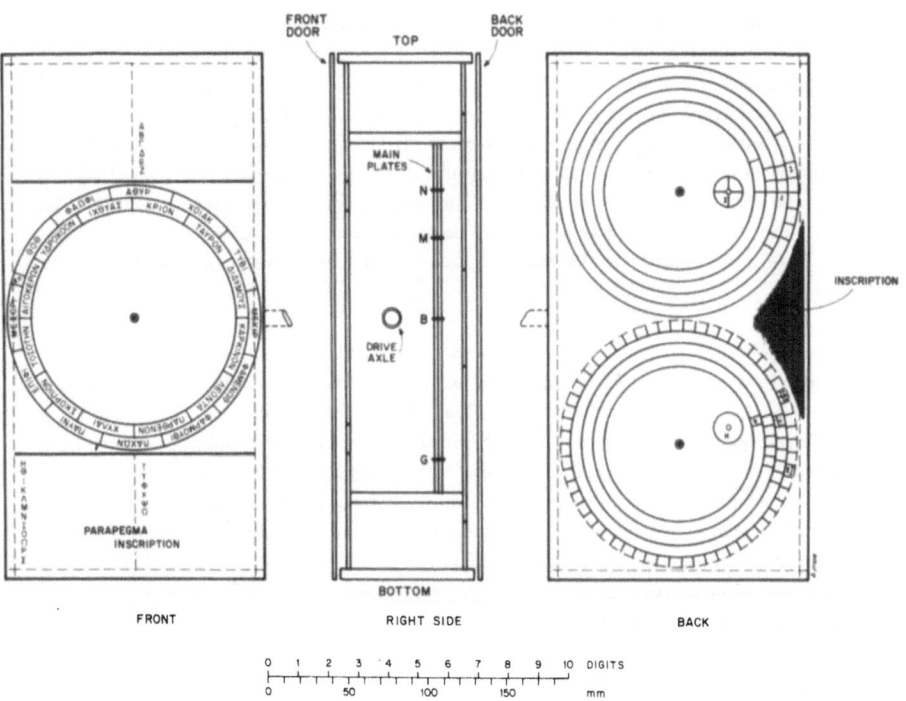

from the region of Rhodes to Rome. I like to think it possible that it·was Cicero himself who lost his baggage in the Antikythera Channel.

As now reconstructed, the device was originally a box measuring exactly one Roman foot in height, half a foot in width, and a quarter of a foot in thickness. The side casing was of wood, and the front and back plates were of brass, inscribed with the dials and covered by doors on which the various settings of the instrument and its general operation were written, together perhaps with the dedicatory and makers' inscriptions.

The front dial of the instrument showed the circle of the zodiac, and on a slip ring around it there was the normal Greco-Egyptian calendar containing each year 12 months of 30 days and 5 epagomenal days. Having no leap-year, the calendar and therefore this ring had to be rotated manually one-quarter of a division each year, or one whole division every four years. It is from this scale that the instrument has been dated. Within this dial there is a pointer or a little model to indicate the place of the Sun and another to indicate that of the moon. Possibly the Moon is shown in its phases, and though no trace of such mechanism exists there could also have been a unit geared to show the mean motions of the five planets. Along the zodiac scale there are also placed at irregular intervals the 26 letters of the Greek alphabet, and an accompanying inscription tells letter by letter, as it is reached by the Sun pointer, the various heliacal risings and settings of major bright stars and constellations which are customarily recorded in a standard Greek *parapegma* calendar — a record which usually gives also the equinoxes and solstices and a set of weather warnings appropriate to each indicated time of year, e.g. the Dog Days, the Halcyon Days.

The back plate of the instrument contains an upper and lower dial of which only the latter can be elucidated. Its main pointer rotates in exactly one synodic month over a series of scales which each contain 59 divisions, each of which therefore corresponds to a half-day, i.e. a night or a day. Inscribed in some of these divisions appear to be numbers giving some data at various points in the cycle of lunar phases. A subsidiary small dial, set like a seconds dial on a modern watch, records numbers of synodic months in cycles of twelve such months which constitute a lunar year. The upper back dial also has a main dial and a small subsidiary, but at this point the gear trains are incomplete right at the end and one can only guess that a longer cycle of perhaps a Metonic cycle of exactly 19 years, or an eclipse cycle of a little more than 18 years may be involved together with some multiple or sub-multiple of this.

Both the gearing and the inscription attest to the fact that what is involved is a mechanization of the Metonic cycle in which 19 solar years corresponds exactly to 235 synodic months and (necessarily) to 254 (i.e. 19 + 235) sidereal revolutions of the Moon. If required to make such a mechanism one would perhaps suppose that gears with 19, 235 and 254 teeth, or various multiples and factors of these, would be involved. For example, the ratio of 235/19 might in fact be accomplished by successive pairs of gears giving ratios of 47/19 and 5/1 working in series and therefore multiplying. What is spectacular about the gearing and the reason for its extremely unexpected sophistication is that this is not done. Instead, the ratio of 254/19 is produced in this way, but that relating to the 235 months is obtained by means of an epicyclic differential gear that mimics the astronomical explanation and actually subtracts the motion of the Sun from that of the Moon in their paths around the zodiac.

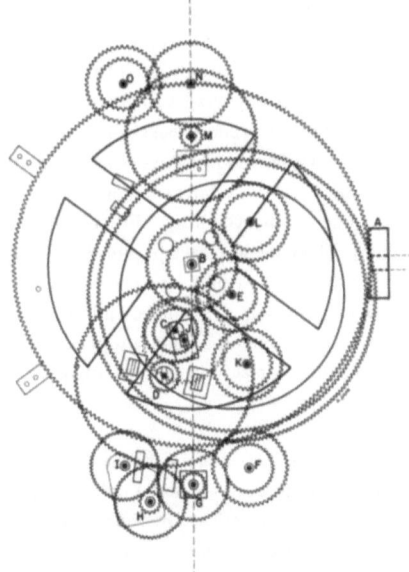

Fig. 5. Plan of the gearing of the Antikythera Mechanism.

Fig. 6. Schematic Diagram of the gear trains of the Antikythera Mechanism.

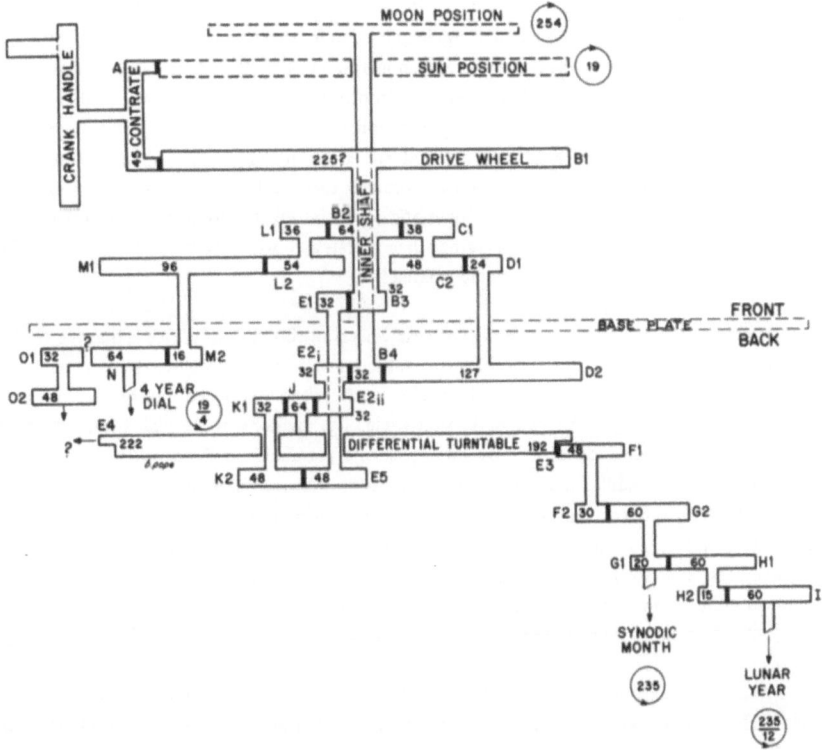

Such a differential gear constitutes one of the most non-obvious devices in the armory of fine mechanics, and its existence at this early date forces a revolutionary upgrading of our entire evaluation of the technological competence of the classical civilizations. The accident that only one of many such devices has been preserved and the fact that the Greeks did not write any classics about such machinery have enabled us to get away with the false assumption that because some Greeks disdained technology as menial work in a slave society, all must have done so. We have in fact a device and a manuscript from 11th/13th century Islam in the same line of evolution, and the differential gear itself turns up in the same context of astronomical clock gearing in 16th century Germany. From here it is adapted by a clockmaker to early textile spinning machinery, where it helps make possible the beginning of the Industrial Revolution in England, and most recently it was adapted by a textile engineer to the first steam motor cars and thence to the back axle of every modern automobile. This saga now has been shown to owe its origin to first century B.C. Greeks and moreover to belong to the ancient tradition of astronomical model-making, and design of *protoclocks* which do not primarily tell the time but rather illustrate the machinery of the cosmos.

Instead of beginning the history of machinery with very simple folk devices and with the disappointingly simple toys described by Heron and Vitruvius, we now have to posit a series of rather complicated geared devices which include the planetariums known to have been designed by Archimedes and lead through the Antikythera Mechanism to the later mechanical astronomical clocks, the later timetelling clocks, and the entire tradition of fine mechanics and scientific technology which has molded the modern world. The line begins then with Archimedes, and I have a strong suspicion that his invention of gear trains which we find flowering in the Antikythera Mechanism was motivated in the first place by the very special requirement of modelling calendrical-astronomical ratios. It has long been familiar in the case of the meso-American Mayan astronomy that the entire calendar revolves like a series of interlocking gear wheels that never slip a tooth, but inflexibly turn over in their sequences as do our modern days of the week. This applies equally to the Metonic cycle and the other interlocking sequences of Greek calendrical art that are so ingeniously described in such texts as the *Isagoge* of Geminos of Rhodes. When one has to satisfy these rational ratios in long series by means of gears between fixed axes the number-theoretic problems are far from trivial, and it is probably this type of mathematical computation rather than metal- and wood-working that was contained in the lost treatise of Archimedes, "On Sphere-making".

A most intriguing point in the general use of gear wheels in this type of device is that the continuity of motion is actually a flaw in the mechanism, rather than a sought property. What is required, for example, is a step by step, tooth by tooth, congruence between the sequence of the days of the week or the month and those of the year or 19-year cycle. The divisions of the dials on the Antikythera Mechanism are numbered or lettered or they bear data pertaining to the division as a whole rather than its endpoint or middle, and one gets the impression that reading them in a continuous fashion, e.g. to 17¼ days or degrees, is as inappropriate as it would be to speak of quarter past Saturday or quarter to January. The model shows a step by step sequence of discrete omens and events in the various cycles, and it is only the tradition of the protoclocks with their geometric dials which later combined with this to become very familiar and readily available as a means of showing and telling the passage of a continuous flow of time and steadily moving cycles.

At all events, it becomes more clear, thanks to this new evidence of the Tower of Winds, the Antikythera Mechanism, and the corpus of sundials, that we are dealing with a tradition of astronomical model-making that became of such sophistication at the very beginning of our era that it was readily available for the new purpose of telling the time in a utilitarian fashion. The techniques came first and with great elaboration and much improvement before the utility first arose. Indeed, so great was the development that these protoclocks seem to have engendered a special reverence for their ingenious precision, and it seems now far less accidental that astronomical showpiece clocks should have adorned the cathedrals of the medieval West and that the calling of precise prayer times should have become the chief externalization of the faith of medieval Islam. It is also to this story of the proto-clock that we seem to owe a large part of the tradition of model-making which constitutes the philosophical kernel of scientific theory, and the fine mechanics which was the crucial craft provoking the scientific revolution and the industrial revolution. The origin of timekeeping is therefore not just one of many possible topics but the main thread that has produced the most notable characteristics of modern civilization.

NOTES

1. Derek de Solla Price: "Automata and the Origins of Mechanism and Mechanistic Philosophy," *Technology and Culture* V (1964), 9-23.

    Derek de Solla Price: "On the Origin of Clockwork, Perpetual Motion Devices and the Compass," *United States National Museum Bulletin 218: Contributions from the Museum of History and Technology,* Paper 6, Washington, D.C. 1959, 81-112.

    Derek de Solla Price: *Science Since Babylon,* New Haven: Yale University Press 1961, chapter 2: 23-44.

2. Sharon Gibbs: "Greek and Roman Sundials," Doctoral Dissertation, Department of History and Medicine, Yale University, University Microfilms Dissertation No. 73-14334, 1972.

3. e.g. See C. Hose and W. McDougall: *Pagan Tribes of Borneo,* London: Macmillan 1912.

4. Joseph V. Noble and Derek de Solla Price: "The Water Clock in the Tower of Winds," *American Journal of Archaeology* 72 (1968), 345-55.

5. "The ✡ , ✡ , and ✧ and Other Geometrical and Scientific Talismans and Symbolisms," *Changing Perspectives in the History of Science, Essays in Honour of Joseph Needham,* (eds. Mikulas Teich & Robert Young) London: Heinemann Educational Books *1973,* 250-64.

6. Early accounts of my researches on this instrument have appeared in "An Ancient Greek Computer," *Scientific American* 201 (1959), 60-67, and *Science Since Babylon,* New Haven: Yale University Press 1961, chapter 2: 23-44; the new evidence will be published in a full monograph later in 1974 in *Transactions of the American Philosophical Society.*

Monasticism and the First Mechanical Clocks

J.D. NORTH

Had I been speaking in an ancient Athenian law court rather than in modern Japan, my address would have been timed, not by a chess clock but by a clepsydra. Aristophanes[1] and Aristotle[2] both testify to its use in the courts, and the custom was still remembered in the time of Lucian, or a little later, when he or an imitator reported that Demades had made fun of Demosthenes for preferring water to wine. 'Others spoke to water, but Demosthenes wrote to it'.[3] Roman senators timed their discourse by means of the clepsydra, as Pliny, Cicero and others bear out; and Cicero indicates that the very acts of asking and giving leave to speak were described, respectively, as ' seeking the clock ' and ' giving the clock '.[4] The clepsydra was used in ancient Greece for timing military watches[5] and for astronomical measurement.[6] It is said that according to Lucian it was used – and if this were not so dubious a reference it would be the oldest historical reference to such a use – for sounding a bell.[7] There is nothing intrinsically surprising about a Greek hydraulic automaton capable of sounding a bell at regular intervals, for Ctesibius had previously, by means described in some detail by Vitruvius, that is, by hydraulic timepieces, caused figures to move, pillars to turn, stones and eggs to fall, trumpets to sound, and other displays *(parerga)*.[8] Other winter timepieces described by Vitruvius, driven likewise by water power, required a measure of astronomical understanding if they were to be used to yield the time, having as they did an astrolabe dial.[9]

I mention these early literary references not merely as an introduction to the medieval scene, but as a reminder that a timepiece is much more than a mechanism. To attempt to understand it in isolation from its human setting is to forget that it was made in the first place in response to specific human needs. As those needs altered, its form tended to change, and there might be times when its very survival was in jeopardy. It is doubtful, for example, whether in a simple agricultural society much meaning would be found in an Athenian clepsydra, let alone in a Vitruvian astrolabe dial. However dark or bright a historian might be inclined to find the first few bucolic centuries following the final collapse of the Roman Empire, one thing seems certain: if the Vitruvian tradition was handed down within the West, it was by the chance preservation of text or artefact in a society hostile to the civilization which gave birth to the underlying rationale. In Islam conditions were ostensibly more favourable, and the tradition of hydraulic automata was positively enriched, helped by a religion that encouraged astronomical expertise of a high order,[10] not to say by a climate that seldom encouraged water to freeze. In due course, Europe became more conscious of the need to learn from the past, and something significantly new was added to the ancient tradition of anaphoric clocks, this without much by way of the mediation of Islamic thought. The institution chiefly responsible, both for preservation and for development, was the Church.

This should not be very surprising. The Church was rich and powerful. It controlled almost all academic education. It could afford to employ the best available craftsmen, and numbered among the lay brethren attached to the monasteries must have been some of the most skilful artisans of the time.[11] The monastic orders had played an important part in the development of the many mills and contrivances which could be powered by a waterwheel, which thus made more time available to the monks for their proper vocation of prayer and meditation. (The Cistercian rule enjoined monks to build near rivers so as to make best use of water power.) The administrative machinery for such a vast enterprise as the construction of a large mechanical clock was available nowhere outside the Church and the courts of princes. The transmission of power by means of gearing, rope, and pulley, was admittedly a part of a common stock of craft-knowledge from ancient times. One must be wary of exaggerating the conceptual difficulty inherent in these ideas. The principle of the rope drive, for example, should be immediately evident to anyone who lets slip the windlass as he hoists up a bucket from a well.[12] But the mathematics of gear-trains − and this is especially true of astronomical trains − were the province of the well educated alone. And education of the appropriate sort was the monopoly of the Church.

The Church was a feudal force, and through the close regulation of the monastic day a measure of regularity was imposed on society at large. With or without automatic control, the canonical hours of the monastic life were struck eight times daily on a tower bell which, in summoning the monks to prayer by day and by night, was heard far beyond the confines of the cloister. The rules for the market of Salisbury, for example, in the early years of the fourteenth century, refer specifically to the striking of the cathedral clock,[13] and there are even cases where the Church obliged the townsmen to maintain a church clock. I do not want to give the impression that people outside the churches had no wish to keep their own time: in Cologne, for instance, a guild of water-clock makers was already in existence in 1183, while by 1220 they occupied a whole street, the *Urlogingasse*.[14] Within the Church, however, timekeeping − which had at first been no more than an aid to the regulation of worship − soon became almost a necessary ingredient of ritual. In due course, timekeeping was to encompass the drama of a mechanical cosmos, combined with a wide range of more earthly amusements: striking jacks, jousting knights, wheels of fortune, and in fact all that the Vitruvian word *parerga* might have signified. Not that we should exaggerate the element of drama, or melodrama, to the exclusion of all else. If the concept of precision in timekeeping had been unknown in the fourteenth century, Chaucer could hardly have written, as he did in the *Nun's Priest's Tale*, of the cock Chaunticleer thus:

'Wel sikerer was his crowyng in his logge
Than is a clokke or an abbey orlogge.'[15]

In lands where the Sun does not always shine, and in a community for which the first cockcrow of the day was not early enough for the call to matins, some sort of clock was sorely needed. A classic list of references to early examples has gradually accumulated, and I have no wish to tread more than absolutely necessary on already well-trodden ground, but it is impossible to appreciate the mechanical clock without a knowledge of the role played by its predecessors. There is no good reason to suppose that knowledge of the water clock in the West depended on its importation from Islam. Considerable doubt attaches to the story that there were diplomatic relations between Charlemagne and Harun al-Rashid, as are presupposed by the legend that the Caliph sent the Christian emperor a water-clock, a silk tent, and so on.[16] The oldest detailed account of the construction of a water-driven alarm is in a tenth or eleventh-century manuscript, now unfortunately incomplete, from the Benedictine monastery of Santa Maria de Ripoll, at the foot of the Pyrenees.[17] The text does not appear to be a translation from an Arabic original. There is every indication that the hydraulic driving mechanism, of which the description is lost, did not turn any astronomical dial, but merely a dial to help in setting the alarm. The weight-operated striking mechanism was very simple: an ordinary rope-and-weight drive turned on an axle which acted as a flail on small bells hanging from a rod. This very primitive device had to be re-set after each use. This re-setting was perhaps done by the sacristan, as is explained in the Cistercian Rule which dates from the early part of the twelfth century.[18] In Rule XCIV, the sacrist was instructed to set the clock *(horologium temperare)* and cause it to sound *(facere sonare)* on winter weekdays before lauds, unless it was daylight. He was to use it to awaken himself before vigils each day, before lighting up the church. In Rules LXXIV and LXXXIII, the brethren and sacrist were told to ring the larger bell *(signum* or *campana)* on hearing the clock *(horologium).* These are not the earliest constitutions known that relate to the subject. In the eleventh century, William, abbot of Hirsau, gave similar instructions to the sacristan, using words which echoed the Cluniac rule and also the ancient customs of the monastery of St Victor in Paris, where the registrar *(matricularius),* the sacristan's companion, was to adjust the clock.[19] Such adjustment was necessitated by the use of unequal hours.[20] Similar commentary on the Benedictine Rule confirms that the same customs were adopted generally.

There can be little doubt that mechanisms were operated by water. The supporting evidence is well known, and has been ably summarized by C.B. Drover.[21] There was the fire of 1198 at Bury St Edmund's abbey, which the clock doubly helped to extinguish, first by rousing the master of the vestry, and secondly by providing water. There are the fragments of slate dating from 1267 or 1268 from Villers Abbey near Brussels, relating to the method of setting a water-clock by the Sun. There are the Alfonsine books of the next decade, which describe both a water-clock and a mercury-clock.[22] There is the clock case of about 1250, drawn in the sketch book of Villard de Honnecourt. And then there is the illustration to which it was the aim of Drover's article to draw attention, an illustration of almost exactly the same period as Villard's, showing a similar clock-case.

The illustration is from a moralized Bible of extraordinary richness.[23] The prophet Isaiah is shown giving the sick king Hezekiah a sign from the Lord that fifteen years will be added to his days. This was done 'that the Sun would be moved back ten degrees in the clock'.[24] I shall discuss this illustration briefly here since it seems to me that it has been misinterpreted by Lynn White Jnr. in a work which is so well known and generally so accurate on points of detail that his version is likely to become canonical.[25]

The medallion illumination illustrates 2 Kings XX.5-11. The same story is told in Isaiah XXXVIII.8. The Bodleian manuscript is one of three needed to complete the original Bible, and the Isaiah passage is to be found in the complementary manuscript in the Bibliothèque Nationale, MS Lat. 11560, f.120r, where, sad to say, there is no comparable illustration, owing to the different wording of the Vulgate.[26] Both illustrations show a symbol denoting the Sun, which is emphatically not a 'fan-escapement to slow the action of the chime, at the striking of the hours, by friction with the air'[27] but is a representation in accordance with a perfectly standard convention.[28] The fifteen divisions of the only visible wheel might be significant, but in view of thirteenth-century artistic conventions, it is unlikely that the number does more than pick up the '15' of the years mentioned in the text, as seems to have happened in the Isaiah illustration. It is certainly rash to conclude that since 15 degrees represent an equinoctial hour, therefore the wheel was probably meant to turn once every hour. In any case, the wheel seems to have about 24 teeth, which rather suggests that it might have been meant to turn once in a day.

The clock clearly has some sort of rope drive, but there are several mechanical reasons for thinking that it did not work on the same principle as the Alfonsine mercury clock. (In the latter, a couple created by viscous forces, as mercury flows between radially divided compartments of a wheel, is opposed by a couple created by a rope drive, thus establishing dynamic equilibrium in the wheel, which turns slowly.) Not the least of the objections to this interpretation of the biblical painting is that water is there shown clearly gushing forth from an animal's head spout into a cistern below the clock. There are unsolved problems of interpretation, certainly,[29] but it is difficult to avoid the general conclusion that, however it worked, the clock was mechanically a simple affair, offering little more than encouragement to the men who would make the first purely mechanical clocks.

Within a century, however, two clocks were begun, by Richard of Wallingford and Giovanni de' Dondi respectively, which were so extraordinarily complex that in the sixteenth century the first could be described as even then surpassing all others in Europe,[30] while the second was so intricate that Charles V could find only one technician, Gianello Torriano, who was capable of repairing it, others having failed.[31] How may we explain such a technological advance, for which there were very few parallels in the Middle Ages and few indeed before the industrial revolution of the eighteenth century?

I must first remind you of that remarkable passage to which Lynn Thorndike first drew attention, in the commentary written by Robertus Anglicus in 1271 on the most widely used of all medieval astronomical textbooks, the *De Sphera* of Sacrobosco.[32] After a discussion of equal and unequal hours,[33] Robert goes on at some length:

> Nor is it possible for any clock *(horologium)* to follow the judgment of astronomy with complete accuracy. Yet clockmakers *(artifices horologiarum)* are trying to make a wheel which will make one complete revolution for every one of the equinoctial circle [i.e. the celestial equator], but they cannot quite perfect their work. If they could, it would be a really accurate clock and worth more than an astrolabe or other astronomical instrument for reckoning the hours if one knew how to do this according to the method aforesaid.

> The method of making such a clock would be this, that a man make a disk *(circulum)* of uniform weight in every part, as far as could possibly be done. Then a lead weight should be hung from the axis of that wheel, and this weight should move that wheel so that it would complete one revolution from sunrise to sunrise, minus approximately as much time as it takes about one degree to rise.[34]

This all suggests that no form of mechanical escapement was known to the writer in 1271, and the simple arrangement he describes is not incompatible with the water-clock illustration discussed earlier. Within a few years, however, the number of documentary references to *horologia* grows so very rapidly that we can only suppose the mechanical escapement to have been found at last. C.F.C. Beeson is persuasive in arguing that the earliest of all European records of a clock with such a control is that of 1283, in the Annals of Dunstable Priory in Bedfordshire.[35] This was a house of Austin canons. The clock was set up alongside a great painted crucifixion scene, with attendant images of Mary and John, on the rood-screen and loft, or gallery. Beeson follows with records from Exeter Cathedral (1284), Old St Paul's, London (1286), Merton College, Oxford (1288?), Norwich Cathedral Priory (1290), Ely Abbey, a house of Benedictine monks in Cambridgeshire (1291), and Christchurch Cathedral,

Canterbury (1292), all before the turn of the century. Taken singly, the records are easy to view with scepticism, but taking them together, and noting especially that relatively large sums of money are involved in payment for the materials used, they persuade us that the mechanical clock had indeed arrived on the scene. When the *orologiarius* Bartholomew drew 281 rations for three quarters and eight days in 1286 at St Paul's, he was surely not building either a sundial or a water clock.

Although it is possible to be reasonably precise as to the time of the invention, the place of origin of the mechanical clock is entirely unknown. Italy has been canvassed, mainly, one suspects, on the grounds that Italy was always a century in advance of the rest of Europe. The earliest acceptable Italian record of which I am aware, however, relates to the year 1309, when an iron clock *(horologium)* was set up in Sant' Eustorgio in Milan.[36] A bell on the bridge at Caen was in 1314 associated with a clock *(l'orloge)* there and according to its inscription served the common people, but this need not have been a purely mechanical clock, and in the absence of further documentation its watery surroundings do not encourage the idea that it was so. The known early English records are at the present time much the richest in Europe, and I am obliged to give most of my attention to them; but I certainly do not suppose that the mechanical escapement was for this reason an English invention. I find it hard to believe, nevertheless, that any early centre of clock-building could have been more advanced than that which took in Norwich and St Albans, and most of what I have to say will relate to the remarkable work done in these two places.

The Sacrist's Rolls of Norwich Cathedral from 1321 to 1325 contain the first extensive financial records concerning the construction and installation of a large mechanical clock.[37] The man in charge of the work was one Roger Stoke, who later worked at St Albans, and who was in both places assisted by Laurence Stoke. The clock had a very large astronomical dial — it was of iron plate and weighed 87 lb — with models of the Sun and Moon, automata, including 59 sculpted images (done by one Adam, a wood-carver), and a choir or procession of monks. There was much colouring and guilding. Smiths, carpenters, masons, plasterers and bell-founders were engaged over a period of three years. The competence of most of the craftsmen concerned seems to have been equal to the occasion, but the making of the main astronomical dial went less smoothly than the rest. In 1323 the fabrication of the large plate was entrusted to Robert of the Tower (Robert de Turri) in London, but in his hands the whole work was ruined. The man was himself ruined *(depauperatus),* and only 10 of the 18 shillings advanced to him could be recovered. Other artisans proved to be equally ineffectual, ruining the material in their attempts. Men were sent from Norwich to London for news of progress, but at length it was necessary for Roger Stoke himself to ride to London to supervise the engraving of the plate. The total cost of the clock was in excess of £52. There are many ways of working out a modern monetary equivalent, none very satisfactory; but in terms of the salary of the best craftsmen of the time, this amounts to around $250,000, in modern American terms.

Complex as the Norwich clock obviously was, it was as nothing by comparison with that designed by Richard of Wallingford, however similar the two might possibly have appeared outwardly. Before considering the mechanical intricacies of the St Albans clock, I should like to consider that aspect of the clock which gave so much trouble to the Norwich builders, namely the dials. These had the same general appearance as an astrolabe, showing the daily rotation of the sky — the stars, and perhaps the Sun and Moon — against a second 'map' of the observer's coordinate lines and unequal hour lines, the horizon and the meridian being chief of these. We know that the Vitruvian anaphoric clock had such a dial,[38] and it is of some interest to ask whether the ancient tradition was ever completely lost.

I shall offer two tentative arguments for supposing that the tradition had some sort of continuity. The first rests on the fact that there is documentary evidence for the survival from Roman, as opposed to Islamic, sources of the tradition of mapping the sky in stereographic projection in the manner required for the main dial — the most difficult theoretical aspect of the construction. Parts of two discs from anaphoric clocks have been found, one at Salzburg, and one at Grand in the Vosges, both of the 2nd century A.D.[39] Judging by the Salzburg disc, the Sun was not moved mechanically with respect to the stars, but was simply plugged into one of 360 or 365 holes distributed around the ecliptic. Other astrolabe dials are those of the water-clocks of Islam, but here one must be cautious and remember the fallacy of *post hoc ergo propter hoc*. In the Islamic world the dial generally resembled the conventional astrolabe, with its pierced star map placed between observer and the plate of local coordinates. The same seems to have been true of the thirteenth-century Alfonsine dials, which were presumably inspired by Islamic sources.[40] In the European tradition it seems to have been far more often the other way round. In this way, it was possible to paint the constellations in as much detail as was thought desirable. The flimsy overlay of wires representing a horizon, meridian, unequal hour lines, and so forth, offered no great obstacle to vision.[41] Bearing the typical European arrangement in mind, there is a certain class of manuscript illustrations dating from the ninth century onwards which closely parallel the anaphoric clock tradition in the West, as I shall explain in a short digression.

By the ninth century, encyclopaedists like Scot Erigena and Isidore of Seville had brought the contents of several late Roman authors to the notice of the educated few, while Lupitus, Hermann and possibly Gerbert, in the tenth century, drew on Arabic sources for their writings on the astrolabe.[42] It is in association with the ancient writings of Aratus (as transmitted by Germanicus), Hyginus and Macrobius on Cicero, however, that we find a certain class of extremely interesting manuscript illustration of this period.[43] One fine example included with scholia to Aratus is in a manuscript of the early ninth century now in Munich.[44] At first sight this is merely a planisphere, showing the constellations pictorially, with one very clearly marked circle. This proves to represent not the ecliptic, but the galaxy, the Milky Way. On closer examination, however, no fewer than eight circles are revealed within the outer pair bounding the planisphere, all in approximate — and not accidental — stereographic projection. Whoever painted this diagram seems to have known the principles of astrolabe projection well. He was certainly not merely painting an ordinary astrolabe rete, since the Milky Way is never — at least to my knowledge — found on an astrolabe, while there is also on the painting an uncharacteristic circle with north polar distance of about 37°. This could be a zenith-track for geographical latitude 53° N,[45] although it is more probably an arctic circle set to a conventional 36° (the height of the Pole at Rhodes). The stars are, moreover, drawn with the signs in a clockwise order, and not as on an astrolabe. I shall return to this point later.[46]

A second, but somewhat different illustration of the same period, also associated with scholia to Aratus, is to be found in the Berlin Codex Phillippicus 1830 at ff. 11 and 12. It is redrawn, perhaps not absolutely accurately, by Georg Thiele,[47] and on the evidence of his drawing it is impossible to know whether the original is as carelessly drafted as it appears to be from Thiele's version. Now, however, we note that this diagram has the sense of rotation of the zodiacal signs as found on an ordinary astrolabe rete, unlike that of the Munich diagram. Yet another example of this type of illustration, which can now be examined in colour in a modern printed source, occurs in a twelfth-century Spanish manuscript.[48] Thiele's *Antike Himmelsbilder* is a useful source of precise information on the texts with which these illustrations are associated, and the book is in fact prefaced with a list of 27

Fig. 1. Construction lines taken from the
constellation map of C.L.M. 210, f. 113v.

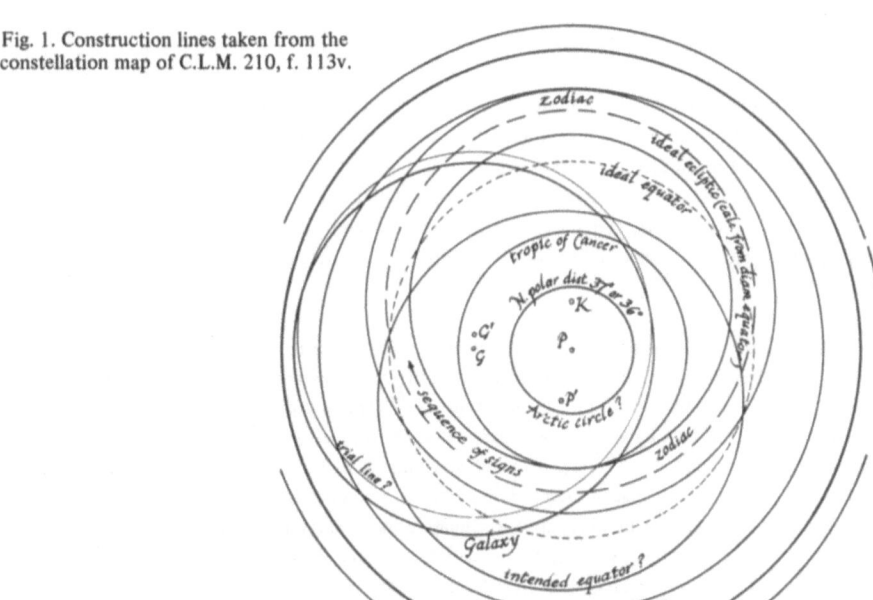

Notes on the constellation map of C.L.M. 210, f. 113v.

1. The projection is from the *south* pole, as on an astrolabe, but the stars are depicted *as seen,* and not with the order to the signs reversed, as on known astrolabes.

2. The tropics are not drawn with our convention, for one touches the northern limit of the zodiacal band, the other the southern limit.

3. The hesitant way in which the circle centred at G' is drawn, together with the equality of its radius and that of the circle centred at G, and their proximity, suggests that the former is a trial line, or a mistake. Neither can be the projection of a great circle, whether from P, P' or K. See below.

4. The circle centred at P' appears to be the intended equator, since no other line will serve, and since the diameter is correct. In stereographic projection, however, the equator would be centred at P; see the dotted line. (The tropics must be concentric with the equator.)

5. The small circle centred at P has a north polar distance of about 37°, and could mark the arctic circle in keeping with one ancient convention (36° = latitude of Rhodes). Alternatively, it could be the locus of the zenith point for an observer in a latitude of approximately 53°. (The MS was for some time in the library of St Emmerammus, in Ratisbon.)

6. The catalogue entry on C.L.M. 210 has

> 'f. 114 Excerptum de astrologia . Adiecta est picta caeli tabula. Inc.: Duo
> sunt extremi vertices mundi quos appellant polos ... '

This text was edited by E. Maass in 1898.

manuscripts (but not all of them illustrated).

I have no wish to suggest that these diagrams were copied from an anaphoric clock plate, but they show that at least from the time of the ninth century the necessary skills for making such a plate were once again available in Europe, and that Pacificus, Archdeacon of Verona, could well have had on the clock mentioned in his epitaph a 'song of the heavens' in the form of an astrolabe dial of one sort or another.[49] It is quite possible, of course, that the medieval illustrations were copied from ancient examplars which were themselves taken from anaphoric dials, or were even done by craftsmen skilled at both arts.

Coming down for a moment to the thirteenth century and another record of an astrolabe dial, this time from northern Italy, we find what is probably the oldest extant detailed description of the sequence and counts of the wheels of a planetary model.[50] This was probably water-driven, if it was ever built, and if so it must have demanded many advanced technological skills.[51] The device had an astrolabe dial in a vertical plane,[52] and it exhibited a number of Ptolemaic planetary motions. It is not at all unlikely, however, that the description related to a non-European device, perhaps that given to the Emperor Frederick II, in 1232, by the ambassadors of al-Ashraf, Sultan of Damascus. This highly valued gift was presumably lost in the seige of Parma in 1248, with the Emperor's treasury, his insignia, some of his ministers and, no less significantly, his harem.

I now come to my second argument — and one which is, I think, much stronger than the first — for supposing that there was a tenuous European continuity in the transmission of the ancient tradition of putting astrolabe dials on clocks. If we look at the astrolabe dials of the great European cathedral and abbey clocks of later centuries, we find that they are for the most part in stereographic projection from the north pole, rather than — as with almost every known portable astrolabe — from the south. This is true of the clocks at Valenciennes, Münster, Prague, Bourges, Doberan, Lübeck, Lund, Stralsund and Ulm, for example. Notre Dame at St Omer, and Berne, were relatively rare specimens in south projection, and like the Alfonsine mercury clock and the De' Dondi clock, they had a rete of stars which was turned by the mechanism. In the typical anaphoric clock, however, and in the Salzburg fragment, and in Richard of Wallingford's clock at St Albans, while the projection was from the south pole, the rete was fixed and was a rete of hours, rather than of stars. If the Phillippicus and Osma manuscripts had been drawn from a moving plate, this would have been in the same tradition.[53] I would therefore like to suggest, very tentatively, that Richard of Wallingford was following an ancient tradition perpetuated, whether by artefact or document, within the monasteries of northern Europe. And lest it seem that the probabilities are high that a fourteenth-century author should hit on the ancient arrangement by chance, it should be noted that there are sixteen possible arrangements of a rete and plate, and that no fewer than five of these are actually found in use among the clocks I have mentioned.[54]

The first English clocks seem to have been made almost wholly of iron, and to have been of large dimensions: a frame three or four feet across was not unusual, and the frame of the St Albans clock was probably more than twice as great as this. De' Dondi's clock, of the mid-fourteenth century, was of brass and much smaller, while in the inventory of Charles V it is recorded that Philippe le Bel, who died in 1314, possessed a clock of an even more costly metal, silver, 'une reloge d'argent, avec deux contre poix d'argent empliz de plomb'.

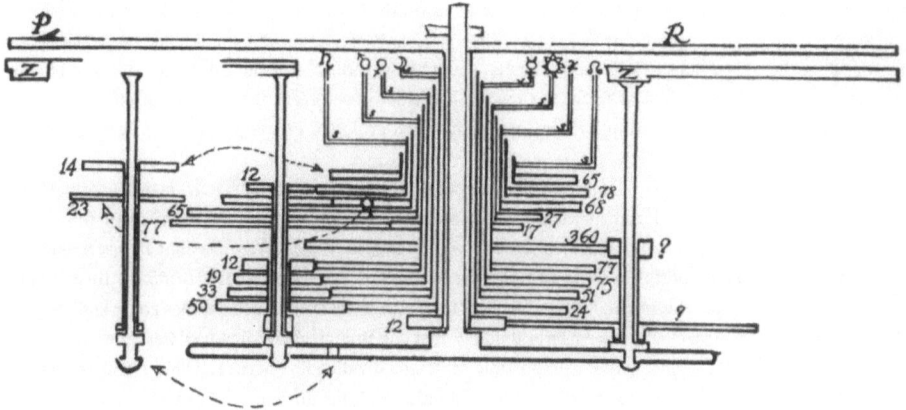

Fig. 2. One possible reconstruction of the 'device of certain remarkable wheels'
described in the manuscripts from northern Italy (13th cent.).

R is the rete, Z the zodiac, P the pointer, and C the supporting column.
The models of the planets are indicated by their symbols, and the numbers indicate the numbers of teeth on the adjacent wheel. For clarity,
one train of gears has been drawn out of position. It is to be inserted in
a way indicated by the dotted arrows.

The Church cannot perhaps claim much of the responsibility for the rapid improvement in the quality of metalwork or for the rapidly growing numbers of metal-workers at the end of the thirteenth century and the beginning of the fourteenth. Chain armour was giving way to intricately hinged plate, and cannon were for the first time being employed in European warfare.[55] There was social mobility enough, however, for ecclesiastic and artisan to come together on such a venture as that of making the most elaborate device in metal designed at that time. Some clocks, to be sure, were relatively simple devices, but a high proportion of those early clocks of which we have records were much more. Some of them showed jacquemarts to amuse or astonish the onlooker, and others perhaps a simple astrolabe dial, but far more significant were those which displayed the daily and annual movements of the Sun, the Moon (and possibly the planets), the phases of the Moon and other astronomical phenomena. At least in its highest form, the mechanical clock was largely the product of an intellectual movement going back to antiquity, and one with which Professor Price's paper very effectively deals.

The tradition of the geared planetarium simply could not have begun or have been maintained without considerable astronomical and mathematical knowledge. The single diurnal movement of the water-clock was something which could be altered at will by adjusting the water-flow; but to interrelate the complex movements of the planets and the cosmos requires a far from trivial ability to compute trains of gears. Since the same ability is needed to compute even the much simpler going and striking trains of a mechanical clock, it seems not unlikely that the first mechanical clocks were the product of the academic world — and here it is impossible to draw a dividing line between Church and University — rather than of the unschooled artisan. We need only recall the difficulties encountered in the much simpler matter of engraving an astrolabe dial for the Norwich clock.

Richard of Wallingford was in an excellent position to unite the craft tradition with the academic. As I mentioned earlier, he was the son of a blacksmith of Wallingford in Berkshire. Orphaned at an early age, he was sent by the Prior of the Benedictine house at Wallingford to study at Oxford, where he stayed with one short break from about 1308, when he was roughly 16, until he took the monastic habit at St Albans at the age of 23. After his double ordination, first as deacon (1316) and then as priest (1317), he was sent back to Oxford by his abbot. There he studied philosophy and theology for nine more years, in the course of which he wrote some remarkable pieces on trigonometry and astronomy, as well as astrology — which was of course almost *de rigeur* at that time. In 1327 he visited St Albans at a time which coincided with the abbot's death. He was himself elected abbot, duly visited Avignon for the papal confirmation, and at length returned to rule his monastery, which was in several ways the most important in England. He had a difficult time. He contracted leprosy on his overseas journey; his abbey was saddled with numerous debts; the townsmen were in revolt against the feudal privileges of the abbey; and there was an inquisition to investigate his fitness to rule. Despite his firmness towards them, however, he appears to have inspired enormous confidence in most of his monks, and when he died after only nine years in office, he left behind him a far greater secular inheritance than did his predecessor. At St Albans he was remembered above all else for his clock.

Leaving aside, for the present, the nature of the going and striking trains, I will first say something of the general character of the astronomical section, which was built into an adjacent frame. It seems, judging from the manuscripts, that Richard began to write a thoroughly formal treatise on the whole subject of making an astronomical clock, that he completed a first section on the arithmetical techniques to be used in calculating trains of gears for general astronomical use, but that he died before editing his several drafts of chapters describing the actual construction of a clock – the St Albans clock. A single manuscript of these drafts survives, not in the hand of the author, but copied out for the sacristy at St Albans by a man who had a very poor idea of the best order, and who mixed the drafts with other quite irrelevant texts. The manuscript was later misbound, and a whole gathering is missing, presumably with details of the actual planetary trains hinted at by certain surviving data.

In successive parallel drafts we can see how Richard improves upon his first thoughts until the accuracy of the gear trains he advocates is exemplary. By using a transported correction train to drive the Moon, for example, the theoretical error in its mean motion is only 7 parts in 10^6. But accuracy is not only achieved by the careful determination of gear ratios. The problem of driving the Sun round the zodiac at the correct rate is simply that its speed should be variable. Richard solves this problem by computing an oval gear wheel, with contrate teeth, linked to a long pinion the axis of which is parallel with the plane of the oval wheel. The result is highly accurate, but more to the point, there is absolutely no known historical precedent for gears of this design, which until the discovery of the St Albans manuscript would have been generally thought beyond the potential of medieval technology. (I should say here that the oval wheels of the later De' Dondi clock are of an entirely different sort, and serve a different purpose.) The intricacy of the gear work of the St Albans clock may be judged from another type of wheel, which transmitted power from conchoidal contrate teeth to a worm gear. This offers an excellent solution to the problem of smooth transmission, and – unlike the oval wheel – is still much used.

Among the clock's other refinements, I may mention a half-blackened Moon-globe which, rotating on its axis as the globe moved round the dial, showed the lunar phase. But this was not all. The lunar globe was so placed that it was automatically drawn under a small eclipsing disc whenever the time was ripe for an eclipse of the Moon. I hardly need say that this was an extraordinary feat of design.

Very little about the arrangement of the planets of the clock is now known, but when John Leland saw the clock at the time of the dissolution of the English monasteries, he reported that it included planetary trains and a tidal dial.(English manuscripts of the period are frequently found which tabulate the time of the tides at London Bridge, and a tidal dial would have been a relatively simple affair, following the lunar motion.) And this is a suitable point at which to record the placing of the clock. Like so many abbey clocks, it was put into the south transept of the church, and was therefore rather better protected from the weather than the average tower clock.

The St Albans clock was not completely finished during Richard's lifetime, but there is no reason to suppose that those who finished the work altered his design.[56] Judging by a remark made in chronicles of the abbey after his death, there was no one capable of doing comparable work – as he had indeed foretold, in excusing his lavish spending on the work on the occasion of his being upbraided by the king who pointed to the poor state of repair of the fabric of the abbey. There was, however, an established clockmaking craft, and it is of the greatest interest that the very Stoke family who had worked at Norwich were also engaged over a long period of time in work on the St Albans clock. Roger and Laurence Stoke were professional *horologiarii,* but they were lay brethren,[57] who must have done other monastic work in the area of England north-east of London. How tight the monastic hold on their profession was, at this period, is difficult to determine for lack of evidence, but the only stylistic parallels I have found with the St Albans clock are with the west country series at Exeter, Wells, Wimbourne and Ottery St Mary. All of these, however, are much later. The resemblances are slight, and mostly concern the simplest aspect of the lunar mechanism.

Finally, I come to the mechanical escapement, that characteristic of the truly mechanical clock which distinguishes it from what went before. It has always been supposed that the first mechanical escapement was the familiar verge and foliot, that is to say, the sort found on the De' Dondi clock, and found on large church clocks throughout Europe, until at length it was replaced by the pendulum. It cannot be said with certainty that this was not the first; but the first escapement of which we have certain knowledge, namely that of the St Albans clock, was substantially different. The clock had two similar escapements, one to control the going train and one to sound the bell at the hour — and on a twenty-four hour system, with a number of strokes equal to the hour. A double-edged pallet was thrown first one way and then the other, each edge being acted upon by one of several radially disposed pins on one or the other of a pair of similar wheels fixed together in parallel planes a suitable distance apart. (The pins on one wheel came mid-way between those on the other.) A vertical verge with cross-bar, carrying weights to adjust the moment of inertia of the verge assembly, and hence the period of swing, oscillated accordingly, and on the striking side of the clock it was these oscillatory swings which caused the bell to be struck. Needless to say, by working with one pallet it was not necessary to traverse the arbor of the crown wheel (or pin wheel) with another arbor. On the other hand, the St Albans mechanism required an extra wheel, and it might well have quickly fallen out of favour on the grounds of expense. I am convinced, however, that it was mechanically more efficient than the common verge and crown wheel. To distinguish this type of escapement, I shall use the word 'strob', which is that given to it in the manuscript. In dictionary language, strob is 'etym.dub.' The pallet is simply called a 'semicircle'.

There is no drawing of the strob wheels and verge in the St Albans manuscript, and the reconstruction is one which was achieved by painstakingly piecing together a whole series of measurements of things given in words, many of which were unknown to the dictionaries. Having, after many attempts, arrived at a successful solution, its correctness seemed to be confirmed when I found drawings of the very same escapement in Carlo Pedretti's *Studi Vinciani*,[58] reproduced, need I say, from the work of Leonardo da Vinci. These drawings are from Codex Atlanticus and date from about 1495. There are similar drawings in Codex Madrid. It is not for me to correct here those who have ascribed to Leonardo the invention of this ingenious device, but rather to suggest that if an escapement is known from two places at opposite extremes of Europe and from periods 160 years apart, there is a strong likelihood that the escapement was once widely diffused. There is in fact one other manuscript, of the early fifteenth century, in which the same escapement appears to be the subject of discussion.[59]

The two types of verge and foliot escapement — they may be distinguished as single- and double-wheel types — are obviously functionally related. The double-wheel is mechanically superior and more expensive, and on both scores it might therefore be thought wise to place it historically earlier than the single-wheel. It is the earlier, as far as extant records go, and moreover when Richard of Wallingford gave some numerical data for it, between 1328 and 1336, he wrote as though its construction was generally known. In the customary scholastic manner, he went into minute detail on the astronomical and arithmetical side of his draft treatise, but when he came to the going and striking trains, he took their general construction for granted and contented himself with measurements and gear counts. Some of the most technologically interesting parts of his work are thus the most difficult to disentangle, and this is especially true of the part which concerns the striking mechanism. I hasten to add, however, that I have not the slightest doubt that the clock had hour-striking of the sort mentioned earlier, with some sort of locking barrel, even though this directly conflicts with the well-known statement by a certain monk of Malmesbury, who dates the invention to 1373.

On this subject of the oscillatory mechanism for striking the bell, I wish to end. This, in the St Albans clock, works exactly as the main escapement, and it is not inconceivable that such an oscillatory striking device triggered at suitably chosen intervals by a hydraulic clock, pointed the way to the first mechanical escapement proper. (Perhaps my title should have echoed Professor Price's, and read simply 'The escapement before the escapement'.) The bell to be struck in this manner could have been large and the noise impressive, unlike the tintinnabulation to be expected of a water clock of the type illustrated in the thirteenth-century moralized bible. Can it be that the more ponderous tones of a clock sounded in this way were those which, somewhere between 1235 and 1260, it was ordained should be heard at the installation of every new abbot of St Albans? There are water-clocks (*orologia ollarum,* 'clocks of earthernware vessels') briefly mentioned in the St Albans chronicles under the rule of John de Maryns (1302-8), but the earlier reference is more revealing by far. The prior shall enter the church from the chapter house, the rule of John of Hertford tells us, to be presented to God and the Holy Martyr Alban at the high altar, with the striking of a summons, the shawms[60] sounding with the horologe, the tapers lighted round the altar, and the throne uncovered. Very few of those treatises from the centuries following, in which are given precise and complete descriptions of actual clocks, can convey quite as much about medieval humanity as can this passing reference to one way in which a great abbey clock of the thirteenth century was absorbed into the high ritual of the church. Nothing could have been more fitting.

NOTES

1.   *Vespae*, 93. 857.

2.   *Athen. Polit.*, 67. 2.

3.   *Demosthenis Encomium*, 15. The work is probably of the second or third century A.D.

4.   *De Oratore*, 3. 34. 138.

5.   Aeneas Tacticus (4th cent. B.C.) ed. R. Schone, Leipzig 1911, 22-24.

6.   See, for example, Proclus: *Hypotyposis astronomicarum positionum*, ed. C. Manitius, Leipzig 1909, 4-74.

7.   This statement is made in the *Enc. Brit.*, 11th edn., art. "Bell", by H.M. Ross, who probably took it from the best-selling book by A[lfred] G[atty]: *The Bell; its Origin, History and Uses* London, 1847, p.16. In neither place is a reference given, but the source was very probably chapter 6 of Hieronumus Magius [or Maggi]: *Anglarensis de tintinnabulis,* many editions; but that of Amsterdam 1689 is illustrated with an imaginary reconstruction, at p.31. A rope from a clepsydra float trips a weight which operates a bell once only (through a crank!). Maggi, writing from prison, gives no precise references, but says that his relative Johannes Nicolaus Justus made him a copy from an old book in which the Lucian clock was delineated.

8.   *De Architectura*, IX. viii. 4-7. Vitruvius was most probably writing a few years before 27 B.C.

9.   *Ibid.*, 8-14.

10.   Astronomy served not only astrology, a subject to which Islamic scholars added significantly, but among other things it was required by any who would master the lunar calendar adopted by the faith.

11.   The Abbot of St Albans, Richard of Wallingford, whose work I shall shortly describe, was the son of a blacksmith.

12.   One might imagine that the principle of the water-driven wheel was arrived at by the same sort of accidental discovery of reversibility, in this case of such a water-raising device as, for example that which appears in Vitruvius: X.4 (the *tympanum*).

13.   The sources relating to the Salisbury clock are to be found in C.F.C. Beeson: *English Church Clocks, 1280 to 1850,* Antiquarian Horological Society 1971, p.16.

14.   Quoted from E. Volkmann by Lynn White Jnr.: *Medieval Technology and Social Change,* Oxford 1962, p.120.

15.   *Canterbury Tales,* Fragment VII, lines 2853-4.

16. For the historical controversy see E.A. Belyaev: *Arabs, Islam, and the Arab Caliphate in the Early Middle Ages,* tr. from the Russian, London 1969, p.221. Belyaev is perhaps too extreme in his criticism of the 'traditional opinion inspired by Christian pietism'.

17. A full discussion and translation of the surviving part of the text was given in F. Maddison, B. Scott and A. Kent: 'An early medieval water-clock', *Antiquarian Horology,* 3 (1962), 348-53.

18. I quote from C.B. Drover: 'A medieval monastic water-clock', *Antiquarian Horology,* 1 (1954) 54-8.

19. These earlier references and those next following are given in greater detail in John Beckmann: *A History of Inventions, Discoveries and Origins,* tr. W. Johnston, 4th edn., London 1846, pp.346-9. This is a fundamental source of information on the history of the clock. The section was actually written originally by Hamberger.

20. See 33 below. Dante has something to say on equal and unequal hours, and the difficulty of regulating church services, in *Convivio* IV.23, near the end.

21. *Ibid.,* passim.

22. Mr. Francis Maddison, who is preparing an English translation (with commentary) of the text describing the Alfonsine mercury clock, points out that the inspiration for this clock is stated in the text to be a work by 'Iran el filosofo', namely Hero, more specifically where he explains ways of lifting heavy weights.

23. Bodleian Library, Oxford, MS Bodley 270b, f.183v. For a facsimile of the entire MS and its missing parts (B.N. Lat. MS 11560, and B.M. MS Harley 1527) see A. de Laborde: *La Bible moralisée,* Paris 1911-27, 5 vols.

24. '...ut Sol x gradibus retrorsum in orologio reuerteretur'.

25. *Op cit.,* pp.120-1.

26. There is no mention of a *horologium,* but the mention of *gradus* prompts the illustrator to paint a flight of 15 or 16 stairs!

27. White, *op.cit.,* p.121.

28. The same convention is to be seen on numerous occasions in the manuscripts (Bodley 270b, ff. 10r, 16r, 34r, 57v etc.), especially in connexion with crucifixion scenes. It was still being followed more than a century later. See, for example, Bodleian Library, MS Ashmole 1522, ff. 27r, 39v, 40r.

29. Drover was puzzled at the cranked form of the arm supporting the wheel, but a comparison with the astronomical instruments of ff.11r, 24r and 27r suggests that this was the standard way of supporting a wheel, resulting in a 'handle' in the plane of the wheel.

30. The antiquary, John Leland. No references to the material on Richard of Wallingford are given here. My complete edition of his writings is awaiting publication at the Clarendon Press, Oxford.

31. The authority is Bernardo Sacco, 1565. A somewhat different report comes from the notoriously unreliable Cardano. For the texts, see S.A. Bedini and F.R. Maddison: 'Mechanical

universe: the astrarium of Giovanni de' Dondi', *Trans. of the American Philosophical Society,* N.S., 66, part 5 (1966), 37-9.

32. The name Robertus Anglicus points to a man of English family, but the commentary was given as a course of lectures at the university of either Paris or Montpellier.

33. Sometimes called 'equinoctial' and 'canonical' or 'seasonal' respectively. The former are in accordance with our modern convention, each being one twenty-fourth part of a day. The latter are each one twelfth part of day or night. This convention goes back to the ancient world and survived in Japan until the last century at least. A night hour is obviously approximately equal to a day hour only twice a year, near the equinoxes. It should not be thought that the concept of equal hours ('horae de clock', in the late fourteenth century) had to wait for the invention of the mechanical clock, as is sometimes suggested. Astronomers made use of the idea in Antiquity, and it was known in the West in the early Middle Ages from the writings of Martianus Capella, Leontinus, Gerbert and many others. William of Hirsau's *naturale horologium* was probably so called because it showed the equal hours of the 'natural day' (24 hours), rather than of the 'artificial day' (sunrise to sunset).

34. Taken, in a form not significantly altered, from L. Thorndike: *The Sphere of Sacrobosco and Its Commentators,* Chicago 1949, pp.180 (text) and 230 (translation).

35. *English Church Clocks,* Antiquarian Horological Society, 1971, pp.13-14.

36. Mentioned in the Chronicle of Galvano Fiamma. See L.T. Belgrano: *Degli antichi orlogi pubbici d'Italia,* Archivo Storico Italiano, 3rd series. 7, Florence 1868. The clock was restored in 1333 and 1555, and renovated in 1572. See J. Drummond Robertson: *The Evolution of Clockwork,* 1931, p.31.

37. Mr. C.B. Drover kindly provided me with transcripts and photographs of the relevant sections. The Roll for 1323-4 is missing. There are extracts printed in *Archaeological Journal,* 12,1855, and in the Centenary Volume of the Norfolk and Norwich Archaeological Society, 29, 1946.

38. On the nature of the dial, its relation to the portable astrolabe, and the possibility that it was the invention of Hipparchus and came before the astrolabe (which of course is much the better known of the two), see A.G. Drachmann: 'The plane astrolabe and the anaphoric clock', *Centaurus,* 3 (1954), 183-9.

39. For references to the original literature, and for further details, see article by D.J. de S. Price in *A History of Technology,* ed. C. Singer, et al., vol.3, Oxford 1957, pp.604-5.

40. The fundamental studies of Islamic clocks are by E. Wiedemann and F. Hauser. See especially their *Uber die Uhren im Bereich d. islamischen Kultur,* Nova Acta. Abhandl. d. konigliche Leopoldinisch-Carolinsche deutsche Akademie der Naturforscher zu Halle, 100, no. 5, 1914.

41. The multitude of lines on the plate of an ordinary astrolabe (almucantars, etc.) were superfluous on an instrument which was not meant to be matched to observations, while the fine star pointers of an astrolabe rete would have been difficult to identify on a large dial high above the observer's head in a dark church.

42. Gerbert, who was to become Pope Sylvester II, was a highly practical man and it is worth noting that there is mention of a clepsydra in one of his letters (ed. Julien Havet, epist. 153).

Mrs. Harriet Lattin has pointed out to me in a note by Oldoin in Alphonso Ciacconius – Augustinus Aldoinus: *Vitae et resgestae Pontificum Romanorum,* Rome 1677, col. 756 – that Gerbert, as Archbishop of Ravenna, constructed a water-clock there: Horologii aquatilis, seu clepsidrae figura est Ravenna in Herculis regione, quam Gerbertus construxit Archiepiscopus tunc Ravennas. This would have been between April 998 and April 999.

43. Aratus of Soli, of the 3rd century B.C., wrote an astronomical poem based on Eudoxus, which was admired by some Roman writers. Cicero, Caesar, Germanicus and Aviennus translated it, and much still survives. There are at least four English translations. The circles which interest us (galaxy, equator, ecliptic and the tropics) are discussed at p.202 of the Loeb edition. Gaius Julius Hyginus was a prolific Latin author, living in Spain or Alexandria, whose elementary astronomical treatise drew heavily on the poem of Aratus. The commentary by Macrobius on Cicero's *Somnium Scipionis* is, of course, much better known.

44. CLM 210, f.113v. This is reproduced in several places, perhaps the best illustration being that in D. Bullough: *The Age of Charlemagne,* London 1965, plate 50 (in colour).

45. There is a very different diagram, but one based on an astrolabe plate for approximately this latitude, in Bodleian MS F.1.9, f.88r. The latter was drawn at Worcester, lat. 52⁰ 10′, in circa 1130.

46. Further notes on the circles will be found with the diagram, in which they are redrawn.

47. *Antike Himmelsbilder,* Berlin, 1898, p.164.

48. The illustration is at f.92v. of an Osma cathedral MS of Ciceronian pieces. It is reproduced in colour in G. de Champeaux and Dom. Sebastian Sterckx, O.S.B., *Le Monde des symboles,* 2nd edn., no place of publication, 1972, p.66.

49. Beckmann, *op.cit.,* p.344, gives as part of the epitaph of this man: 'Horologioque carmen spherae coeli optimum,/ Plura alia graviaque prudens invenit.'

50. I gave the text and translation of the work, with detailed discussion and two potential reconstructions, in 'Opus quarundam rotarum mirabilium', *Physis,* 8 (1966), 337-72.

51. One of my suggested reconstructions required it to have no fewer than ten concentric arbors (tubes). This might seem improbable, and yet we do know that the St. Albans clock used multiple tubes (*caligae*), which were therefore not beyond 14th century technological resources.

52. See the note added in proof, *ibid.,* p.368. Note that on p.362 I was wrong to repeat a claim that the stereographical projection of the Salzburg fragments is from the north pole.

53. A plate in the style of the Munich MS would merely have been required to turn in an anticlockwise sense.

54; The projection may be north or south; the stars may be as seen, or as they would be seen by an observer outside the star sphere, as it were; the rete may be of hours or stars; and it may be fixed or moving.

55. There are actually many late medieval examples of men who were both armourers and clockmakers.

56. Abbot Thomas de la Mare (ruled 1349-96) saw to it that 'the upper dial and the wheel of fortune were perfected' by Laurence Stoke and a monk who was skilled in woodcarving, William Walsham by name. Richard of Wallingford 'first arranged' them, but they were left off the clock on account of his early death. Almost all of the clock was therefore completed before 1336.

57. Laurence was important enough to accompany abbot Thomas to the papal court when he sought confirmation of his election.

58. Geneva 1957, pp.103-4.

59. Cracow, MS 551, ff.44v -49r. There is a further class of Italian double-wheeled escapements used in alarm clocks, as well as cognate devices from later periods. These are cited in my forthcoming edition of the works of Richard of Wallingford.

60. The chronicler adds that they are sometimes called 'mules'!

The Cathedral Clock and the Cosmological Clock Metaphor

F.C. HABER

INTRODUCTION

A new world picture was developed in the seventeenth century around the mechanistic philosophies of leading men of science, particularly those of Galileo, Descartes, Boyle and Newton, but it was a development grounded in religion as well as science. The scientists themselves showed concern with making their picture of the world as a machine harmonize with religion, and some even felt that they were making a contribution to religion by showing the structure of the divine workmanship in the Creation.

It was natural that the new mechanical philosophers would adopt the clock as their favorite explanatory model for the illustration of a world machine. The clock was the most complex and scientific machine at the beginning of the seventeenth century, but more important, the astronomical clock had evolved as a mechanical model to represent the motions of the heavens, as Derek J. de Solla Price has convincingly shown.[1] Furthermore, from the origin of the mechanical clock when the escapement was invented some time around the beginning of the fourteenth century, the astronomical clock had been closely associated with religion.

One of the earliest documented astronomical clocks was that of Richard of Wallingford, Abbot of St. Albans, built between 1327 and 1330, but the construction of clocks in religious buildings was by then already well established.[2] In 1324, for instance, the Treasurer of Lincoln Cathedral offered a donation for a new *horologium* for the Cathedral because, as he said, "the Cathedral was destitute of what other cathedrals, churches and convents almost everywhere in the world are generally known to possess."[3] References to the *horologium* or clock in religious establishments go back much earlier and before 1280 probably indicate water clocks, but that the prototype of the monumental astronomical cathedral clock with its trains of automatons had come into existence early in the fourteenth century seems clear from the records of the Norwich Cathedral Priory which describe a clock built between 1322 and 1325 that had a large astronomical dial and automata with 59 images and a choir or procession of monks.[4]

A clock that was often imitated was the first clock of the Strasbourg Cathedral, begun in 1352 and completed in 1354, almost a century before the last stone of the Cathedral itself was laid in 1439. It had an automated astrolabe, a perpetual calendar, a carillon that played tunes from hymns, a Virgin holding the Christ child before whom the three Magi presented themselves, a magnificent mechanical cock that flapped its wings and crowed, and a tablet showing the body parts and their correlation with the zodiac for the favorable and unfavorable times for bloodletting.[5]

From a *Leidensuhr* (Clock of the Passion) written about the time the clock was completed, it appears that the Passion plays given in the Cathedral were coordinated with the striking of the hours of the clock and the performance of its automatons.[6] When the second Strasbourg Cathedral clock was built in 1574, its architect, Conrad Dasypodius, had the old cock cleaned and restored. He said of it: "This poultry cock itself was skillfully made two hundred years ago and placed on the old clock, and since at that time it was customary to commemorate the Passion of Christ in the Christian church, this cock by its crowing warned men of the denial of Peter."[7] It seems possible that the clock played its part in the performance of the Passion plays.

It is clear that from the outset of the mechanical astronomical clocks monks were deeply involved in their design and construction. Indeed, the desire to automate the figure of the Virgin Mary may have been as strong a motivation in the invention of the mechanical escapement as the more pragmatic desire to make a clock to regulate daily life.[8] The design and construction of the complicated astronomical church clocks involved the close interaction of scholarly learning and the craft of the artisan, and these were often combined in one person, as in the notable instances of Richard of Wallingford, Abbot of St. Albans, and, a century later, Johann Stoeffler von Justingen (1452-1531), monk, professor, humanist and clockmaker.[9] Both were learned in mathematics and astronomy, and both were skilled craftsmen. For the more complex astronomical cathedral clocks, knowledge from the entire quadrivium of the Liberal Arts — arithmetic, geometry, astronomy and music — might be utilized in the design, while a knowledge of the mechanical arts was needed for their execution. In this area of clockmaking there was a close collaboration between theoretical knowledge and practical knowledge for some three centuries before Galileo visited the workshops of the artisans, which has sometimes been held to be an important step in the scientific revolution of the seventeenth century. There were also three hundred years of familiarity with representing the operations of the heavens with a machine powered by gravitational force before the machine of the world had been reduced to the laws of gravitation. A mechanical working model of the world system had come into existence long before there was a mechanical philosophy to go with it.

The monumental astronomical cathedral clocks were rich with cosmological significance, but as a class, they have been largely neglected and their importance has been obscured, at least in part, by the high seriousness of modern utilitarianism in which the clock is seen merely as a prototype machine of modern industrialism or as a timekeeper whose principal function is to produce units of time.[10]

The invention of the mechanical escapement around the beginning of the fourteenth century made possible the transformation of the accelerating progress of a descending weight into a regular and steady motion. This has been recognized as one of the important innovations for the development of automatic machines, as well as for the regular production of uniform units of time. When the pendulum was adapted to the mechanical clock in the middle of the seventeenth century, a timekeeper was developed of sufficient precision to be used in observational astronomy, experimental physics and mapping. The utilitarian histories of the mechanical clock tend to leave a barren plateau between these two innovations, a period covering the Renaissance and Reformation, the very period when the monumental astronomical cathedral clocks were flourishing. The reduction of the history of the pre-Galilean clock to the single function of timekeeping makes these clocks seem redundant and a miscarriage of effort. Their trains of automatons have been viewed as "mechanical puppet shows",[11] and their ornamentation has been regarded as irrelevant to clockmaking, so they have been left to the interest of the antiquarian or to the amusement of tourists.

The neglect of the larger and multi-functional aspects of the church clocks in their culture has been so complete that I shall have to confine myself to the culmination of the development and work from the second Strasbourg clock of 1574 for which there is available documentation. This clock retained some of the tradition reaching back to the early fourteenth century and also incorporated influences from the Renaissance. As the most famous clock in the early seventeenth century, it was known to some of the mechanical philosophers and was probably the specific clock from which the cosmological metaphors began to proliferate.

THE STRASBOURG CLOCK

The fourteenth–century Strasbourg Cathedral clock had ceased to function by the early sixteenth century and plans were made to build a new one around 1547. The construction of the clock housing was begun, but Strasbourg was overtaken by the religious conflicts of the Reformation and the work was abandoned. It was taken up again in 1571 and the Senate of Strasbourg appointed Conrad Dasypodius (ca.1530-1600), Professor of Mathematics at the Strasbourg Academy, to undertake the direction of its construction. It was his injunction to make a "magnificent, splendid, and artistic work" that would be an ornament to the Cathedral and bring honor to the Senate and people of Strasbourg.[12] He designed the clock, actually helped make the celestial globe which was used in it, engaged the Habrecht brothers, clockmakers, to execute the clockwork, and put Tobias Stimmer, a leading German painter of the time, in charge of the art work. He was assisted by David Wolkenstein, a professor of music, in the design of the musical parts of the clock and also in the supervision of the work as a whole. The clock was finished in 1574.

The clock was probably the largest of its kind ever built, twenty-five feet wide at the base and about sixty feet high. It was covered with art work. On the weight-tower there were paintings of the three Fates: Clotho spinning the thread of life, Lachesis setting its length, and Atropos cutting it; a painting of Urania, the Muse of Astronomy; a painting of Colossus; and a copy of the portrait of Copernicus. Panels on the base portrayed by means of representative warriors the four monarchies of the ancient world — Assyria, Persia, Greece and Rome — which were associated with the apocalyptic vision in the Book of Daniel. There were scenes of the Creation, Christ Judging the World, Resurrection, Last Judgment, Vice and Innocence. Also on the clock were *putti* and other statuary, including the armorial lions of Strasbourg.

The automated astronomical devices included a large calendar dial with the holy days, a clock giving local time, an astrolabe with delicately wrought signs of the zodiac and planets, a mechanism that showed the phases of the moon, and a celestial sphere supported by a pelican mounted on the floor in front of the clock. The principal trains of automatons were: the tutelary gods of the days of the weeks being borne around in elaborate chariots; figures of the four ages of man that struck the quarter hours in their circuit; and the figures of Christ and Death that dueled at the stroke of the hour with Death winning all hours except the last. Also, the old restored cock was mounted on the crest of the weight-tower and at midday flapped its wings, raised its head, opened its beak, shook its tail and crowed, to the accompaniment of a carillon playing bits of music.

HOROLO-
gium Argento-
ratense.

Fig. 1. The Astronomical Clock of the Strasbourg Cathedral built in 1574, from a contemporary woodcut drawn by Tobias Stimmer and printed in Nikodemus Frischlin, *Operum Poeticarum* (Strasbourg, 1598). Photograph courtesy of the Warburg Institute, University of London.

Fig. 2. The Astronomical Clock of the Strasbourg Cathedral today. The 1574 clock was rebuilt and modernized in 1842 by Jean-Baptiste Schwilgué. Photograph courtesy Photo Service d'Architecture de l'Oeuvre Notre-Dame, Strasbourg.

All the movements were powered by trains hidden from view that were driven by a descending weight inside the weight-tower. "What one saw was a Theater of the World with a morality play taking place through mechanisms and automatons... The position of the clock inside the Cathedral was fitting indeed, for it was a huge visual aid, or mechanized teaching machine, rehearsing the meaning of life and the epitome of the microcosm-macrocosm relationship."[13] The so-called "puppetry" of the clock was an integral part of the symbolic art of religion whose purpose was to illustrate and memorialize the complete meaning of time in a Christian world.

A vision of this concept of total time "keeping" was presented by Dasypodius himself:

> And on this clock we exhibit eternity, the century, the orbits of the planets, the yearly and monthly revolutions of the sun and moon, the divisions of the week, days, hours, parts of hours, minutes; all these I say, we exhibit to be seen. We have added also, for the sake of adornment, splendor, admiration, various contrivances, pneumatic, sphaeropoetic and automatic, everything from history and the tales of the poets, and also from sacred and profane writings in which there is or can be some delineation of time. And we show these things by paintings, pictures, statues and other works similar to these.[14]

The art work by Stimmer was notable even by traditional standards of art.[15] It was all subordinated to the expression of time and this was achieved by a kind of coding. The symbol was a coded sign for part of a myth which the viewer could understand and complete for himself. It was like other mnemotechnic devices popular in the theater of the sixteenth and seventeenth centuries.[16] The assemblage of symbols was the key to the whole story of the Divine Creation, so that from the code the clock could be read like a book and watched like a play to the accompaniment of mechanically produced hymns.

Without entering into a detailed description of the coding, one example will illustrate the technique. The pelican which supported the celestial globe was shown pecking at its stomach to draw blood to feed its young in a nest at its feet. The pelican was a symbol of Christ, a piece of typology from the medieval moralizing of nature. It was described in the popular cosmic poem of Du Bartas, *La Semaine, ou Création du Monde* (1578) as follows:

> [The Pelican], kindly for her tender Brood
> Teares her owne bowells, trilleth out her blood
> To heale her young, and in wondrous sort
> Unto her Children doth her life transport:
> For, finding them by some fell Serpent slaine,
> She rents her brest, and doth upon them raine
> Her vitall humour; whence, recovering heat,
> They by her death, another life doo get:
> A Type of *Christ*, who, sinne-thrall'd man to free,
> Became a Captive; and on shamefull Tree
> (Self-guiltless) shed his blood, by's wounds to save-us,
> And salve the wounds th' old Serpent firstly gave-us:
> And so became, of meere immortall, mortall;
> Thereby to make, fraile mortall man, immortall.[17]

Dasypodius said of the pelican supporting the celestial globe:

> But we have attached this Pelican so that it should be in place of Atlas and represent a symbol of eternity, or even of our Redeemer and Saviour. For

Fig. 3. Mechanical cock from the 1354 Strasbourg clock which was re-used on the 1574 clock and is now preserved in the Strasbourg Museum of Decorative Arts in the Rohan Castle. Photograph courtesy Photo Service d'Architecture de l'Oeuvre Notre-Dame, Strasbourg.

Fig. 4. Detail from the woodcut of the 1574 clock showing the pelican supporting the celestial globe at the base of the clock.

all individual details were so ordered and arranged by us that they have a definite meaning and one worthy of note, taken either from things sacred or the pagans, or the stories of the poets, or writings of historians and annalists, such things, indeed, commend the magnificence, elegance, and benefit of our work, and impress a greater admiration upon the minds of men skilled as well as unskilled. Finally, they render our invention, arrangement and efforts more splendid.[18]

The long poem of Du Bartas put forward a summation of the biblical world view as it had been conventionalized by the sixteenth century. Often translated and reprinted, it was widely read down to the middle of the seventeenth century. It presented a vision of reality that was to lose its place before the rise of modern science, history and secularism, and with the loss of that vision the meaning of the symbolism about time and eternity on the Strasbourg clock would fade away until its automatons could be viewed merely as redundant "puppetry."

The Strasbourg clock continued traditions that reached back to the Middle Ages, but it was also a product of the late Renaissance and Reformation standing on the brink of the seventeenth century scientific revolution (Galileo was ten years old when the clock was finished). There were emphases connected with the clock that continued into a new world view.

Strasbourg had been one of the centers of northern humanism. It was here that Erasmus published many of his works and the humanist tradition remained strong in the city during the early Reformation. When the educational institutions were reformed, a humanist, Jean Sturm, was brought to Strasbourg to head the newly established Academy.[19] His ideal was piety and learning. Greek and Latin were rigorously taught and the classics took a strong place in the curriculum. The Academy attracted students from Europe's nobility as well as townsmen and achieved a considerable reputation for the quality of its education. The clock was in some respects an extension of the teaching functions of the Academy. When the astronomical parts of the clock are considered, the public exhibition of the clock is comparable to a planetarium for illustration and comprehension of astronomy.

Conrad Dasypodius, as professor of mathematics and astronomy at the Academy, emphasized the teaching value of the clock, not only for astronomy, but also for the learning of the theoretical aspects of mechanics. When he published descriptions of the clock, he actually gave a small presentation on mechanics and its history. If the artisan guilds tried to keep their skills cloaked in secrecy, the scholar-clockmakers made their knowledge public.

The lifetime ambition of Dasypodius was to publish all the mathematical and mechanical treatises of the Greeks, a task only partially completed. Building the clock was an interruption to his scholarly work, but he took the occasion to sum up the achievements of the ancients in mechanics and to confront his own age with the question of the superiority of the ancients. He thought that God distributed talents to various peoples at various times in an Aristotelian way and that some things in his own age could equal those of the ancients. Clockmaking was such an instance. The mechanical clock had not been known to the Greeks, but the making of the Strasbourg clock represented an interaction of theory and practical knowledge in the tradition of the best Greek mechanicians. He felt that his own age had not excelled the ancients in mechanics as well as it might because of the ignorance of artisans in a theoretical knowledge of mechanics. The construction of the clock and his description of it were conceived by him as a step towards educating the public in rational mechanics and its importance for chiurgical, or applied, mechanics. His *Description* was published in Latin for the learned and in German for the general citizenry.[20]

Above all, the treatise of Dasypodius was a glorification of the role of the architect who understood both theory and the artisan skills and who could plan the design of a work in his mind and then see that it was properly executed. Dasypodius saw the role of the architect fulfilled in Archimedes, Hero of Alexandria, and Vitruvius. He saw his own role in the construction of the clock as modeled upon Archimedes and Vitruvius, although with the false modesty that was usual at the time, pointed out that he fell far short of their achievements. The architect as a creative genius can be discerned in his description of the nobler natures in the following account of the mechanician, or architect of contrivances:

> There is a very great variety of arts and sciences, not only those of a theoretical sort, but also those which contrive and accomplish something. Among those needed in doing and making, mechanics does not hold the last place. Those who in former times especially mastered this art were reckoned among the philosophers. For just as the philosophers examine by observing the nature, force, and effect of things, so do mechanicians bring about with the work of their hands, their industry, talent, and skill those things which are either necessary for life, or made for pleasure, or benefit daily use.

> For nature has provided that wise and ingenious men, men equipped with a nobler nature and an understanding of skills, above all in their hearts and minds, after comprehending many precepts and scrutinizing things at length, conceive ideas. When these have been conceived, they afterwards fashion and present models of them in some material which, so far as is possible, simulates them. Each one is eager to render perfect his own work which has been devised as much as he is able with intelligence and skill to make that work itself pretty, useful, and sound.[21]

One of the developments of the Renaissance had been the efforts of artists to take their work out of the artisan crafts and elevate it into the liberal arts. They made the case that it was necessary to have a theoretical knowledge for a proper comprehension of their art and they also glorified the artistic genius. When Dasypodius took the same path, he articulated in Renaissance terms a relationship that had existed in fact for centuries in the making of monumental astronomical clocks between theoretical knowledge and artisan skills but had probably passed unnoticed. It is doubtful if Dasypodius would have made his appeal to the tradition in any case, for he was trying to put his case in the mainstream of the ancient mechanical philosophers. If the idea of the architect was slow in being spelled out in connection with clockmaking, it was an outgrowth of a central conception of Christianity about God as Maker and the world as a work of art.[22] It was this idea of the clockmaker as designer and architect which was to be the most important analogy in the various uses of the clock metaphor from the seventeenth century to the nineteenth century and it was around this idea that the religious grounding of the new mechanical philosophy was to turn.

THE CLOCK AS A WORK OF ART

From the dim past of mythology through the entire tradition of Western thought, there has been a strong strain of technomorphism. God has been extolled as potter, poet, musician, geometer, playwright and architect. All of these analogies, and many more, drawn from technics have found expression in the Western tradition with varying degrees of popularity. The idea of God the Maker has been central to Christianity. The first verses of Genesis present God as a Maker, and if his manner of making in these first verses is by word and spirit, more artisan-like conceptions of making appear elsewhere in the Old Testament. God comes close to being an architect in the Psalms of David, and by the sixteenth century the Psalms were being used to suggest that man should stand back in awe and admire God's handiwork as though God had put on an exhibition to astonish man with His artistry.

The Platonic tradition and other sources in antiquity supported the idea of God the Maker and these were fused with the Old Testament ideas by the Church Fathers. The Platonist ideas were strongly emphasized in St. Augustine's conception of God the Maker, an Intelligence and a Designer who planned the whole Creation in his mind before executing it. St. Augustine's use of musical harmony metaphors to extol the order of the universe as well as ideas of the Creation being a stage on which the salvation drama was being played enjoyed great popularity through the Middle Ages and Renaissance. Both conceptions reinforced the idea of God as an artist and the world as a work of art.

The idea of God as Architect or Artist was developed in natural theology in connection with the Argument from Design to prove the existence of God. "There cannot be a design without a designer; contrivance without a contriver"; the argument went forward in an unbroken liturgy from the Middle Ages to William Paley's *Natural Theology* in the nineteenth century.[23] It was an easy step to make the Designer a Mechanician once Nature began to be explained in terms of mechanisms.

The Argument from Design was also used to support the idea of divinity in man. God as Maker served as a role model for the creature who was made in God's image. The history of this idea is complex, and it is enough for our purposes to see how it was given expression by Calvin in his *Institutes*. He argues that our ability to comprehend the works of God has little use for our bodily needs, therefore it must be a function of our soul, and so, too, is our ability to devise things.

> Of what concern is it to the body that you measure the heavens, gather the
> number of the stars, determine the magnitude of each, know what space lies
> between them, with what swiftness or slowness they complete their courses,
> how many degrees this way or that they decline? I confess, indeed, that astro-
> nomy has some use; but I am only showing that in this deepest investigation
> of heavenly things there is no organic symmetry, but here is an activity of the
> soul distinct from the body. I have put forth one example, from which it will
> be easy for my readers to derive the rest. Manifold indeed is the nimbleness
> of the soul with which it surveys heaven and earth, joins past to future, re-
> tains in memory something heard long before, nay, pictures to itself what-
> ever it pleases. Manifold also is the skill with which it devises things incredible,
> and which is the mother of so many marvelous devices. These are unfailing
> signs of divinity in man.[24]

It is probable that astronomical clocks were among Calvin's "many marvelous devices", for no machine illustrated so well the necessity of design in the making of a contrivance. When the Calvinist poet Du Bartas was developing this, he wrote:

> But who would think, that mortall hands could mold
> New Heavens, new stars, whose whirling courses should
> With constant windings, though contrary wayes,

Marke the true mounds of Yeares, & Months, & Dayes![25]

In a marginal note was written: "Admirable Dialls and Clockes, namely, at this day, that of Strasbourg." But even before Du Bartas, the divine art of the architect of the clock had begun to be celebrated by German Humanists.

Philip Nicodemus Frischlin (1547-1590), a noted neo-Latin poet, wrote a long poem about the clock and its maker in 1575, shortly before he was named poet laureate of Germany. In the poem, he exclaimed: "Oh, divine inventions of the human hand! What work does either God or Nature do anywhere which we do not imitate with our thumb, a people rivalling our Father?" He lavished his praises upon Dasypodius, concluding one long section extolling his genius with the following:

> He augments the praises of the ancients with his own virtues, and he surpasses all the works of the men of the past with his new inventions. He challenges the outstanding contrivers of this age with the new praise of his construction, imitating divine Olympus. And yet all the others whom these lofty studies tire out, how small a part of you are they, Conrad? What of such a sort has great Purbachius or thrice greatest Stoeffler done? Or has Copernicus done anything like this with the unmoved orb of heaven? If what was illustriously formed by all these men should be joined into one, you alone will outstrip them all in rank.[26]

Frischlin's poem was dedicated to the Magistrates and Senate of Strasbourg. He justified undertaking such a work on the grounds of extending the usefulness of the clock far and wide in teaching morals and astronomy, and also to commend such an admirable work to posterity. To achieve these goals, he felt that poetry was more fitting than prose, but writing a poem in Latin about a clock presented special difficulties. Since mechanical clocks were not known to the ancients, the subject matter was new, but so too was the present astronomical terminology, much of which had been taken from Arabic. As a result, he had no ancient Latin models to follow, and he saw his exercise as an entirely new venture in literary art.

In the poem there were extravagances, and his goal of being useful sometimes ran counter to his desire to be elegant, but he did tell the story evoked by the parts of the clock with skill and sometimes dramatically. He also put the Strasbourg clock forward as the work of art of a genius that excelled anything the ancients had done in mechanics. Although some of Frischlin's literary works have retained their reputation down to the present, the work on the clock is hardly one of them. Nevertheless, at that time, Frischlin firmly established the clock in the literary *exempla*, and there was a rash of attempts to write poems about the Strasbourg clock. His poem was accompanied by a woodcut of the clock, as was the later *Description* of Dasypodius, and the fame of the clock spread rapidly.

The Strasbourg clock became the paradigm of the clock in the early seventeenth century, a paradigm that carried with it associations of modern genius equaling the ancients in mechanics, the imitation of the Divine Creation through mechanics and a descending weight hidden from view, "the unfailing signs of divinity in man," and the glorification of the mechanical arts when raised to the theoretical level. It was also a convincing model of a machine that would run indefinitely.

If the Strasbourg clock attempted to symbolize the many dimensions of time in the world of the Renaissance, it was not to be despised as a conventional timekeeper. It has been estimated that if the central works were properly wound and maintained, they would vary only about nine hours in 600 years.[27] It had been set to have some of its astronomical calculations redone at one hundred year periods, and this confidence in a long future operation was emphasized both by Dasypodius and Frischlin. Interestingly, this model of stability and the exuberant powers of man that it symbolized were put before the age at a time when Europe was going through a crisis of faith about the future of the world.

THE DECAY OF NATURE

In the period roughly from 1570 to 1660, there was an extensive literature on the decay of nature, the decay of man, and the decay of the world. It was an old idea and one that had been discouraged by the Catholic Church, but the Reformation and the availability of the printing presses to prophets of despair accelerated its influence. It was also spread by the prophets of hope in millenarian movements who saw the end of the world coming in the immediate future and the ushering in of the heavenly state.

It was a European phenomenon as well as an English one, but Hiram Haydn writing about England has said of this image of a world dissolving into chaos:

> ... when one becomes acquainted with all the different ways in which the
> general theory was treated in the last decade of the sixteenth century —
> the theological, philosophical, astrological, political and even scientific
> versions — one discovers that there is almost no one among the Elizabethans
> who did not at one time or another take up the question of the decay of
> nature and the imminent disintegration of the universe.[28]

And along with the idea of decay, there was intermittently, and often in the same person, the expression of a rampant Prometheanism.

This was also a period in which the idea of a personal God was widely held, and not just a benevolent God, but at times an angry Moralist, and the fires of Hell loomed large in the perception of reality. Nature was filled with spirits, witches and mysterious forces, most of which were malevolent, and it was not uncommon still to see warfare, plagues, famine and social chaos as punishment visited upon man for his sins. It was a heavy burden to bear, and all of this may have made a de-personalized world seem rather attractive.

A world whose invisible powers were merely a descending weight behind the phenomena was not only benign in contrast to the world animated with spirits, but it was also intelligible. The emergence of the "new mechanical philosophy" of the early seventeenth century was not only a positive development on the road to modern science; it also cut with one mighty stroke through that enormous burden of a highly personalized world.

Descartes knew the Strasbourg clock well as early as 1629, and although there were many elegant automatons other than those on clocks, it is possible that he may have had the cock on the Strasbourg clock in mind when he developed the idea that animals were merely automatons without souls. In presenting his mechanical philosophy, Descartes used the clock analogy on a number of occasions, but primarily as a structure in which the parts were interdependent and functionally disposed. In describing the circulation of the blood, for instance, he wrote, "...this movement which I have just explained follows as necessarily from the very disposition of the organs, as can be seen by looking at the heart, and from the heat which can be felt with the fingers, and from the nature of the blood of which we can learn by experience, as does that of a clock from the power, the situation and the form, of its counterpoise and of its wheels."[29] More important than any particular use of the clock as an analogy, however, was the assumption of Descartes that all the material parts of the world functioned on mechanical principles. Any machine could be invoked as an illustration, but the clock was especially apt.

It was the Honourable Robert Boyle who most thoroughly and effectively brought the Strasbourg clock into the "new mechanical philosophy" to explain the operations of nature. Of the several references he makes to the clock and its maker, the following emphasizes the difference between the spectator of the phenomena and the philosopher who goes behind the scene to examine the works of the machinery:

> The curious works of famous artificers are wont to invite the visits, and excite the wonder of the generality of inquisitive persons. And I remember, that in my travels, I have often taken no small pains to obtain the pleasure of gazing upon some masterpiece of art; but now, I confess, I could not with more delight look upon a skilful dissection, than the famous clock at *Strasburgh.*
>
> But if the bare beholding of this admirable structure, is capable of pleasing men so highly; how much satisfaction, Pyrophilus, may it be supposed to afford to an intelligent spectator, who is able both to understand and to relish the admirable architecture and skilful contrivance of it: for the book of nature is to the ordinary gazer, and a naturalist, like a rare book of hieroglyphics to a child, and a philosopher; the one is sufficiently pleased with the oddness and variety of the curious pictures that adorn it; whereas the other, is not only delighted with those outward objects, that gratify his sense, but receives a much higher satisfaction, in admiring the knowledge of the author, and in finding out and inriching himself with those veiled truths dexterously hinted in them.[30]

Boyle neatly severed a world run by secondary causes from a personal God and left God as merely the Designer of the World Clock, much as Dasypodius had designed the Strasbourg clock and then left it to run by itself. And the little statues on the clock, Boyle emphasized, were not like puppets that were manipulated by a hand pulling strings or wires, but instead, the little statues performed without the "interposing of the artificer."[31]

The idea that a world clock designed by God was a better illustration of the art of the clockmaker the less a finger had to be put to it to make it run properly had a wide currency in the seventeenth century. Its most famous example was the charge of Leibniz against Newton that God had to wind up his clock periodically,[32] but the idea had by then become a cliché. Bishop John Wilkins in 1640 had popularized the idea in defending the Copernican system. He also developed the idea that no watchmaker would put anything superfluous in his mechanisms, that the Clockmaker always designed his works with the principle of parsimony in mind. "We allow every watchmaker so much wisdom as not to put any motion in his instrument, which is superfluous, or may be supplied an easier way: and shall we not think that nature has as much providence as every ordinary mechanic?"[33]

Around the principle of parsimony in clockmaking, the idea of pure mechanical utilitarianism would develop against the manifold purposes of a Strasbourg clock and bring the whole class of cathedral clocks into contempt. This can already be seen taking shape in Bishop Wilkins when he attacked the Ptolemaic system. "Those antiquated engines that did consist of such a needless multitude of wheels, and springs, and screws, (like the old hypothesis of the heavens) may be compared to the notions of a confused knowledge, which are always full of perplexity and complications, and seldom in order; whereas inventions of art are more regular, simple, and perspicuous, like the apprehensions of a distinct and thoroughly-informed judgment."[34]

The use of the clock analogy to remove God from the daily operations of nature received an additional refinement in the early eighteenth century from Benoît de Maillet when he envisioned a world clock with the ability to automatically replace its own worn parts. He maintained that it was more worthy of the Creator to have designed an order of nature that was self-renovating.

> What comparison could we make between a clock-maker, who had skill
> enough to make a clock so curiously, that by the disorder which time
> should produce on her parts and movements, there should be new wheels
> and springs formed out of the pieces, which had been worn and broken;
> and another artist of the same profession, whose work should every day,
> every hour, and minute, require his attention to rectify its errors, and
> eternal variations?[35]

There were many examples of the analogy of the clock or watch being used to de-personalize God's role in nature, getting his finger out of the operation of the works. In the usage, the motivation of shoring up the world against decay and imminent destruction was often implicit, but the motivation was made explicit by Henry Power. He saw a great discouragement to the promotion of the arts and sciences in "The Universal Exclamation of the World's decay and approximation to its period."[36] Observing that it was a conceit that had possessed all ages, "yet the Clamour was never so high as it is now;" so he offered something to abate the influence of the conception. By placing the beginning of the world 5,000 years ago with the sun's apogee set in Aries, he thought that in his own time (1663) there would be at least 15,000 years left to complete the cycle of the apogee. He reasoned:

> Now in all likelihood, he that made this great Automaton of the world,
> will not destroy it, till the slowest Motion therein has made one Revolution.

> For would it not even in a common Watchmaker (that has made a curious
> Watch for some Gentleman or other, to shew him the rarity of his Art) be
> great indiscretion, and a most imprudent act, and argue also a dislike of his
> own work, to pluck the said Watch in pieces before every wheel therein
> had made one revolution at least?[37]

CONCLUSION

By means of the clockwork and automatons of the cathedral clocks, the idea had become familiar of making an imitation of the Divine Creation through the art of man. By the end of the seventeenth century, the Divine Creation had become an imitation of the art of man. Analogies based on clocks, watches and automatons were extended to brutes, plants, the economy of nature, the system of the universe and man himself. The mind, through association psychology, assumed the pattern of a clockwork system; the brain was often referred to as a clock, as well as the heart and the whole system of the body. The passions had their mainspring; the economy of society took on the systematics of clockwork; and John Adams had the clock analogy in mind when he emphasized checks and balances in the American constitutional system.

The observation has been made that in the use of models in science:

> A model only becomes fertile by its own impoverishment. It must lose some of its own specific singularity to enter with the corresponding object into a new generalization. When some kind of machine becomes a valid model for an organic function it is not the machine in its entirety that becomes the model, but only the pattern of its operations such that it can be expressed in mathematical terms.[38]

Certainly, with the work of Newton one part of the clock model, its simulation of the world system, had lost all of its particularity and had been reduced to a mathematical system. Many of the other areas in which the clock model was applied, man the machine and society as a machine, have yet to find as full an expression in mathematics, but the clockwork analogy did provide an aesthetic sense about order which exerted a powerful influence on how reality was perceived at least down to Darwin's *Origin of Species* in 1859, which advanced a system that did not require a completed clockwork Design in the mind of the Creator.

NOTES

1. Derek J. de Solla Price: "On the Origin of Clockwork, Perpetual Motion and the Compass," *U.S. National Museum Bulletin 218: Contributions from the Museum of History and Technology,* Paper 6 (Washington 1959), pp. 82-112; and "Automata and the Origins of Mechanism and Mechanistic Philosophy," *Technology and Culture* V (1964) 9-23. See also in the same volume, Silvio A. Bedini: "The Role of Automata in the History of Technology," pp. 24-42.

2. John D. North, Oxford University, is publishing a major study on Richard of Wallingford.

3. C.F.C. Beeson: *English Church Clocks, 1280-1850: History and Classification,* (London: Antiquarian Horological Society 1971), p. 18.

4. *Ibid.,* p. 16.

5. Alfred Ungerer: *Les Horloges astronomiques et monumentales les plus remarquables de l'Antiquité jusqu'à nos jours,* (Strasbourg 1931), p. 165.

6. Theodore Ungerer and l'abbé André Glory: "L'astrologue au cadran solaire de la Cathédrale de Strasbourg (1493)," *Archives alsaciennes d'histoire de l'art,* XII (1933), 73-108; and Joseph Walter: "Le Mystère 'Stella' des trois mages joué à la Cathédrale de Strasbourg au XIIe siècle," *ibid.,* VIII (1929), 39-50.

7. Conrad Dasypodius: *Heron mechanicus: seu De mechanicis Ejusdem Horologii astronomici,* (Strasbourg 1580). Dr. Bernard Aratowsky, Professor of History, State University College of New York at New Paltz, is preparing translations for publication of this text, a German edition of the same work, *Warhafftige Ausslegung und Beschreybung des Astronomischen Uhrwercks zu Strassburg* (1580), and Nicodemus Frischlin: *Carmen de Astronomico Horologio Argentoratensi* (1575). He has kindly given me permission to use his translations from Dasypodius and Frischlin in this paper.

8. The rudimentary escapement sketched in the *Album* of Villard de Honnecourt about 1250 was a device to turn an angel so that its finger would point towards the sun.

9. J.C. Albert Moll: *Johannes Stoeffler von Justingen,* (Lindau 1877).

10. For example, Joseph Needham: *Clerks and Craftsmen in China and the West,* (Cambridge 1970), p. 204; Lewis Mumford: *Technics and Civilization,* (New York 1963), Chapter One; and Abbott Payson Usher: *A History of Mechanical Inventions,* (Boston 1959), Chapter Eight.

11. Usher, p. 209.

12. Dasypodius, *Heron mechanicus.*

13. F.C. Haber: "The Darwinian Revolution in the Concept of Time," *Studium Generale* 24 (1971), p. 298; and in *The Study of Time, Proceedings of the First Conference of the International Study of Time*, (New York: Springer-Verlag 1972), p. 392.

14. Dasypodius, *Heron mechanicus.*

15. Max Bendel: *Tobias Stimmer, Leben und Werke*, (Zurich/Berlin 1940).

16. Frances A. Yates: *The Art of Memory*, (Chicago 1966) and *Theatre of the World*, (Chicago 1969).

17. *Bartas, His Devine Weekes and Works* (1605), translated by Joshua Sylvester, (Gainesville, Fla.: Scholars' Facsimiles and Reprints 1965), pp. 180-1.

18. Dasypodius, *Heron mechanicus.*

19. Charles Schmidt: *La vie et les travaux de Jean Sturm*, (Strasbourg/Paris 1855); Henri Strohl: *Le Protestantisme en Alsace*, (Strasbourg 1950).

20. Dasypodius, *Warhafftige.*

21. Dasypodius, *Heron mechanicus.*

22. Edgar Robert Curtius: *European Literature and the Latin Middle Ages*, (New York 1963) pp. 544-46; Leo Spitzer: *Classical and Christian Ideas of World Harmony*, (Baltimore 1963); Rudolf and Margot Wittkower: *Born Under Saturn*, (London 1963); Edgar Zilsel: *Die Entstehung des Geniebegriffes*, (Tübingen 1926); and Paolo Rossi: *Philosophy, Technology, and the Arts in the Early Modern Period*, (New York 1970).

23. William Paley: *Natural Theology. or, Evidences of the Existence and Attributes of the Deity, Collected from the Appearances of Nature*, (London 1802), Chapter Two.

24. John Calvin: *Institutes of the Christian Religion* (1560), Book I, Ch. 5, par. 5.

25. *Bartas*, p. 222.

26. Frischlin, *Carmen.*

27. The Clockmakers' Company of Copenhagen, 1934, in Otto Mortensen: *Jens Olsen's Clock: A Technical Description*, (Copenhagen 1957), p. 32. The 1574 clock ceased to function in the late 18th century. A new set of clockworks and a restoration of the outer works was completed by Jean-Baptiste Schwilgué in 1842, and it is this clock which is now to be seen in the Cathedral.

28. Hiram Haydn: *The Counter-Renaissance*, (New York 1960), p. 22. See also Victor Harris: *All Coherence Gone*, (Chicago 1949).

29. René Descartes: *Discourse on the Method* (1637), Part V, in E.S. Haldane and G.T.R. Ross: *The Philosophical Works of Descartes*, 2 vols. (Cambridge 1955), I, 112.

30. *The Works of the Honourable Robert Boyle*, ed. Thomas Birch, 6 vols. (London 1772), II. 7.

31. *Ibid.*, V, 163.

32. Correspondence with Samuel Clarke, 1715, in *Leibniz: Philosophical Writings*, tr. Mary Morris (Everyman's Library 1934), p. 192.

33. John Wilkins: *Discourse concerning a New Planet; tending to prove, that it is probable our Earth is one of the Planets*, (London 1640), in *Mathematical and Philosophical Works* 2 vols. (London 1802), I, 239.

34. *Mathematical Magic* (1648), *ibid.*, II, 183.

35. Benoît de Maillet: *Telliamed* (1748, first posthumous edition), (Baltimore 1797), p. xxvi.

36. Henry Power: *Experimental Philosophy*, (London 1664), p. 188.

37. *Ibid.*, p. 189.

38. Georges Canguilhem: "The Role of Analogies and Models in Biological Discovery," in *Scientific Change*, ed. A.C. Crombie (New York 1963), p. 515.

The Development of the Pendulum as a Device for Regulating Clocks Prior to the 18th Century

S.G. ATWOOD

INTRODUCTION

The pendulum, according to *Webster's Third International Dictionary,* is: (a) "a body suspended from a fixed point so as to swing to and fro under the action of gravity and commonly used to regulate the movements of clockwork and other machinery;" or (b) "a suspended body that vibrates not by swinging but by rotating, with alternate twisting and untwisting (as the balance wheel of a watch) — called also *torsion pendulum.*" In the case of the swinging pendulum, the period is constant and more or less independent of the amplitude and of the swing, angular velocity of the swing, and of the mass of the bobbin. Because the pendulum swings with a regular motion of equal periods without regard to amplitude or mass, it is useful as a timekeeper.

As background to the history of the pendulum in horology, it may be helpful to define the pendulum more completely. There are four kinds of pendulums: the simple pendulum, the physical pendulum, the torsion pendulum, and the vertical pendulum. The simple pendulum is an idealized body consisting of a point mass suspended inextensibly, by a thread, for example, from a point. When pulled to one side of its equilibrium position and released, the device swings in a vertical plane under the influence of gravity. The motion is periodic and oscillatory. A physical pendulum is any rigid body mounted so that it can swing in a vertical plane about an axis that does not pass through its center of gravity. It is a generalization of the simple pendulum, since no point mass can be held by a weightless device. A torsional pendulum might be described as a disc suspended by wire and attached to the center of the mass of the disc. The wire is securely fixed to this disc and at the other end of the wire to a solid support. If the disc is rotated in a horizontal plane, the wire will be twisted. The twisted wire will exert a torque on the disc and return it to a central position (see Figure 1). A watch spring attached to the balance wheel of a watch provides angular harmonic motion, and the formula in physics for this motion satisfies the definition of a pendulum. The vertical pendulum is so named because a vertically oscillating coiled spring will function as a pendulum (see Figure 2).

In addition to the above pendulums, there is the so-called "conical pendulum", defined as a mass hung from a point swinging in such a way that the mass describes a circle and the thread member between the mass and the fixed point of support describes the surface of a cone. A conical pendulum does not fit the definition of a pendulum under the terms and conditions that prevail for pendulums. That is, it will not maintain a relatively regular period except under conditions of constant velocity. In other words, the period will change with the velocity and the distance between the suspension point on one

*The author and the editors wish to express their appreciation to Prof. F.C. Haber for his assistance in abridging and editing the manuscript.

Fig. 1. Torsion Bar Pendulum Year Clock by the Year Clock Company of New York City, under the patents of Aaron D. Crane of Caldwell, New Jersey, patented March 18, 1829.

Fig. 2. Bouncing Doll Clock by the Ansonia Clock Company, New York. Patented December 14, 1886.

end, and the center of the circle described by the circular swinging mass at the other correspondingly changes. Therefore, a "conical pendulum" is essentially a brake or governor, and not a true pendulum. In spite of this variation in the height of the conical pendulum, this difference is slight enough over a long distance with a small radius swing, that for practical purposes, a conical swinging weight may be used as a timekeeping, regulating device (see Figure 3).

Pendulums are subject to a great many variables. The acceleration of gravity will vary with altitude, that is, with distance from the center of the earth, and is also affected by forces resulting from the rotation of the earth. Another variable is introduced in the rate of the pendulum by temperature changes. These do not only affect the expansion or contraction of the materials used to construct the pendulum, but could also affect the viscosity of oil or other lubricants, and the clearances of gears in the trains, leading to minute variations in the period. Compensation must also be made for changes in barometric pressure.

Under all conditions, isochronism demands that the rate of the pendulum be equal for any amplitude. So far, we have been talking about the simple harmonic motion of the swinging pendulum, and it has been assumed that the rate of this pendulum will remain constant as long as the length of the pendulum remains constant. Experiments show that this statement is incorrect and that such a pendulum is not isochronous. For the simple or physical pendulum to be isochronous, the restoring force should be directly proportional to the angular displacement from the equilibrium point. However, the force of gravity causes the restoring force to be proportional to the line of the angular displacement. For very simple amplitudes, the difference between the angular displacement and the line of the angular displacement is very small.

In order to bring the periods of freely swinging pendulums of different amplitudes and large oscillations into agreement, the length of the pendulum must be shortened in proportion to the amplitude. This means that the curvature of the correct curve is greater than that of a circle, the radius of which is equal to the length of the perpendicular pendulum.

To attain an isochronous motion, a mechanical device must be related to the pendulum which will cause the pendulum to swing through a cycloid which is a tautochronous curve, that is, the descents and ascents of a body or a pendulum mass from any point on the curve that the mass swings shall take equal times. Since clock pendulums are relatively simple pendulums, they have this so-called minor circular error. One of the problems that relates to developing the perfect pendulum is that it is nearly impossible to determine the center of oscillation in the case of a pendulum suspended in such a way as to achieve a cycloid arc of swing, (to eliminate circular error). The center of oscillation must be determined as a pendulum changes its length continuously as it moves. As will be mentioned later, Huygens' solutions to isochronism were geometric. While Huygens dealt with the pendulum in a particular geometric way, Newton approached pendulums in a more general algebraic-calculus description, which is still valid today.

In summary, it is possible to define the pendulum in rather precise terms and to describe the variables which will affect its rate. A truly isochronous pendulum must obey Newton's laws and yet we find the application of these laws and the theoretical attempt to design a truly isochronous pendulum on paper thwarted by the variables described.

Fig. 3. Porcelain Conical Pendulum Clock, circa 1880. The fisherman's line is rotated by a 3/16 in. diameter wheel of a special watch movement.

For the clockmaker, an isochronous-designed pendulum is unnecessary because he provides for a nearly constant restoring force. The final adjustment of a pendulum in a clock is today, as it was when pendulums were first used, a matter of trial and error adjustment in comparison with some time reference when the clock is in position.

HISTORY OF THE THEORY OF THE PENDULUM

No one knows where the first theory of the pendulum arose. There is apparently no reference to the pendulum, as such, in ancient literature. It could have occurred in China or in ancient Greece. No one knows when the first "talented tinkerers" used levers of various sorts weighted at one end and pivoted, i.e., a pendulum, so that it could swing to and fro to furnish simple reciprocating power. Probably long before the 16th century, a pendulum was used for power without theoretical comprehension of its principles.

Piero E. Ariotti points out that Nicole of Oresme (1330-1382) makes one of the first direct references to the pendulum.

> ... if an opening were made from here to the center of the earth and beyond and a heavy object fell through this opening or hole, upon reaching the center, it would pass beyond and begin to go upward by reason of this accidental and acquired property; then it would fall back again and come and go several times just as we can observe in the case of a heavy object hanging from a beam by a long cord. Therefore, since this property causes a heavy body to move upward, it is definitely not the same as weight or heaviness. And such property is present in all motion, both natural and violent, whenever the speed is increased, save only in the motion of the heavens.[1]

By the end of the 15th century, Leonardo da Vinci (1452-1519) and others were deeply interested in both clockwork and the pendulum. Leonardo's notebooks have drawings of ingenious mechanical devices involving many kinds of motion. The notebooks include a drawing of a weight-driven clock with a possible suggestion of a pendulum (Manuscript II Codex Atlanticus 378 Rb); a pendulum (Manuscript II Codex Atlanticus 257 Ra); a clock escapement labelled "tempo d'orlogio" (Codex Madrid, folio 157, verso, dated 1493); a discussion of how temperature, friction, and humidity affect the rate of a clock (Codex Madrid K, folio 48, verso); and an analysis of the variations of force on geared wheels to explain the lack of uniform motion of the escapement of clocks (Codex Madrid I, folio 1, verso).[2]

There is more than a hint that Leonardo recognized that the pendulum would be a useful device *in connection with clockwork*. Whether he noted the periodicity of the pendulum or whether a clock with a pendulum was in existence during his lifetime is unknown. It is the writer's opinion in the light of the Madrid Codex drawings, that Leonardo da Vinci recognized that a pendulum might be useful as a controller in clockwork.

In 1578, Francois Beroald published in French an expanded translation of Jacques Besson's *Theatrum Instrumentorum et Machinarum* (1569). This is one of the early compendiums of mechanical inventions. Illustrations in this book show the pendulum at work as a stone cutting and polishing device, as a mechanism to actuate a pump and as a device for actuating forge bellows (see Figure 4).[3]

Fig. 4. Jacques Besson's Pendulum Powered Well Bucket System, 16th Century.[3]

Fig. 5. Modern reproduction from Galileo's sketch of an escapement and pendulum mechanism for a clock.

The 17th century opens with many public clocks in place and an awakening interest on the part of the mechanically talented and scholars to probe for the reasons why things move and operate as they do. As the 17th century progressed, the demand was felt for an understanding of geometry and astronomy in order to improve techniques of navigation and surveying. The Spanish, Portuguese, English and the Dutch ranged the oceans in their ships. In this connection, the King of Spain in 1598, offered a prize of 1,000 crowns for a means of finding longitude at sea. In 1602, the States General of the Netherlands offered 10,000 florins for a similar result. A better measurement of time would win the prizes offered.

In the early part of the 17th century, Isaac Beeckman (1588-1637) clearly conceived of applying the pendulum to a clock. In spite of the fact that he apparently did not understand the value of the pendulum's periodicity, some time before 1634 he wrote as follows:

> A clock could also be fashioned from such a string arrangement. Let there be hung a large weight (maximum pondus) such that would go back and forth all day, or better still, [take] a small staff of suitable length (the longer it is, the longer the weight will go to and fro) and [let it be] attached to a transverse [bar] which is inserted in bearings [foraminibus] at both ends so that its motion is like that of bells. As often as the staff is perpendicular, let it engage a certain gear. When this gear is released, a certain small wheel is put in motion which, when the gear is again engaged, immediately stops. (*Quoties vero baculus perpendicularis est tangat pinnacidium quoddam, quo remoto, rotula quaedam movetur statimque, pinnacidio recidente, quiescat.*) In this fashion by means of wheels connected to one another after the fashion of clocks, there would be a new kind of clock and indeed, unlike our present ones which are subject to the perturbations of the air, most useful in astronomical observations.[4]

The work of Galileo Galilei (1564-1642) is of particular importance as it relates to the development of the theory of the pendulum as a timekeeping device. Apparently, neither Galileo nor his last disciple, Vincenzo Viviani, claimed that Galileo was the first to conceive of the idea of applying a pendulum to clockwork. The pendulum may have been used more or less by accident by clockmakers in clocks in or before the first quarter of the 17th century. Nothing, however, should detract from the fact that through Galileo's scientific analysis of the motion of the pendulum, he put the world on notice that the pendulum could be used as an horological device of substantial accuracy. Thereby, he stimulated the mechanical practitioners, who were building clocks, and scientists who were interested in motion from a scientific point of view.

There has been controversy over whether Christiaan Huygens (1629-1695) conceived of applying a pendulum to clockwork before Galileo. Although Huygens claims in his *Horologium* (1658), a short treatise describing the application of the pendulum to the clock escapement, that he designed and invented a practical pendulum clock, he does not say that he was the first to think of the idea. An account of the relationship of Galileo to Huygens concerning the development and first applications of the pendulum has been researched and described by J. Drummond Robertson.[5] It is clear from Robertson's study that Galileo had priority over Huygens in developing the idea of a pendulum clock.

A second very complete story of Galileo in relation to the pendulum has been written by Silvio A. Bedini.[6] Bedini indicates that the basis for Galileo's interest in the pendulum was his alleged discovery of the periodicity of the pendulum from a swinging lamp in the Cathedral at Pisa, when he was a student. Viviani reported that Galileo verified by experiment that the pendulum did in fact swing through various arcs in the same period of time.

On June 5, 1637, Galileo wrote a letter to Admiral Lorenzo Realio, president of the committee of the States General of the Netherlands for examining proposals to accurately determine longitude. In this letter Galileo states:

> ... It is also possible to adapt around the center of the first toothed wheel, another wheel having a lesser number of teeth, which engages another larger toothed wheel and from the motion of which it is possible to determine the number of internal revolutions of the first wheel, compartmenting the number of teeth in such a manner that, for example, when the second wheel has given a conversion, the first has already given 20, 30, 40 or as many as may be desired. But to bring this to the attention of Their Excellencies, who have men of the most exquisite and ingenious ability in constructing clocks and other admirable machines, would be a superfluous act, because on this new foundation, knowing that the pendulum, moving in large or small arcs, makes its oscillations most equally, they would find consequences much slighter than those which I could possibly imagine. And inasmuch as the fallacy of clocks consists principally in its not having been possible to manufacture until now those which we call the balance (tempo) of the clock, so adjusted that it will have equal vibration, thus in this my most simple pendulum which is not subject to any alteration, is contained the method of maintaining always most equal the measure of time.[7]

In 1639, Galileo published in Paris a booklet about the pendulum.[8] According to Robertson, Galileo deals with using the pendulum in a clock for means of ascertaining longitudes and quotes from the publication as follows: "This method was valueless as long as the pendulum was not used for regulating the movement of clockwork."[9] Clearly, in 1637 and again in 1639, Galileo relates the pendulum to clockwork long before Huygens' *Horologium* of 1658. It seems clear that Galileo should be credited with the first comprehensive, scientific, theoretical examination of the pendulum, and except for his lack of understanding of the cycloid curve as a truly isochronic curve, he developed the basic laws of motion relating to a physical pendulum. From the viewpoint of horology, Galileo's great contribution was that he clearly pointed the way to others to construct clocks utilizing his ideas. This horological stimulation bore fruit in Italy as well as in Holland.

As has been noted by Bedini[10] and H. Alan Lloyd,[11] not only did Galileo contribute to horology the fundamental ideals of the pendulum, but he also contributed an elegant concept of an escapement which unfortunately lay dormant for years (see Figure 5). This escapement serves to indicate the genius of Galileo in designing things mechanical and his deep interest in horology. Ironically, in spite of further improvements in the theory of how a more accurate, i.e., isochronic, pendulum should be constructed, in clockmaking we find that this cycloidal curve is not of practical value for more accurate timekeeping, and that Galileo's original conception of a simple physical pendulum with circular error is quite adequate indeed. He never lived to see his design made and tested.

The mathematical work of Rene Descartes (1596-1650) in linking algebra to geometry to define curves by coordinate equations in 1637 had a bearing on later research related to the pendulum. So, too, did the work of Pierre Fermat (1601-1665), who independently invented analytical geometry and extended it to three dimensions, and Blaise Pascal (1623-1662) who solved many of the problems of the cycloid. These three men, Descartes, Fermat and Pascal, represent the great pre-calculus geometric mathematicians. They set the stage for Christiaan Huygens, the Dutch mathematician, physicist and astronomer, who made an important contribution to the theory of the pendulum.

Christiaan Huygens proved that a cycloid curve is a tautochrone. An object such as a ball placed at any point on an upside-down bowl of cycloid curve shape will roll or slide to rest in the same time interval. It is also to Christiaan Huygens that full credit must be given for developing *the theory* of an *isochronous pendulum.* He showed that if a pendulum could be made to swing in the arc of a cycloid, its period, regardless of amplitude, would be truly equal. This discovery and the idea of applying it to the pendulum was not made by him prior to 1657.[12]

Gian Battista Riccioli (1598-1671), Isaac Beeckman, Descartes, and Mersenne were all concerned that Galileo's simple pendulum was not truly isochronous. Mersenne, in a letter dated December 8, 1646 to Christiaan Huygens, describes the state of the art relative to the pendulum, and asked him to look into the problem of isochronism.[13] The results of Huygens' work on the cycloid are not contained in his *Horologium,* and the clock illustrated in this monograph is a clock designed to limit the swing of the pendulum in order to maximize isochronism without a mechanical system that provides for swinging the pendulum in a cycloidal arc and there is no mention in the text of cycloidal cheeks. The results of Huygens' work on the cycloid and cycloidal pendulum are contained in his *Horologium Oscillatorium* of 1673. In this work, he shows that a cycloid is a tautochronous curve, i.e., the descents and ascents of a body from any point on the curve take equal times, hence a cycloidal pendulum should be isochronous.

Huygens' pendulum clock ideas show quite clearly that his work on the application of the pendulum to clockwork was independently conceived and certainly was not a copy of the work of Galileo. Huygens certainly must be credited with furnishing a clear, analytic, geometric statement of an ideal clock pendulum. Further, Huygens stimulated clockmakers to develop pendulum clocks in a very direct and personal way, as will be described later.

Following Huygens, it became clear to many mathematicians that Newton's work in differential calculus furnished a much more precise tool for reasoning and solving problems about the theory of the pendulum than was previously available. Newton in his *Principia* in 1687 completed the work of Huygens by general statements concerning the oscillations of all pendulums and cycloid curves and linking them to the laws of gravity.[15]

It is now evident that Huygens' important contribution to the theory of the pendulum was an unimportant contribution as far as the clockmaker was concerned. The theory of the thermal properties of materials and the effect of variations in temperature, barometric pressure, and mechanical friction as they relate to the pendulum were not formulated or expressed during the 17th and early 18th century so as to be of any vlaue to the clockmakers. It is, of course, true that problems relating to these matters were considered by Newton, Boyle, and the Bernoullis, Pascal and Torricelli, but the practical clockmaker was far ahead of the scientist and observed that nature and the simple mechanics of the clock caused variations in timekeeping. The clockmaker's approach to the solution of problems relating to temperature and pressure was an experimental cut-and-try workbench solution which was practically successful.

EXAMPLES OF THE DEVELOPMENT OF THE PENDULUM BY CLOCKMAKERS

Fig. 6   Before 1500, the talented Volpaia family of Florence, Italy, constructed fine complicated clocks.[16] Benvineto Volpaia, perhaps the most talented member of the family, made a drawing of a clock mechanism about 1520 (see Figure 6).[17]

Our interest in this drawing centers on the bent bar with the forked element at the lower end which would, by definition, act as a pendulum if the toothed wheel is the driven wheel. It is possible that the fork was intended to give impulse to the rod of a free hanging pendulum. If this was the intention, the design represents a very modern clock design indeed. On the other hand, the fork at the end of the member may imply that this is only part of a mechanism to convert rotary motion to a more or less linear motion to some other vertically pivoted member. Later in the 16th century, it was relatively common to actuate automata with a linkage of this sort. Even if this forked member was used in conjunction with some other motion, and if the toothed wheel in the drawing is the driven wheel, a pendulum action would result unless there was some kind of banking in this linkage system. Further knowledge of the Volpaias is needed. The history of horology must be rewritten if the Volpaias constructed pendulum clocks.

Fig. 7   An early Italian clock pictured in Figure 7 is perhaps the most interesting of all extant pendulum clocks to be discussed. As can be seen, the clockmaker has taken the usual adjustable cross bar or foliot member and rotated it 90°. In doing so, it was necessary to eliminate the adjustable hanging weights. In their stead, he has placed two brass bobs on either end of this oscillating member which can be adjusted by sliding them up and down. An interesting point to notice is that the lower bob or weight is somewhat larger and heavier than the upper bob and, more importantly, the lower bob is further from the center of oscillation than the upper bob. This is clearly a physical pendulum and therefore a pendulum clock. The brass movement, weight-driven, three-wheel train clock has a fixed pointer and the circular brass dial revolves about this fixed hour hand. Pins are set in the rim of the great wheel that provide for the bell to be struck at the quarters, one blow for the first quarter and one in addition for each succeeding quarter, including the hours. The system of tying the foliot or pendulum directly to the pallet shaft, the shape and position of the bell, and the general layout of the construction mark this clock as an Italian clock.[18] The height of this clock is 12 inches. Numerous horological books have illustrations of early monastic clocks of this general pattern which date back as early as 1500. This clock is typical of the early Italian monastic clocks.

Horologists looking at this undated clock will invariably say, if asked to date it, that it's probably the last quarter of the 17th century, and this is because they recognize that it has a pendulum. Everything about it except the pendulum would lead one to believe that it was earlier. This could be a sport pendulum clock made before 1600. No one knows for certain. The writer recalls looking at a particular clock in the Ilbert Room of the British Museum with the late Philip Coole, the horological curator, and asking him whether a certain clock was a 16th *or* 17th century clock. His appropriate reply was, "Yes."

Fig. 8   A German clock is shown in Figure 8. It is a rare and fine example of the beautiful workmanship that was done in the Augsburg, Germany, area from about 1590 until 1635. Although this

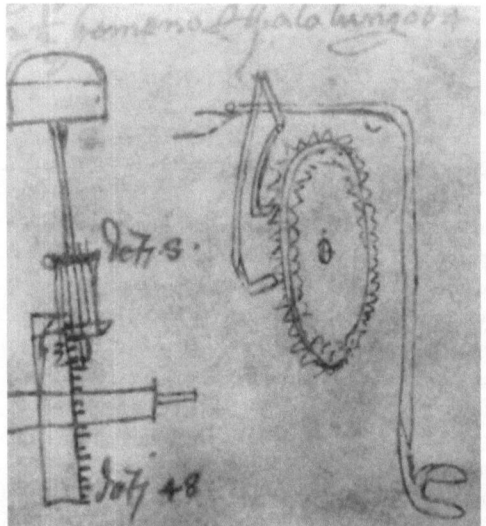

Fig. 6. Benvineto Volpaia, Florence, Italy, circa 1520, clock mechanisms.[17]

Fig. 7. Unsigned early Italian pendulum clock.

Fig. 8. Unsigned masterpiece, Augsburg, Germany, circa 1625.

masterpiece is neither dated nor signed, horologists date it circa 1625. The overall height of this clock is 23-½ inches. The chased fire gilt figure of the elephant is comparisoned in silver. The engraved gilt metal howdah contains the striking and quarter chiming movement. This movement is of 30-hour duration as is the main movement within the body of the elephant. The bell is housed in an arcaded gallery over the movement about which, at the hours and quarters, the three knights revolve around the turret. The time dial has an engraved landscape at its center and silver chapter ring indicating the hours and minutes. The reverse side of the howdah has a quarter chime indicator dial. The elephant stands on a repousse fire gilt dais and the front of the ebonized octagonal base is a recessed stage on which at the hours, three animals move around a central figure at the striking of the hours, and a small dog also revolves between the front legs of the elephant. Clocks from this area at this time are usually fire gilt and beautifully executed in terms of decorative finishing. They also exhibit a high quality of clockwork. Augsburg was an important mecca for fine workmanship, not only in clocks, but in instruments and silver work. It was on the path of the trade route between the East and the West, but its importance as a mecca for fine craftsmanship died out during the Thirty Years War. This clock is important because it may have been an accidental pendulum clock. When the clock is running, the elephant's tail swings and its eyes oscillate. It would have been a very natural thing for the clockmaker to have these motions occur in view of the general automata motions of the clock as described. Further, it was quite common to incorporate oscillating eyes in lion and dog clocks made in Augsburg at this time. The crown wheel of the movement is placed horizontally and the pallet shaft is directly connected to the tail. In this condition it is a pendulum clock. Some horologists feel that this linkage is a later conversion. If this is so, it was most expertly done indeed, since there is no indication within the clock of extra holes or a rearrangement of the movement. Let us assume, however, that the elephant's tail was not originally connected to the pallet shaft, and that typically, there was a bent forked member connected to the pallet shaft similar to the Volpaia illustration shown previously in this paper. This forked member would have engaged a vertical member inside the elephant's body which was attached to the tail. If this was the case, the assembly in toto would constitute a pendulum clock. Perhaps none of these assumptions are correct, and the tail did not act as a pendulum; no one can be certain.

Fig. 9   The plausibility of pendulums at this early period is strengthened by an anonymous drawing of a verge escapement and the pendulum of Italian origin discovered by Enrico Morpurgo (see Figure 9). He states that this is an early 17th century drawing but, of course, we do not know whether clocks embodying this idea were constructed. In this same article, another illustration (Figure 10) of an early 17th century pendulum is shown which may or may not be a clock.

There is a pendulum clock in the Science Museum, London, signed Camerini. This clock is dated 1656, which is one year earlier than the clock made by Solomon Coster after Huygens' design. Camerini was a Turin clockmaker, whose name does not appear in the literature related to clockmaking or the pendulum. The writer has seen several clocks by Camerini, and it appears that they have been converted to a pendulum motion by placing a swinging pendulum in front of the dial which is a typical conversion practice. The writer believes that the Camerini clocks are conversions to pendulum and this opinion is shared by some other horologists.

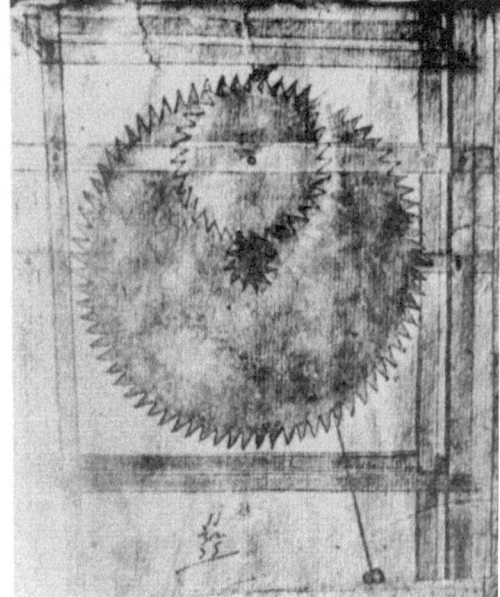

Fig. 9. Early Italian verge escapement with pendulum.[19]

Fig. 10. Is this an early drawing of clockwork with pendulum?[19]

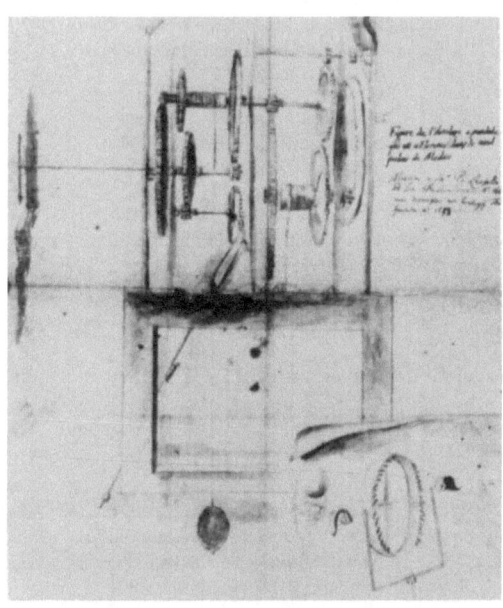

Fig. 11. Drawing by Treffler made circa 1659, showing how Prince Leopold's palace clock was converted to pendulum regulation.

Neither Huygens nor Galileo is entitled to take credit for first conceiving the idea of applying a pendulum to clockwork. It is unlikely that either of them could claim priority to pendulum application, that is, the actual construction of a clock utilizing a pendulum. We can be quite certain that the earliest complete clock constructed after Huygens' design was made by Solomon Coster in 1657. The writer believes that Galileo stimulated the construction of a pendulum clock before that date. It should not be concluded, however, that these applications occurred other than independently.

Figs. 5, 11 We know that Galileo did not construct a clock during his lifetime, and records indicate that Vincenzio Galileo, his son, attempted to build a clock with a pendulum prior to his death in 1649. Viviani, as executor of the estate, in an inventory of effects made the following entry: "An iron clock unfinished with pendulum; the first invention of Galileo." (See Figure 5.) Prince Leopold received a letter from Ismael Boulliau in Paris, a friend and correspondant of Huygens in February 1659, saying that he was sending the Prince a Huygens clock. The Prince replied in March that, "One must not rob Galileo of his glory." He goes on to state that he had discovered a model already made by Vincenzio Galileo and that three years ago (i.e. 1656) a clock was made from it by an expert.[20]

From the research of Silvio A. Bedini, it seems clear that the expert referred to was Johanne Philipp Treffler. The Treffler family was from Augsburg, Germany, and Treffler's father was a turner in silver. He was born in 1625, and while in his early twenties migrated to Florence to become a turner, instrument maker, and clockmaker to Archduke Ferdinand II of Tuscany.

The following is a quotation from Bedini:

> A few months later Prince Leopold wrote another letter to Bouiliau, on August 21, 1659, telling him that he was forwarding the design of Galileo's clock "drawn with the same roughness as is the model made from it, which is now in my room.

> Becher supplies a further note of interest in this matter in a statement which appears in Chapter XIV, page 15 of his second work entitled *Narrische Weisheit* (9):

> "An artistic clockmaker of Augsburg who for a long time was in Florence the art clockmaker to the former Archduke, by name of Treffler, told me that they had a perpendicular clock long before Zulichem [Huygens]."

> The claim that Treffler actually completed a model pursuant to Galileo or Vincenzo Galilei's design is confirmed.[21]

Bedini also points out the importance of the drawing shown in Figure 11 [22] in relation to the Galileo-Huygens controversy.

Again, quoting from Bedini:

> This sketch in pencil was made for Prince Leopold in 1659 specifically with the intention of having it sent to Huygens as evidenced by Leopold's statement in his letter to Bouiliau dated August 21, 1659, to the effect that "... I shall also have a drawing made showing the manner in which we adapted the pendulum to our

clocks particularly of a very large striking clock in the Piazza of our Palace and shall send it to you ..." This sketch accompanied a copy of Viviani's Letter Report and a drawing of the Galilei clock which Bouiliau was to forward to Huygens.

One of the inscriptions on the sketch is written in French in Bouiliau's hand and reads "Drawing of the pendulum clock which is in Florence in the old palace of the Medici." The second inscription is in Latin and not completely legible and seems to be "missum a Sr. Pr. Leopold ad Zm Buillialdum a° 1660 cum 2 r scripti mir (?) Eorology dilv (?) a° 1658."

The clock pictured in the sketch was undoubtedly the work of Treffler. The outstanding feature deserving note is the placement of the crown wheel and verge at the base of the movement with the great wheel at the top. This placement is characteristic of the very few works in existence signed by Treffler. The use of this unusual placement has been otherwise noted only in the movements of some of the earlier night clocks by Guiseppe Campani and his brother Pietro Tommaso.

It is somewhat difficult to understand why this sketch has not previously been inter-related with the text of Viviani's Report of 1659, and it is just as difficult to understand why Treffler's application of a pendulum to the Pitti Palace clock has not been given more consideration in the history of the pendulum clock, since the reconstruction of the Pitti clock occurred between 1655 and 1658 — that narrow interim of time which is of the most critical importance in the story of the pendulum clock.

In the light of this evidence, one can say that from the stimulation of Galileo there was application of a pendulum to a clock a year or two before the work of Huygens. Huygens made his model of a pendulum clock on December 25, 1656.[23] There is no indication that either copied the other, as will be shown after describing the construction of Huygens' clocks.

Figs. 12, 13 It is clear that Treffler was an important person in the history of the application of the pendulum to clockwork. Treffler returned to Augsburg in 1664 and continued to work in part for the Medicis in Augsburg until his death in 1697.

A rare Treffler clock is shown in Figures 12 and 13. The reader will notice his placement of the escapement and pendulum at the base of the movement. This is peculiar to his clocks and quite different from the practice of the Dutch and the English. The clock is a day and night clock, with a 30-hour movement and inverted verge pendulum escapement. The time is registered by the revolving disc with pierced hour numerals transversing the semi-circular lunette in the painted dial. The hours are also registered in the normal manner. This clock was made in Italy between 1656 and 1664. The case is walnut and in a style of this period. Note the Italian form of signature which reads, "Gio Philipe Treffler Augusto". The interior of this case is lined in metal, so that an oil lamp could be set in the case to allow light to shine through the perforations, thus registering each hour at night in relation to quarters. There is a ventilation hole at the top of the case and a small door at the rear. The overall height of the clock is 19-½ inches.

Figs. 14, 15 In this exciting horological period marked by the beginnings of the application of the pendulum to clockwork, surprisingly enough we find a totally independent development and appli-

Fig. 12. Rare night clock by George Philip Treffler made between 1656 and 1664 while he was in Italy.

Fig. 13. View of Treffler clock movement from the rear showing Treffler's unique underhung escapement.

435

cation to clockwork by the Campani brothers in Rome. Their work seems to be nearly coincident with that of Treffler and Coster. Matteo Campani related that after his own accidental invention of a pendulum clock, which occurred at some time during the year 1656 or early 1657, he had applied the same principle of a pendulum regulator to a neck watch. After having observed the equal motion of the pendulum applied to the neck watch for several days, he journeyed to Florence to bring the watch to his patron, Archduke Ferdinand II.

> Since this (invention) appeared to be such a thing of wonder and most
> useful, I brought it to His Serene Highness the Grand Duke of Tuscany;
> but his Serene Highness, to my marvelling unbelief, exhibited to me a
> clock with a similar artifice which had been constructed from his own
> personal invention and he made me understand that his Serene Highness
> as a result of his own personal studies and application, had most ingeni-
> ously invented this himself, which I had discovered so accidentally. Then
> he showed me a print (published work) of a similar one, although it was
> in part a very different artifice discovered in Holland by Signor Christian
> Huygens; and finally, His Highness, in order to honor me even more,
> allowed me to see a large old chamber clock which had been constructed
> for Signor Galilei, which likewise had a pendulum for its balance (tempo);
> and which, although it was easily motioned it was not so perfect as was
> His (Highness') own, and it was also enough different from the method of
> the Dutch clock, nevertheless, it could not be denied that this also was a
> pendulum clock. And in consequence, it is necessary to confess that
> Galilei was the first inventor, of the pendulum, and of the application of
> these same pendulums to ordinary clocks, and that furthermore, His
> Serene Highness the Prince (Leopold?) was the first to give (these pendu-
> lum clocks) their necessary disposition and simplicity which was lacking
> in the mechanism of Signor Galilei for the perfection of the clock and
> the accuracy of its movement.[24]

The Campani brothers were clearly excellent workmen. Guiseppi Campani was noted for his fine telescopes. The *Philosophical Transactions* of the Royal Society of England, No. 2 (1665) has an account of some observations by Guiseppi Campani on Saturn's ring, which tended to confirm Huygens' theory of the movement of Saturn's ring. Matteo Campani, in 1655, in a well documented story, stated that Pope Alexander VII had asked him to apply the pendulum to clocks and that he wanted a silent clock, so that he could sleep at night without interruption. Morpurgo felt that Pietro Tommaso was the most talented of all of the brothers, and it was he who invented the silent escapement for a night clock.[25] In this device, the pendulum is designed in such a way as to allow through linkage an eccentrically placed disc to continuously rotate rather than give the intermittent motion of the usual escapement mechanism.

Shown in Figures 14 and 15 is a rare night clock signed in an arc on the barrel, "Petrus Thomas Campanus, inventor, Romae, in Via Julia ad Ponte Sixtu 1664." This clock contains his silent crank escapement and pendulum which has been restored as closely as possible to its original design. The problem of illumination was solved by placing in the case behind the copper dial, a candle on a holder with an appropriate flue through the case. Behind the copper dial revolves a metal disc which has two shaped openings through which pierced hour numerals show. The movement revolves the disc in a clockwise direction carrying the current hour numeral in an arc so that it traverses the semicircular lunette in the dial. In doing so, the

Fig. 14. A rare Italian night clock by Petrus Thomas Campanus, dated 1664. Apparently, an independent development of the pendulum.[25]

Fig. 15. A view of the movement of the Campani clock showing the pendulum. Hidden from view is the silent crank escapement invented by Campani.

minutes may be gauged in relation to the appropriate quarter hours. The light from the candle shows through the pierced numerals. The copper painted dial shows the rape of Ganymede. The ebonized case is of architectural design and the base is molded in the classical manner. The overall height is 40-¾ inches. Campani's silent escapement pendulum clock design never became popular and, in fact, it is quite difficult to keep these clocks running properly.

Figs. 16, 17   Huygens' independent development of the pendulum for clockwork was applied by Solomon Coster in 1657. The clock shown in Figure 16 is inscribed on the cartouche, "Solomon Coster-Haghe met privileg 1657." *Met privileg* indicates that Huygens had assigned his rights to the invention to Coster and a patent was granted to him on June 16, 1657, for 21 years. This clock is of interest because it is one of three extant Coster clocks which are thought to be the earliest surviving pendulum clocks. In the author's opinion, pendulum clocks were made earlier than 1657 by accident and by design. This early clock did contain an undeveloped form of cycloidal cheeks, which Huygens abandoned in making his recommendations concerning the construction of a clock with a pendulum in his *Horologium* (1658). In 1659, he evolved the theoretical form required to build cycloidal cheeks and he went back to recommending this construction. The clock shown in Figure 17 limits the amplitude of the pendulum as a result of interposing a gear on the pallet shaft and a second meshing gear on the pendulum clutch shaft. Huygens' primary motivation was that his clock design would solve the problem of longtude. He continued during his lifetime to experiment with pendulum marine clocks. His experiment to hang the pendulum clock from a ball joint quickly revealed that a pendulum clock even in a ball joint did not work at sea.

Fig.18   Four other people are known to have worked directly with Coster prior to his death in December, 1659, namely Visbach, Hanet, Pascal and Fromanteel. Peter Eraerts was born in 1634 and died circa 1700. He is better known as Peter Visbach, the name of his stepfather. In 1646 he started working part-time with Solomon Coster and on Coster's death in 1659, he bought Coster's business. In conversation with Enrico Morpurgo, the writer learned that Visbach probably did not work directly with Coster since he moved from Middleburg to the Hague in 1660, when he bought a house on the east side of Toorenstraat from Coster's estate. Visbach was clearly among the earliest of the Dutch makers to apply a pendulum to a clock, and it is known that he had direct association with Christiaan Huygens. He was an official in the Guild at the Hague. A very early clock by Visbach is shown in Figure 18. If Morpurgo's assumption is correct that Visbach did not work directly with Coster, it may explain why this clock is signed on the gilten metal cartouche *Pieter Visbach Secit Hagae met privilege*. This signature seems to imply that although the patent was assigned to Coster, Huygens gave Visbach the right to construct clocks to Huygens' design as well. It will be noted that this clock is encased in a different style from that of Coster. It is faced in red and black tortoise shell with split door pilasters and a plumb type indicator is fixed to the right hand side. One is led to believe that this is a very early Huygens type clock. This is an 8-day clock with a going barrel movement and a verge pendulum escapement with Huygens' cycloidal cheeks. The pinned back plate is also signed "Pieter Visbach Hagae." The chapter ring is 5-½ inches in diameter and the height is 13 inches. There is no way to determine whether this clock was made during Coster's lifetime or shortly thereafter.

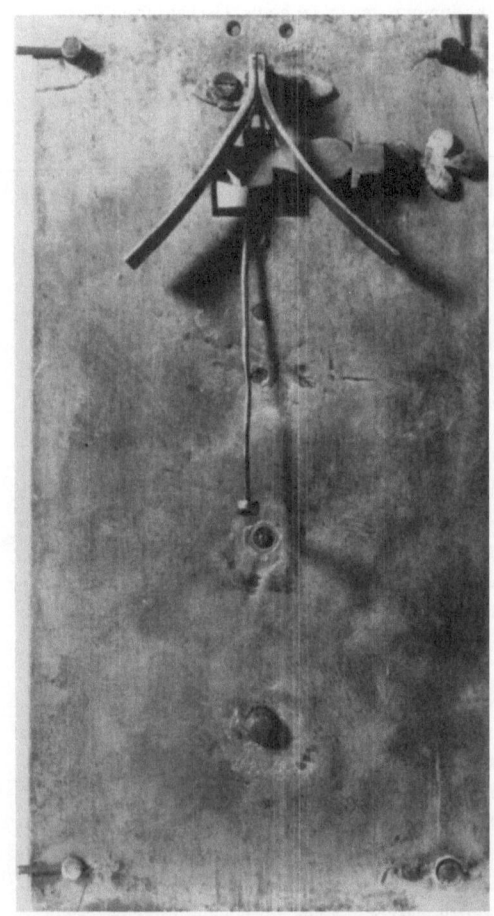

Fig. 16. These two photographs are of a
recently discovered Solomon Coster clock.
The upper photograph of the movement
from this clock shows the pinned back
plate and the rudimentary form of the
cycloidal cheeks. (The string suspension
pendulum is missing). The lower photograph is of the cartouche, which is hung
on the lower portion of the dial plate.
The cartouche is, for all purposes, identical on all three Coster clocks.

Fig. 17. Modern reproduction of pendulum clock mechanism utilizing Huygens' design for a pendulum clock.

Fig. 18. Clock by Peter Visbach. One of four men to work directly with Huygens and Coster. He took over Coster's business upon Coster's death in 1659. This Huygens-type pendulum clock is not dated, but it is probably circa 1660.

Fig. 19. Clock by Nicholas Hanet, Coster's agent in Paris. Another fine maker who worked with Coster in Holland. The movement shows the cycloidal cheeks for the pendulum.

Fig. 20. A beautifully made clock by Claude Pascal. This clock was undoubtedly made by Pascal while he was working with Coster in Holland. The striking mechanism is a rack striking device which predates Barlow's rack striking patent.

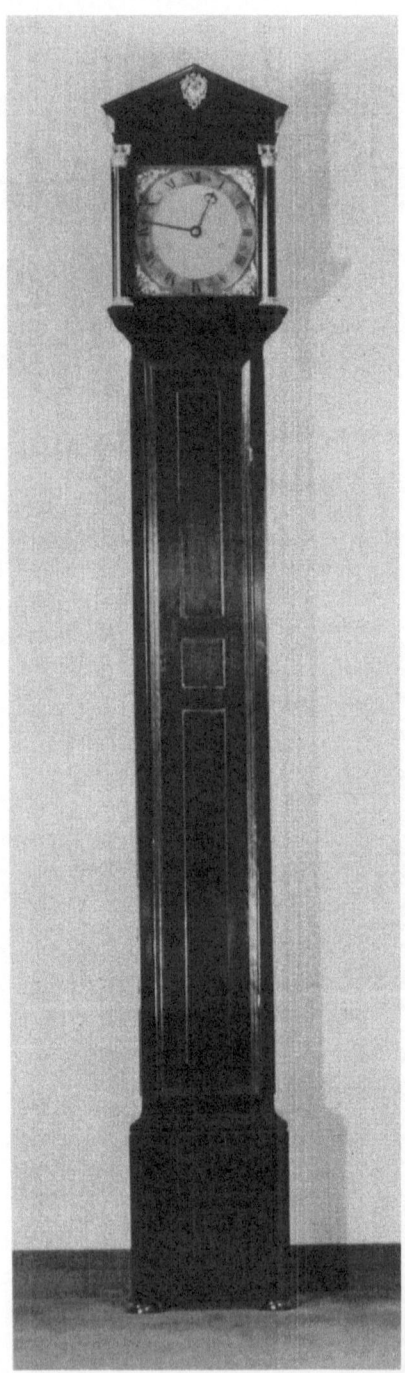

Fig. 21. A longcase clock with a short bob pendulum by Ahasuerus I. Fromanteel. This clock was made between 1662 and 1666 in London. It has a 6 - ¼ in. pendulum attached directly to the pallet arbor. Ahasuerus' son, Johannes, worked directly with Coster in Holland and the Fromanteels introduced the pendulum clock into England in the early 1660's.

Fig.19 It is known that Hanet of Paris worked with Coster from September 1657 to May 1658 in Holland, and was his agent in Paris. It is also known that Huygens and Coster arranged at a very early date to have clocks shipped to Paris. Le Duc de Luynes received a clock in September, 1658. Other clocks were shipped in 1659, but after Coster's death, few were shipped to Paris since Huygens was unable to obtain a patent on these clocks in France and French clockmakers freely copied Huygens' type clocks. Winthrop Edey[26] points out that between 1650 and roughly 1660, clockmaking was practically nonexistent in France. Huygens' clocks caused, as Edey puts it, "an explosion of clockmaking in France." This is in contrast to the situation in Holland, where from 1659 until 1678 Visbach alone had the right to produce Huygens' pendulum clocks. The photo shown is a clock by Hanet, and because of his relationship with Huygens and Coster, he certainly was among the earliest to construct pendulum clocks on the Huygens principle in the French capital. Examples by this maker of religeuse cycloidal cheek clocks are of considerable rarity. The clock in Figure 19 is 17-¾ inches high. Again, it is very difficult to date this clock, but it might be circa 1661. The construction of the clock follows very closely Coster's design in that it has one drum for the going and striking.

Fig.20 The third of the four makers to work with Coster was Claude Pascal. Notice that the clock illustrated in Figure 20 is very similar in appearance to the clock by Coster. The plate below the dial is inscribed "C. Pascal Hagae Hollandiae", which indicates the clock was made while he was in Holland prior to his death in 1671. From the horologist's point of view, this clock is extremely interesting, since it contains a rack striking mechanism. Edward Barlow of London is generally credited with inventing rack striking in 1676, some years after Pascal's application.[27]

Fig.21 The only other maker that we know of who worked directly with Coster during that critical period when he was building clocks under the Huygens plan is Johannes Fromanteel, who was in the service of Coster in 1657 and 1658. Dr. Plomp points out that the Fromanteels may have migrated to England from the French speaking part of the low countries in the 16th century and settled in Colchester and/or Norwich, England.[28] We know that Ahasuerus I was born in Norwich about 1605 and that he migrated to London in 1629. His son, Johannes, who was an apprentice to Coster, was born in 1638 or 1639. This family was clearly mechanically inclined. The *Commonwealth Mercury*, November 25, 1658, points out that Ahasuerus I Fromanteel not only announced the new invention of the pendulum, but also advertised for sale engines made in a way of his own invention for quenching fire. It is Dr. Plomp's opinion that Ahasuerus I Fromanteel was a very brilliant man indeed. Clearly, Ahasuerus I Fromanteel and his son, Johannes, were the first to build pendulum clocks in England. Ronald A. Lee illustrates perhaps the only dated and only cycloidal cheek clock made in England on plates 2 and 3 in his booklet largely devoted to the Fromanteels. It is signed: "A. Fromanteel, London, FECIT 1658." This plain portable table clock is, as Lee states, "the foundation stone of all pendulum clocks made in England."[29]

The fertile minds of the Fromanteels are evident in other clocks by them. Various features are introduced and as near as we can tell from surviving examples, the Fromanteels immediately dropped the idea of cycloidal cheeks and attached the short bob pendulum directly to the pallet shaft. In order to reduce the amplitude of the swing, the pallets which meshed with the crown wheel were set at 70°. The beautiful long case clock illustrated in Figure 21 by Aha-

Fig. 22. A beautiful silver decorated clock by Johannis Van Ceulen of the Hague utilizing Huygens' pendulum design made after Huygens' patent in Holland expired in 1678.

suerus I Fromanteel represents the basic design of all long case clocks made in England through the 17th and 18th centuries. The cabinet work in these clocks is of the very highest quality and one is immediately struck by the perfect proportions of the Renaissance style. The movement of this clock is typical of his work between 1662 and 1666. It has a 6-¼ inch long bob pendulum attached directly to the pallet arbor. This clock features, among other things, one of Fromanteel's contributions to clockmaking, i.e., bolt and shutter maintaining power.

Fig.22 The beautiful workmanship of Johannis Van Ceulen of the Hague is shown in Figure 22. After the expiration of Huygens' patent in Holland in 1678, a few other Dutch makers began to make clocks after the Huygens' pattern. For a time, Van Ceulen was closely associated with Huygens for whom he made a number of watches incorporating a balance spring. He also constructed a planisphere to Huygens' design circa 1682.

It is clear that by 1675, there was a widespread preference through England and Europe for clocks made with pendulums, and the knowledge of how to make a clock was easily obtained. From this date until the middle of the 18th century, English clockmaking was significantly refined and improved. The English, through many improvements, some of which will be detailed hereafter as they relate to the pendulum, set the standard for the rest of the world to copy.

Fig.23 The next clock, illustrated in Figure 23, is by Joseph Knibb. It is important in relation to the pendulum, not only because of refinements in adjustment relating to the pendulum, but also because it incorporates something new in escapements which improved the accuracy of timekeeping. In order to appreciate the statement previously made concerning the importance of English clockmaking starting in 1675, the following is a complete quotation of a description of this clock made by R.K. Foulkes, an expert horologist from England.

> An outstanding long case clock by the celebrated Joseph Knibb of London, the movement of the month duration having striking on the Roman Notation Principle, the ten inch square dial signed at the base, the finely proportioned case in ebony veneer typical of Knibb's case maker, the whole embodying the best in English clockmaking of ca. 1680.

> Joseph Knibb, 1640-1711, was the most illustrious member of the famous Knibb family of clockmakers. England owes much to the marvellous workmanship of the Fromanteels, the Knibbs, Tompion, East, Jones, and Quare, who with others, perhaps less well known, all contributed to placing England, horologically in a supreme position during the last quarter of the seventeenth century.

> Joseph Knibb owes much of his fame to his uniformly high quality, delicate work and inventive skill. He was among the first to use the true anchor escapement in long case clocks. Joseph and his brother, John, concerned themselves with another form of escapement known as the tic-tac which was fitted to spring clocks about 1673-78. Joseph applied this escapement to a hanging clock supplied in 1673 to the astronomer Gregory.

> The Roman notation, Joseph's invention, probably in 1677, was a system of striking to economise in the use of power. It provided for thirty blows instead of the usual seventy-eight in twelve hours. The system was accomplished with two bells, the smaller bell striking the I and the larger bell the V in the appro-

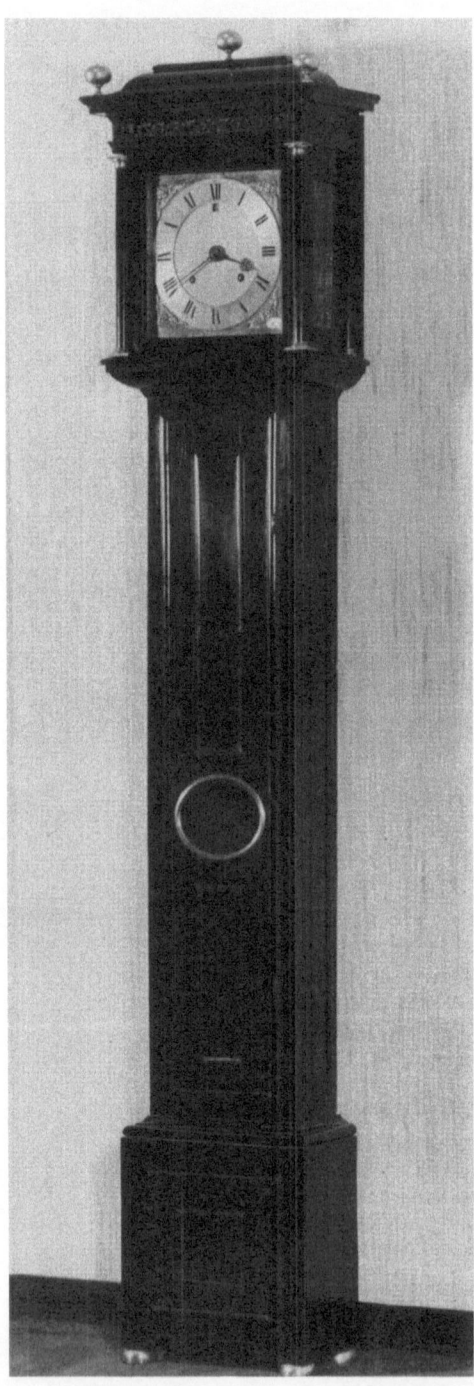

Fig. 23. English longcase clock by the famous Joseph Knibb, circa 1681. One of the very early clocks to incorporate a long (i.e. one second) pendulum. The long pendulum was made possible by discarding the verge escapement and utilizing the anchor or tic-tac escapement.

priate sequence for the Roman hour numerals. Thus IV was one blow on the smaller bell followed by one on the larger. At X there were two blows on the larger bell. Such a plan for power economies was undoubtedly precipitated by his research into duration periods of one month in spring clocks and three months and more in weight clocks.

The movement has a duration of one month. The front plate is latched and the dial plate pinned. There are brass pulleys and brass cased weights.

The going train, planted on the left is conventional for the duration, with an intermediate wheel before the centre (third). This is an anchor escapement clock with seconds pendulum. The back cock is separate to the pendulum suspension, the latter being on Knibb's principle with a butterfly net and the block for the crutch hooked to the bottom of the suspension spring. To regulate, the whole pendulum bob is turned on the rod which is threaded. The fixed butterfly above is held to prevent the pendulum rod turning. Small regulation adjustments are performed by using the suspension butterfly.

The striking train planted on the right has locking plate striking mechanism on the Roman notation principle, the locking plate being squared on to the second wheel arbor outside the back plate. This is the pin wheel and has pins both sides that engage with the tails on the two hammer arbors. The two bells are positioned, one inside the other above the movement.

The ten inch square gilt metal dial has an engraved feathered edge and bears the signature along the base, "Joseph Knibb Londini Fecit". These are winged cherubs head spandrels, a silvered chapter ring and date aperture. The original blued steel hands are of high quality, the hour hand being delicately pierced and the minute hand tapered in a spiral design.

The case has an oak carcase and is veneered in ebony. The lift-up hood (now cut for a door) has the caddy top and flattened gilt brass finials associated with Knibb. There are finese frets and fine pilasters with gilt metal cappings. A convex moulding unites the hood with the truck which has fielded panels bordered by delicate scratch mouldings. The door is provided with a large oval lenticle. The plinth base is raised on four gilt metal bun feet.

The dial 10-inches square, height 6 feet 9½-inches English ca. 1680.

The importance of this clock in relation to this paper lies in the fact that it is one of the very early clocks to incorporate a pendulum which swings at a one-second rate. This is accomplished by discarding the verge escapement and employing an anchor escapement. The anchor escapement allowed the clockmakers to avoid a pendulum with a wide amplitude (about 40°). as required by the usual 90° pallet position on a verge escapement. It is generally stated that William Clement (1639-1704), an anchorsmith and locksmith, invented and used this escapement in a large turret clock made for Kings College, Cambridge, in 1671.[30] This invention is truly important to horology, since with it the amplitude of the swing of the pendulum can be restricted to 3° or 4°, and it is the basis in essence of all subsequent escapements. There are some horologists who feel that there is a possibility that Joseph Knibb was the first to use an anchor escapement. Be that as it may, it seems clear that Knibb was one of the very first people to employ this escapement in a long case clock.

CONCLUSION

The year is 1700. The precision clock has been born. Subsequent years find the clockmaker pragmatically experimenting with devices to improve the accuracy of clocks by minimizing friction and counteracting the effects of temperature and barometric changes. The need for even more precise timekeepers continues to the present.

NOTES

1. Nicole Oresme: *Le livre du ciel et du monde,* cited in Piero E. Ariotti: "Aspects of the Conception and Development of the Pendulum in the 17th Century," *Archive for History of Exact sciences,* VIII (1972), 340.

2. See Antonio Simoni: "Oddities of Certain Leonardian Codices," *La Clessidra* (1971), 20-24.

3. Jacques Besson: *Theatre des instrumens mathematiques & mecaniques,* ed. Francois Beroald (1578), cited in Edwin A Battison: *Technology and Culture,* VII (1966), 202-5.

4. Isaac Beeckman: *Journal,* cited in Ariotti, *op. cit.,* p. 375.

5. J. Drummond Robertson: *The Evolution of Clockwork,* (London: Cassell & Co. 1931), Chapter V.

6. Silvio A. Bedini: "Saggi Su Galileo," in *Galileo Galilei and the Measure of Time,* ed. G. Barbera (Florence, Italy: Comitato'Nazionale per Le Manifestazioni Celebrative del IV Centario della Nascita di Galileo Galilei 1967).

7. Letter from Galileo to Realio, 5 June 1637, cited in Bedini, "Galileo Galilei," *ibid.*

8. G. Galilei: *L'usage du cadran ou de l'Horloge physique universal,* (Paris 1639), cited in Robertson, *op. cit.,* p. 93.

9. Robertson, *Ibid.,* p. 82 n2.

10. Bedini, *op. cit.*

11. H. Alan Lloyd: *The Collectors Dictionary of Clocks,* (London: Country Life Ltd. 1964), 53.

12. Christiaan Huygens: *Horologium,* tr. E.L. Edwardes in *Antiquarian Horology, VII* (1970), 35-55.

13. Ariotti, *op. cit.,* pp. 372-80.

14. Cited in *ibid.,* p. 381. See also Robertson, *op. cit.,* p. 76.

15. E.g., *Principia,* ed. Florian Cajori, Book I, Prop. LII, Prob. XXIV (Vol. I, p. 157).

16. Enrico Morpurgo: "Alcuni appunti sugli orologiai Della Volpaia," *La Clessidra* (1959).

17. *Ibid.*

18. Enrico Morpurgo: "L'Orologia e il Pendolo," *La Clessidra* (1950).

19.   *Ibid.*

20.   Robertson, *op. cit.*, p. 101.

21.   Silvio A. Bedini: *Johann Philipp Treffler, Clockmaker of Augsburg,* (National Association of Watch and Clock Collectors, Inc. 1956-57), and a supplement, "Agent for the Archduke," *Physis,* III (1961).

22.   *Ibid.*

23.   *Oeuvres completes de Christiaan Huygens,* (Den Haag: Martinus Nijhoff 1889-93), Oeuvres 2, p. 109.

24.   Enrico Morpurgo: "Gli Oralogi Notturni E. Guisette Compani," *La Clessidra* (1959).

25.   *Ibid.*

26.   Winthrop Edey: *French Clocks,* (New York: Walker & Co. 1967), 30-1.

27.   G.A. Baille: *Watchmakers and Clockmakers of the World,* (London: Finwell House 1963), 16.

28.   R. Plomp: "The Dutch Extraction of the Fromanteel Family," *Antiquarian Horology,* VII (1971), 324. The writer has had the opportunity to review the work and influence of Solomon Coster with Dr. R. Plomp of Holland, who has been most helpful indeed, and has furnished in writing some notes about early Dutch pendulum clocks, which he painstakingly obtained from the correspondence of Christiaan Huygens and other persons.

29.   Ronald A. Lee: *The First Twelve Years of the English Pendulum Clock or the Fromanteel Family and their Contemporaries,* (Guildford, England: Seven Corners Press 1969).

30.   Lloyd, *op. cit.*, p. 58.

Unless otherwise indicated, all photographs are from the author's collection of clocks on public display at the Time Museum, Rockford, Illinois.

Oriental Concepts of the Measure of Time

S.E. BEDINI

Time and its measurement have been preoccupations of mankind since the most primitive civilization in all parts of the world, and as his needs and his knowledge increased, so did man's awareness and concepts of time develop. In the Orient as in the Western World, the first timekeepers were such simple phenomena of nature as the sun, the moon, and the stars. The shadow cast by a tree or vertical object led to the invention of the sundial, and then the measured flow of water was discovered to provide another unit of time measure, as indeed did subsequently the flow of sand.

With his increasing awareness man realized that in everyday life time provided the means by which order was furnished to all organic and human experience. He learned to consider time in terms of two basic units, the interval and the epoch, the first to measure duration of, and the second to measure location in, time. Thus, time became a subject of man's philosophical consideration long before he attempted to harness it for practical use. Observation of such celestial phenomena as the periodicity of the sun and moon brought even greater cognizance, which led to the creation of myths and superstitions, which in turn gave rise to cults and religions evolving from special rites designed to keep these deities benign.

The Chinese, for example, evolved the legend of their first Adam, P'an Ku, who created the universe by chiselling the stars and planets from the cliffside of Chaos, an undertaking which required eighteen thousand years and from the initiation of which the Chinese measured the beginning of time.[1] Another was the legend of the Ten Chariots of the Sun, which raced across the sky after each other, each drawn by six dragon horses and driven by the mother of the suns.[2] Each of the suns represented one hour of the day, and it was an evil omen if more than one was to be viewed at the same time. Celestial phenomena were explained in the stories which described them, such as the tale of the Excellent Archer who shot down one of the nine suns with his magic bow, and the legend of the Cowherd and the Weaving Girl which explained the phenomenon of the Milky Way. Then also there was the great dragon, whose waking and sleeping determined the duration of day and night, and whose breathing regulated the winds and the seasons.[3]

Such myths and legends became interwoven with the religions that evolved and influenced and subsequently became part of the philosophic considerations of scholars. They permeated the recording of early history leading to the evolution of a literature, and played a role in the development of early science.

Fig. 1. P'an Ku came into life endowed with perfect knowledge. He was the legendary "great architect of the universe" and was usually depicted with a hand chisel carving the planets and the stars from the cliff of Chaos. The inseparable companions of his toils were the dragon, phoenic and tortoise. His efforts continued for eighteen thousand years during which the heavens rose and the earth spread out and thickened. P'an Ku himself grew in stature six feet a day until his labors ended and he died giving birth to the details of the existing material universe. His head became the mountains, his breath the wind and the clouds and his voice was transmuted into the thunder. His left eye became the sun and his right eye the moon, his beard was transformed into the stars, his limbs the four quarters of the globe, his blood the rivers and his flesh became the soil. His legend is said to have been invented in the 4th Century A.D. by Ko Kung.

The primitive tools of time measurement were gradually improved and transformed into devices useful to court astrologers and astronomers as the primitive concern with heavenly deities became gradually transformed into the beginning of science. The water clock in particular proved to be adaptable for astronomical measure and made possible the invention of sand clocks from which progressed gearwork for time measurement. These preoccupations culminated with the sophisticated clock tower of Su Sung in the eleventh century as an achievement combining philosophy and technology.[4] Throughout this evolution of time measurement in China, time remained the concern of the philosopher and the astronomer, however, and historians relate that it was not until the second half of the fourteenth century that the function of time measurement became separated from astronomical apparatus, resulting in the development of the first independent clockwork which could be utilized for common needs.[5]

Meanwhile, approximately the same history of evolution had taken place in Western civilization during approximately the same period. The sundial and water clock had been introduced into Greece and Rome from the Arabic world at early periods and paved the way for a separate astronomical technology culminating in the first century B.C. with a geared planetary device of considerable sophistication.[6]

It was not until the late thirteenth century that the mechanical clock made its appearance in the Western World, and considerable speculation persists among scholars that it may have been introduced from China, possibly by traders through one of the seaports of northern Italy. The first appearance of mechanical clockwork, probably water-powered instead of weight-powered, is confirmed in northern Italian cities by the beginning of the fourteenth century, whence the invention was gradually transmitted westward throughout Europe.[7]

Meanwhile, as the mechanical clock emerged and evolved in Europe, it was simultaneously lost to history in China. Natural and political events, furthered by the wear of Time itself, effectively erased the great achievements of Chinese astronomers and technologists, leaving surviving evidence only in the written word of forgotten manuscripts and published works. So well did the passage of centuries blot out the memory of these accomplishments, such as the clocktower of Su Sung, that by the advent of the first European missionaries to the East in the mid-century, the Oriental world had returned to the use of the simplest forms of timekeeper in the community, in the temple and even in the palaces of sovereigns.

The first Jesuit missionaries who penetrated the Orient found a new world in which time and its measurement were little more than philosophical preoccupations. It was with the utmost naivité that among their gifts to Oriental potentates to encourage the propagation of the Christian faith they brought mechanical clocks among other representations of the finest Western craftsmanship. The first missions arrived in the Orient approximately two and one-half centuries after the mechanical clock had made its first appearance in Europe. In the intervening period this invention had developed into little more than a toy for the cabinets of curiosities of princes and prelates, and it remained only a symbol of status in Europe until it was produced in substantial numbers in the eighteenth century. In the centuries that followed the establishment of the first missions in the East, more and more European missionaries, diplomats and traders directed their attention to the Orient, and although the mechanical clock played a critical role in the acceptance of the foreigners by the Eastern potentates, it contributed little to the Oriental awareness of time and its measurement.

The European mechanical clock was first introduced to the Orient in Japan in 1549, shortly after the arrival of the first Portuguese traders in the Japanese port of Kui-Shiu in 1542.[8] When reported in Europe, the success of the traders aroused considerable interest in the religious centers, and plans were promptly made for propagation of the faith in Japan and China, particularly by the Jesuits. The first Jesuit mission was undertaken under Father Francis Xavier who arrived in Japan in 1549 with plans to establish himself at Yamaguchi. Travelling from Kyoto to Yamaguchi, Xavier was provided with appropriate gifts by the Portuguese Viceroy to India. These he presented to Yoshitaka Ouchi, governor of Yamaguchi, in 1550 and most prominent among them was a clock described in a sixteenth century biography of Ouchi as "... an instrument ... which controls the twelve 'hours' and rings during day and night."[9] Additional details have not survived, and even less is known about the second clock to make its way into Japan, the gift of a Jesuit missionary to General Hideyoshi, then the ruler of all Japan.[10] Clocks may have also been among the gifts presented to the Japanese embassy of four envoys who visited the Vatican in 1591. The best documented gift of a clock was one presented by a Jesuit missionary to Iyeyashu a short time later. This was a spring-driven brass lantern clock made by Hans de Evalo, a Madrid clockmaker, in 1581, which has survived to the present.[11] The clocks were particularly well received and contributed greatly to the success of the missionaries in establishing themselves. Joan Rodriguez, a Portuguese missionary assigned to the mission at Kyoto, was subsequently employed to work as a clockmaker to both Iyeyashu and Hideyoshi, to keep these precious gifts in repair.[12] The Jesuits proved to be more zealous than discreet, however, and in 1624 the Spanish Jesuits were expelled from Japan, and the Portuguese suffered the same fate in 1638. Thereafter, Japan remained open to only Chinese and Dutch traders, who were, furthermore, limited to trading only in the port city of Nagasaki.

The first travellers to Japan reported that the Nipponese seemed to "... have great need to mark in their history, not only the day on which events occurred, but even the hour and the part of the hour."[13] The Japanese proved to be curious about all things, and soon applied themselves to imitation of the European timepieces brought into their country by the Portuguese. With the assistance of a few of the religious missionaries who were skilled in the arts, native craftsmen soon established workshops in which others were trained. Even after all missionaries had been proscribed from Japan in 1637, the making of clocks was continued without interruption by Japanese craftsmen. Probably because this continuity of craftsmanship persisted without Jesuit supervision, the clocks produced by the Japanese began to evolve a character entirely their own. Although clocks continued to be occasionally imported into Japan in small numbers by Dutch traders, Japanese clockmaking gradually assumed an artistic character and mechanical form that became recognized as indigenous to the country, retaining only vestigial evidence of their European inspiration.[14]

At first Japanese artisans, while working under Jesuit supervision, produced copies of European clocks with native decoration, but without making any serious attempt for their utilization as timekeepers because of the disparity of the systems of time division.

Among the first problems which Japanese clockmakers had to overcome was the representation of Japanese time division by mechanical means. The Japanese day was reckoned in hours of variable length, and the periods from sunrise to sunset and from sunset to sunrise were each divided into six hours. The length of the hours varied from summer to winter, being long in summer daytime and short during the summer night, and vice-versa during the winter period. This system of timetelling was

Fig. 2. Portrait of Father Matteo Ricci shown with Paul Zi, from a drawing by the Chinese Jesuit, Father Hoang.

Fig. 3. Chinese Weight driven bracket shelf clock, circa 1680-1700, possibly a product of the clockmaking atelier of the Imperial Palace Manufactories at Peking. The dial plate is in four-color enamel on copper, the hour figures shown as Roman numerals, with a sweep seconds hand that requires 1½ minutes to complete a revolution. The rosewood case is made in two parts, with the hood having a hinged glass door, gilt brass terminals, and mother-of-pearl inlay on the front panel. Collection of the author.

Fig. 4. View of movement of the Chinese weight-driven shelf clock, showing the characteristic thick plates. Inscribed in ink on the front plate and on several locations inside the case are the words in Chinese "Thick plate No. 1." The movement has a crown wheel and verge and knife-edge suspension pendulum regulator. The weights are cast in lead, covered in gilt brass, and shaped as gourds. The clock strikes the hours and quarters on two bells. The height is 23 inches and the width overall is 11 inches.

difficult to incorporate in a European lantern clock of the period, which was designed to measure the hours equally. Various means were developed by the Japanese clockmakers to overcome these difficulties. Early in the seventeenth century the European dial was replaced with one which incorporated the twelve symbols of the Chinese zodiac to represent the hours with corresponding numerals. The movement remained little changed. Later in the century the fixed dial with revolving index, which was common in European clocks, was revised by the Japanese into a revolving dial ring with a fixed index and having adjustable hour plates. The last of the major innovations came later in the seventeenth century with the invention of the double-escapement, one of which was used for the measure of the daylight hours and the other serving for the hours of the night. The verge and foliot escapement was retained for some time to come and eventually was replaced first by the circular balance and later still by the pendulum regulator.[15]

In due course the Japanese evolved four separate types of clocks, each of which was designed to serve a specific purpose in the Japanese home. The first, of course, was the weight-driven lantern clock made to resemble its European prototype, and designed to be suspended from a wall support or to be placed upon a wall-shelf. The weight-driven lantern clock was also made in varying sizes in a floor model, supported upon a pyramidal structure placed upon the floor. The Western table clock was duplicated with appropriate modifications in the timekeeping mechanism as well as its decoration, and produced in a variety of sizes and forms. Finally, the Japanese evolved a new form of clock which did not resemble any European prototypes but which was designed specifically for the Japanese home. This was the pillar clock, made in a vertical or "stick" form which could be easily suspended from a roof pillar inside a room. These various clock forms continued to be produced by the Japanese until the national adoption of Western time division at the end of 1872. After that date, the importation of Western timepieces became widespread, and the Japanese clock became obsolete.

No certain date is known for the production of the first clocks that were uniquely Japanese, and records relating to early Japanese craftsmen trained in clockmaking are fragmentary and inconclusive. Some evidence is provided, however, by Japanese artwork and literature.

Among the earliest of the Japanese craftsmen known to have made clocks was Tsudo Sukezaiema, a late sixteenth century smith who was claimed to have made copies of the de Evalo clock presented to Iyeyashu.[16]

The earliest known graphic representation of a Japanese type clock occurred in a work by Hishikawa Moronobu dated 1681 (Tenna 1) and entitled *Bokuyo Kyoka Shu*. A weight-driven lantern clock on a wall support is shown in a simple line drawing, but without sufficient specification to determine whether the clock is of European origin or of Japanese craftsmanship. In an illustration of a literary work entitled *Nippon Yeitai Kura* by Ihara Saikaku dated 1688 (jokyo 5), the Governor of Nagasaki is depicted receiving a Korean envoy bearing gifts which included a lantern clock on a pyramidal stand. A clockmaker is shown at his work in a print dated 1690 (Genroku 3) by Makieshi Genzaburo entitled *Jinrin Kummo Dzui*, clearly evidence that the craft was already well established in Japan in that period. A Japanese lantern clock is depicted in a work by Nishikawa Sukenobu having the title *Honcho Shin Kanninki* and dated 1708 (Hoyei 5) while a clockmaker is shown working in his shop in a print by Tanchosai Masanobu entitled *Ehon Shomotsu Hajime* produced in 1723 (Kyoho Rabbit Year). An inscription relating to the illustration noted that "The first recorded mention of clocks occurs in a book of the T'ang Dynasty (A.D. 618-906) where it is referred to as a Jimeisho or 'Timetelling Bell'.

Its invention is said to have resulted from profound study of the action of the human body. Functioning independently of weather changes — indifferent to rain, sunshine or cloudiness — the Jimeisho was regarded as one of the most wonderful and admirable things in the world."

Another illustration, dated 1740, depicting a clockmaker holding a measure or square and seated before a weight-driven clock with pyramidal base was published in *Haikai Futawarai*, a work attributed to Hasegawa Mitsunobu. A seventeen syllable poem (a *haiku*) by Sigyu which accompanied the print noted that *"Hidokei no hari suru kagemo toji kana,"* which may be rendered as

> "The craftsman's lengthened shadow on the mat
> Gives warning that the hours for work are short."[17]

The chief market for the clocks produced in Japan was provided by members of the local nobility and regional officials as well as the provincial sovereigns. Consequently, the number of craftsmen at work at any one time was relatively small, consisting primarily of skilled smiths who produced timepieces occasionally and on a limited scale. The names and addresses of some of the better known Japanese clockmakers were listed in several works published in Japan before the beginning of the eighteenth century. The listings included makers working in Kyoto, Osaka, Edo (now Tokyo), Nagasaki and later in Ise, Nagoya, Sendai and other centers. Among the earliest of the master clockmakers noted was Hokyo Gensa, who maintained a shop near the Imperial Palace in Edo.[18]

It was not until forty-two years after the first European clock had been brought into Japan that the first clock was introduced into China, by Italian Jesuits seeking to establish missions on the mainland. After numerous attempts to obtain approval for their mission, the Chinese Viceroy, Ch'en Jui, issued an invitation in 1583 to Father Michele Ruggieri, head of the Jesuit mission at Macao, to come to Chaoching. Ruggieri was accompanied by a young Jesuit priest, Matteo Ricci, and on land provided by the Viceroy they built the first Christian mission in the Chinese interior. In the course of their stay, they were informed by the Viceroy that he would be interested in acquiring one of the "self-ringing bells", as clocks were described by the Chinese, which he had learned might be available from Macao. The Jesuits lost no time in sending to Macao to seek such a clock but could find none for sale. Instead, they discovered an Indian blacksmith who had been trained as a clockmaker by Europeans, and they brought him back to work in Chaoching (Schiaochin). There, with the assistance of two Chinese smiths hired for the purpose, he managed to complete an iron clock for the Viceroy by 1584. The official was extremely pleased with the achievement, but some years later, when the clock had ceased to function and he was unable to find anyone to repair it, he returned it to the mission for the use of the missionaries.[19]

During the next six years Father Ricci and his companions attempted to consolidate their position by developing a discreet Sino-Christian community. Ricci sought to propagate the faith more by means of discussion of common concerns and the practical sciences than by urging religious conversion. In 1589 he moved northward to Shaochow as the next stage towards Peking, his eventual destination. By 1595 he had managed to establish himself first at Nanking and then at Manchang. It was slow progress towards the Chinese capital, but during these years Ricci was not idle, and Chinese officialdom had become increasingly aware of his great learning and of his many scholarly attainments. Finally, in September 1598 he reached Peking after having experienced innumerable difficulties en route, only to discover that a war with Japan that was then in progress made the presence of foreigners unwelcome.

Fig. 5. Chinese screen clock with rosewood case, chased gilt brass dial plate, and Western hour numerals on a white enamel dial. Height overall 24 inches. Crown wheel and verge escapement, with pendulum regulator. Private Collection.

Fig. 6. Musical bracket clock with gilt bronze case and white enamel dials made for the Chinese court by Robert Philp of London c. 1780-1790. Lunaision dial, and originally equipped with revolving globe depicting the phases of the moon. Two-bell striking for the quarters, and plays either a jig or a minuet following the hour striking, determined by one of the smaller dials. Formerly in the collection of the Imperial Palace at Peking, and described in the *Catalogue* of Simon Harcourt-Smith on page 24. Each of the three spring barrels bears an inscription in Chinese. In the collection of the author.

Fig. 7. Left. View of the reverse side of the dial plate. The inscription in Chinese may be translated as "In the Kuei-ch'ou year, the 4th month and the 16th day, Wang Chiu-chieh. Sold" (or "To be sold").

Fig. 8. Right. View of the back plate, showing the elaborate decoration and the maker's signature, "ROBT. PHILP LONDON". Philp worked in London between 1775 and c. 1800, specializing in musical bracket clocks for the Oriental market, some with elaborate automata.

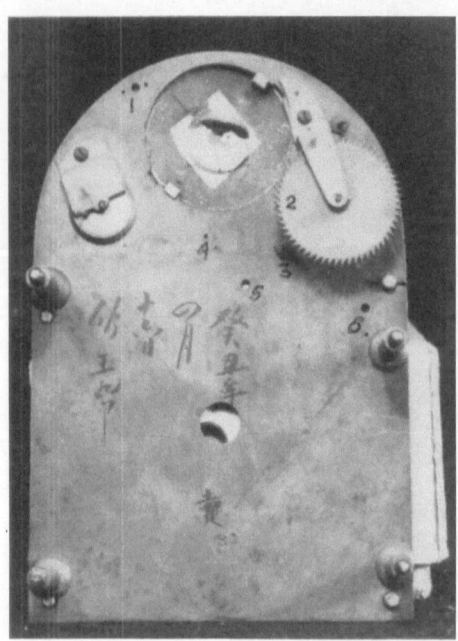

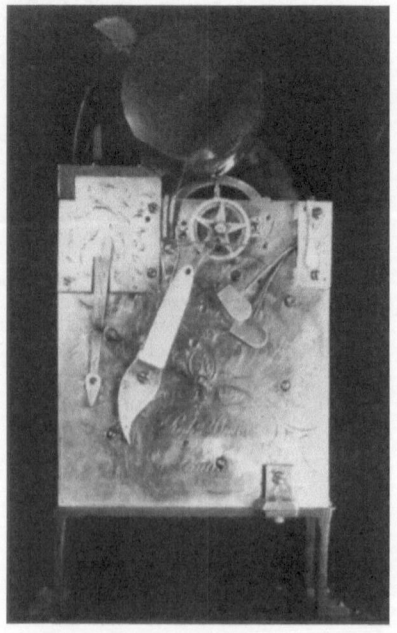

Fig. 9. View of back plate of another Chinese made movement which is apparently copied from a European clock. Private Collection.

Ricci and his companions were forced to turn southward once more to Nanking, to await a more propitious time.[20]

In anticipation of making another attempt to reach Peking, Ricci sent members of his party to Macao to select gifts worthy of the Emperor's interest. Among the objects assembled was a clock which had been originally acquired in 1581 by the Jesuit provincial at Goa for the mission at Macao.[21]

Ricci's successful move to Peking did not come until the spring of 1600, at which time he negotiated passage with a eunuch in charge of a flotilla of ships bearing silks for the court. It was a grueling journey, each step of the way having to be carefully negotiated to avoid conflicts with local officials. The rumor that the Jesuits were bearing valuable gifts for the Emperor led them into numerous difficulties, however. At Lintsing they fell into the clutches of a local official named Ma -t'ang who attempted to acquire the gifts for himself, which led to another delay of six months while awaiting approval of their request to come to the court. Finally, an imperial order arrived early in 1601 permitting the foreigners to proceed to Peking.

Arriving at the Imperial City at last on January 24, 1601, Ricci and his fellow travellers found themselves once more at the mercy of political factions. After enforced delays, their gifts were presented in due time to the Emperor by the court eunuchs. Those that won the Emperor's particular attention were two iron clocks, one of which was very large, in the form of a public clock. The other was a domestic clock having a wooden case surrounded with sculptured gilt dragons. The time was indicated by the beak of an eagle upon a dial having the hour figures engraved as Chinese characters. The clock also struck the hours and quarters on bells.[22]

The Emperor appointed four of his eunuchs from the College of Mathematicians to be taught by missionaries how to maintain and regulate the clocks, and the Jesuits were furthermore required to assign names to each clock part and to provide a description of its function, so that the eunuchs might thereafter maintain the clocks independently.

Father Ricci then presented a gift for the Emperor from the Father General of the Jesuit order. This was a spring-driven table clock encased in gilt metal, measuring a palm in height. The Emperor was extremely pleased with this portable timepiece, and thereafter kept it always beside him.[23]

Meanwhile, the large iron weight-driven clock proved to be too large for installation within the palace, and in the following year the Emperor ordered the construction of a wooden tower to house it in one of the gardens outside the second wall of the palace. Ricci and his companions designed the tower and supervised the installation of the clock.

The Jesuits soon discovered that their problems at the court were not yet over. It was determined that they had erred in submitting their gifts for the Emperor through the eunuchs instead of through the Tribunal of Rites. Once more they were jailed and brought to trial. Ricci had no difficulty in proving his innocence of political dealing, and they were subsequently set free. Later, when the Jesuits planned to leave Peking, the court eunuchs opposed their departure because they feared they would be unable to maintain the clocks without their assistance. The "clocks that struck of themselves" had made a tremendous impact on the court, and the Emperor was so insistent that the clocks should be kept in constant operation that he offered a pension to any of the missionaries who agreed to remain at the court to maintain the clocks. He then wished to meet these foreigners, but the suggestion was opposed by the Tribunal of Rites. Instead, arrangements were made for the best court painters to produce portraits of the missionaries so that the Emperor would have some knowledge of their appearance.[24]

With his awareness of the value of timepieces as items for negotiation, Ricci in 1603 reclaimed from the Vice-Provincial of the Philippines a Flemish clock which the General had acquired with the intention of providing it to the Chinese mission. It had been lost en route and later found. In the same period the Bishop of Manila sent Ricci a small clock which struck the hours and quarters on three bells which Ricci kept for his own use in Peking. Meanwhile, the energetic Ricci personally constructed for the Emperor a brass clock which struck the twelve double-hours twice daily in the Chinese fashion.[25]

These early timepieces were of immeasurable value to the Jesuits in establishing their Chinese missions, but contributed only peripherally to the achievement of a new awareness of time and its measurement by the Chinese. Ricci and his associates attempted to supplement it by assembling an impressive library of European works in all fields of knowledge at their headquarters in Peking for the use of the mission. By means of these works they were able to impart some knowledge of Western technology including timekeeping. The library had been collected in Europe by Johann Schreck (Father Terrentius) prior to his embarkation for China. It was he who composed, and rendered into Chinese with the assistance of Mandarin Wang Tcheng, a work describing and illustrating the mechanical marvels of the Western world. The text and illustrations were subsequently included in the *K'in -ting Kou -kin T'ou -chou Tsi -tch'eng*, the great Chinese encyclopedia published in 1728.[26]

Other talented European missionaries found their way to the Peking mission in the years that followed, frequently deliberately assigned because of the favorable response to mechanical skills. Among them was Father Gabriel Magalhaes, who presented the young Emperor K'ang Hsi with an automaton in the form of a soldier holding a sword in one hand and a shield in the other, motivated by interior clockwork. Later he produced a clock which played a musical melody after each hour had been struck.[27] Another who contributed mechanical skills was Father Thomas Pereira, who installed in a church in Peking a carillon of bells which played Chinese airs.[28] Clocks as well as automata were constructed in Peking by the French missionaries, Frère Jacques Brocard and Father Charles Slaviczek.[29] Among this group was Father Valentin Chalier who produced horizontal clocks set into lacquer tables for the Emperor's palace. He subsequently also constructed a magnificent timepiece which indicated the Western hours, minutes and seconds on one dial, and the Chinese mansions of the night upon another dial having the numerals in the form of the Chinese zodiac, and the clock also struck the mansions on bells. Four months of work were required for the project. The Emperor was so delighted with it that he spared no expense in its decoration. He frequently exhibited it to princes and nobles of the court, implying that it might in fact have been his own invention.[30]

Among the French missionaries whose horological work was recieved with enthusiasm at the court was Father Jean Matthieu De Ventavon who produced an automaton in the form of two men carrying a vase as they walked, motivated by clockwork. It was De Ventavon who later modified a Chinese-writing automaton created by Jacques-Droz so that it was enabled to write in Manchu, Mongolian and Tibetan.[31]

The French mission included among its numbers at Peking two professional clockmakers, the Frères coadjuteurs Gilles Thebault and Francois-Louis Stadlin of Zoug. Stadlin received many favors from the

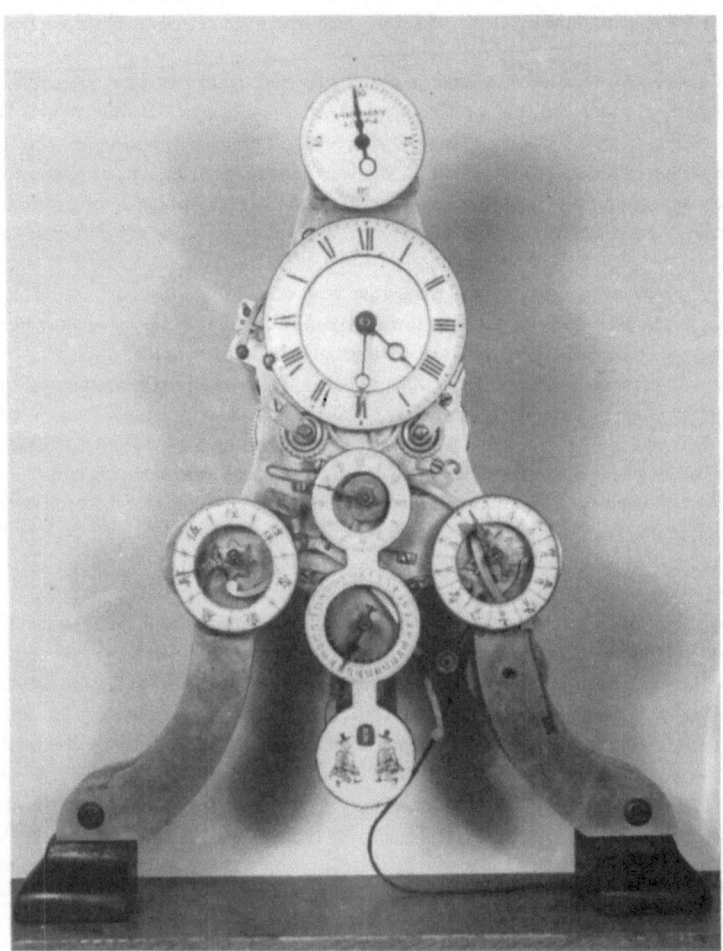

Fig. 10. Chinese skeleton clock with enamel dials for seconds, hours and minutes, days of month, days of the week, lunaison months and moon phases. Moon phase dial features figures of two Europeans shown with the customary cariacatured long noses, green jackets and red trousers. Hour strike only. Uppermost dial is inscribed "Bver Patnet Londno."

Emperor in acknowledgement of his numerous horological achievements. Among his most successful were two automata created in the forms of a lion and a tiger which realistically walked a distance of about one-hundred paces.[32]

The example of the Jesuits, who relied on the Oriental preoccupation with these toys of time to further their religious aims, inspired other religious groups to send missionaries who had been professionally trained as clockmakers in Europe. Notable among them was Gaspar-Francois Guety of the French Missions Etrangeres. He came from Paris with the Papal legation of Cardinal Maillard d'Tournon and worked for a few years for the Emperor K'ang Hsi first at Peking and then at Jehol. Another was Sigismondo di San Nicola, who spent thirty years in the service of Emperor Ch'ien L'ung as a mechanician. Most notable among his many achievements were a clock with automata which he constructed in cooperation with Father Thebault in 1752 and another automaton which he made independently two years later.[33] Other clockmaker missionaries working in Peking were the Discalced Carmelite Arcangelo-Maria di Sant'Anna and the Augustine Pietro Adeodato, the last of the missionary clockmakers to be employed in the service of the Chinese Emperor. In 1789 Adeodato cooperated with the Lazariste missionary, Father Charles Paris, in the production of a writing androide, which was probably inspired by a similar work made by Jacques-Droz and brought to the Chinese court several years previously. Adeodato continued to work in Peking until 1811.[34]

Following the persecution of Western missionaries in Peking, which had begun in 1723, they were no longer permitted to remain freely in the Chinese provinces. They could find haven only at the court at Peking, and then only if their particular skills were considered to be useful in the service of the court. An Italian friar, Carlo Orazi of Castorano, managed to remain in Peking by claiming to know how to make sundials. It was only a pretext and was soon discovered, but he succeeded in compiling a small treatise on the subject. Other Italian Franciscan friars and priests, skilled as mechanicians and clockmakers, were premitted to remain at Peking between 1720 and 1800, but always in the Imperial service.

Papal envoys made their way to Peking in 1705 and again in 1720, bearing valuable gifts for the Emperor. Since his proclivity for horological curiosities had become well known in Europe, these were featured among the papal gifts. In 1720, for example, Monsignor Mezzabarba presented to the young Emperor K'ang Hsi " ... a clock which kindles fire (*l'orologio che accende il fuoco*)" which was in fact a tinder lighting clock. Mezzabarba was accompanied by an Italian friar, Fratre Angelo Maria Pavese de San Siro, who had been well trained as a clockmaker.[35]

So successful had been the horological gifts of the early missionaries in achieving their religious aims in China, and so well publicized were these successes in Europe, that European diplomatic missions to the Orient followed suit. When the ambassador of the King of Spain arrived in China in 1581, he did not reach Peking, although he bore with him gifts including clocks[36] for the Emperor. The British ambassador Lord MacCartney arrived in China in 1794 accompanied by two Swiss clockmakers as members of his party to care for the gifts of clocks and scientific instruments which he brought as gifts from King George III to the Emperor.[37] Upon the advice of the clockmakers in the MacCartney mission, the ambassador from the Netherlands purchased gifts of complicated clocks for the Emperor from the shop of Beale, the successor to James Cox in Canton. The timepieces were damaged during the journey; Charles H. Petitpierre, who accompanied the Dutch diplomatic mission, was unable to

466

Fig. 11. Illustration depicting interior of a Chinese home with a weight-driven wall clock and an incense seal. From *Peking, Histoire et Description* by Alphonse Favier, Lille, 1900.

Fig. 12. Chinese painting on glass depicting girl with monstrance type clock having Western hour numerals. The poem inscribed on the fan is by the Sung dunasty poet, Shao Yung (1011-1077) and may be translated as

"When the moon reaches the heart of the heavens

And when the wind comes over the water's surface,

Generally there is a flavor of pure concepts

Such is as known by few people."

Fig. 13a. Chinese wall and table clock, from *L'Horloge* by M. Planchon.

Fig. 13b. Illustration of a Chinese wall clock and a table clock from Planchon's *L'Horloge*. Neither of these clocks can presently be identified in the collections of the Louvre or other public Parisian collections.

repair them. Accordingly, the task was assigned to Frere Charles Paris, then serving as official clock-maker to the Imperial court, and he accomplished the repairs successfully.[38]

Clocks, watches and automata of great complexity of mechanism and decoration continued to figure prominently among the Western gifts of religious, diplomatic and trade missions to China throughout the eighteenth century in response to the Chinese preoccupation with these mechanical curiosities. The interest of the Chinese of the privileged class in these objects never waned and led to the establish-ment of a considerable trade in such items during the eighteenth and nineteenth centuries.

Meanwhile, elaborate timepieces and automata continued to be brought into China for the delectation of the privileged class; clocks were also being produced by native Chinese craftsmen as early as the first half of the 17th century. The earliest reference to this activity was published by a Jesuit missiona-ry, Father Alvaro Semedo, who had spent twenty-eight years in Nanking, Hangchow and Shanghai. In an account of his travels published after his return to Europe in 1637, Samedo wrote that " ... the Chi-nese, who greatly admire European mechanisms, are already making table clocks and they also make them of the smallest size, and the workers are as well paid as those of the West."[39] Further published confirmation of this activity was provided by G. Brusoni in a work published in Venice in 1659, in which he noted that " ... the Chinese craftsmen are very skilled in manual work and especially in working ivory, ebony and amber ... . In mathematical things they are inferior to the Europeans. How-ever, they are capable of making mechanical table clocks and they would make also small ones if they are paid as our craftsmen are."[40]

The first Chinese clockmakers were native craftsmen of demonstrated superior skills, recruited by the Imperial government to work at first at the Jesuit mission and later at the clock atelier in the Imperial Manufactories at Peking, to which the Emperor invited craftsmen from all over his domain. In this offer of employment favor was shown to those Chinese craftsmen who had been converted to Chris-tianity and to whom the missionaries had taught the art of clockmaking. It was forced labor in the sense that the individual was given no choice. As well as can be determined, such artisans were not treated with the consideration that their skills required, and received less acknowledgement than was prevalent in Europe or even in Japan, where the crafts were favored and encouraged.

The clockmaking atelier at which the first formal training in clockmaking was provided to native craftsmen was another of the Imperial Manufactories (*tsao pan chu*) established at the Imperial Palace at Peking under the Board of Works (*Kung Pu*) by the Emperor K'ang Hsi. The training of apprentices and the work of the atelier were supervised by Jesuit missionaries. They emphasized repair and main-tenance of the Court's numerous timepieces, and subsequently the duplication of some of them. The clockmaking manufactory was the fourth of twenty-seven Imperial ateliers which remained active for more than a century through the reigns of K'ang Hsi and then of his son and later the latter's heir, Ch'ien Lung, to almost the end of the eighteenth century. The clock manufactory expanded substan-tially under the reign of Ch'ien Lung, becoming professionalized and serving also as a training center for clockmakers later sent out to work in the provinces. After Ch'ien Lung's death in 1796, the ateliers were closed down, and what remained of the buildings was destroyed by fire in 1869.[41]

The Imperial clockmaking enterprise was among the favorite preoccupations of K'ang Hsi. In his "In-structions Sublime and Familiar" he confirmed that Emperor Chouen -tche had attempted to have copied the clocks and automatons he had received as gifts from the West but that these duplications operated poorly because his craftsmen were not successful in manufacturing clock springs that were sufficiently supple and elastic. The Emperor went on to note that after having obtained the necessary information about spring-making from European sources, he proceeded to manufacture hundreds of them with such success that he could have made thousands of clocks "which indicated the time with precision."[42] Surviving evidence is in conflict with the Imperial report, and it is doubtful that the atelier was indeed capable of producing such timepieces in such numbers as he had described. During the period of K'ang Hsi's reign, the atelier was supervised by Frère Francois-Louis Stadlin, the only professional clockmaker residing at Peking at that time. He was extremely competent, and a substantial number of clocks were produced for the Emperor and his court during this period entirely in the palace manufactory, although many of them may have been weight-powered because of the difficulties encountered in producing acceptable springs.[43]

K'ang Hsi was so proud of his atelier that in his "Instructions" he could not forbear boasting about the timepieces produced under its auspices, and he commented to his "children" that " ... do not you think you are very fortunate? Because of my initiative, you can play with ten and twenty self-ringing bells."[44]

By the end of K'ang Hsi's reign the atelier was well manned, and Father Valentin Chalier, who supervised the manufactory at that time, stated that in 1736 he had at his command one hundred "slave" workers who did not as much as hammer a nail, he reported, without an order and the specification as to how it was to be hammered.[45] In 1767 Father Jean Mathieu Ventavon reported that the missionaries were then working in the atelier with tranquility and that they had numerous helpers at their command. It was in that same year that the Emperor had issued a public appeal to the artisans of Macao, inviting them to come to Peking to establish shops of their own such as already existed at Macao.[46] The existence of independent clockmaking shops in various cities was confirmed by the Dutch ambassador, Van Braam, who expressed surprise to find two such shops in Hangchow when he passed through that city in 1793.[47]

During the late eighteenth and early nineteenth century native Chinese clockmakers flourished in such trade centers as Soochow, Canton and Nanking. It was reported that in the latter two cities as many as two hundred craftsmen were engaged in commercial horological production at one time.

By the early nineteenth century clockmaking had become established as a Chinese craft and occupation. By 1809 the first manual in clockmaking in the Chinese language was produced by Siu Tch'ao-tsiun, a scholar of Shanghai who noted that he was the fifth generation of his family to work as a professional clockmaker, and that he had devoted his early life to the craft. His modest little treatise was presented in eight chapters for the guidance of clockmakers and named, described and illustrated each of the parts of the clock, and explained their functions and provided instruction for their assembly and disassembly as well as for their repair and regulation. The dials shown in the illustrations were marked in each instance with the Western symbols for the hours and minutes.[48]

By the time clockmaking had become well established in China travellers frequently noted the activity. In a report compiled by the Abbot Grosier in Peking in 1820, he remarked that clockmaking had become a familiar art to the Chinese and that they not only manufactured clocks and watches but were capable of repairing them.[49]

470

Fig. 14. Figures 14 - 21 are pages with woodcut illustrations from a clock-making manual in the Chinese language published at Shanghai in 1809. From the Bibliotheque National, Paris. Note the use of Western hour numerals on the dial plate.

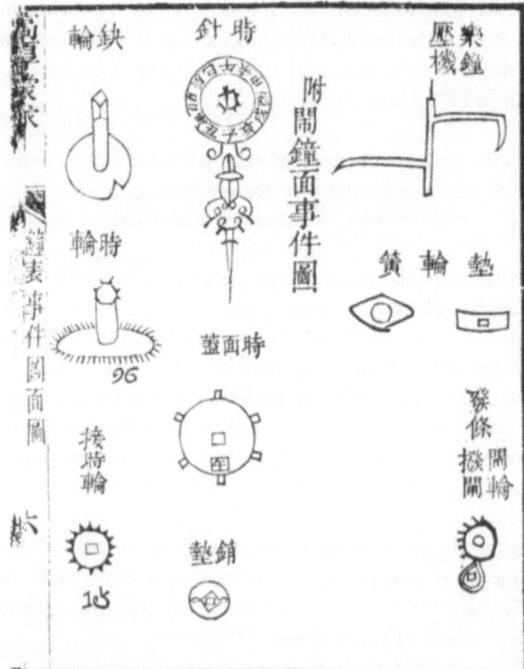

Fig. 15.

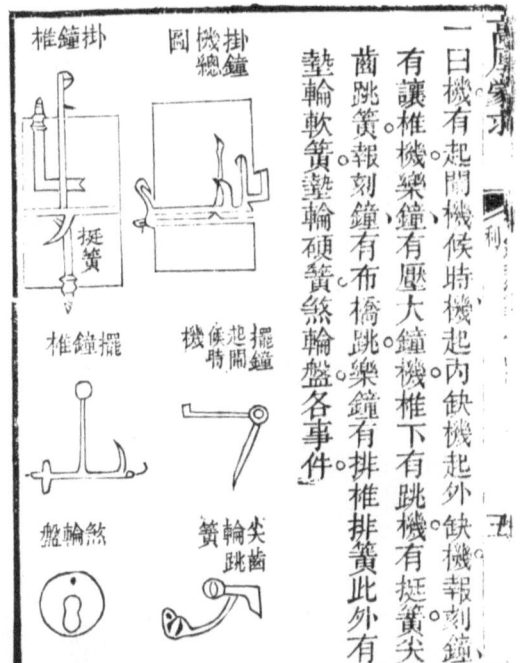

Fig. 16.

椎鐘掛　掛鐘機總圖

挺簧

椎鐘擺　擺鐘起間候時機

煞輪盤　尖齒輪跳簧

一曰機有起間機候時機起內缺機起外缺機報刻鐘
有讓椎機樂鐘有壓大鐘機椎下有跳機有挺簧尖
齒跳簧報刻鐘有布橋跳樂鐘有排椎排簧此外有
墊輪軟簧墊輪硬簧煞輪盤各事件

萬厘掌測

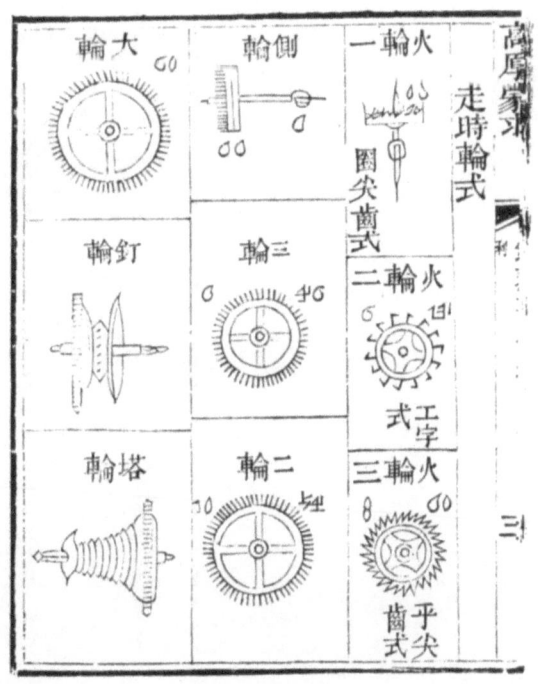

Fig. 17.

大輪　側輪　火輪一　圓尖齒式
釘輪　三輪　火輪二　工字輪式
塔輪　二輪　火輪三　平尖齒式

走時輪式

萬厘掌測

Fig. 18.

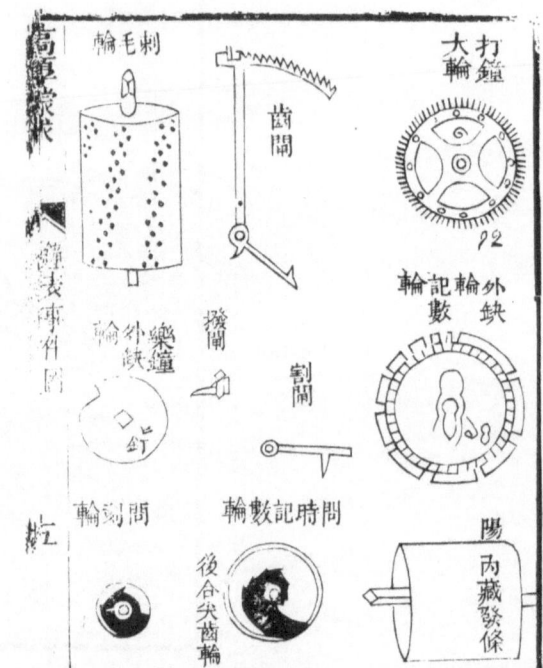

Fig. 19.

473

Fig. 20.

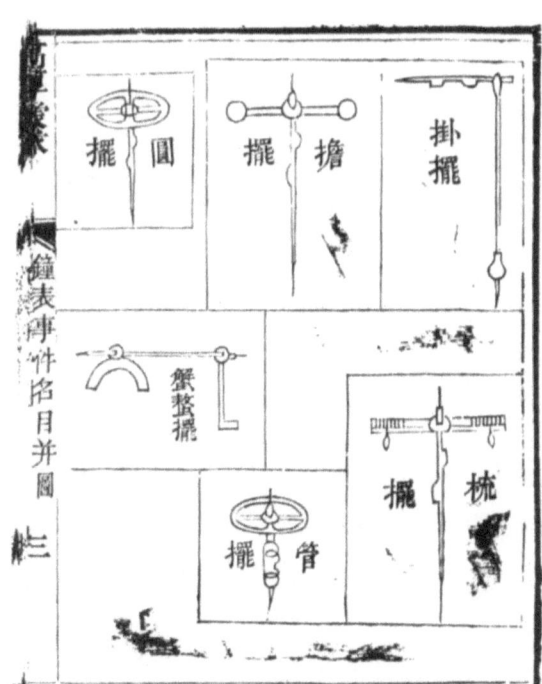

Fig. 21.

Fig. 22. Emperor K'ang-Hsi. Illustrated
from *Peking, Histoire et Description,* by
Alphonse Favier, Lille, 1900.

康熙皇帝

L'EMPEREUR K'ANG-SI.

At the same time that the Chinese were establishing their own capability in the production of time-pieces, a trade specifically in such items was being developed between China and England and Europe. The so-called Chinese taste in timepieces, of pieces of considerable ornateness and complication, had become well known in Europe and paved the way for a commerce which was at first occasional and then of constantly increasing proportions. M. de Lange, a French traveller in the Orient, reported in 1721 that clocks and watches could be sold with profit in China. The India Company in 1747 confirmed the existence of such a market, and other travellers of the period noted that income to be gained from such trade was sufficient so that adequate profit could be made even after customs and tariffs.[50]

By the second half of the eighteenth century, the Chinese market for timepieces had swelled to such a degree that the English turned their attention to it and developed a new type of export, jewelled horological toys rendered in gold, silver and precious stones with mechanical works of considerable complexity. Among the English artisans who specialized in their production were the London clock-makers, William Anthony, Hughes William and James Cox. Featured were bronze figures which performed as automatons, and other combinations of wheelwork which boggled the Oriental mind. So successful were the English in their new venture that the son of James Cox established an outpost in Canton. He had obtained permission from the India Company to go to Canton for presumed reasons of health, while in fact he became agent in residence for his father's products.[51]

French clockmakers also discovered the new market and created dazzling pieces for the China trade, a trade which in turn influenced the decorative arts of Switzerland, England and France throughout this period. As the commerce in "sing songs", as the elaborate timepieces and automata were known in China, developed, the market expanded beyond the Imperial Palace and the Court. Many of the imports were sold to Chinese merchants, who in turn sold them to mandarins who in turn presented them as gifts to superiors and officials for the gain of favors.

This trade, which had reached major proportions during the late eighteenth century, was faced with strong competition from native craftsmen by the end of that period. In a letter written from China to his clockmaker father in London in 1815, Charles Magniac reported the difficulties he encountered in disposing of his clocks and watches in Canton due to competition from Chinese-made timepieces, which were sold at but half the prices at which he was able to offer his.[52]

By this time the European trade to China in timepieces and automata began to diminish abruptly and substantially, partly for the reason reported by Magniac and partly for other reasons as well. The Chinese were capable of making faithful copies of the European products, although not of the same mechanical quality, and at far less cost. At the same time the Emperor, Chia -Ch'ing,and his court had begun to take a dim view of the craze for these relatively useless toys since they were no longer rarities, and he imposed major restrictions upon their importation. European merchants who had established sales outlets at Canton found it no longer possible to continue selling their products under the same favorable conditions that had existed before. The Emperor's restrictions, combined with native competition, forced them to extend credit and increase their prices. At the trade diminished, the English lost interest and left the market to Swiss clockmakers, who had recently discovered the China trade. They organized outlets for their mass-produced timepieces directly in the Chinese port cities. Their production methods enabled the Swiss to manufacture clocks and watches of modest price which reached a new and different Chinese market, produced in their own manufactories established in southern Chinese port cities, such as Canton and Macao, until other ports were opened to European trade.[53]

By the mid-nineteenth century, the native Chinese clockmaking activity came to an abrupt halt chiefly as a result of the Opium War, which left the country impoverished, and which also effectively reduced the importation of Swiss and other European timepieces. A report compiled in 1851 for the U.S. Commissioner of Patents by D.J. MacGowan, an American medical missionary at Ningpo, provided various statistics which may or may not have been based on careful compilation. He noted that there were forty clockmakers' shops in Nankin, thirty at Suchau, seventeen at Hangchau and seven in Ninpo at that time, each employing an average of less than four employees, and that they were for the most part engaged in the repair of clocks and watches and not their manufacture. A manufacturer, he reported, estimated that the total output did not exceed fifteen hundred clocks annually for all of China, and that they were sold at prices ranging from seven to one hundred dollars.[54]

Despite the documentary evidence that clocks and watches were produced in relatively substantial numbers by Chinese craftsmen over a period of one and one-half centuries, these timepieces have remained relatively unknown to the historian and the collector alike. A comprehensive survey of private and public collections has brought a relatively small number of these works to light and leads to further speculations concerning their apparent rarity. One factor may have been due to the fact that although Chinese craftsmen were capable of duplicating and developing original decorative elements of European timepieces, they did not in general succeed in adequately mastering the mechanical aspects. Those surviving timepieces of Chinese origin which have been examined rarely demonstrate competent mechanical production. Furthermore, the deliberate attempts to reproduce European timepieces in a quality which was not comparable to the originals may have led to the assumption that they were in actuality poor examples of European worksmanship and for that reason did not survive or come to notice.

Since the Western numerals and numbers had been adopted in China in an early period, they were used as commonly as the symbols of the Chinese zodiac. Consequently, there was not the need to accommodate a special system of time division such as had occurred in Japan. The measure of the unequal hours, although it had at one time prevailed in China, had been abandoned before the beginning of the sixteenth century. The system of twelve equal double hours had been stabilized from at least the Han Dynasty at the beginning of the second century B.C. It was, therefore, a simple matter for the Chinese to adopt the European dial of twelve hours, use the western hour numerals, or convert them to the signs of the Chinese zodiac with the relative numbers, and use it in that form. Meanwhile, the European hour numerals had become familiar in China and clocks produced by native craftsmen just as frequently utilized the European dial as not.[55]

Just as the Japanese had, the Chinese evolved their own clock forms which were quite distinctively their own, soon after they began to produce timepieces. Early clocks made by Chinese craftsmen clearly demonstrate Jesuit and other Western influences in the style of the case, frequently housing the movements in cases of fine Oriental woods having the altar shapes of the Italian, Spanish or French prototypes on which they were based, often decorating the cases with mother-of-pearl or jade inlay in decorative designs. The pagoda form gradually replaced the altar and other European shapes; the clocks eventually evolved into an Oriental appearance of subtle forms.

One of the clock forms which the Chinese developed, and apparently produced in substantial numbers, was based on the traditional scholar's screen, in which a round clock dial was centered within a rectangular panel of gilt damascened brass or copper having the movement attached to its reverse side. The panel was generally framed simply in fine rosewood or other Oriental woods, with an appropriate carved footed stand, frequently with glazed panels in front of the dial and a wooden door in the rear. Most frequently the chapter ring was of white enamel with Roman hour numerals, and with the quarters marked in the appropriate segmentation in arabic digits, as in the clocks of the Western world. The Chinese developed other shapes and forms for their clocks, sometimes based on a European prototype, or else derived from some object in the Chinese home.

On the basis of available examples, the concensus is that the Chinese were not successful as mechanicians, and in general the movements of clocks they produced left much to be desired. Of examples examined, the movements frequently had excessively thick brass plates. A crown wheel and verge escapement with knife-edge suspension of a pendulum regulator was the one most commonly used. Almost without exception, the movements of Chinese clocks appear to be literal copies of brass movements made by European clockmakers, rendered with a determination to represent the work as European. This is evidenced in many ways, even to attempts to copy the names of European makers, with occasional inevitable slips in spelling or formation of the letters, due to lack of familiarity with the language or alphabet.

The mechanical clock and related horological curiosities played a significant role in opening up the Oriental countries to Western religious, diplomatic and trade missions, and for a time provided a new preoccupation for the privileged classes of Japan and China. Although in Japan these timepieces subsequently influenced the evolution of a clockmaking tradition which had practical and useful results, with the evolution and production of new forms of timekeepers which were practical and appropriate to the national traditions, the same was not true in China. Although the imported timepieces also led to the evolution of a clockmaking tradition in China, neither the intent nor the product served the needs to measure time, but remained always titillating toys for the edification of the privileged. In his report MacGowan noted that clocks were not commonly seen in China even in that period except in public offices where it was usual to find as many as half a dozen of them all in a row, and few if any operative![56] Although the trade developed with China by various European countries for the importation of "sing songs" for more than a century constituted an important chapter in the history of world trade, it had little if any effect on the Oriental concepts of the measure of time, either temporarily or permanently.

478

Fig. 23. Chinese screen clock with elaborately carved rosewood case, white enamel dial on a gilt brass chased dial plate, having Western hours. Crown wheel and verge escapement with a pendulum regulator. Courtesy the Time Museum, Rockford, Ill. (Seth G. Atwood, Jr. Collection).

Fig. 24. Elephant and pagoda clock, probably eighteenth century. Courtesy the Time Museum, Rockford, Ill. (Seth G. Atwood, Jr. Collection).

Fig. 25. Automaton clock of extraordinary elaboration representing the Sacred Mountain of the Chinese. The mountain itself is sculpted in wood, with a waterfall represented with glass twisted bands, with a horse poised on a rock bearing on his back a gilt disk inlaid with red stones, with priests and other figures carved in ivory. The base is bronze with a blue enamel background. The clock strikes the hours and quarters, which set into motion the figures, the cascade and the disk on the horse's back. This timepiece was undoubtedly produced in London for the Chinese Imperial Court between 1760 and 1780, by a maker unknown, but possibly attributable to Stephan Rimbault who worked in London during this period. Courtesy the Time Museum, Rockford, Ill. (Seth G. Atwood, Jr. Collection).

480

Fig. 26. Automaton clock made for the Chinese market, probably English eighteenth century. Gilt bronze figure of court jester shown above dial, and automaton
of a Chinese lady, etc. below the dial. Courtesy the Time Museum, Rockford, Ill.
(Seth G. Atwood, Jr. Collection).

NOTES

1. Charles A.S. Williams: *Outlines of Chinese Symbolism,* Peiping: Customs College Press 1931, pp. 273-4.

2. *Ibid.,* p. 347.

3. *Op. cit.,* p. 121.

4. Joseph Needham, Wang Ling and Derek J. Price: *Heavenly Clockwork, The Great Astronomical Clocks of Medieval China — A Missing Link in Horological History,* Cambridge: Cambridge University Press 1960, pp. 16-59.

5. *Ibid.,* pp. 133-42.

6. Derek J. De Solla Price: "An Ancient Greek Computer," *Scientific American,* Volume 200, June 1959, pp. 60-7.

7. Silvio A. Bedini and Francis R. Maddison: *Mechanical Universe,* [Transactions of the American Philosophical Society], New Series — Volume 56, Part 5, 1966, pp. 8-10.

8. Chester W. Howard: "A De Evalo Returned," *Bulletin of the National Association of Watch and Clock Collectors,* Volume VII, No. 5, October 1956, pp. 296-300.

9. Hyoe Takabayashi: *Tokei Hattatsu -shi,* Tokyo: Toyo Shuppansha 1924; Sir Ernest Satow: "The Church of Yamaguchi From 1550 to 1586," *Transactions of the Asiatic Society of Japan,* Volume II, p. 135.

10. Takabayashi, *op. cit.,* pp. 51-2. See also Drummond Robertson: *The Evolution of Clockwork,* London: Cassells & Co. 1932, pp. 196-7.

11. Paulina Junquera: *Relojeria Palatina,* Madrid: Roberto Carbonell Blasco 1956, pp. 14-15; 17-18; see also Note 8.

12. George H. Dunne: *Generation of Giants,* Notre Dame: University of Notre Dame Press 1962, pp. 215-18.

13. Pere Charlevoix: *Histoire du Japon,* quoted in Mathieu Planchon: *L'Horloge, Son Histoire Retrospective, Pittoresque et Artistique,* Paris: Henri Laurens 1898, p. 209.

14. Robertson, *op. cit.,* pp. 285-7.

15. For a detailed description of the various forms of Japanese clocks, see Robertson, *op. cit.,* pp. 217-84.

16. Ryuji Yamaguchi: *The Clocks of Japan,* Tokyo: Nippon Hyoron -Sha Publishing Co. 1950, p. 6.

17. N.H.N. Mody: *Japanese Clocks*, Tokyo: Kegan Paul, Trench, Trubner & Co., Ltd. 1932, pp. 42-6, pl. 128-35.

18. *Op. cit.*, pp. 128-35.

19. Pasquale D'Elia: *Fonte ricciane, Documenti originati concernenti Matteo Ricci e la storia della prima relazioni tra Europa e la Cina 1579-1615*, Roma: 1942-1949, Volume I, pp. 201-12.

20. Nigel Cameron: *Barbarians and Mandarins, Thirteen Centuries of Western Travellers in China*, New York: Walker Weatherhill 1970, pp. 149-94; Georges Bonnant: "L'Introduction de l'horlogerie occidentale en Chine," *La Suisse Horlogere*, 27 Aout 1959, pp. 767-8.

21. D'Elia, *op. cit.*, Volume I, pp. 147, 166, 216.

22. D'Elia, *op. cit.*, Volume II, pp. 29, 91, 99, 124, 126.

23. D'Elia, *loc. cit.*, Volume II, p. 149.

24. D. Enshoff: "Riccis Uhren," *Die Katolischen Missionen*, Band 65, 1937, pp. 190-4; R. Sarreira: "Horas boas e horas mas para a civilizacao Chinesa," *Broteria*, Volume 36, 1943, pp. 518-28.

25. D'Elia, *op. cit.*, Volume II, p. 149.

26. Guiseppe Gabrieli: *Giovanni Schreck linceo, gesuita, e missionario in Cina e le sue lettere dall' Asia*, Roma 1937, passim; Fritz Jaeger: *Das Buch von den wonderlichen Maschinen in Asia Major*, Leipsig 1944, Band I, pp. 78-96; Henri Bernard: "Les adaptations chinoises d'ouvrages europeens 1514-1688," *Monumenta serica*, Peking 1945, pp. 1-57; 309-88.

27. Dunne, *op. cit.*, pp. 325-38; Louis Pfister: *Notices biographiques et bibliographiques sur les Jesuites des anciens missions de Chine, 1553-1773*, Shanghai 1932-1934, p. 253.

28. Pfister, *op. cit.*, p. 384.

29. Pfister, *op. cit.*, pp. 592, 655.

30. Pfister, *loc. cit.*, p. 718; Alfred Chapuis: "Montres et Pendules 'chinoises' et chinoises," *Journal Suisse d'Horologerie et de Bijouterie*, Nr. 12, Decembre 1932, pp. 248-9; Jean Rispaud: "Le Jesuite Valentin Chalier de Briancon ... ," *Le Petit Dauphinois*, 20 Aout 1939, pp. 1-2.

31. Pfister, *op. cit.*, pp. 913-4; Henri Cordier: "Les correspondants de Berlin," *T'oung Pao*, Leyden 1913, Volume XIV, p. 239.

32. Pfister, *op. cit.*, pp. 793, 620.

33. Henri Cordier: "La suppression de la Compagnie de Jesus," *T'oung Pao*, Leyden 1916, Volume XVII, p. 272. *Lettres edificantes et curieuses*, Paris 1781, Volume XXII, pp. 176, 361; Volume XXIV, pp. 284, 387.

34. D'Elia, *op. cit.*, Volume I, pp. 201, 211.

35. Personal correspondence with Fr. George Mensaert, O.F.M., Director of Editions Sinica Franciscana, October 25, 1959.

36. D'Elia, *op. cit.,* Volume I, p. 168.

37. E.H. Pritchard: "Letters from Missionaries at Peking Relating to the MacCartney Embassy," *T'oung Pao,* Volume XXXI, 1935, p. 39; John Barrow: *Some Account of The Public Life and a Selection from the Unpublished Writings of the Earl of MacCartney,* London 1807, p. 219.

38. Andre Everard Van Braam Hougckest: *Voyage de l'Ambassade de la Compagnie des Indes orientales hollanaises vers l'Empereur de la Chine dans les annees 1794 et 1795,* Philadelphia 1797-1798, Volume II, p. 377.

39. Alvaro Semedo: *Relatione della Grande Monarchia della Cina,* Roma 1643, p. 38.

40. G. Brusoni: *Varie osservazioni sopra le Relazioni Universali de G. Botero,* Venezia 1659, p. 13.

41. Stephen W. Bushell: *Chinese Art,* London 1906, Volume I, p. 108; Alfred Chapuis: *La Montre Chinois,* Neuchatel: Attinger Freres 1919, pp. 42-4; Matthieu Planchon, *op. cit.,* pp. 261-2.

42. "Instructions sublimes et familieres," in *Memoires concernant l'histoire, les sciences, les arts, les usages, ecc ... des Chinois par les missionaires de Pekin,* Paris 1776 and 1814, Volume IX, p. 179.

43. Chapuis, *op. cit.,* p. 45; Pfister, *op. cit.,* pp. 619-20.

44. See Note 41,

45. Pfister, *op. cit.,* pp. 718-720; D'Elia, *op. cit.,* Volume I, p. 29.

46. *Lettres edificantes, op. cit.,* Volume XXIV, p. 110; *Ta -ts'ing Hui -tien Cheu -li,* Paris 1886, Volume 512, p. 19.

47. Van Braam Hougckest, *op. cit.,* Volume II. p. 212.

48. Siu Tch'ao'Tsuin: *Tseu -ming -tchong Piao -t'ou Fa,* Shanghai 1809. (Copy in the Bibliotheque Nationale, Department des Manuscrits, Chinois 4941).

49. J.B. Grosier: *De la Chine ou description generale de cet empire redigee d'apres les memoires de la mission de Pekin,* Paris 1818-1820, 3rd edition, Tome VII, p. 165.

50. Michael Greenberg: *British Trade and the Opening of China, 1800-1842,* Cambridge University Press 1952, p. 87.

51. Chapuis, *Montre Chinois,* pp. 28-32; 61-64; Faith Dennis: "Some Jewelled 'Toys' of Georgian London," *Bulletin of the Metropolitan Museum of Art,* February 1947, Volume V, No. 6 pp. 164-8; John Hayward: "English Clocks for China," *Antiques,* July 1959, pp. 46-9; H. Alan Lloyd: "English Clocks for the Chinese Market," *Antique Collector,* February 1951, pp. 25-9; Simon Harcourt-Smith: *A Catalogue of Various Clocks, Watches, Automata, and Other Miscellaneous Objects of European Workmanship Dating from the XVIIIth and the*

Early XIXth Centuries, In the Palace Museum and the Wu Ying Tien, Peiping, Peking: The Palace Museum 1933, pp. 1-2 and passim.

52. Greenberg, *op. cit.,* p. 87; Chapuis, *Montre Chinois,* pp. 76-9.

53. Chapuis, *Montre Chinois,* pp. 87-130.

54. D.J. MacGowan: "Modes of Timekeeping Known Among the Chinese," *Patent Office Report,* Washington, Government Printing Office, 1851, pp. 335-42.

55. Leopold de Saussure: "L'Horometre et le Systems Cosmologique des Chinois," in Chapuis, *Montre Chinois,* pp. 1-18; Needham, Ling and Price, *op. cit.,* pp. 199-205.

56. Silvio A. Bedini: "Chinese Mechanical Clocks," *Bulletin of the National Association of Watch and Clock Collectors,* Volume 7, 1956, pp. 211-21; MacGowan, *op. cit.,* p. 336.

LIST OF PARTICIPANTS

International Society for the Study of Time,
Second World Conference

Piero E. Ariotti, Verrazzano College, Saratoga Springs, New York, U.S.A.

Seth G. Atwood, The Time Museum, Rockford, Illinois, U.S.A.

Silvio E. Bedini, Smithsonian Institution, The National Museum of History and Technology, Washington, D.C., U.S.A.

Peter Bieri, Universität Heidelberg, Heidelberg, Germany.

Dorothea Watanabe Dauer, University of Hawaii, Honolulu, Hawaii, U.S.A.

Hubert Dreyfus, University of California, Berkeley, California, U.S.A.

J.T. Fraser, International Society for the Study of Time, Westport, Connecticut, U.S.A.

Norio Fujisawa, Sakyo, Kyoto, Japan.

James J. Gibson, Cornell University, Ithaca, New York, U.S.A.

Helen B. Green, Middlesex College, Middletown, Connecticut, U.S.A.

John Gunnell, State University of New York, Albany, New York, U.S.A.

Francis C. Haber, University of Maryland, College Park, Maryland, U.S.A.

José Huertas-Jourda, Wilfred Laurier University, Waterloo, Canada.

Kodi Husimi, Vice President, Science Council of Japan.

Shiro Imai, Sapporo, Japan.

Hide Ishiguro, University College, London, England.

S. Kamefuchi, Tokyo University of Education, Tokyo, Japan.

Robert Kastenbaum, University of Massachusetts, Boston, Massachusetts, U.S.A.

Dale R. Koehler, Bulova Watch Company, New York City, U.S.A.

Helmut Lamm, Universität Mannheim, Mannheim, Germany.

Nathaniel Lawrence, Williams College, Williamstown, Massachusetts, U.S.A.

Masao Matsumoto, Keio University, Tokyo, Japan.

Wolfe Mays, University of Manchester, Manchester, England.

John A. Michon, State University, Groningen, Netherlands.

Otoya Miyagi, Tokyo Institute of Technology, Tokyo, Japan.

Gert Heinz Müller, Universität Heidelberg, Heidelberg, Germany.

William Newton-Smith, Balliol College, Oxford, England.

J.D. North, Oxford College, Oxford, England.

Helga Nowotny, Universität Wien, Vienna, Austria.

Yoichiro Murakami, University of Tokyo, Tokyo, Japan.

Shozo Ohmori, Tokyo, Japan.

Ken-ichi Ono, University of Tokyo, Tokyo, Japan.

David Park, Williams College, Williamstown, Massachusetts, U.S.A.

Derek de Solla Price, Yale University, New Haven, Connecticut, U.S.A.

Ricardo J. Quinones, Claremont Men's College, Claremont, California, U.S.A.

Curt P. Richter, Johns Hopkins University School of Medicine, Baltimore, Maryland, U.S.A.

George Rochberg, University of Pennsylvania, Philadelphia, Pennsylvania, U.S.A.

Georges Schaltenbrand, Leutpold Hospital, Würzburg, Germany.

Manfred E.A. Schmutzer, Vienna, Austria.

Charles M. Sherover, Hunter College, New York City, U.S.A.

Masanao Toda, Hokkaido University, Sapporo Hokkaido, Japan.

A. Toffler, Ridgefield, Connecticut, U.S.A.

Gisela Trommsdorff, Universität Mannheim, Mannheim, Germany.

Waldemar Voisé, Polish Academy of Sciences, Warsaw, Poland.

Michael S. Watanabe, University of Hawaii, Honolulu, Hawaii, U.S.A.

Makoto Yamamoto, University of Tokyo, Tokyo, Japan.

Takahiko Yamanouchi, Tokyo, Japan.

Michael M. Yanase, Sophia University, Tokyo, Japan.

Natsuhiko Yoshida, Kanagawa-ken, Japan.

Jiri Zeman, Czechoslovak Academy of Sciences, Prague, Czechoslovakia.